Lecture Notes in Mathematics

Edited by A. Dold, J. M. Morel, F. Takens and B. Teissier

Editorial Policy
for the publication of monographs

1. Lecture Notes aim to report new developments in all areas of mathematics – quickly, informally and at a high level. Monograph manuscripts should be reasonably self-contained and rounded off. Thus they may, and often will, present not only results of the author but also related work by other people. They may be based on specialized lecture courses. Furthermore, the manuscripts should provide sufficient motivation, examples and applications. This clearly distinguishes Lecture Notes from journal articles or technical reports which normally are very concise. Articles intended for a journal but too long to be accepted by most journals, usually do not have this "lecture notes" character. For similar reasons it is unusual for doctoral theses to be accepted for the Lecture Notes series.

2. Manuscripts should be submitted (preferably in duplicate) either to one of the series editors or to Springer-Verlag, Heidelberg. In general, manuscripts will be sent out to 2 external referees for evaluation. If a decision cannot yet be reached on the basis of the first 2 reports, further referees may be contacted: the author will be informed of this. A final decision to publish can be made only on the basis of the complete manuscript, however a refereeing process leading to a preliminary decision can be based on a pre-final or incomplete manuscript. The strict minimum amount of material that will be considered should include a detailed outline describing the planned contents of each chapter, a bibliography and several sample chapters.
Authors should be aware that incomplete or insufficiently close to final manuscripts almost always result in longer refereeing times and nevertheless unclear referees' recommendations, making further refereeing of a final draft necessary.
Authors should also be aware that parallel submission of their manuscript to another publisher while under consideration for LNM will in general lead to immediate rejection.

3. Manuscripts should in general be submitted in English.
Final manuscripts should contain at least 100 pages of mathematical text and should include
– a table of contents;
– an informative introduction, with adequate motivation and perhaps some
  historical remarks: it should be accessible to a reader not intimately familiar
  with the topic treated;
– a subject index: as a rule this is genuinely helpful for the reader.

Dedication to Albrecht Dold

Springer-Verlag and the junior editors of this series dedicate this volume to their senior colleague Albrecht Dold who has chosen to retire from active service as editor at the close of this millennium: this volume is the last of the Lecture Notes in Mathematics for which he shares the responsibility of publication.

A more profound reason for this dedication is our wish to record the unique service that Albrecht Dold has rendered to these Lecture Notes for which he has been an editor from the very beginning until the present, i.e. for a total of 1750 volumes, that appeared over a period of 36 years. The "LNM" series is the oldest of the Lecture Notes series. It was the brainchild of Konrad Springer and Beno Eckmann in 1964. Albrecht Dold was the first editor, joined, from Volume 3 by Beno Eckmann: they worked in tandem on more than 1500 volumes until Beno Eckmann retired in 1994.

The series bridges the gaps between longer research papers, advanced graduate textbooks and classical "hard-cover" research monographs.

The subsequent emergence of many other such lecture notes series, both in mathematics and in other sciences demonstrates the success of the LNM formula.

Over all these years the main constant factor in the Lecture Notes in Mathematics has been the great energy that Albrecht Dold has dedicated to the series. In this period he amassed an astounding wealth of knowledge on the research activities and the questions driving research in essentially all branches of mathematics. Extending well beyond the mere cataloguing of open questions, main results and protagonists, his appreciation of their intrinsic significance was often a decisive factor in complicated decision-making situations.

In addition to Albrecht Dold's outstanding scientific qualities, his fine human features were of immense if much less conspicuous importance, and consistently provided the right counterpoise when difficult decisions had to be reached and communicated.

Over the years the mathematical world has changed and the editorial policy of the Lecture Notes has evolved in line with these changes. One notable such change resides in the phenomenal expansion of quantity and diversity of the applications of mathematics. This in turn motivated Springer to propose inviting to the series a new editor, whose scientific profile might raise the visibility of LNM in these new areas. As a result, Jean-Michel Morel is joining Floris Takens and Bernard Teissier as editor of the LNM series, as of this very volume.

The hope of the editors and the publisher is that our continuing work on the Lecture Notes in Mathematics, evidenced by the future volumes to be published in the 21st century, will be viewed by Albrecht Dold and Beno Eckmann as maintaining the standards and the spirit that have been the driving force of the LNM for the first 36 years.

Floris Takens · Bernard Teissier
Catriona Byrne

Lecture Notes in Mathematics 1750

Editors:
A. Dold, Heidelberg
J.-M. Morel, Cachan
F. Takens, Groningen
B. Teissier, Paris

Springer
Berlin
Heidelberg
New York
Barcelona
Hong Kong
London
Milan
Paris
Singapore
Tokyo

Brian Conrad

Grothendieck Duality and Base Change

Springer

Author

Brian Conrad
Department of Mathematics
Harvard University
1 Oxford Street
Cambridge, MA 02138, USA

As of Sept. 1, 2000:

University of Michigan
Department of Mathematics
2074 East Hall
525 East University Ave.
Ann Arbor, MI 48109, USA
E-mail: bdconrad@math.lsa.umich.edu

Cataloging-in-Publication Data applied for

Die Deutsche Bibliothek - CIP-Einheitsaufnahme

Conrad, Brian:
Grothendieck duality and base change / Brian Conrad. - Berlin ;
Heidelberg ; New York ; Barcelona ; Hong Kong ; London ; Milan ; Paris
; Singapore ; Tokyo : Springer, 2000
  (Lecture notes in mathematics ; 1750)
  ISBN 3-540-41134-8

Mathematics Subject Classification (2000): 14A15

ISSN 0075-8434
ISBN 3-540-41134-8 Springer-Verlag Berlin Heidelberg New York

Springer-Verlag Berlin Heidelberg New York
a member of BertelsmannSpringer Science+Business Media GmbH

© Springer-Verlag Berlin Heidelberg 2000
Printed in Germany

Typesetting: Camera-ready T$_E$X output by the author
SPIN: 10734287   41/3142-543210 - Printed on acid-free paper

Preface

Grothendieck duality theory on noetherian schemes, particularly the notion of a dualizing sheaf, plays a fundamental role in contexts as diverse as the arithmetic theory of modular forms [**DR**], [**M**] and the study of moduli spaces of curves [**DM**]. The goal of the theory is to produce a trace map in terms of which one can formulate duality results for the cohomology of coherent sheaves. In the 'classical' case of Serre duality for a proper, smooth, geometrically connected, n-dimensional scheme X over a field k, the trace map amounts to a canonical k-linear map $t_X : H^n(X, \Omega^n_{X/k}) \to k$ such that (among other things) for any locally free coherent sheaf $\mathscr{F}$ on X with dual sheaf $\mathscr{F}^\vee = \mathscr{H}om_{\mathscr{O}_X}(\mathscr{F}, \mathscr{O}_X)$, the cup product yields a pairing of finite-dimensional k-vector spaces

$$H^i(X, \mathscr{F}) \otimes H^{n-i}(X, \mathscr{F}^\vee \otimes \Omega^n_{X/k}) \to H^n(X, \Omega^n_{X/k}) \xrightarrow{t_X} k$$

which is a perfect pairing for all i. In particular, using $\mathscr{F} = \mathscr{O}_X$ and $i = 0$, we see that $\dim_k H^n(X, \Omega^n_{X/k}) = 1$ and t_X is non-zero, so t_X must be an isomorphism. Grothendieck duality extends this to a relative situation, but even the relative case where the base is a discrete valuation ring is highly non-trivial. The foundations of Grothendieck duality theory, based on residual complexes, are worked out in Hartshorne's *Residues and Duality* (hereafter denoted [**RD**]). These foundations make the duality theory quite computable in terms of differential forms and residues, and such computability can be very useful (e.g., see Berthelot's thesis [**Be**, VII, §1.2] or Mazur's pioneering work on the Eisenstein ideal [**M**, II, p.121]).

In the construction of this theory in [**RD**] there are some essential compatibilities and explications of abstract results which are not proven and are quite difficult to verify. The hardest compatibility in the theory, and also one of the most important, is the base change compatibility of the trace map in the case of proper Cohen-Macaulay morphisms with pure relative dimension (e.g., flat families of semistable curves). *Ignoring* the base change question, there are simpler methods for obtaining duality theorems in the *projective* CM case (see [**AK1**], [**K**], which also have results in the projective non-CM case). However, there does not seem to be a published proof of the duality theorem in the general proper CM case over a locally noetherian base, let alone an analysis of its behavior with respect to *base change*. For example, the rather important special case of compatibility of the trace map with respect to base change to a geometric fiber is not at all obvious, even if we restrict attention to duality for projective *smooth* maps. This was our original source of motivation in this topic and (amazingly) even this special case does not seem to be available in the published literature.

The aim of this book is to prove the hard unproven compatibilities in the foundations given in [**RD**], particularly base change compatibility of the trace map, and to explicate some important consequences and examples of the abstract theory. This book should be therefore be viewed as a companion to [**RD**], and is by no means a logically independent treatment of the theory from the very beginning. Indeed, we often appeal to results proven in [**RD**] along the way, rather than reprove everything from scratch (and we are careful to avoid any circular reasoning). More precisely, we will give the definitions of most of the basic constructions we need from [RD] (aside from a few cases in which the definitions are very elaborate, in which case we refer to specific places in [RD] for the relevant definitions), and we will sometimes refer to [RD] for proofs of various properties of these basic constructions. It is our hope that by providing a detailed explanation of some of the more difficult aspects of the foundations, Grothendieck's work on duality for coherent sheaves will be better understood by a wider audience.

There is a different approach to duality, and particularly the base change problem for the trace map, which should be mentioned. In [LLT], Lipman works out a vast generalization of Grothendieck's theory, using Deligne's abstract construction of a trace map [RD, Appendix] in place of Grothendieck's 'concrete' approach via residual complexes (as in the main text of [RD]). Lipman's theory requires a lot more preliminary work with derived categories than is needed in [RD], but it yields a more general theory without noetherian conditions or boundedness hypotheses on derived categories (though the 'old' theory in [RD] is adequate for nearly all practical purposes). In these terms, Lipman says that he can deduce the base change compatibility of the trace map in the proper Cohen-Macaulay case. However, it seems unwise to ignore the foundations based on residual complexes, because of their usefulness in calculations.

In any case, it is unlikely that Lipman's powerful abstract methods lead to a much shorter proof that the trace map is compatible with base change. The reason is that ultimately one wants to have statements in terms of sheaves of differentials or (at least in the projective case) their $\mathcal{E}xt$'s, with *concrete* base change maps. Translating Deligne's more abstract approach into these terms is a non-trivial matter which cannot be ignored, and this appears to cancel out any appearance of brevity in the proofs. Either one builds the concreteness directly into the foundations (as in [RD] and this book) and then one needs to check a lot of commutative diagrams, or else one uses abstract foundations and has to do a lot of hard work to make the results concrete. To quote Lipman on the issue of the choice of foundations,

> " ... The abstract approach of Deligne and Verdier, and the more recent one of Neeman, seem on the surface to avoid many of the grubby details; but when you go beneath the surface to work out the concrete interpretations of the abstractly defined dualizing functors, it turns out to be not much shorter. I don't know of any royal road ... "

Contents

CONTENTS

CHAPTER 1

Introduction

1.1. Overview and Motivation

Let $f : X \to Y$ be a proper, surjective, smooth map of schemes, with all fibers equidimensional with dimension n, and let $\omega_{X/Y} = \Omega^n_{X/Y}$. Grothendieck's duality theory [**RD**, VII, 4.1] produces a trace map

$$(1.1.1) \qquad \gamma_f : \mathrm{R}^n f_*(\omega_{X/Y}) \to \mathscr{O}_Y$$

which is an isomorphism when f has geometrically connected fibers. When $n = 0$, this is just the usual trace map $f_*(\mathscr{O}_X) \to \mathscr{O}_Y$.

One aspect of the 'duality' is that if $\mathscr{F}$ is any locally free sheaf of finite rank on X with dual $\mathscr{F}^\vee = \mathscr{H}om_{\mathscr{O}_X}(\mathscr{F}, \mathscr{O}_X)$, the cup product pairing

$$\mathrm{R}^i f_*(\mathscr{F}) \otimes \mathrm{R}^{n-i} f_*(\mathscr{F}^\vee \otimes \omega_{X/Y}) \xrightarrow{\cup} \mathrm{R}^n f_*(\omega_{X/Y}) \xrightarrow{\gamma_f} \mathscr{O}_Y$$

(where $\cup$ denotes cup product) induces a map

$$(1.1.2) \qquad \mathrm{R}^{n-i} f_*(\mathscr{F}^\vee \otimes \omega_{X/Y}) \to \mathscr{H}om_{\mathscr{O}_Y}(\mathrm{R}^i f_*(\mathscr{F}), \mathscr{O}_Y)$$

which is an isomorphism if $\mathrm{R}^j f_*(\mathscr{F})$ is locally free (necessarily of finite rank) for all j. To make the proofs work, one actually establishes a more general isomorphism on the level of derived categories in the locally noetherian case.

An important property of the trace map is that it is of formation compatible with arbitrary (e.g., non-flat) base change. More precisely, recall that if

$$
\begin{array}{ccc}
X' & \xrightarrow{u'} & X \\
{\scriptstyle f'}\downarrow & & \downarrow{\scriptstyle f} \\
Y' & \xrightarrow{u} & Y
\end{array}
$$

is a cartesian diagram of schemes, then there is a natural base change morphism

$$u^* \mathrm{R}^n f_*(\omega_{X/Y}) \to \mathrm{R}^n f'_*(u'^* \omega_{X/Y}) \simeq \mathrm{R}^n f'_*(\omega_{X'/Y'}).$$

Since f is proper and finitely presented with n-dimensional fibers, it follows from direct limit arguments and Grothendieck's theorem on formal functions that $\mathrm{R}^n f_*$ is a right exact functor on quasi-coherent sheaves, so this base change morphism is an isomorphism. The compatibility of the trace map with base change means that the diagram

$$(1.1.3) \qquad\qquad u^* \mathbf{R}^n f_*(\omega_{X/Y}) \xrightarrow{\;\simeq\;} \mathbf{R}^n f'_*(\omega_{X'/Y'})$$

$$u^*(\gamma_f) \downarrow \qquad\qquad\qquad \downarrow \gamma_{f'}$$

$$u^*(\mathscr{O}_Y) =\!\!=\!\!=\!\!=\!\!=\!\!=\!\!=\!\!= \mathscr{O}_{Y'}$$

commutes. One of the main goals of this book is to prove this commutativity. This is needed over certain bases in [**RD**] in order to *define* the trace map γ_f over an arbitrary base. More importantly, the commutativity of (1.1.3) is crucial in the proof that γ_f is an *isomorphism* when f has geometrically connected fibers.

The standard references [**RD**], [**Verd**] ignore the verification that (1.1.3) commutes. In [**RD**] this is left to the reader and [**Verd**] only checks the case of *flat* base change. From the point of view of either of these references, the analysis of (1.1.3) is a non-trivial matter. The very definition of the trace map γ_f in [**RD**] involves a series of intermediate steps for which general base change *makes no sense*; general base change maps are only meaningful for the 'outer pieces' $\mathbf{R}^n f_*(\omega_{X/Y})$ and $\mathscr{O}_Y$ left at the end of the construction. This makes the commutativity of (1.1.3) seem like a miracle. The methods in [**Verd**] take place in derived categories with "bounded below" conditions. This leads to technical problems for a base change such as $p : \mathrm{Spec}(A/\mathfrak{m}) \hookrightarrow \mathrm{Spec}(A)$ with $(A, \mathfrak{m})$ a non-regular local noetherian ring, in which case the right exact p^* does not have finite homological dimension (so $\mathbf{L}p^*$ does not make sense as a functor between "bounded below" derived categories). Moreover, Deligne's construction of the trace map in [**RD**, Appendix], upon which [**Verd**] is based, is so abstract that it is a non-trivial task to relate Deligne's construction to the sheaf $\mathbf{R}^n f_*(\Omega^n_{X/Y})$. However, a direct relation between the duality theorem and differential forms is essential for many important calculations (e.g., [**M**, §6, §14(p.121)]).

I initially tried to verify the commutativity of (1.1.3) by a direct calculation with $\mathscr{E}xt$ sheaves in the smooth projective case. This approach quickly gets stuck on the fact that base change maps for $\mathscr{E}xt_{\mathscr{O}_X}$'s [**AK2**, §1] are only defined when the sheaves involved satisfy certain flatness and quasi-coherence conditions, and, more importantly, this definition is local on X and *not* local over Y (unless X is *finite* over Y). When I asked Deligne about this difficulty, he agreed that the projective smooth case seemed puzzling if one tried to analyze it directly via $\mathscr{E}xt$-sheaves. A subsequent discussion about the general case of (1.1.3) with some other experts was also inconclusive. Despite the fact that everyone believes that (1.1.3) commutes, no published proof seems to exist. Nevertheless, it is widely used. A proof of the commutativity of (1.1.3) will be given in this book; it makes essential use of a generalization to the proper Cohen-Macaulay case, using the foundations of duality theory in [**RD**].

Many of the unverified compatibilities in [**RD**] are not hard to check (and compatibilities with respect to translations, flat base change, and composites of scheme morphisms are often trivial to verify), but some unverified compatibilities in [**RD**] are genuinely difficult to prove and their truth depends in an essential way on a correct choice of sign conventions. Thus, in order to construct a global theory and to make explicit calculations, we must fix once and for all a correct and consistent choice of sign conventions in the main constructions of Grothendieck duality theory (e.g., for Koszul complexes, residues, etc.). Anyone who has ever used an argument of the form "$x = -x$, so therefore $x = 0$" (in a $\mathbf{Z}[1/2]$-module) can appreciate the importance of eliminating sign ambiguities in the foundations, even if one admits that a global theory ought to exist.

Prior to the statement and proof of the main duality theorem [**RD**, VII, 3.4], nearly all of the difficult compatibility problems in [**RD**] are in the foundational chapter [**RD**, III]. Most of the remaining omitted proofs and omitted compatibilities are quite straightfoward to fill in. Thus, we devote approximately the first half of the book (Chapters 1 and 2) to justifying certain difficult compatibilities which arise without proof in [**RD**, III]. Chapters 3 and 4 are devoted to developing the theory of the dualizing sheaf for Cohen-Macaulay morphisms and proving that the base change diagram (1.1.3), as well as a generalization to the proper Cohen-Macaulay case, commutes. We conclude in Chapter 5 and the appendices by giving some important consequences and examples of the general theory.

There are two observations that enable us to successfully analyze the base change question for the trace map. First of all, for a technical reason to be explained shortly, we relax the smoothness condition to the condition that the proper map f be a (proper) CM map — that is, locally finitely presented and flat with Cohen-Macaulay fibers (so if f is finite and finitely presented, then f is CM if and only if f is flat) — and then $\omega_{X/Y} = \Omega^n_{X/Y}$ has to be replaced by a 'relative dualizing sheaf' ω_f. Second and more importantly, we *ignore* a direct treatment of the projective case. Instead, we study the definition in [**RD**, VI, VII] of ω_f and the trace map γ_f for proper morphisms $f : X \to Y$ to noetherian schemes Y which admit a dualizing complex (this includes Y of finite type over $\mathbf{Z}$ or over a field, or more generally over a complete local noetherian ring). This definition of (ω_f, γ_f) uses the theory of *residual complexes*, to be discussed in Chapter 3, and the definition of γ_f is built up from derived category trace maps associated to certain *finite* morphisms which are 'supported' at closed points in the fibers of f. The auxiliary *finite* maps arising in this 'residual complex' definition of γ_f cannot generally be chosen to be smooth (i.e., étale) if f is smooth, but they can be chosen to be CM (i.e., flat) if f is CM. Thus, our goal is to use the 'residual complex' definition of the trace map and the base change theory of $\mathscr{E}xt$'s [**AK2**, §1] in order to formulate a base change theory for (ω_f, γ_f) in the CM case and to reduce the base change problem for proper CM maps to the special case of *finite flat maps*, in which case a direct calculation is possible. We emphasize that this method forces us to go outside of the category of smooth maps, but one wants a duality theorem for possibly non-smooth Cohen-Macaulay maps anyway (e.g., for the study of flat families of semistable curves).

We now describe the basic idea that makes reduction to the finite case plausible. Recall the following fact [**EGA**, IV$_4$, 19.2.9]: if $f : X \to Y$ is a CM map of schemes and $x \in X_y \overset{\text{def}}{=} f^{-1}(y)$ is a *closed* point in the fiber over y, there exists a commutative diagram of locally finitely presented maps

in which g is quasi-finite, separated, and flat (hence CM) and i is an immersion which passes through x. Moreover, i can be chosen to contain *any* desired infinitesimal thickening $\text{Spec}(\mathcal{O}_{X_y,x}/\mathfrak{m}_x^n)$ of $\{x\}$ along X_y. To define i, we simply choose a system of parameters in the local Cohen-Macaulay ring $\mathcal{O}_{X_y,x}$ which lie in $\mathfrak{m}_x^n$ and we lift these to a small neighborhood of x in X; these lifted functions then define the subscheme Z. Note that if we restricted ourselves to the case of smooth Y-schemes, such a quasi-finite *smooth* g exists if and only if the finite extension $k(x)/k(y)$ is separable, and such a Z *never* contains the higher infinitesimal neighborhoods of x in $f^{-1}(y)$.

The theory of residual complexes (used in the definitions of ω_f, γ_f) is a priori compatible with base change to a henselization of the local ring at any point in Y, so it is easy to reduce the general base change question to the case of a local henselian Y. When Y is local *henselian* with closed point y, it follows from Zariski's Main Theorem [**EGA**, IV$_4$, 18.12.13] that Z as above breaks up as a disjoint union of a part Z_{fin} which is *finite* (flat) over Y and a part Z' which has empty closed fiber (so Z' does not contain x and therefore can be ignored). By letting x run through the closed points in the fibers of f and choosing such Z's as above containing larger and larger infinitesimal neighborhoods of x, we are *almost* able to reduce to the study of the Z_{fin}'s in place of X, which would reduce us to the finite flat case. To be more precise, we will use such i's as above to reduce the general base change question (1.1.3) to the following two cases:

- Y, Y' are local artin schemes and f is *finite* flat,
- $Y = \text{Spec}(A)$ with A a local noetherian ring admitting a dualizing complex (e.g., a complete local noetherian ring or a local ring of a finite type **Z**-algebra) and the base change is to an artinian quotient of A.

When the details are carried out in Chapter 4, we will use the Krull Intersection Theorem to bypass the need to use henselizations or Zariski's Main Theorem, but the above idea was our source of motivation.

In §3.1–§3.4, we review the construction of duality theory in terms of residual complexes and in §3.5–§3.6 we use this to formulate the base change question in terms of residual complexes and $\mathcal{E}xt$-sheaves rather than in terms of sheaves of relative differentials. The advantage of this reformulation is that it makes sense for proper CM maps, not just proper smooth maps. In particular, we can work with finite flat maps, which are "usually" not smooth. It suffices to consider the two special cases above, and these are treated by direct calculations with residual complexes in §4.1–§4.2. This gives a duality theorem for proper Cohen-Macaulay maps $f : X \to Y$ with pure relative dimension over a base Y which is noetherian and admits a dualizing complex, and we prove that the corresponding trace map is compatible with any base change $Y' \to Y$ where Y' is also noetherian and admits a dualizing complex. In §4.3–§4.4, we use this base change compatibility of the trace map to obtain a 'derived category' duality theorem for proper CM maps with pure relative dimension over *any* locally noetherian base, along the lines suggested in [**RD**, p.388] (though the proof of this duality theorem in the CM case is significantly more difficult than in the smooth case). In Chapter 5, we unwind the abstract machinery to recover two of the most widely used consequences of Grothendieck's theory: duality for high direct image sheaves (as in (1.1.2)), and the explicit description of duality on proper reduced curves over an algebraically closed field in terms of 'regular differentials' and residues (for which some delicate questions of compatibility arise).

In Appendix A, we explain how to use the general duality theory to establish the basic properties of the residue symbol, Grothendieck's higher-dimensional generalization of the classical notion of residue of a meromorphic differential at a point on a smooth curve. These results are stated in [**RD**, III, §9], but proofs are not given. A detailed analysis of the residue symbol can also be found in [**L**], based on a completely different foundation for the theory of the residue symbol, via Hochshild homology. Appendix B explains the relation between the abstract theory and classical duality on a smooth curve via residues. This example deserves special attention for two reasons. First of all, the relationship between residues and Grothendieck's abstract trace map on a proper smooth curve rests on the unproven result [**RD**, VII, 1.2] (for which we give a proof, in Theorem B.2.1). Second, the theoretical foundations of the general theory rely upon this example because the proof that the 'residual complex trace map' for *proper* morphisms is a map of complexes ultimately involves reduction to the case of proper smooth curves over an artin local ring A with algebraically closed residue field (and even just $\mathbf{P}^1_A$), and in this case the desired result is exactly the residue theorem (thanks to the relation already "established" between Grothendieck's theory and residues in [**RD**, VII, 1.2]).

1.2. Notation and Terminology

Our notation and terminology is almost identical to that in [**RD**], and is fairly standard. For example, if $(X, \mathscr{O}_X)$ is a ringed space, the derived category of the category of $\mathscr{O}_X$-modules is denoted $\mathbf{D}(X)$ and if X is a scheme (resp. locally noetherian scheme), we denote by $\mathbf{D}^+_{qc}(X)$ (resp. $\mathbf{D}^b_c(X)$)the full subcategory in $\mathbf{D}(X)$ consisting of complexes whose cohomologies are all quasi-coherent (resp. coherent) and vanish in sufficiently negative degrees (resp. sufficiently negative and positive degrees). It is always assumed that complexes in an abelian category are cochain complexes (i.e., the differential increases degree by 1). We denote by $\mathbf{D}(X)_{fTd}$ (resp. $\mathbf{D}(X)_{fid}$) the full subcategory of $\mathbf{D}^b(X)$ consisting of complexes with finite Tor-dimension (resp. finite injective dimension) that is, complexes $C^\bullet$ quasi-isomorphic to a bounded complex of flat sheaves [**RD**, II, 4.3] (resp. quasi-isomorphic to a bounded complex of injective sheaves, which can all be taken to be quasi-coherent if X is a locally noetherian scheme and $C^\bullet$ has quasi-coherent cohomologies [**RD**, I, 7.6; II, 7.20]). For a ring A, we denote by $\mathbf{D}(A)$ the derived category of the category of A-modules. If A is noetherian, we denote by $\mathbf{D}_c(A)$ the full subcategory of $\mathbf{D}(A)$ consisting of objects with finitely generated cohomology modules. The full subcategories $\mathbf{D}_c^+(A)$, etc. are defined in the obvious manner.

Whenever we form the derived category $\mathbf{D}(\mathscr{A})$ of an abelian category $\mathscr{A}$, it is assumed that the multiplicative system of quasi-isomorphisms in the homotopy category $K(\mathscr{A})$ of complexes in $\mathscr{A}$ is 'locally small' in the sense of [**W**, 10.3.6, p.381], so we may construct the derived category without needing universes. This hypothesis is satisfied for all derived categories we will need to consider. As a general reference for detailed proofs of the basic facts about derived categories, we recommend [**RD**, I, II] and [**W**, Ch 10], but we generally follow the sign conventions in [**BBM**, 0.3] (as discussed in §1.3 below). In particular, we define $C^\bullet[m]$ to be the complex whose pth term is C^{p+m} and pth differential is $(-1)^m d_{C^\bullet}^{p+m}$ (beware that [**W**] writes $C^\bullet[-m]$ for what we have called $C^\bullet[m]$). To make this a functor, for any map of complexes $f^\bullet : C^\bullet \to C'^\bullet$ we define $f^\bullet[m] : C^\bullet[m] \to C'^\bullet[m]$ to be f^{n+m} in degree n. If $\mathbf{F}, \mathbf{G} : \mathbf{D}(\mathscr{A}) \to \mathbf{D}(\mathscr{B})$ are two δ-functors (i.e., maps of triangulated categories) with the same variance, we say that a natural transformation $\varphi : \mathbf{F} \to \mathbf{G}$ is δ-functorial if it is compatible with translations.

It is important to keep track of translations because we will sometimes be able to reduce ourselves to studying a complex which is concentrated in a single degree. In such a situation, reduction to the analysis of an object concentrated in degree 0 requires that we have kept track of translation-compatibility throughout the theory; this occurs in the proof of Lemma A.2.1, for example.

For a ringed space X, we usually write $\mathscr{H}om_X$ or $\mathscr{H}om$ instead of $\mathscr{H}om_{\mathscr{O}_X}$, and we usually write $\otimes$ instead of $\otimes_{\mathscr{O}_X}$. If $\varphi : \mathscr{F} \to \mathscr{G}$ is a map between sheaves on a ringed space, we denote by $\mathrm{im}(\mathscr{F})$ the image sheaf of φ in $\mathscr{G}$. If $f : X \to Y$ is a locally finite type map of schemes, we define the *relative dimension of f at $x \in X$* to be $\dim \mathscr{O}_{X_{f(x)},x}$. If this function is identically equal to $n \geq 0$, we say that f has *pure relative dimension n* (i.e., all non-empty fibers of f are equidimensional with dimension n). When $f : X \to Y$ is a smooth map of schemes, then $\omega_{X/Y}$ denotes the top exterior power of the locally free finite rank sheaf $\Omega^1_{X/Y}$ on X. The relative dimension of such smooth f is a locally constant $\mathbf{Z}$-valued function on X.

Somewhat less standard notation and terminology (mostly taken from [**RD**]) is the following. For a locally free $\mathscr{O}_X$-module $\mathscr{F}$ on a ringed space X, we denote the dual sheaf $\mathscr{H}om(\mathscr{F}, \mathscr{O}_X)$ by $\mathscr{F}^\vee$. A map $i : Y \hookrightarrow X$ of schemes is said to be a *local complete intersection* map, or an *lci* map, if i is a closed immersion and the associated quasi-coherent ideal sheaf $\mathscr{I}_Y$ on X is locally generated by a regular sequence, in the sense of [**EGA**, 0_{IV}, 15.2.2] (one could alternatively use the notion of regular sequence as defined in [**SGA6**, VII, 1.1], which behaves well for permutations of the sequence, but the definition in [**EGA**] suffices for our purposes). This is not a relative notion, but is preserved under composition. As an example, a section to a separated smooth map is an lci map. By [**EGA**, 0_{IV}, 15.1.9; IV_4, 16.9.2], for an lci map $i : Y \hookrightarrow X$ the $\mathscr{O}_Y$-module $i^*(\mathscr{I}_Y/\mathscr{I}_Y^2)$ is locally free with finite rank (called the *codimension of i*, and said to be the *pure* *codimension of* i if it is a constant) and if the rank is n near $y \in Y$, then any n generators of $\mathscr{I}_Y$ near $i(y)$ induce a basis of $i^*(\mathscr{I}_Y / \mathscr{I}_Y^2)$ near y. For example, a section to a separated smooth map with pure relative dimension n is an lci map with pure codimension n (and therefore remains so after any change of base on X). For an lci map $i : Y \hookrightarrow X$, the *dual* of $i^*(\mathscr{I}_Y / \mathscr{I}_Y^2)$ is called the *normal bundle* to Y in X, and the top exterior power of the normal bundle is denoted $\omega_{Y/X}$.

We say that a map of schemes $f : X \to Y$ is *Cohen-Macaulay* (or *CM*) if it is locally of finite presentation, flat, and all fibers are Cohen-Macaulay schemes (in this definition, it is equivalent to work with geometric fibers [**EGA**, IV_3, 6.7.1(i)]). A composite of CM maps is CM and a base change of a CM map is CM. If S is a scheme, X is a locally finitely presented flat S-scheme, and Y is a finitely presented closed subscheme of X, we say that the closed immersion $i : Y \hookrightarrow X$ is *transversally regular over* S if Y is S-flat and the fiber maps $Y_s \hookrightarrow X_s$ are local complete intersection maps for all $s \in S$. This property is preserved by any base change over S and sections to separated smooth S-maps have this property. If i is transversally regular over S, then i is an lci map. See [**EGA**, IV_4, 19.2.2, 19.2.4] for more details and equivalent formulations.

If $\mathscr{F}^{\bullet\bullet}$ is a double complex, then we denote by $\mathrm{Tot}^{\oplus}\mathscr{F}^{\bullet\bullet}$ the usual direct sum total complex with degree n term

$$(\mathrm{Tot}^{\oplus}\mathscr{F}^{\bullet\bullet})^n = \bigoplus_{p \in \mathbf{Z}} \mathscr{F}^{p,n-p}.$$

For any complex $\mathscr{G}^{\bullet}$, we define the *canonical truncation in degrees* $\leq n$ to be the subcomplex $\tau_{\leq n}(\mathscr{G}^{\bullet})$ obtained by replacing $\mathscr{G}^p$ with 0 if $p > n$ and with $\ker(\mathscr{G}^n \to \mathscr{G}^{n+1})$ if $p = n$. There is a canonical map $\tau_{\leq n}(\mathscr{G}^{\bullet}) \to \mathscr{G}^{\bullet}$ inducing an isomorphism on H^i's for $i \leq n$. If $\mathscr{F}^{\bullet\bullet}$ is a double complex, we define the *canonical truncation in rows* $\leq n$ to be the double subcomplex obtained by replacing each *column* with its canonical truncation in degrees $\leq n$, and we similarly define the *canonical truncation in columns* $\leq n$. Analogues for '$\geq n$' are defined in the obvious way by using cokernels in place of kernels. If we are given a commutative upper-half plane diagram in which the rows and columns are complexes, we define the *associated double complex* to be the upper-half plane double complex obtained by multiplying differentials in the pth column by $(-1)^p$. A typical example is to have a resolution of a complex by other complexes, and to make this into a double complex.

We sometimes write $A = B$ to denote the fact that A is canonically isomorphic to B (via an isomorphism which is always clear from the context). One notational deviation from [**RD**] is that we denote the total derived tensor product by $\overset{L}{\otimes}$ rather than by $\underline{\otimes}$.

1.3. Sign Conventions

We define the general isomorphism

(1.3.1) $$\mathrm{H}^n(C^{\bullet}[m]) \simeq \mathrm{H}^{n+m}(C^{\bullet})$$

without the intervention of signs. Variations on this isomorphism will play a prominent role in many explications of the general theory.

For the definition of the mapping cone we follow [**BBM**, (0.3.1), (0.3.2)], which is slightly different from [**RD**], [**W**]. More precisely, if $u^\bullet : A^\bullet \to B^\bullet$ is a map of complexes, we define the *mapping cone complex* $\mathrm{cone}^\bullet(u^\bullet)$ to have nth term $\mathrm{cone}^n(u^\bullet) = A^{n+1} \oplus B^n$ and nth differential

$$d^n_{\mathrm{cone}^\bullet(u^\bullet)} = \begin{pmatrix} -d_{A^\bullet}^{n+1} & 0 \\ u^{n+1} & d_{B^\bullet}^n \end{pmatrix}.$$

Following [**BBM**], we say that a diagram

$$A^\bullet \xrightarrow{u} B^\bullet \xrightarrow{v} C^\bullet \xrightarrow{w} A^\bullet[1]$$

in $\mathbf{D}(\mathscr{A})$ is a *distinguished triangle* if there is a commutative diagram in $\mathbf{D}(\mathscr{A})$

$$
\begin{array}{ccccccc}
A^\bullet & \xrightarrow{\ u\ } & B^\bullet & \xrightarrow{\ v\ } & C^\bullet & \xrightarrow{\ w\ } & A^\bullet[1] \\
\downarrow{\scriptstyle f}\,\simeq & & \downarrow{\scriptstyle g}\,\simeq & & \downarrow{\scriptstyle h}\,\simeq & & \downarrow{\scriptstyle f[1]} \\
A'^\bullet & \xrightarrow[u']{} & B'^\bullet & \xrightarrow[v']{} & \mathrm{cone}^\bullet(u') & \xrightarrow[-p]{} & A'^\bullet[1]
\end{array}
$$

with vertical isomorphisms (in $\mathbf{D}(\mathscr{A})$), where the bottom row is a diagram of complexes with $v'(b_n) = (0, b_n)$ and $-p(a_{n+1}, b_n) = -a_{n+1}$. Note the sign! This is equivalent to requiring that

$$A^\bullet \xrightarrow{-u} B^\bullet \xrightarrow{v} C^\bullet \xrightarrow{w} A^\bullet[1]$$

is a triangle (resp. an exact triangle) in the sense of [**RD**] (resp. [**W**]). The definition in [**BBM**] recovers the snake lemma as follows. Let

$$0 \to A^\bullet \xrightarrow{u} B^\bullet \xrightarrow{v} C^\bullet \to 0$$

be a short exact sequence of complexes and consider the quasi-isomorphism

(1.3.2) $\mathrm{cone}^\bullet(u) \to C^\bullet$

defined by $(a_{n+1}, b_n) \mapsto v(b_n)$ and let $p : \mathrm{cone}^\bullet(u) \to A^\bullet$ denote the canonical projection $(a_{n+1}, b_n) \mapsto a_{n+1}$. Via (1.3.1) and the quasi-isomorphism (1.3.2), the long exact cohomology sequence arising from the distinguished triangle

(1.3.3) $A^\bullet \xrightarrow{u} B^\bullet \to \mathrm{cone}^\bullet(u) \xrightarrow{-p} A^\bullet[1]$

(using $b_n \mapsto (0, b_n)$ in the middle) is exactly the usual long exact sequence from the snake lemma, without any extraneous signs in the coboundary maps (e.g., there is no sign as in [**W**, Exer 1.5.6, p.24]).

If $F : \mathscr{A} \to \mathscr{B}$ is a covariant left-exact functor and $\mathscr{A}$ has enough injectives, the δ-functor $\mathbf{R}F : \mathbf{D}^+(\mathscr{A}) \to \mathbf{D}^+(\mathscr{B})$ has its translation-compatibility defined *without* the intervention of signs, and likewise for any object A in $\mathscr{A}$ there is no intervention of signs in the definition of the isomorphism

(1.3.4) $\mathrm{H}^i(\mathbf{R}F(A[n])) = \mathrm{H}^i(\mathbf{R}F(A)[n]) = \mathrm{H}^{i+n}(\mathbf{R}F(A)) \simeq \mathrm{R}^{i+n}F(A),$

where this requires computing $\mathbf{R}F(A) = F(I^\bullet)$ if we choose to compute the derived functors $\mathrm{R}^\bullet F(A)$ via a specified injective resolution $I^\bullet$ of A. When $n = 0$, this convention makes (1.3.4) δ-*functorial* in A, due to the mapping cone and distinguished triangle conventions from [**BBM**] as in (1.3.3)! We adopt similar conventions for covariant total left-derived functors (and for $\mathbf{R}F : \mathbf{D}(\mathscr{A}) \to \mathbf{D}(\mathscr{B})$ if F has finite cohomological dimension, ...). The only contravariant δ-functors we have to deal with arise from Hom in the first variable, and this situation is carefully discussed below.

A source of much confusion (at least for the author) is the interpretation of (1.3.4) when translating abstract derived category results into the 'classical' language of ordinary derived functors (higher direct image sheaves, Ext sheaves, etc.). Fix an object A of $\mathscr{A}$ and choose an injective resolution $I^{\bullet}$ of A (in degrees ≥ 0), in terms of which we compute the derived functors

$$\mathbf{R}^j F(A) = \mathrm{H}^j(F(I^{\bullet})).$$

The definition of (1.3.4) requires us to compute $\mathbf{R}F(A[n]) = F(I^{\bullet}[n])$ (i.e., use $I^{\bullet}[n]$ as an injective resolution of $A[n]$ in the derived category), transforming (1.3.4) into the isomorphism

$$\mathrm{H}^i(F(I^{\bullet})[n]) = \mathrm{H}^{i+n}(F(I^{\bullet}))$$

without the intervention of signs, exactly as in (1.3.1). The headaches begin when we confront the fact that several derived category isomorphisms that pervade Grothendieck duality theory are defined in terms of Cartan-Eilenberg resolutions.

More precisely, recall that if $C^{\bullet}$ is a bounded below complex and $I^{\bullet\bullet}$ is a Cartan-Eilenberg resolution of $C^{\bullet}$, then the standard conventions are that the column $I^{p,\bullet}$ is taken as the injective resolution of C^p to be used in the computation of derived functors $\mathbf{R}^j F(C^p)$ (as in the construction of the Grothendieck spectral sequence), and likewise the canonical quasi-isomorphism

(1.3.5) $$C^{\bullet} \to \mathrm{Tot}^{\oplus}(I^{\bullet\bullet})$$

induced by augmentation is taken as the injective resolution to be used in the computation of the total derived functor $\mathbf{R}F(C^{\bullet})$. We now explicate these spectral sequence conventions when $C^{\bullet} = A[n]$. A Cartan-Eilenberg resolution of $C^{\bullet}$ is just an injective resolution $I^{\bullet}$ of A placed in the $-n$th column (to be used in the computation of ordinary derived functors $\mathbf{R}^j F(A)$) and forming the total complex leads us via (1.3.5) to compute $\mathbf{R}F(A[n]) = F(I^{\bullet+n})$, so (1.3.4) computes $\mathrm{H}^i(\mathbf{R}F(A[n])) = \mathrm{H}^{i+n}(F(I^{\bullet+n}[-n]))$, without the intervention of signs. But the (unique up to homotopy) isomorphism between $I^{\bullet+n}[-n]$ and $I^{\bullet}$ as injective resolutions of A is given by multiplication by $(-1)^{nm}$ in degree m, so the identification

$$\mathrm{H}^{i+n}(F(I^{\bullet+n}[-n])) = \mathrm{H}^{i+n}(F(I^{\bullet}))$$

without the intervention of signs corresponds to multiplication by $(-1)^{(i+n)n}$ on $\mathbf{R}^{i+n}F(A)$!

We will be faced with exactly this situation on several occasions, often with $i = 0$ and n equal to some geometrically significant parameter. In such cases, we must be vigilant in order to avoid a lot of confusion about powers of $(-1)^n$ when we attempt to explicate concrete consequences of the abstract derived category theory in terms of the 'classical' language of higher direct image sheaves and $\mathscr{E}xt$ sheaves. For example, the relationship between residues and the Grothendieck trace map on proper smooth curves, as in Appendix B, involves issues of the above sort. Fortunately, the relevant integer n will usually be additive with respect to composites of scheme morphisms, so the identity $(-1)^n(-1)^m = (-1)^{n+m}$ and the compatibility of linear maps with respect to multiplication by -1 will help to eliminate many potential difficulties along these lines. Alas, in general it seems that no matter which sign convention we choose to explicate results in concrete terms, we cannot avoid the occasional interference of signs in the statements. The reader should keep in mind that this sign nuisance is essentially just a problem in the explications of the abstract theory and has no influence on what is happening in the derived category formulations.

Let X be a ringed space (e.g., a 1 point space recovers commutative algebra situations). If $\mathscr{F}^\bullet$ is a complex of $\mathcal{O}_X$-modules, the sign conventions on $\mathscr{H}om^\bullet(\cdot,\cdot)$ should be such that the complex $\mathscr{H}om^\bullet(\mathcal{O}_X[0],\mathscr{F}^\bullet)$ is isomorphic to $\mathscr{F}^\bullet$ via the canonical isomorphism $\mathscr{H}om_{\mathcal{O}_X}(\mathcal{O}_X,\mathscr{F}^n) \simeq \mathscr{F}^n$ in degree n, without the intervention of any signs. Under the conventions in [**RD**], this is not the case. This can cause problems in the study of duality of sheaves. Moreover, some of the important (unchecked) diagrams which are claimed to commute in [**RD**] only commute up to slightly complicated signs. It is important to eliminate this type of error, because otherwise there are difficulties with globalization. We use different sign conventions from those in [**RD**] in order to avoid all such problems. This difference in conventions will force us to restate some results in [**RD**] in a 'corrected' form which agrees with our conventions. In general (with a notable exception mentioned in the definition of (1.3.16) below), our sign conventions agree with the ones in [**BBM**, 0.3] (which are the same as in [**SGA4**, XVII]). For example, $\mathscr{F}^\bullet[n]$ denotes the complex with pth term $\mathscr{F}^{p+n}$ and pth differential $(-1)^n d_{\mathscr{F}^\bullet}^{p+n}$ and we define $H^i(\mathscr{F}^\bullet[n]) \simeq H^{i+n}(\mathscr{F}^\bullet)$ without the intervention of signs.

Let us make explicit the important examples of the total tensor product and total $\mathscr{H}om$ complexes (whose definitions in [**RD**] involve different signs). Let $\mathscr{F}^\bullet$ and $\mathscr{G}^\bullet$ be two complexes of $\mathcal{O}_X$-modules on a ringed space X. The total tensor product and total $\mathscr{H}om$ complexes are defined in degree n to be

$$(\mathscr{F}^\bullet \otimes \mathscr{G}^\bullet)^n = \bigoplus_{p+q=n} \mathscr{F}^p \otimes_{\mathcal{O}_X} \mathscr{G}^q,$$

$$\mathscr{H}om^n(\mathscr{F}^\bullet,\mathscr{G}^\bullet) = \prod_{q\in\mathbf{Z}} \mathscr{H}om_{\mathcal{O}_X}(\mathscr{F}^q,\mathscr{G}^{n+q}),$$

and the differentials are respectively defined as follows:

$$d_{\mathscr{F}^\bullet \otimes \mathscr{G}^\bullet}^n|(\mathscr{F}^p \otimes_{\mathcal{O}_X} \mathscr{G}^q) = d_{\mathscr{F}^\bullet}^p \otimes 1 + (-1)^p \otimes d_{\mathscr{G}^\bullet}^q,$$

$$d_{\mathscr{H}om^\bullet(\mathscr{F}^\bullet,\mathscr{G}^\bullet)}^n((f^q)) = (d_{\mathscr{G}^\bullet}^{n+q} \circ f^q + (-1)^{n+1} f^{q+1} \circ d_{\mathscr{F}^\bullet}^q).$$

Note the $(-1)^{n+1}$ in the case of $\mathscr{H}om^\bullet$. The signs in the definitions of $\mathbf{R}\mathscr{H}om^\bullet$ and $\overset{\mathbf{L}}{\otimes}$ in [**RD**] are adjusted accordingly. Although this convention agrees with

Cartan-Eilenberg [**CE**, IV, §5] for the total $\otimes$, it disagrees with Cartan-Eilenberg for the total $\mathscr{H}om$ (in fact, [**RD**] and [**CE**] have the same sign convention for the total $\mathscr{H}om$).

We define the 'associativity' of $\otimes$ and the compatibility of $\otimes$ (resp. $\mathscr{H}om^{\bullet}$) with respect to translation in the first (resp. second) variable without the intervention of signs. The isomorphism

$$(1.3.6) \qquad \mathscr{F}^{\bullet} \otimes (\mathscr{G}^{\bullet}[m]) \simeq (\mathscr{F}^{\bullet} \otimes \mathscr{G}^{\bullet})[m]$$

is defined by using a sign of $(-1)^{pm}$ on $\mathscr{F}^{p} \otimes \mathscr{G}^{q}$ (cf. [**BBM**, (0.3.5.5)]) and the isomorphisms

$$(1.3.7) \qquad \alpha_1 : \mathscr{H}om^{\bullet}(\mathscr{F}^{\bullet}[1], \mathscr{G}^{\bullet}) \simeq \mathscr{H}om^{\bullet}(\mathscr{F}^{\bullet}, \mathscr{G}^{\bullet})[-1],$$

$$\alpha_{-1} : \mathscr{H}om^{\bullet}(\mathscr{F}^{\bullet}[-1], \mathscr{G}^{\bullet}) \simeq \mathscr{H}om^{\bullet}(\mathscr{F}^{\bullet}, \mathscr{G}^{\bullet})[1]$$

are defined by using respective signs of $(-1)^n$, $(-1)^{n+1}$ in degree n (cf. [**BBM**, (0.3.5.3), (0.3.5.1)] in case $\mathscr{G}^{\bullet}$ is supported in degree 0). These two isomorphisms are *inverse* to each other, in the sense that $T^{-1}(\alpha_{-1}) \circ \alpha_1$ and $T(\alpha_1) \circ \alpha_{-1}$ are the identity maps, where $T = [1]$ denotes the translation functor. This allows us to uniquely define isomorphisms

$$(1.3.8) \qquad \alpha_m : \mathscr{H}om^{\bullet}(\mathscr{F}^{\bullet}[m], \mathscr{G}^{\bullet}) \simeq \mathscr{H}om^{\bullet}(\mathscr{F}^{\bullet}, \mathscr{G}^{\bullet})[-m]$$

for all $m \in \mathbf{Z}$ such that $\alpha_0 = 1$, we recover (1.3.7) for $m = 1$, and

$$(1.3.9) \qquad T^{-n}(\alpha_m) \circ \alpha_n = \alpha_{n+m}$$

for all $n, m \in \mathbf{Z}$ (so we also recover the first definition of α_{-1} when $m = -1$). To be explicit, α_m involves a sign of $(-1)^{pm+m(m-1)/2}$ in degree p (as a safety check, this is compatible with (1.3.9)). We warn the reader that

$$(1.3.10) \qquad \mathscr{H}om^{\bullet}(\mathscr{F}^{\bullet}[m], \mathscr{G}^{\bullet}[m]) \xrightarrow[\simeq]{\alpha_m} \mathscr{H}om^{\bullet}(\mathscr{F}^{\bullet}, \mathscr{G}^{\bullet}[m])[-m]$$

$$\Big\| \qquad\qquad\qquad\qquad\qquad \Big\|$$

$$\mathscr{H}om^{\bullet}(\mathscr{F}^{\bullet}[m], \mathscr{G}^{\bullet})[m] \xrightarrow[\alpha_m]{\simeq} \mathscr{H}om^{\bullet}(\mathscr{F}^{\bullet}, \mathscr{G}^{\bullet})$$

only commutes up to a sign of $(-1)^m$.

With these conventions, $\mathbf{R}\mathscr{H}om^{\bullet}$ and $\overset{\mathbf{L}}{\otimes}$ are δ-functors in each variable; for complexes supported in degree 0 this recovers the classical (covariant) δ-functor structure on $\mathscr{T}or$ in the both variables, thanks to (1.3.1) and our conventions for mapping cones and translations. Also, we define

$$\mathscr{F}^{\bullet} \otimes \mathscr{G}^{\bullet} \simeq \mathscr{G}^{\bullet} \otimes \mathscr{F}^{\bullet}$$

using a sign of $(-1)^{pq}$ on $\mathscr{F}^{p} \otimes_{\mathcal{O}_X} \mathscr{G}^{q}$; this is compatible with translation in either variable. For applications in the study of dualizing sheaves, it will be convenient to also introduce the isomorphisms

$$(1.3.11) \qquad \mathscr{H}om^{\bullet}(\mathscr{G}^{\bullet}[n], \mathscr{H}^{\bullet}[n]) \simeq \mathscr{H}om^{\bullet}(\mathscr{G}^{\bullet}, \mathscr{H}^{\bullet})$$

which are defined with a sign of $(-1)^{nm}$ in degree m. In general, this isomorphism is *not* equal to going either way around (1.3.10), and has the advantage that it is compatible with respect to the identification $T^{n_1+n_2} = T^{n_1} \circ T^{n_2}$ (where T is the translation functor); this is why we must use (1.3.11) later on.

The case of δ-functoriality of $\mathscr{E}xt$ in the first variable presents a subtle sign issue which we must explain in order to remove any possibility of confusion. First of all, for any two $\mathcal{O}_X$-modules $\mathscr{F}$ and $\mathscr{G}$, we officially *define*

$$(1.3.12) \qquad \mathscr{E}xt_X^p(\mathscr{F}, \mathscr{G}) \overset{\text{def}}{=} H^p(\mathbf{R}\mathscr{H}om_X^\bullet(\mathscr{F}[0], \mathscr{G}[0])),$$

$$\operatorname{Ext}_X^p(\mathscr{F}, \mathscr{G}) \overset{\text{def}}{=} H^p(\mathbf{R}\operatorname{Hom}_X^\bullet(\mathscr{F}[0], \mathscr{G}[0]))$$

for all p. This "is" the standard definition of $\mathscr{E}xt$ and Ext via injective resolutions in the second variable, and we use (1.3.1), (1.3.7), (1.3.12), and the δ-bifunctor structure on $\mathbf{R}\mathscr{H}om^\bullet$ and $\mathbf{R}\operatorname{Hom}^\bullet$ to *define* the δ-bifunctor structure on $\mathscr{E}xt$ and Ext. For the second variable, this gives the usual coboundary maps via injective resolutions and the snake lemma, thanks to the equalities of complexes

$$\mathscr{H}om_X^\bullet(\mathscr{F}[0], \mathscr{G}^\bullet) = \mathscr{H}om_X(\mathscr{F}, \mathscr{G}^\bullet),$$

$$\operatorname{cone}^\bullet(\mathscr{H}om_X(\mathscr{F}, u^\bullet)) = \mathscr{H}om_X(\mathscr{F}, \operatorname{cone}^\bullet(u^\bullet))$$

and our earlier remarks concerning (1.3.3) and the snake lemma.

What about the first variable? Following [**Tohoku**, II, 2.3, p.144], one can use injective resolutions in the second variable to directly define a δ-functor structure in the first variable (and this is the only option for most ringed spaces, where there aren't enough projectives). It is an exercise with mapping cones to check that this recovers the same first variable δ-functor structure as via $\mathbf{R}\mathscr{H}om$'s and $\mathbf{R}\operatorname{Hom}$'s above. This uses the fact that the sign $\varepsilon(m) = (-1)^{m(m-1)/2}$ of (1.3.8) in degree 0 satisfies

$$(1.3.13) \qquad \varepsilon(-m)\varepsilon(1-m) = (-1)^m$$

and the differential $\mathrm{d}_{\mathscr{H}om^\bullet}^{-1}$ involves *no* sign. In the module setting, one needs to be careful when comparing this with the "classical" coboundary maps via projective resolutions. This is crucial, because we will often calculate some Ext's using Koszul complexes in order to define some important isomorphisms in duality theory.

To be precise, if $P^\bullet \to M[0]$ is a projective resolution and $N[0] \to I^\bullet$ is an injective resolution, then the diagram of quasi-isomorphisms

$$(1.3.14) \qquad \operatorname{Hom}_A^\bullet(P^\bullet, N[0]) \to \operatorname{Hom}_A^\bullet(P^\bullet, I^\bullet) \leftarrow \operatorname{Hom}_A^\bullet(M[0], I^\bullet)$$

is what gives the identification of our Ext's "computed" via projective (resp. injective) resolutions in the first (resp. second) variable. Consequently (as one checks via mapping cones and (1.3.13)), for a fixed A-module N our coboundary maps on $\text{Ext}_A^\bullet(\cdot, N)$ can be computed by using projective resolutions

$$P^\bullet \overset{\text{def}}{=} \cdots \to P^{-1} \to P^0 \to 0$$

of the first variable and then applying the snake lemma to complexes of the form $\text{Hom}_A^\bullet(P^\bullet, N[0])$. This is *not* the same as the "classical" (chain) complex $\text{Hom}_A(P^\bullet, N)$, but is instead the result of negating the indices in $\text{Hom}_A(P^\bullet, N)$ (to make it a cochain complex) and then multiplying the differentials by $(-1)^{n+1}$ in degree n (which is compatible with [**BBM**, (0.3.3.2)] for the contravariant functor $F = \text{Hom}_A(\cdot, N)$). Note that the 'canonical' isomorphism of complexes

(1.3.15) $$\text{Hom}_A(P^{-\bullet}, N) \simeq \text{Hom}_A^\bullet(P^\bullet, N)$$

lifting the identity in degree 0 is given by multiplication by $(-1)^{n(n+1)/2}$ in degree n.

For our purposes, the significance of this is that when we compute Ext's via projective resolutions in the first variable, certain *universal* signs in each degree must be introduced if we wish to correctly compute the coboundary maps. However, if we are trying to make a term-by-term construction or prove such a construction is δ-functorial, then these universal signs are sometimes irrelevant because all linear maps commute with multiplication by -1.

Using the general bifunctorial isomorphism

$$\text{Hom}_{\mathbf{D}(\mathscr{A})}(C^\bullet, C'^\bullet) \simeq \text{H}^0(\mathbf{R}\,\text{Hom}_{\mathscr{A}}^\bullet(C^\bullet, C'^\bullet))$$

for bounded below $C'^\bullet$, it follows essentially by *definition* that the isomorphism

$$\text{Hom}_{\mathbf{D}(X)}(\mathscr{F}[0], \mathscr{G}[n]) \overset{\simeq}{\longrightarrow} \text{H}^0(\mathbf{R}\,\text{Hom}_X^\bullet(\mathscr{F}[0], \mathscr{G}[n]))$$

$$\|$$

$$\text{Ext}_X^n(\mathscr{F}, \mathscr{G}) = \!\!=\!\!=\!\!= \text{H}^0(\mathbf{R}\,\text{Hom}_X^\bullet(\mathscr{F}[0], \mathscr{G}[0])[n])$$

is δ-functorial in the $\mathscr{O}_X$-module $\mathscr{G}$ (using the mapping cone mechanism to convert a short exact sequence of $\mathscr{O}_X$-modules into a distinguished triangle in $\mathbf{D}(X)$), and likewise we have δ-functoriality in $\mathscr{F}$ for the isomorphism

$$\text{Hom}_{\mathbf{D}(X)}(\mathscr{F}[-n], \mathscr{G}[0]) \overset{\simeq}{\longrightarrow} \text{H}^0(\mathbf{R}\,\text{Hom}_X^\bullet(\mathscr{F}[-n], \mathscr{G}[0]))$$

$$\|$$

$$\text{Ext}_X^n(\mathscr{F}, \mathscr{G}) = \!\!=\!\!=\!\!= \text{H}^0(\mathbf{R}\,\text{Hom}_X^\bullet(\mathscr{F}[0], \mathscr{G}[0])[n])$$

(in which the middle isomorphism involves an intervention of signs).

There is one sign convention in [**BBM**] which is wrong, and which we use in a corrected form. This is quite important for duality. Suppose X is a ringed space and $\mathscr{F}^\bullet$ and $\mathscr{G}^\bullet$ are two complexes of $\mathscr{O}_X$-modules. We want to define a 'double duality' map of complexes

(1.3.16) $$\mathscr{F}^\bullet \to \mathscr{H}om^\bullet(\mathscr{H}om^\bullet(\mathscr{F}^\bullet, \mathscr{G}^\bullet), \mathscr{G}^\bullet).$$

In degree p, this should be a map of $\mathcal{O}_X$-modules

$$\mathcal{F}^p \longrightarrow \prod_{q \in \mathbf{Z}} \mathscr{H}om(\prod_{p' \in \mathbf{Z}} \mathscr{H}om(\mathcal{F}^{p'}, \mathcal{G}^{p'+q}), \mathcal{G}^{p+q})$$

$$\downarrow \simeq$$

$$\prod_{(p',q)} \mathscr{H}om(\mathscr{H}om(\mathcal{F}^{p'}, \mathcal{G}^{p'+q}), \mathcal{G}^{p+q})$$

which projects to 0 in all factors with $p' \neq p$ and which projects to the canonical 'double duality' map $\mathcal{F}^p \to \mathscr{H}om(\mathscr{H}om(\mathcal{F}^p, \mathcal{G}^{p+q}), \mathcal{G}^{p+q})$ multiplied by some suitable sign $\epsilon_{p,q}$ in the factors with $p' = p$. It is easy to check that (1.3.16) is a map of complexes if and only if $\epsilon_{p+1,q} = (-1)^{p+q}\epsilon_{p,q+1}$ and $\epsilon_{p,q} = (-1)^p \epsilon_{p,q+1}$, which is equivalent to $\epsilon_{p,q} = (-1)^{pq}\epsilon_{0,0}$. Since we want to recover the usual double duality of sheaves when $\mathcal{F}^\bullet = \mathcal{F}[0]$, $\mathcal{G}^\bullet = \mathcal{G}[0]$ for $\mathcal{O}_X$-modules $\mathcal{F}$, $\mathcal{G}$, we must set $\epsilon_{0,0} = 1$. Thus, we must define $\epsilon_{p,q} = (-1)^{pq}$, and this ensures that (1.3.16) is a map of complexes. In [**BBM**, (0.3.4.2)], it is mistakenly claimed that the definition $\epsilon_{p,q} = (-1)^p$ gives a map of complexes. It should be noted that (1.3.16) is functorial in $\mathcal{F}^\bullet$ and is functorial with respect to *isomorphisms* in the '$\mathcal{G}^\bullet$-variable'.

As special cases of (1.3.16), $\mathcal{F}^\bullet \to \mathscr{H}om^\bullet(\mathscr{H}om^\bullet(\mathcal{F}^\bullet, \mathcal{O}_X[0]), \mathcal{O}_X[0])$ is given in degree n by the usual double duality map

$$\mathcal{F}^n \to \mathscr{H}om(\mathscr{H}om(\mathcal{F}^n, \mathcal{O}_X), \mathcal{O}_X)$$

multiplied by $(-1)^{-n^2} = (-1)^n$, while

(1.3.17) $\mathcal{O}_X[0] \to \mathscr{H}om^\bullet(\mathscr{H}om^\bullet(\mathcal{O}_X[0], \mathcal{G}^\bullet), \mathcal{G}^\bullet)$

is given in degree 0 by the canonical map

$$\mathcal{O}_X \to \prod_{q \in \mathbf{Z}} \mathscr{H}om(\mathscr{H}om(\mathcal{O}_X, \mathcal{G}^q), \mathcal{G}^q) \simeq \prod_{q \in \mathbf{Z}} \mathscr{H}om(\mathcal{G}^q, \mathcal{G}^q)$$

without any intervention of signs. If $\mathcal{E}^\bullet$ is a complex of locally free finite rank $\mathcal{O}_X$-modules, then we denote by $\mathcal{E}^{\bullet\vee}$ the 'dual' complex $\mathscr{H}om^\bullet(\mathcal{E}^\bullet, \mathcal{O}_X[0])$. There is a natural isomorphism of complexes $i_{\mathcal{E}^\bullet} : \mathcal{E}^\bullet \to \mathcal{E}^{\bullet\vee\vee}$ which involves a sign of $(-1)^n$ in degree n and the maps $i_{\mathcal{E}^\bullet\vee}$ and $i_{\mathcal{E}^\bullet}^\vee$ are inverses. When $\mathcal{E}^\bullet$ is bounded as well, then for any complex $\mathcal{G}^\bullet$ there is a natural isomorphism

(1.3.18) $\mathcal{G}^\bullet \otimes \mathcal{E}^{\bullet\vee} \simeq \mathscr{H}om^\bullet(\mathcal{E}^\bullet, \mathcal{G}^\bullet)$

which involves no intervention of signs. Similarly, if $\mathcal{E}'^\bullet$ is another bounded complex of locally free sheaves with finite rank, then the isomorphism

(1.3.19) $(\mathcal{E}^\bullet \otimes \mathcal{E}'^\bullet)^\vee \simeq \mathcal{E}'^{\bullet\vee} \otimes \mathcal{E}^{\bullet\vee}$

involves no intervention of signs (note the 'flip'!). Moreover, it is easy to check that both (1.3.18) and (1.3.19) are compatible with translation in any of the variables. Using (1.3.18) and (1.3.19) is usually a convenient formal way to guess which signs arise in general isomorphisms involving $\overset{\mathbf{L}}{\otimes}$ and $\mathbf{R}\mathscr{H}om^\bullet$, by 'reduction' to an assertion about $\overset{\mathbf{L}}{\otimes}$'s.

We should make some remarks concerning the interaction of (1.3.16) with translations. Fix complexes of $\mathcal{O}_X$-modules $\mathcal{R}^\bullet$ and $\mathcal{F}^\bullet$. Using (1.3.16), we have a natural transformation of functors of $\mathcal{F}^\bullet$

(1.3.20) $$\eta_{\mathcal{R}^\bullet} : \mathcal{F}^\bullet \to \mathcal{H}om^\bullet(\mathcal{H}om^\bullet(\mathcal{F}^\bullet, \mathcal{R}^\bullet), \mathcal{R}^\bullet).$$

In applications to duality theory (on suitable schemes), $\mathcal{R}^\bullet$ is a dualizing complex, unique up to translation and tensoring with an invertible sheaf (these two operations commute without the intervention of signs). There is no issue of signs when identifying $\eta_{\mathcal{R}^\bullet}$ and $\eta_{\mathcal{L}[0]\otimes\mathcal{R}^\bullet}$ for an invertible sheaf $\mathcal{L}$. Moreover, we identify $\eta_{\mathcal{R}^\bullet}$ and $\eta_{\mathcal{R}^\bullet[n]}$ by using the canonical isomorphism (1.3.11) which involves multiplication by $(-1)^{nm}$ in degree m. Thus, the identification of $\eta_{\mathcal{R}^\bullet}$ and $\eta_{\mathcal{L}[n]\otimes\mathcal{R}^\bullet}$ for an invertible sheaf $\mathcal{L}$ and integer n is compatible with the isomorphisms $\mathcal{L}[n] \otimes \mathcal{R}^\bullet \simeq (\mathcal{L}[0] \otimes \mathcal{R}^\bullet)[n]$ and $\mathcal{L}_1[n_1] \otimes (\mathcal{L}_2[n_2] \otimes \mathcal{R}^\bullet) \simeq (\mathcal{L}_1 \otimes \mathcal{L}_2)[n_1 + n_2] \otimes \mathcal{R}^\bullet$, both of which are defined without the intervention of signs.

Now fix $\mathcal{R}^\bullet$ and define the 'dual' complex $D(\mathcal{F}^\bullet) = \mathcal{H}om^\bullet(\mathcal{F}^\bullet, \mathcal{R}^\bullet)$. We want to consider the natural transformation $\eta = \eta_{\mathcal{R}^\bullet} : \mathrm{id} \to DD$ arising from (1.3.20). The preceding discussion makes clear the sense in which η is independent of replacing $\mathcal{R}^\bullet$ by any $\mathcal{L}[n] \otimes \mathcal{R}^\bullet$ (with $\mathcal{L}$ an invertible sheaf and $n \in \mathbf{Z}$). It is very important for some compatibilities in the Grothendieck Duality Theorem [**RD**, VII, 3.4] that η respects 'triple duality' in the sense that the composite

(1.3.21) $$D(\mathcal{F}^\bullet) \xrightarrow{\eta(D(\mathcal{F}^\bullet))} (DD)(D(\mathcal{F}^\bullet)) = D(DD(\mathcal{F}^\bullet)) \xrightarrow{D(\eta(\mathcal{F}^\bullet))} D(\mathcal{F}^\bullet)$$

is the *identity*. This is a straightfoward calculation. A more delicate point is the fact that η commutes with translation. More precisely, let $T(\mathcal{F}^\bullet) = \mathcal{F}^\bullet[1]$ denote the translation functor on complexes. By (1.3.7), we have natural isomorphisms $\alpha : DT \simeq T^{-1}D$, $\beta : TD \simeq DT^{-1}$ which involve the respective signs $(-1)^n$ and $(-1)^{n+1}$ in degree n. The translation-compatibility of η amounts to the claim that the diagram

(1.3.22)

$$\begin{array}{ccc} T & \xrightarrow{\ \eta_T\ } & DD \circ T \\ {\scriptstyle T(\eta)}\big\downarrow & & \big\uparrow{\scriptstyle D(\alpha)} \\ T \circ DD & \xrightarrow[\ \beta_D\ ]{} & DT^{-1}D \end{array}$$

commutes. This is ultimately because for any p and q,

$$(-1)^{p(q+1)} = (-1)^{(p+1)q}(-1)^{p+1}(-1)^{q+1}.$$

The commutativity of (1.3.22) is needed in some of the translation-compatibility proofs for isomorphisms in Grothendieck's duality theory. When $\mathcal{R}^\bullet$ is bounded below, it follows from (1.3.22) that the natural map of functors on $\mathbf{D}^+(X)$

$$\eta_{\mathcal{R}^\bullet} : 1 \to \mathbf{R}\mathcal{H}om^\bullet(\mathbf{R}\mathcal{H}om^\bullet(\cdot, \mathcal{R}^\bullet), \mathcal{R}^\bullet)$$

is δ-functorial (i.e., is also compatible with translations), and likewise as functors on $\mathbf{D}(X)$ if $\mathcal{R}^\bullet$ has *finite* injective dimension.

We will occasionally need to use Čech theory and the "Čech to derived functor cohomology" spectral sequence, so we recall here the basic definitions. Quite generally, if X is any ringed space, $\mathscr{F}$ is any $\mathscr{O}_X$-module, we define $\underline{\mathrm{H}}^q(\mathscr{F})$ to be the *presheaf*

$$U \mapsto \mathrm{H}^q(U, \mathscr{F})$$

on X. Also, if $\mathfrak{V} = \{V_i\}_{i \in I}$ is any *ordered* open covering of X, for each $\underline{i} = (i_0, \ldots, i_p) \in I^{p+1}$ with $i_0 < \cdots < i_p$ we define

$$V_{\underline{i}} = V_{i_0} \cap \cdots \cap V_{i_p}$$

and let $j_{\underline{i}} : V_{\underline{i}} \hookrightarrow X$ denote the canonical open immersion. If $\mathscr{F}$ is any $\mathscr{O}_X$-module *presheaf*, we define the Čech complex $\mathscr{C}^\bullet(\mathfrak{V}, \mathscr{F})$ to have degree p term

$$\prod_{i_0 < \cdots < i_p} j_{\underline{i}_*}(\mathscr{F}|_{V_{\underline{i}}}),$$

and a typical section of this presheaf is denoted $s = (s_{\underline{i}})$. The differential is given by the habitual formula

$$\mathrm{d}^p(s)_{(i_0, \ldots, i_{p+1})} = \sum_{j=0}^{p+1} (-1)^j s_{(i_0, \ldots, \widehat{i_j}, \ldots, i_{p+1})}|_{V_{(i_0, \ldots, i_{p+1})}}$$

and the augmentation from $\mathscr{F}$ in degree 0 is defined in the obvious way without the intervention of signs. Observe that the signs in the differential are $(-1)^j$, not $(-1)^{j+1}$. Naturally this choice must be remembered whenever we explicitly compute cohomology using Čech theory (cf. the analysis of (B.4.8)).

When $\mathscr{F}$ is a sheaf (rather than a presheaf), $\mathscr{C}^\bullet(\mathfrak{V}, \mathscr{F})$ is a complex of sheaves and is a resolution of $\mathscr{F}$. The above is the standard definition of the Čech complex. The complex of global sections of $\mathscr{C}^\bullet(\mathfrak{V}, \mathscr{F})$ is denoted $\check{\mathrm{C}}^\bullet(\mathfrak{V}, \mathscr{F})$ and the cohomology of this complex is denoted $\check{\mathrm{H}}^\bullet(\mathfrak{V}, \mathscr{F})$. As is well-known, we can make the same definition for non-ordered covers by omitting the condition $i_0 < \cdots < i_p$ in the products, and this defines a complex equipped with a functorial quasi-isomorphism to the above definition (upon choosing an ordering of I). On the level of cohomology, this permits us to pass to direct limits over covers, with respect to refinements. We write $\check{\mathrm{H}}^\bullet(X, \mathscr{F})$ for this limit.

The "Čech to derived functor cohomology" spectral sequence is a special case of the Grothendieck spectral sequence. More precisely, if we denote by $\mathscr{O}_X$−mod the abelian category of $\mathscr{O}_X$-modules, $\Gamma(X, \mathscr{O}_X)$−mod the abelian category of $\Gamma(X, \mathscr{O}_X)$-modules, and $\mathrm{Ch}_{\geq 0}$ the abelian category of complexes of $\Gamma(X, \mathscr{O}_X)$-modules in degrees ≥ 0, then we can consider the left exact functors

$$\check{\mathrm{C}}^\bullet(\mathfrak{V}, \cdot) : \mathscr{O}_X -\mathrm{mod} \to \mathrm{Ch}_{\geq 0}, \quad \mathrm{H}^0 : \mathrm{Ch}_{\geq 0} \to \Gamma(X, \mathscr{O}_X) -\mathrm{mod}$$

whose composite functor is $\Gamma(X, \cdot)$. These categories all have enough injectives. By universal δ-functor arguments, the derived functors are respectively

$$\check{\mathrm{C}}^\bullet(\mathfrak{V}, \underline{\mathrm{H}}^q(\cdot)), \quad \mathrm{H}^q$$

for $q \geq 0$. The construction of the Grothendieck spectral sequence can be applied here, yielding a spectral sequence of the form

$$E_2^{p,q} = \check{H}^p(\mathfrak{V}, \underline{H}^q(\mathscr{F})) \Rightarrow H^{p+q}(X, \mathscr{F}).$$

This is compatible with the 'non-ordered' Čech construction and with respect to refinements in $\mathfrak{V}$.

In particular, we get a functorial edge map

$$(1.3.23) \qquad E_2^{p,0} = \check{H}^p(\mathfrak{V}, \mathscr{F}) \to H^p(X, \mathscr{F})$$

compatible with refinement in $\mathfrak{V}$. This edge map is used in some calculations (see (B.4.8)ff), so it is crucial to have an 'explicit' construction of this map. To give such a construction, let $\mathscr{I}^\bullet$ be any injective resolution of $\mathscr{F}$. There is a map of resolutions

$$(1.3.24) \qquad \mathscr{C}^\bullet(\mathfrak{V}, \mathscr{F}) \to \mathscr{I}^\bullet,$$

unique up to homotopy. Applying $\Gamma(X, \cdot)$, the induced map on homology in any degree p is *exactly* the map (1.3.23). The verification of this fact proceeds by first immediately reducing to the case $\mathscr{O}_X = \mathbf{Z}$ (i.e., abelian sheaves). In this setting, the global sections of an injective sheaf form an injective abelian group, and an unfortunately lengthy analysis of double and triple complexes based on this fact leads to the result. The reader who prefers to bypass this can use (1.3.24) to directly define (1.3.23), since the properties we need for (1.3.23) later on — functoriality, pullback compatibility, and δ-functorial isomorphism results for H^0 and H^1 in general and for $H^\bullet$ with quasi-coherent sheaves on quasi-compact separated schemes — can be directly proved on the basis of this more concrete definition (using the *existence* of the above Grothendieck spectral sequence, without needing an explication of it).

Finally, we address the important topic of Koszul complexes. If A is a ring, M is an A-module, and $f_1, \ldots, f_n \in A$ is an ordered sequence of n elements (with $n \geq 1$), we define the Koszul complex $K_\bullet(\mathbf{f})$ to have terms $K_{-p}(\mathbf{f}) = \bigwedge^p(A^n)$ in degree $-p$ for $0 \leq p \leq n$ (we use the chain complex notation '$\bullet$' for this cochain complex purely for psychological reasons). If $e_1, \ldots, e_n \in A^n$ is the standard basis, the differential $d_{-p} : K_{-p}(\mathbf{f}) \to K_{-(p-1)}(\mathbf{f})$ satisfies

$$d_{-p}(e_{i_1} \wedge \cdots \wedge e_{i_p}) = \sum_{j=1}^{p} (-1)^{j+1} f_{i_j} e_{i_1} \wedge \cdots \wedge \widehat{e_{i_j}} \wedge \cdots \wedge e_{i_p}$$

When $n = 0$, we define $K_\bullet(\mathbf{f}) = A[0]$. Note that if $g_1, \ldots, g_m \in A$ is an ordered sequence of m elements, we have a canonical isomorphism of complexes

$$K_\bullet(\mathbf{g}, \mathbf{f}) \simeq K_\bullet(\mathbf{g}) \otimes K_\bullet(\mathbf{f})$$

which does not involve the intervention of signs.

For reasons which will become clear in the proof of Lemma 2.6.3, we define

$$(1.3.25) \qquad K^\bullet(\mathbf{f}, M) \stackrel{\text{def}}{=} \operatorname{Hom}_A(K_{-\bullet}(\mathbf{f}), M)$$

(supported in degrees from 0 to n). Explicitly, $K^\bullet(\mathbf{f}, M) = M[0]$ when $n = 0$, and when $n > 0$ the differential d^p in degree p satisfies

(1.3.26)

$$\mathrm{d}^p(\alpha)(e_{i_1} \wedge \cdots \wedge e_{i_{p+1}}) = \sum_{j=1}^{p+1} (-1)^{j+1} f_{i_j} \alpha(e_{i_1} \wedge \cdots \wedge \widehat{e_{i_j}} \wedge \cdots \wedge e_{i_{p+1}});$$

For example, for $\alpha \in K^0(\mathbf{f}, M) = \mathrm{Hom}_A(A, M)$, we have

$$\mathrm{d}^0(\alpha)(e_i) = f_i \alpha(1).$$

In general, (1.3.25) is *not* the same as the complex $\mathrm{Hom}_A^\bullet(K_\bullet(\mathbf{f}), M)$, for which the analogue of (1.3.26) has $(-1)^{j+1}$ replaced by $(-1)^{j+p}$. The sign convention for $K^\bullet(\mathbf{f}, M)$ in [**EGA**, III$_1$, 1.1.2.3] coincides with our definition (1.3.25), and this is also adapted to δ-functorial computations of Čech cohomology [**EGA**, III$_1$, (1.2.2.3)]. The use of the definition (1.3.25), rather than $\mathrm{Hom}_A^\bullet(K_\bullet(\mathbf{f}), M)$, is crucial in the proof of Theorem 2.5.1 (it is surprising to the author that the "wrong" definition $\mathrm{Hom}_A^\bullet(K_\bullet(\mathbf{f}), M)$ for $K^\bullet(\mathbf{f}, M)$ is the one that a priori seems better-adapted to derived category considerations).

We define $\mathrm{H}^n(\mathbf{f}, M) = \mathrm{H}^n(K^\bullet(\mathbf{f}, M))$. The discussion near (1.3.15) shows that the 'canonical' (unique-up-to-homotopy) isomorphism

(1.3.27) $\mathrm{Hom}_A^\bullet(K_\bullet(\mathbf{f}), M) \simeq K^\bullet(\mathbf{f}, M)$

lifting the identity map on M in degree 0 is given in degree p by multiplication by $(-1)^{p(p+1)/2}$. Suppose that $\{f_1, \ldots, f_n\}$ is a regular sequence in A generating an ideal J, so $\mathrm{Spec}(A/J) \hookrightarrow \mathrm{Spec}(A)$ is a local complete intersection morphism with pure codimension n. Using the derived category isomorphism $K_\bullet(\mathbf{f}) \simeq (A/J)[0]$ induced by the canonical surjection $A \to A/J$ in degree 0, via the isomorphism of complexes (1.3.27) we get δ-*functorial* isomorphisms

$$\mathrm{Ext}_A^\bullet(A/J, M) \simeq \mathrm{H}^\bullet(\mathbf{f}, M)$$

for variable A-modules M, recovering in degree 0 the canonical identification of $\mathrm{Hom}_A(A/J, M)$ with J-torsion in M. In degree n this isomorphism plays a rather special role in Grothendieck's theory, so we give it a name:

(1.3.28) $\psi_{\mathbf{f}, M} : \mathrm{Ext}_A^n(A/J, M) \simeq \mathrm{H}^n(\mathbf{f}, M) = M/JM$

(there is no intervention of signs in the definition of the equality on the right). Keep in mind that, due to the isomorphism (1.3.27), the definition of (1.3.28) involves a sign of $(-1)^{n(n+1)/2}$, and of course it depends on the *ordered* sequence $\{f_1, \ldots, f_n\}$.

Acknowledgements I would like to thank deJong, Deligne, and Emerton for helpful discussions, Grothendieck, Hartshorne, Kleiman, and Lipman for inspiration, and Berthelot, Kisin, Lang, Messing, and Ogus for their encouragement. I am grateful to the Institute for Advanced Study and the University of Sydney for their hospitality while a preliminary version of this book was written, to Joe Harris and Silvio Levy for their generous computer assistance, and to the editorial staff at Springer-Verlag for their infinite patience. Finally, I am deeply indebted to Ofer Gabber for his careful reading of an earlier version of this manuscript. It is my belief that there are no mathematical errors in this book; Gabber's comments, corrections, and suggestions were invaluable in (hopefully) making this goal a reality. All responsibility for the accuracy of statements in this book rests with me. This work was partially supported by an NSF grant.

CHAPTER 2

Basic Compatibilities

After reviewing some preliminary general nonsense, we discuss the basic formalism of the functor $f^\sharp : \mathbf{D}^+_{qc}(Y) \to \mathbf{D}^+_{qc}(X)$ (resp. $f^\flat : \mathbf{D}^+_{qc}(Y) \to \mathbf{D}^+_{qc}(X)$) for a smooth (resp. finite) map $f : X \to Y$ between locally noetherian schemes. A 'projective trace' map Trp_f is defined in case $X = \mathbf{P}^n_Y$ and f is the canonical projection, and a 'fundamental local isomorphism' η_f is defined in case f is a closed immersion which is a local complete intersection morphism of pure codimension. Most of this chapter is concerned with verifying several important non-trivial properties of these constructions. At the end of the chapter, these compatibilities are used to 'glue' the definitions of $(\cdot)^\sharp$ and $(\cdot)^\flat$ in the case of more general maps which are neither smooth nor finite.

2.1. General Nonsense

There are a series of rather general isomorphisms among composites of hyperderived functors given in [**RD**, I, II], and these are frequently used when one works with derived categories. Such isomorphisms should involve little or no intervention of signs. We want to begin by modifying some of these general isomorphisms so as to avoid sign problems later on (and to ensure compatibility with the conventions in §1.3). Since the proofs in [**RD**] carry over with very minor changes, in this section we often list which isomorphisms are to be changed and leave it to the reader to consult the cited places in [**RD**] for details.

The following extremely important lemma is the source of many isomorphisms in [**RD**], such as (2.5.3) below, and involves some delicate sign considerations.

LEMMA 2.1.1. [**RD**, I, 7.4] *Let $\mathscr{A}$, $\mathscr{B}$ be abelian categories, assume $\mathscr{A}$ has enough injectives, and let $F : \mathscr{A} \to \mathscr{B}$ be a covariant left exact functor with finite cohomological dimension n. Assume that every object in $\mathscr{A}$ is a quotient of an object P for which $\mathrm{R}^i F(P) = 0$ for all $i \neq n$. Let $G = \mathrm{R}^n F$, so $\{\mathrm{R}^i F\}_{i \leq n}$ is an erasable δ-functor and hence a left-derived functor of the right exact functor G (so G has finite homological dimension). Then the hyperderived δ-functors $\mathrm{R}F, \mathrm{L}G : \mathbf{D}(\mathscr{A}) \to \mathbf{D}(\mathscr{B})$ exist and there is a functorial isomorphism $\psi : \mathrm{R}F \simeq (\mathrm{L}G)[-n]$.*

The diagram

(2.1.1)
$$\mathbf{R}F(C^{\bullet}[m]) \xrightarrow[\simeq]{\psi_{C^{\bullet}[m]}} \mathbf{L}G(C^{\bullet}[m])[-n]$$

$$\mathbf{R}F(C^{\bullet})[m] \xrightarrow[\psi_{C^{\bullet}}[m]]{\simeq} (\mathbf{L}G(C^{\bullet})[-n])[m]$$

commutes up to a sign of $(-1)^{nm}$, *where the vertical maps are defined without the intervention of signs (in accordance with our convention for translation-compatibility of total derived functors).*

The sign in (2.1.1) is crucial in several later results, such as the translation-compatibility of the isomorphisms (2.3.5) and (2.5.3), as well as the commutativity of (2.7.17).

Before we recall the definition of the map ψ in Lemma 2.1.1, we note that [**RD**] says "$[-n]$ means shift n places to the right," but the natural construction of the proof also requires multiplying differentials by $(-1)^n$; that is, "$[-n]$" really is the usual $-n$-fold translation functor on complexes. Let us see where the extra sign of $(-1)^n$ comes from. Let $C^{\bullet}$ be a complex of objects in $\mathscr{A}$ and let the double complex $I^{\bullet\bullet}$ be a Cartan-Eilenberg resolution of $C^{\bullet}$. For any fixed $m \geq 0$, we can form the mth 'vertical' cohomology $H^{m,p} = \mathrm{H}_v^m(F(I^{p,\bullet}))$ along each column $F(I^{p,\bullet})$, and this is a complex $H^{m,\bullet}$ in the 'horizontal' direction. Since each column $I^{p,\bullet}$ is an injective resolution of C^p, we have a canonical term-by-term identification $H^{m,p} \simeq \mathbf{R}^m F(C^p)$. However, a Cartan-Eilenberg resolution is a double complex and *not* a commutative diagram, so $\mathrm{d}_{H^{m,\bullet}}^p : H^{m,p} \to H^{m,p+1}$ is $(-1)^m \mathbf{R}^m F(\mathrm{d}_{C^{\bullet}}^p)$. When $m = n$, this yields the desired extra sign of $(-1)^n$. More precisely, in order to describe $\psi : \mathbf{R}F \simeq (\mathbf{L}G)[-n]$, it suffices to consider $\psi(C^{\bullet})$ for a complex $C^{\bullet}$ all of whose terms are G-*acyclic*. If $I'^{\bullet\bullet}$ denotes the canonical truncation of $I^{\bullet\bullet}$ in rows $\leq n$, then $\psi(C^{\bullet})$ is the composite

(2.1.2) $\mathbf{R}F(C^{\bullet}) \mathrel{=\!=\!=} F(\mathrm{Tot}^{\oplus} I'^{\bullet\bullet}) \longrightarrow F(I'^{\bullet-n,n})/\mathrm{im}F(I'^{\bullet-n,n-1})$

$$((\mathbf{L}G)(C^{\bullet}))[-n] \xleftarrow[\simeq]{} \mathbf{R}^n F(C^{\bullet})[-n]$$

From this description, one checks that ψ is natural. We refer the reader to [**RD**, I, 7.4] for the proof that it is a quasi-isomorphism; this uses the coerasability hypothesis on $\mathbf{R}^n F$. The analysis of the sign-commutativity in (2.1.1) is a little tricky, so we now give the justification.

Let $C^{\bullet}$ be a complex of G-acyclics and let $I^{\bullet\bullet}$, $I'^{\bullet\bullet}$ be as above. Define the double complex $\widetilde{I}^{\bullet\bullet}$ by $\widetilde{I}^{p,q} = I^{p+m,q}$ (i.e., shift all columns m units to the left) and

$$\widetilde{\mathrm{d}}_v^{p,q} = (-1)^m \mathrm{d}_v^{p+m,q}, \quad \widetilde{\mathrm{d}}_h^{p,q} = (-1)^m \mathrm{d}_h^{p+m,q}$$

(i.e., multiply all differentials by $(-1)^m$). The canonical truncation $\widetilde{I}'^{\bullet\bullet}$ of $\widetilde{I}^{\bullet\bullet}$ in rows $\leq n$ is similarly related to $I'^{\bullet\bullet}$, and $\widetilde{I}^{\bullet\bullet}$ is a Cartan-Eilenberg of $C^{\bullet}[m]$.

Note that

$$\mathrm{Tot}^\oplus \widetilde{I}'^{\bullet\bullet} = (\mathrm{Tot}^\oplus I'^{\bullet\bullet})[m].$$

Thus, the commutativity of (2.1.1) up to a sign of $(-1)^{nm}$ amounts to the commutativity in the *derived category* of the outside edge of

(2.1.3)

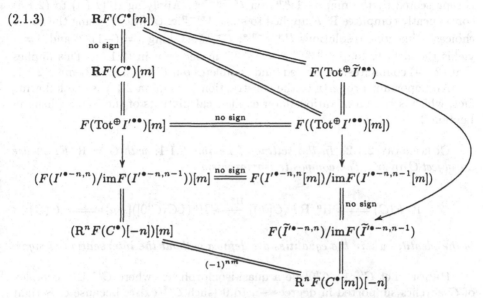

where the unlabelled equalities involve no intervention of signs and the curved arrow is the canonical map. All subdiagrams aside from the bottom part obviously commute on the level of complexes (for the top left part, this uses our convention that translation-compatibility of total derived functors involves no intervention of signs, as in (1.3.1)). For the bottom part, we will check commutativity on the level of complexes by looking in each separate degree. In degree r, the diagram can be written as

(2.1.4)

$$
\begin{array}{ccc}
H^n(F(I^{r+m-n,\bullet})) & =\!=\!= & R^n F(C^{r+m-n}) \\
\| & & \Big\|{\scriptstyle (-1)^{nm}} \\
H^n(F(\widetilde{I}^{r-n,\bullet})) & =\!=\!= & R^n F(C^\bullet[m]^{r-n})
\end{array}
$$

where the horizontal equalities are defined by viewing $I^{r+m-n,\bullet}$ (resp. $\widetilde{I}^{r-n,\bullet}$) as an injective resolution of C^{r+m-n} (resp. $C^\bullet[m]^{r-n}$), while the right vertical equality involves multiplying the identity $C^{r+m-n} = C^\bullet[m]^{r-n}$ by $(-1)^{nm}$ and the left vertical equality uses the term-by-term identity $I^{r+m-n,q} - \widetilde{I}^{r-n,q}$ defined without the intervention of signs.

By *definition*, $\widetilde{I}^{r-n,\bullet}$ is simply $I^{r+m-n,\bullet}$ with all differentials multiplied by $(-1)^m$. Thus, for $\varepsilon \in \{\pm 1\}$, the unique-up-to-homotopy map

(2.1.5)

$$I^{r+m-n,\bullet} \to \widetilde{I}^{r-n,\bullet}$$

over

(2.1.6) $$C^{r+m-n} \xrightarrow{\;\varepsilon\;} C^{r+m-n}$$

is represented by the map $(-1)^{pm}\varepsilon$ on $I^{r+m-n,p}$. Applying $\mathrm{H}^p(F(\cdot))$ to (2.1.5) consequently computes $\mathrm{R}^p F$ applied to $\varepsilon : C^{r+m-n} \simeq C^\bullet[m]^{r-n}$ using the above choices of injective resolutions $I^{r+m-n,\bullet}$, $\widetilde{I}^{r-n,\bullet}$. Taking $\varepsilon = (-1)^{nm}$ and $p = n$ yields the identity map $I^{r+m-n,n} = \widetilde{I}^{r-n,n}$ in degree n in (2.1.5). This implies that (2.1.4) commutes, as desired, and completes our discussion of Lemma 2.1.1.

An important property of the construction in Lemma 2.1.1 is the following fact, which is crucial in various later explicit calculations of the isomorphism in Lemma 2.1.1:

COROLLARY 2.1.2. *In the setting of Lemma 2.1.1, with $G = \mathrm{R}^n F$, choose an object C in $\mathscr{A}$. The composite isomorphism*

$$\mathrm{R}^n F(C) =\!\!= \mathrm{H}^n(\mathbf{R}F(C[0])) \xrightarrow{\;\mathrm{H}^n(\psi)\;} \mathrm{H}^n(\mathbf{L}G(C[0])[-n]) =\!\!= G(C)$$

is the identity, where the equalities are defined without the intervention of signs.

PROOF. Let $C'^\bullet \to C[0]$ be a quasi-isomorphism, where $C'^\bullet$ is a complex of G-acyclics supported in degrees $-n$ to 0 (such $C'^\bullet$ exists because G is right exact with cohomological dimension n). Let $I'^{\bullet\bullet}$ be the canonical truncation in rows $\leq n$ of a Cartan-Eilenberg resolution of $C'^\bullet$, so

$$C'^\bullet \to \mathrm{Tot}^\oplus(I'^{\bullet\bullet})$$

is a quasi-isomorphism to a bounded complex of F-acyclics, supported in degrees $-n$ to n. Let $C[0] \to J^\bullet$ be an injective resolution (with $J^q = 0$ for $q < 0$), so by Lemma 2.1.4 there is a quasi-isomorphism $h : \mathrm{Tot}^\oplus(I'^{\bullet\bullet}) \to J^\bullet$, unique up to homotopy, which makes the diagram

(2.1.7)
$$
\begin{array}{ccc}
C'^\bullet & \longrightarrow & \mathrm{Tot}^\oplus(I'^{\bullet\bullet}) \\
\downarrow & & \downarrow{\scriptstyle h} \\
C[0] & \longrightarrow & J^\bullet
\end{array}
$$

homotopy-commutative. But $C'^\bullet$ is supported in degrees ≤ 0 and $J^\bullet$ is supported in degrees ≥ 0, so this diagram is even *commutative*.

Since $\mathrm{Tot}^\oplus(I'^{\bullet\bullet})$ is supported in degrees $\leq n$, the map h factors through the canonical truncation

$$J'^\bullet = \tau_{\leq n}(J^\bullet)$$

via a map $h' : \text{Tot}^{\oplus}(I'^{\bullet\bullet}) \to J'^{\bullet}$. Since $J'^{\bullet}$ is a complex of F-acyclics, it computes $\mathbf{R}F(C[0])$. Thus, we get a diagram of complexes

(2.1.8)

$$
\begin{array}{ccccc}
F(\text{Tot}^{\oplus}(I'^{\bullet\bullet})) & \longrightarrow & F(I'^{\bullet-n,n})/\text{im}F(I'^{\bullet-n,n-1}) & =\!=\!= & \mathbf{R}^n F(C'^{\bullet})[-n] \\
{\scriptstyle F(h')}\downarrow & & \downarrow & & \downarrow \\
F(J'^{\bullet}) & \longrightarrow & \mathbf{H}^n(F(J'^{\bullet}))[-n] & =\!=\!= & \mathbf{R}^n F(C)[-n]
\end{array}
$$

in which the left square is commutative (since the degree n term of $\text{Tot}^{\oplus}I'^{\bullet\bullet}$ is just $I'^{0,n}$). The right square is also commutative. In order to verify the commutativity of the right square, recall that the *definition* of the functorial structure on $\mathbf{R}^n F$ with respect to $C'^0 \to C$ is via the applying F to a map of F-acyclic resolutions and looking at what happens in degree n homology. In our situation, $C'^0 \to I'^{0,\bullet}$ and $C \to J'^{\bullet}$ are F-acyclic resolutions, so it suffices to check that the map $I'^{0,n} \to J'^n$ between degree n terms inducing the middle column of (2.1.8) can be realized in a map of resolutions $I'^{0,\bullet} \to J'^{\bullet}$ compatible with $C'^0 \to C$.

But $C'^p = 0$ for $p > 0$ and $I'^{0,\bullet}$ is a *subcomplex* of $\text{Tot}^{\oplus}(I'^{\bullet\bullet})$ since $I'^{p,q} = 0$ for $p > 0$, so using the *commutative* diagram (2.1.7) we get a commutative diagram of complexes

$$
\begin{array}{ccccc}
I'^{0,\bullet} & \longrightarrow & \text{Tot}^{\oplus}(I'^{\bullet\bullet}) & \overset{h'}{\longrightarrow} & J'^{\bullet} \\
\uparrow & & \uparrow & & \uparrow \\
C'^0[0] & \longrightarrow & C'^{\bullet} & \longrightarrow & C[0]
\end{array}
$$

The composite map across the top does the job. Applying $\mathbf{H}^n(\cdot)$ to (2.1.8) finishes the proof. $\blacksquare$

We will use many general isomorphisms in [**RD**, II, §5, §6], and most of these remain true with our sign conventions and are compatible with translations. However, we need to make a few modifications to [**RD**, II, 5.14, 5.16], as follows. We replace [**RD**, II, 5.14] with the map

(2.1.9) $$\mathcal{H}^{\bullet} \overset{\mathbf{L}}{\otimes} \mathbf{R}\mathcal{H}om^{\bullet}(\mathcal{F}^{\bullet}, \mathcal{G}^{\bullet}) \to \mathbf{R}\mathcal{H}om^{\bullet}(\mathcal{F}^{\bullet}, \mathcal{H}^{\bullet} \overset{\mathbf{L}}{\otimes} \mathcal{G}^{\bullet})$$

for a scheme X, with $\mathcal{F}^{\bullet}$ in $\mathbf{D}(X)$, $\mathcal{G}^{\bullet}$ in $\mathbf{D}^+(X)$, $\mathcal{H}^{\bullet}$ in $\mathbf{D}^b(X)_{\text{fTd}}$. This is defined in the obvious way by replacing $\mathcal{H}^{\bullet}$ (resp. $\mathcal{G}^{\bullet}$) with a bounded (resp. bounded below) complex of flats (resp. injectives) *without* the intervention of signs, and is compatible with translations.

For a bounded complex $\mathcal{E}^{\bullet}$ of locally free finite rank sheaves on a scheme X and a bounded above (resp. bounded below) complex $\mathcal{F}^{\bullet}$ (resp. $\mathcal{G}^{\bullet}$), we replace [**RD**, II, 5.16] with the following two isomorphisms, which avoid any intervention of signs and are compatible with translation in all variables:

$$\mathscr{E}^\bullet \overset{\mathbf{L}}{\otimes} \mathbf{R}\mathscr{H}om^\bullet(\mathscr{F}^\bullet, \mathscr{G}^\bullet) \simeq \mathbf{R}\mathscr{H}om^\bullet(\mathscr{F}^\bullet, \mathscr{E}^\bullet \overset{\mathbf{L}}{\otimes} \mathscr{G}^\bullet)$$

(this is a special case of (2.1.9)) and

$$\mathbf{R}\mathscr{H}om^\bullet(\mathscr{F}^\bullet \overset{\mathbf{L}}{\otimes} \mathscr{E}^\bullet, \mathscr{G}^\bullet) \simeq \mathbf{R}\mathscr{H}om^\bullet(\mathscr{F}^\bullet, \mathscr{G}^\bullet \overset{\mathbf{L}}{\otimes} \mathscr{E}^{\bullet\vee}).$$

(defined by replacing $\mathscr{G}^\bullet$ with a bounded below complex of injectives). In applications, one links up these last two isomorphisms by using the isomorphisms $\mathscr{E}^\bullet \overset{\mathbf{L}}{\otimes} \mathscr{G}^\bullet \simeq \mathscr{G}^\bullet \overset{\mathbf{L}}{\otimes} \mathscr{E}^\bullet$ and $\mathscr{E}^\bullet \simeq \mathscr{E}^{\bullet\vee\vee}$, both of which involve the intervention of signs (according to our sign conventions).

For a quasi-compact map $f : X \to Y$ between locally noetherian schemes with finite Krull dimension and $\mathscr{F}^\bullet \in \mathbf{D}^-(X)$, $\mathscr{G}^\bullet \in \mathbf{D}^-_{\mathrm{qc}}(Y)$, it should be noted that the useful translation-compatible projection formula [**RD**, II, 5.6]

$$(2.1.10) \qquad \mathbf{R}f_*\mathscr{F}^\bullet \overset{\mathbf{L}}{\otimes} \mathscr{G}^\bullet \simeq \mathbf{R}f_*(\mathscr{F}^\bullet \overset{\mathbf{L}}{\otimes} \mathbf{L}f^*\mathscr{G}^\bullet),$$

which is defined by replacing $\mathscr{F}^\bullet$ (resp. $\mathscr{G}^\bullet$) with a bounded above complex of f_*-acyclics (resp. flats), respects iterated $\overset{\mathbf{L}}{\otimes}$'s without the intervention of signs in the sense that for $\mathscr{H}^\bullet \in \mathbf{D}^-_{\mathrm{qc}}(Y)$, the diagram

(2.1.11)

$$
\begin{array}{ccc}
\mathbf{R}f_*\mathscr{F}^\bullet \overset{\mathbf{L}}{\otimes} (\mathscr{G}^\bullet \overset{\mathbf{L}}{\otimes} \mathscr{H}^\bullet) & \overset{\simeq}{\longrightarrow} & (\mathbf{R}f_*(\mathscr{F}^\bullet) \overset{\mathbf{L}}{\otimes} \mathscr{G}^\bullet) \overset{\mathbf{L}}{\otimes} \mathscr{H}^\bullet \\
\downarrow & & \downarrow \\
\mathbf{R}f_*(\mathscr{F}^\bullet \overset{\mathbf{L}}{\otimes} \mathbf{L}f^*(\mathscr{G}^\bullet \overset{\mathbf{L}}{\otimes} \mathscr{H}^\bullet)) & & \mathbf{R}f_*(\mathscr{F}^\bullet \overset{\mathbf{L}}{\otimes} \mathbf{L}f^*\mathscr{G}^\bullet) \overset{\mathbf{L}}{\otimes} \mathscr{H}^\bullet \\
\simeq \downarrow & & \downarrow \simeq \\
\mathbf{R}f_*(\mathscr{F}^\bullet \overset{\mathbf{L}}{\otimes} (\mathbf{L}f^*\mathscr{G}^\bullet \otimes \mathbf{L}f^*\mathscr{H}^\bullet)) & \overset{\simeq}{\longrightarrow} & \mathbf{R}f_*((\mathscr{F}^\bullet \overset{\mathbf{L}}{\otimes} \mathbf{L}f^*\mathscr{G}^\bullet) \overset{\mathbf{L}}{\otimes} \mathbf{L}f^*\mathscr{H}^\bullet)
\end{array}
$$

commutes. The proof is trivial. Likewise, if $g : Y \to Z$ is another quasi-compact map between locally noetherian schemes with finite Krull dimension, the projection formulas for f, g, and gf are compatible in the sense that for $\mathscr{F}^\bullet \in \mathbf{D}^-(X)$, $\mathscr{G}^\bullet \in \mathbf{D}^-_{\mathrm{qc}}(Z)$, the diagram

(2.1.12)

$$
\begin{array}{ccc}
\mathbf{R}(gf)_*\mathscr{F}^\bullet \overset{\mathbf{L}}{\otimes} \mathscr{G}^\bullet & \overset{\simeq}{\longrightarrow} & \mathbf{R}(gf)_*(\mathscr{F}^\bullet \overset{\mathbf{L}}{\otimes} \mathbf{L}(gf)^*\mathscr{G}^\bullet) \\
\simeq \downarrow & & \downarrow \simeq \\
\mathbf{R}g_*(\mathbf{R}f_*\mathscr{F}^\bullet) \overset{\mathbf{L}}{\otimes} \mathscr{G}^\bullet & \overset{\simeq}{\longrightarrow} & \mathbf{R}g_*\mathbf{R}f_*(\mathscr{F}^\bullet \overset{\mathbf{L}}{\otimes} \mathbf{L}(gf)^*\mathscr{G}^\bullet) \\
\simeq \downarrow & & \downarrow \simeq \\
\mathbf{R}g_*(\mathbf{R}f_*\mathscr{F}^\bullet \overset{\mathbf{L}}{\otimes} \mathbf{L}g^*\mathscr{G}^\bullet) & \overset{\simeq}{\longrightarrow} & \mathbf{R}g_*\mathbf{R}f_*(\mathscr{F}^\bullet \overset{\mathbf{L}}{\otimes} \mathbf{L}f^*\mathbf{L}g^*\mathscr{G}^\bullet)
\end{array}
$$

commutes. The proof is trivial.

An important fact, stated as a question in [**RD**, II, §7] but whose proof is quite easy to give using the results proven there, is the following:

LEMMA 2.1.3. *Let X be a locally noetherian scheme, $\mathscr{I}$ an injective object in the category of quasi-coherent sheaves on X. Then $\mathscr{I}$ is an injective object in the category of $\mathcal{O}_X$-modules.*

PROOF. By [**RD**, II, 7.18], there is an injection $i : \mathscr{I} \hookrightarrow \mathscr{J}$ for some quasi-coherent $\mathcal{O}_X$-module $\mathscr{J}$ which is injective as an $\mathcal{O}_X$-module. By the hypothesis on $\mathscr{I}$, this injection splits. Thus, $\mathscr{I}$ is a direct summand of $\mathscr{J}$ as an $\mathcal{O}_X$-module and so is injective as an $\mathcal{O}_X$-module. ∎

By this lemma, the injective objects in the category of quasi-coherent $\mathcal{O}_X$-modules on a locally noetherian scheme X are exactly the injective $\mathcal{O}_X$-modules which are quasi-coherent as sheaves. We will often invoke this without comment when we construct maps from a very non-quasi-coherent sheaf such as $j_! \mathcal{O}_U$ (for an open subscheme $j : U \hookrightarrow X$) to an injective quasi-coherent sheaf.

As a reference for later, we record some other facts we will frequently need to use. First, there is the well-known

LEMMA 2.1.4. [**W**, Cor 10.4.7] *Let $\mathscr{A}$ be an abelian category, $I^\bullet$ a bounded below complex of injectives in $\mathscr{A}$, $C^\bullet$ a complex in $\mathscr{A}$. Any map $C^\bullet \to I^\bullet$ in the derived category $\mathbf{D}(\mathscr{A})$ is induced by a map of complexes $C^\bullet \to I^\bullet$ which is unique up to homotopy.*

Next, we strengthen Lemma 2.1.3 by recording some results from [**RD**, II, §7] concerning the structure of the category $\mathrm{Qco}(X)$ of quasi-coherent sheaves on a scheme X.

LEMMA 2.1.5. [**RD**, II, 7.13, 7.17] *Let X be a locally noetherian scheme. For $x \in X$, let $\mathscr{J}(x) = i_{x*}J(x)$, where $i_x : \mathrm{Spec}(\mathcal{O}_{X,x}) \to X$ is the canonical map and $J(x)$ is the quasi-coherent sheaf on $\mathrm{Spec}(\mathcal{O}_{X,x})$ associated to an injective hull of $k(x)$ over $\mathcal{O}_{X,x}$. For any set of cardinals $\{\Sigma_x\}_{x \in X}$, the direct sum*

$$(2.1.13) \qquad \bigoplus_{x \in X} \mathscr{J}(x)^{\oplus \Sigma_x}$$

is an injective $\mathcal{O}_X$-module, where $\mathscr{J}(x)^{\oplus \Sigma_x}$ denotes a direct sum of copies of $\mathscr{J}(x)$ indexed by the cardinal Σ_x; moreover, every quasi-coherent injective $\mathcal{O}_X$-module can be written in the form (2.1.13) with unique cardinals Σ_x. In particular, a direct sum of quasi-coherent injective $\mathcal{O}_X$-modules is injective and the stalk of a quasi-coherent injective $\mathcal{O}_X$-module at $x \in X$ is an injective $\mathcal{O}_{X,x}$-module.

We refer the reader to [**Mat**, §18] for an elementary discussion of the commutative algebra analogue of Lemma 2.1.5, giving the structure of injective modules over a noetherian ring in terms of injective hulls at the prime ideals. A related fact which we will use without comment several times later on is that if $\mathscr{I}$ is an $\mathcal{O}_X$-module on a ringed space X and $\mathscr{I}|_{U_i}$ is an injective $\mathcal{O}_{U_i}$-module for an open covering $\{U_i\}$ of X, then $\mathscr{I}$ is an injective $\mathcal{O}_X$-module. This is an easy consequence of Zorn's Lemma and the definition of an injective object.

LEMMA 2.1.6. [**RD**, II, 7.19] *Let X be a locally noetherian scheme, $\mathscr{F}^{\bullet} \in$
$\mathbf{D}_{qc}^{+}(X)$. There is a quasi-isomorphism of complexes $\mathscr{F}^{\bullet} \to \mathscr{I}^{\bullet}$ where $\mathscr{I}^{\bullet}$ is a
bounded below complex of quasi-coherent injectives. In particular, every quasi-
coherent $\mathscr{O}_{X}$-module has a resolution by quasi-coherent injective $\mathscr{O}_{X}$-modules.
Moreover, the 'inclusion' functor $\mathbf{D}^{+}(\mathrm{Qco}(\mathrm{X})) \to \mathbf{D}_{qc}^{+}(X)$ is fully faithful.*

We conclude this section with remarks on why, for any scheme X, the multiplicative system of quasi-isomorphisms in the homotopy category $K(\mathrm{Qco}(X))$
of complexes in $\mathrm{Qco}(X)$ is 'locally small' in the sense of [**W**, 10.3.6, p.381], so
we do not need universes in order to work with $\mathbf{D}(\mathrm{Qco}(X))$ or $\mathbf{D}^{+}(\mathrm{Qco}(X))$.
In order to clarify the nature of this 'local smallness', fix an open affine cov-
ering $\{U_i\}$ of X and an infinite cardinal κ at least as large as the number of
U_i's. Following Gabber, we say that a quasi-coherent $\mathscr{F}$ on X is *of type κ* if,
for all i, the $\mathscr{O}_X(U_i)$-module $\mathscr{F}(U_i)$ is generated by κ elements, in which case
the same clearly holds for the $\mathscr{O}_X(U)$-module $\mathscr{F}(U)$ for all open affines U in
X. Since there is certainly a *set* of isomorphism class representatives for the
quasi-coherent sheaves of type κ, the desired 'local smallness' can be proven by
modifying the proof of [**W**, 10.4.4, pp.386-7], once we know:

LEMMA 2.1.7. *(Gabber) On an arbitrary scheme X, there exists an infinite
cardinal κ such that every quasi-coherent sheaf $\mathscr{F}$ on X is the sum of its quasi-
coherent subsheaves of type κ.*

This lemma implies that the category $\mathrm{Qco}(X)$ admits a set of generators
(in the sense of [**Tohoku**]) consisting of quasi-coherent sheaves of type κ. Since
we only need the locally noetherian case, we do not reproduce Gabber's general
proof of Lemma 2.1.7 (which determines κ in terms of a choice of open affine
covering of X). Instead, we record the following result from EGA which handles
the locally noetherian case:

LEMMA 2.1.8. [**EGA**, I, 9.4.9] *Let X be a locally noetherian scheme. Then
every quasi-coherent sheaf is the sum of its finitely generated quasi-coherent (i.e.,
coherent) subsheaves.*

2.2. Smooth and Finite Maps

In Grothendieck's approach to duality for proper maps $f : X \to Y$ between
schemes, the goal is to construct a δ-functor $f^{!} : \mathbf{D}_{qc}^{+}(Y) \to \mathbf{D}_{qc}^{+}(X)$ and a
δ-functorial trace morphism $\mathrm{Tr}_f : \mathbf{R}f_* \circ f^{!} \to 1$ with various properties. The
method of construction of duality in [**RD**] proceeds by developing a theory of
the δ-functor $f^{\sharp}$ (resp. $f^{\flat}$) for a smooth (resp. finite) map f, and then 'gluing'
these to define $f^{!}$ for more general maps. The first step in this procedure is to
construct general isomorphisms relating canonical bundles for smooth maps and
normal bundles for local complete intersection maps. This section reviews some
of these initial constructions in [**RD**] and their relation with $(\cdot)^{\sharp}$, $(\cdot)^{\flat}$, and we
correct some sign errors along the way.

In general, [**RD**, II, 1.5] defines an isomorphism

$$(2.2.1) \qquad \zeta_{f,g} : \omega_{X/Z} \simeq f^* \omega_{Y/Z} \otimes \omega_{X/Y}$$

for any scheme maps $f : X \to Y$, $g : Y \to Z$ such that each of g, f, and gf is either a separated smooth map or a local complete intersection (lci) map. We will give explicit definitions shortly. Recall that we require lci maps to be closed immersions, that any section to a smooth separated map is an lci map, and that for a smooth (resp. lci) map $f : X \to Y$, we define $\omega_{X/Y}$ to be the top exterior power of $\Omega^1_{X/Y}$ (resp. of the normal bundle). It is incorrectly claimed in [RD, III, 1.6] that (2.2.1) is compatible with triple composites in the sense that for any third $h : Z \to W$ such that each possible composite among f, g, h is either an lci map or a separated smooth map,

$$(2.2.2) \qquad (f^*(\zeta_{g,h}) \otimes 1) \circ \zeta_{f,hg} = (1 \otimes \zeta_{f,g}) \circ \zeta_{gf,h}.$$

Under the definitions in [RD], (2.2.2) is only true up to somewhat complicated sign errors, as one checks by working in local étale coordinates. To be precise, when f, g, hgf are lci maps and h, hg are separated and smooth, or when f is an lci map and all other composites are separated and smooth, (2.2.2) is generally *false*. If one computes the sign errors in terms of relative dimensions and codimensions, one is led to flip around the tensor product and change signs a little to define an isomorphism

$$(2.2.3) \qquad \zeta'_{f,g} : \omega_{X/Z} \to \omega_{X/Y} \otimes f^*\omega_{Y/Z}$$

which satisfies

$$(2.2.4) \qquad (1 \otimes f^*(\zeta'_{g,h})) \circ \zeta'_{f,hg} = (\zeta'_{f,g} \otimes 1) \circ \zeta'_{gf,h}$$

when each composite among f, g, h is either an lci map or a separated smooth map (and $\zeta'_{f,g}$ is the identity when f is the identity or g is the identity). Before giving the definition of (2.2.3), which we regard as a replacement of (2.2.1) from [RD], we remark that we have 'flipped' around the tensor product in (2.2.3) to avoid sign problems in the subsequent theory of $f^\sharp$ on the level of complexes, where f is a smooth morphism.

In each of the four possible cases, we define $\zeta'_{f,g}$ by making some sign changes in the definition of (2.2.1), in addition to 'flipping' the tensor product (*explicit local descriptions will be given below*):

(a) The maps f and g are smooth. Multiply the definition in [RD] by $(-1)^{nm}$ where n and m are the (locally constant) relative dimensions of f and g respectively.

(b) The maps f and g are local complete intersections. Multiply the definition in [RD] by $(-1)^{nm}$ where n and m are the (locally constant) codimensions of f and g respectively.

(c) The map f is a local complete intersection, g and gf are smooth. Multiply the definition in [RD] by $(-1)^{n(n-1)/2}$, where n is the (locally constant) codimension of f.

(d) The maps f and gf are local complete intersections, g is smooth and separated. Multiply the definition in [RD] by $(-1)^{n(n-1)/2}$, where n is the (locally constant) relative dimension of g.

We note that the *method* used to define (2.2.1) in [RD] is completely intrinsic, so to avoid tedious issues of well-definedness it is convenient to have given our definitions in terms of the construction in [**RD**]. But for convenience and clarity, we describe these maps locally. If $f : X \to Y$ is smooth with relative dimension n near $x \in X$, *local coordinates* $x_1, \ldots, x_n$ (*relative to* f) around x are n ordered local sections of $\mathcal{O}_X$ on an open U around x such that $(x_1, \ldots, x_n) : U \to \mathbf{A}_Y^n$ is étale. We will write $\mathrm{d}x$ for the free generator $\mathrm{d}x_1 \wedge \cdots \wedge \mathrm{d}x_n$ of $\omega_{X/Y}$ on U. If $i : Y \hookrightarrow X$ is an lci map with codimension m near $y \in Y$, *local equations* $t_1, \ldots, t_m$ (*relative to* i) around y are m ordered generators of the quasi-coherent ideal sheaf $\mathscr{I}_Y$ over an open $U \subseteq X$ around $i(y)$. We get from this an ordered basis of $i^*(\mathscr{I}_Y / \mathscr{I}_Y^2)$ on $U \cap Y$, and denote by $t_1^\vee, \ldots, t_m^\vee$ the corresponding dual basis. We denote by $t^\vee$ the free generator $t_1^\vee \wedge \cdots \wedge t_m^\vee$ of $\omega_{Y/X}$ over $U \cap Y$. If f is a smooth separated map with a section i, we can regard local equations relative to i around $y \in Y$ as local coordinates relative to f around $i(y)$. Note that if f and g are smooth (resp lci) maps, we can easily view an 'ordered union' of local coordinates (resp. local equations) for f near a point x and local coordinates (resp. local equations) for g near $f(x)$ as local coordinates (resp. local equations) for gf near x.

In these terms (and working locally for the Zariski topology), we can describe all four cases in the definition of (2.2.3) in the following shorthand, whose precise formulation we leave to the reader:

(a) $\mathrm{d}x \wedge \mathrm{d}(f^*(y)) \mapsto \mathrm{d}x \otimes f^*(\mathrm{d}y)$

(b) $u^\vee \wedge (f^*(t))^\vee \mapsto u^\vee \otimes f^*(t^\vee)$

(c) $\mathrm{d}x \mapsto (t_1^\vee \wedge \cdots \wedge t_n^\vee) \otimes f^*(\mathrm{d}t_n \wedge \cdots \wedge \mathrm{d}t_1 \wedge \mathrm{d}x)$

(d) $t^\vee \mapsto (t^\vee \wedge y_1^\vee \wedge \cdots \wedge y_n^\vee) \otimes f^*(\mathrm{d}y_n \wedge \cdots \wedge \mathrm{d}y_1)$.

Note the ordering of the exterior products in the last two cases. It is a straightfoward exercise with these local descriptions to check that (2.2.4) holds and that whenever one of the maps f, g, or gf is both smooth and lci (i.e., an open and closed immersion), there are no inconsistencies. One particular example which will occur repeatedly and which deserves special emphasis is the case of a section i to a smooth separated map $f : X \to Y$ with pure relative dimension n. The isomorphism

$$\zeta'_{i,f} : \mathcal{O}_Y \simeq \omega_{Y/X} \otimes i^*\omega_{X/Y}$$

is given in local coordinates by

$$(2.2.5) \qquad 1 \mapsto (x_1^\vee \wedge \cdots \wedge x_n^\vee) \otimes i^*(\mathrm{d}x_n \wedge \cdots \wedge \mathrm{d}x_1).$$

Since $\mathrm{d}x_n \wedge \cdots \wedge \mathrm{d}x_1 = (-1)^{n(n-1)/2}\mathrm{d}x_1 \wedge \cdots \wedge \mathrm{d}x_n$, (2.2.5) is the source of several sign errors in [**RD**].

If $f : X \to Y$ is a smooth (resp. lci) morphism, we define the locally constant **Z**-valued function $\dim_f$ on X to be the relative dimension (resp. the *negative* of the codimension) of f. If $g : Y \to Z$ is another morphism and each of f, g, gf is either an lci map or a separated smooth map, then $\dim_{gf} = \dim_f + \dim_g$. We can reformulate (2.2.3) as an isomorphism of complexes

$$(2.2.6) \qquad \zeta'_{f,g} : \omega_{X/Z}[\dim_{gf}] \simeq \omega_{X/Y}[\dim_f] \otimes f^*(\omega_{Y/Z}[\dim_g])$$

which is exactly (2.2.3) in degree $-\dim_{gf} = -\dim_f - \dim_g$ (*without* the intervention of any signs). One then gets an analogue of (2.2.4) on the level of complexes in degree $-\dim_{hgf}$, without any intervention of signs.

When $f : X \to Y$ is smooth with relative dimension n (which is a locally constant function on X), we define the δ-functor $f^\sharp : \mathbf{D}(Y) \to \mathbf{D}(X)$ by

$$(2.2.7) \qquad f^\sharp(\mathscr{G}^\bullet) = \omega_{X/Y}[n] \overset{\mathrm{L}}{\otimes} f^*\mathscr{G}^\bullet,$$

whereas in [RD] the definition is $f^*\mathscr{G}^\bullet \overset{\mathrm{L}}{\otimes} \omega_{X/Y}[n]$. Recall that we use the modified definition (2.2.3) in place of the isomorphism (2.2.1) in [RD] in order to have a compatibility such as (2.2.4) without the intervention of signs. In order to use (2.2.6) to define an isomorphism $(gf)^\sharp \simeq f^\sharp g^\sharp$ which is compatible with respect to translations, triple composites, and a global theory later on, we must use the modified definition (2.2.7) instead of the one in [RD]. Note that the isomorphism $f^\sharp(\mathscr{G}^\bullet[m]) \simeq (f^\sharp(\mathscr{G}^\bullet))[m]$ involves an intervention of the sign $(-1)^{nm}$ (according to our sign convention (1.3.6)). This makes $(gf)^\sharp \simeq f^\sharp g^\sharp$ translation-compatible, and the compatibility of this isomorphism with respect to triple composites uses (2.2.4).

For a finite morphism $f : X \to Y$ of locally noetherian schemes, [RD, III, §6] defines the δ-functor

$$(2.2.8) \qquad f^\flat : \mathbf{D}^+(Y) \to \mathbf{D}^+(X)$$

to be $\overline{f}^* \mathbf{R}\mathscr{H}om_Y(f_*\mathscr{O}_X, \cdot)$, where $\overline{f} : (X, \mathscr{O}_X) \to (Y, f_*\mathscr{O}_X)$ is the canonical flat map of ringed spaces. For example, if $\mathscr{I}^\bullet$ is a bounded below complex of quasi-coherent injective sheaves on Y, then the complex of quasi-coherent $f_*\mathscr{O}_X$-modules $\mathscr{H}om_Y(f_*\mathscr{O}_X, \mathscr{I}^\bullet)$ can be viewed as a complex of quasi-coherent sheaves on X, and this represents the complex $f^\flat(\mathscr{I}^\bullet)$. In [RD, III, 6.5] a δ-functorial trace map

$$(2.2.9) \qquad \mathrm{Trf}_f : \mathbf{R}f_* \circ f^\flat \to 1$$

is defined on $\mathbf{D}_{qc}^+(Y)$, which for flat f is just the 'evaluate at 1' map

$$\mathscr{H}om_Y(f_*\mathscr{O}_X, \mathscr{G}^\bullet) \to \mathscr{G}^\bullet.$$

More generally, for a bounded below complex of quasi-coherent injectives $\mathscr{I}^\bullet$ on Y, the map $\mathrm{Trf}_f(\mathscr{I}^\bullet)$ is represented by the natural 'evaluate at 1' map $\mathscr{H}om_Y(f_*\mathscr{O}_X, \mathscr{I}^\bullet) \to \mathscr{I}^\bullet$. If we assume instead that f is a finite map between arbitrary schemes such that, Zariski locally, $f_*\mathscr{O}_X$ has a finite resolution by locally free finite rank $\mathscr{O}_Y$-modules with the resolutions of globally bounded length (e.g., f is an lci map with bounded codimension between arbitrary schemes, or $f_*\mathscr{O}_X$ is locally free of finite rank), then $\mathscr{H}om_Y(f_*\mathscr{O}_X, \cdot)$ has finite cohomological dimension, so the same definition $f^\flat = \overline{f}^* \mathbf{R}\mathscr{H}om_Y(f_*\mathscr{O}_X, \cdot)$ makes sense as a δ-functor $\mathbf{D}(Y) \to \mathbf{D}(X)$. Assuming *in addition* that Y is noetherian of finite Krull dimension (so $\mathbf{R}f_*$ makes sense on unbounded complexes), then (2.2.9) can be defined on all of $\mathbf{D}_{qc}(Y)$, and is again given by 'evaluation at 1' when $f_*\mathscr{O}_X$ is finite locally free.

The basic theory of $(\cdot)^b$ and the δ-functorial trace map $\mathrm{Trf}_f : \mathbf{R}f_* \circ f^b \to 1$ for finite f is worked out in [**RD**, III, §6], and is compatible with our sign conventions. The δ-functor Trf is naturally compatible with flat base change. However, for locally noetherian Y and $\mathscr{F}^\bullet \in \mathbf{D}^b_{\mathrm{qc}}(Y)$, $\mathscr{G}^\bullet \in \mathbf{D}_{\mathrm{qc}}(Y)_{\mathrm{fTd}}$, the top row in the commutative diagram [**RD**, III, 6.9(c)] is wrong; the diagram should be

$$(2.2.10)$$

$$
\begin{array}{ccc}
\mathbf{R}f_* f^b \mathscr{F}^\bullet \overset{\mathbf{L}}{\otimes} \mathscr{G}^\bullet & \longrightarrow & \mathbf{R}f_*(f^b \mathscr{F}^\bullet \overset{\mathbf{L}}{\otimes} \mathbf{L}f^*\mathscr{G}^\bullet) \\
{\scriptstyle \mathrm{Trf}_f \otimes 1} \downarrow & & \downarrow \\
\mathscr{F}^\bullet \overset{\mathbf{L}}{\otimes} \mathscr{G}^\bullet & \underset{\mathrm{Trf}_f}{\longleftarrow} & \mathbf{R}f_* f^b(\mathscr{F}^\bullet \overset{\mathbf{L}}{\otimes} \mathscr{G}^\bullet)
\end{array}
$$

where the top row is the projection formula (2.1.10) and the right column is $\mathbf{R}f_*$ applied to the obvious canonical map [**RD**, III, 6.9(a)] which is defined by replacing $\mathscr{G}^\bullet$ (resp. $\mathscr{F}^\bullet$) with a bounded (resp. bounded below) complex of flats (resp. injectives).

2.3. Projective Space and the Trace Map

The most fundamental proper smooth morphism in Grothendieck's approach to duality theory is the projection $f_Y : \mathbf{P}^n_Y \to Y$ for a fixed integer n. Since the natural edge map (1.3.23) from Čech cohomology to derived functor cohomology respects pullback in the two theories [**EGA**, 0_{III}, 12.1.4.2], by Čech cohomology calculations with $\mathscr{O}_{\mathbf{P}^n_A}(-n-1) \simeq \omega_{\mathbf{P}^n_A/A}$ over an affine base $\mathrm{Spec}(A)$ we know that $\mathbf{R}^n(f_Y)_*(\omega_{\mathbf{P}^n_Y/Y})$ is an *invertible* sheaf on Y and is of formation compatible with the natural base change morphism. By base change from the case $Y = \mathrm{Spec}(\mathbf{Z})$, over which an invertible sheaf is always trivial, it follows that either of the two choices of generator of the free rank one $\mathbf{Z}$-module $\mathrm{H}^n(\mathbf{P}^n_{\mathbf{Z}}, \omega_{\mathbf{P}^n_{\mathbf{Z}}/\mathbf{Z}})$ gives rise to a general isomorphism

$$(2.3.1) \qquad \mathbf{R}^n(f_Y)_*(\omega_{\mathbf{P}^n_Y/Y}) \simeq \mathscr{O}_Y$$

which commutes with base change. In (2.3.3) below, we will make a definite explicit choice of generator over $\mathbf{Z}$ to unambiguously define (2.3.1) in general. The choice of generator over $\mathbf{Z}$ must be correctly made so as to later fit into the global theory (cf. Lemma 2.8.2). Due to sign issues which will emerge in the proof of Lemma 2.8.2, the choice of generator over $\mathbf{Z}$ in [**RD**, III, §3] is not the correct one when using the corrected definition (2.2.3).

We now define the canonical generator of $\mathrm{H}^n(\mathbf{P}^n_{\mathbf{Z}}, \omega_{\mathbf{P}^n_{\mathbf{Z}}/\mathbf{Z}})$. For conceptual clarity (and technical necessity later), we work with any ring A in place of $\mathbf{Z}$. Fix an integer n and let $\omega_A = \omega_{\mathbf{P}^n_A/A}$. Let $\mathfrak{U} = \{U_0, \ldots, U_n\}$ be the standard ordered open affine covering of $\mathbf{P}^n_A$, with U_i the non-vanishing locus of the ith homogenous coordinate X_i. The natural isomorphism

$$(2.3.2) \qquad \check{\mathrm{H}}^n(\mathfrak{U}, \omega_A) \simeq \mathrm{H}^n(\mathbf{P}^n_A, \omega_A)$$

is functorial in A, by [**EGA**, 0_{III}, 12.1.4.2]. Let $U = U_0 \cap \cdots \cap U_n$, $t_j = X_j/X_0$ for $1 \leq j \leq n$. If we calculate in terms of $\mathscr{O}_{\mathbf{P}_A^n}(-n-1) \simeq \omega_A$, we see that the Čech n-cocycle

$$(2.3.3) \qquad (-1)^{n(n+1)/2} \frac{dt_1 \wedge \cdots \wedge dt_n}{t_1 \cdots t_n} \in \check{C}^n(\mathfrak{U}, \omega_A)$$

maps to a generator $c \in H^n(\mathbf{P}_A^n, \omega_A)$ under (2.3.2). We *define* the isomorphism (2.3.1) by using this generator over an affine base $\operatorname{Spec} A$ (in [**RD**, III, §3], the analogous definition omits the sign in (2.3.3)). Our definition is chosen to make Lemma 2.8.2 hold (and this, in turn, is what makes Lemma 3.4.2 hold). When $A = \mathbf{C}$ is the field of complex numbers and we fix a choice of $i = \sqrt{-1} \in \mathbf{C}$ to orient $\mathbf{C}$ (with $1 \wedge i > 0$), there is also an integration isomorphism (with $\mathbf{P}^n$ denoting the complex manifold attached to $\mathbf{P}_{\mathbf{C}}^n$)

$$(2.3.4) \qquad H^n(\mathbf{P}_{\mathbf{C}}^n, \omega_{\mathbf{C}}) \stackrel{\text{GAGA}}{\simeq} H^n(\mathbf{P}^n, \omega_{\mathbf{C}}^{\text{an}}) \simeq H_{\text{dR}}^{2n}(\mathbf{P}^n, \mathbf{C}) \stackrel{\overline{\frac{1}{(2\pi i)^n}} \int_{\mathbf{P}^n}}{\simeq} \mathbf{C}.$$

All maps in (2.3.4) are independent of the choice of i and the middle isomorphism uses the Hodge to deRham spectral sequence. In [**D**, Appendix $(e),(f)$] it is asserted that when $A = \mathbf{C}$, the image of (2.3.3) under the composite of (2.3.2) and (2.3.4) is 1; this is not relevant to our considerations, but is psychologically reassuring (the case $n = 1$ is a classical calculation with a C^∞ bump function supported near $[0,1] \in \mathbf{P}^1$).

Before proceeding, we note that (2.3.1) is 'independent of projective coordinates'. That is, if we choose *any* system of projective linear coordinates L on $\mathbf{P}_A^n$ and repeat the same Čech cohomology construction as above with respect to these new coordinates, we claim that the resulting element $c_L \in H^n(\mathbf{P}_A^n, \omega_A)$ is always the same. To be precise, we choose an A-automorphism $\iota : \mathbf{P}_A^n \simeq \mathbf{P}_A^n$ and define $\mathscr{L} = \iota^* \mathscr{O}(1)$. Let the global sections $X_j' = \iota^* X_j \in \Gamma(\mathbf{P}_A^n, \mathscr{L})$ be the associated 'projective coordinate system' L and let $U_j' = \iota^{-1}(U_j)$ be the open where X_j' generates $\mathscr{L}$. The functions $t_j' = X_j'/X_0'$ for $1 \leq j \leq n$ make sense on $U' = U_0' \cap \cdots \cap U_n'$, so for $\mathfrak{U}' = \{U_0', \ldots, U_n'\} = \iota^{-1}(\mathfrak{U})$, the Čech n-cocycle $(-1)^{n(n+1)/2}(dt_1' \wedge \cdots \wedge dt_n')/(t_1' \cdots t_n') \in \check{C}^n(\mathfrak{U}', \omega_A)$ defines an element $c_L \in H^n(\mathbf{P}_A^n, \omega_A)$.

We claim that c_L is *independent of L*. The proof of this in [**RD**, III, 10.2] is problematic, because it relies upon [**RD**, III, 10.1], whose proof appears to require this 'independence of coordinates' in the first place. More precisely, the argument for *arbitrary* (e.g., non-flat) base change compatibility in step 3 of the proof of [**RD**, III, 10.1] is incorrect, but if one knew a priori that everything was independent of the choice of projective coordinates, then the reduction to step 4 of that proof is not hard to do, since any section of projective space over a local ring is equal to $[1, 0, \ldots, 0]$ in suitable coordinates (the proof of this final step 4 is explained in Lemma 2.8.2). In order to prove that c_L is independent of L, we note that by using the canonical isomorphism $\iota^* \omega_A \simeq \omega_A$, [**EGA**, 0_{III}, 12.1.4.2] ensures the commutativity of the diagram

$$\begin{array}{ccc} \check{H}^n(\mathfrak{U},\omega_A) & \xrightarrow{\simeq} & H^n(\mathbf{P}_A^n,\omega_A) \\ {\scriptstyle \iota^*}\downarrow & & \downarrow{\scriptstyle \iota^*} \\ \check{H}^n(\mathfrak{U}',\omega_A) & \xrightarrow[\simeq]{} & H^n(\mathbf{P}_A^n,\omega_A) \end{array}$$

so it suffices to use the following well-known fact:

LEMMA 2.3.1. *For any scheme* Y, *the natural action of* $\mathrm{Aut}(\mathbf{P}_Y^n/Y)$ *on* $R^n(f_Y)_*(\omega_{\mathbf{P}_Y^n/Y})$ *is trivial, where* $f_Y : \mathbf{P}_Y^n \to Y$ *is the projection.*

Due to lack of an adequate reference, we give a proof.

PROOF. Since $R^n(f_Y)_*(\omega_{\mathbf{P}_Y^n/Y})$ is invertible, the fppf sheaf

$$Y \rightsquigarrow \mathrm{Aut}_{\mathcal{O}_Y}(R^n(f_Y)_*(\omega_{\mathbf{P}_Y^n/Y}))$$

is represented by $\mathbf{G}_m$. The fppf sheaf $Y \rightsquigarrow \mathrm{Aut}(\mathbf{P}_Y^n/Y)$ is represented by the (affine) group scheme PGL_{n+1} over $\mathbf{Z}$, so the action of $\mathrm{Aut}(\mathbf{P}_Y^n/Y)$ on $R^n(f_Y)_*(\omega_{\mathbf{P}_Y^n/Y})$ corresponds to a natural map of smooth affine $\mathbf{Z}$-*group schemes* $\Pi_n : \mathrm{PGL}_{n+1} \to \mathbf{G}_m$. We claim that the only such map is the trivial one. Indeed, we may base change to $\mathbf{Q}$ and since $\mathrm{GL}_{n+1} \to \mathrm{PGL}_{n+1}$ is fppf surjective, the irreducibility of the determinant polynomial and the fact that Π_n respects the group structures forces Π_n to be induced by a power of the determinant. Since Π_n is unaffected by scaling matrices by units, Π_n must be trivial. ∎

Now that the basic projective trace map (2.3.1) is unambiguously defined, we can carry over (without any intervention of signs) the fundamental construction [**RD**, III, 4.3] of the general projective trace morphism: if $f : \mathbf{P}_Y^n \to Y$ is the projection and Y is a locally noetherian scheme, using (2.3.1) in [**RD**, III, 4.3] yields an isomorphism of δ-functors $\mathbf{D}_{\mathrm{qc}}^+(Y) \to \mathbf{D}_{\mathrm{qc}}^+(Y)$

$$(2.3.5) \qquad\qquad \mathrm{Trp} = \mathrm{Trp}_f : (\mathbf{R}f_*) \circ f^\sharp \simeq \mathrm{id}.$$

Because the isomorphism (2.3.5) is important, we want to give an 'explicit' description of it along the same lines as in (2.1.2). By Lemma 2.1.6, any complex in $\mathbf{D}_{\mathrm{qc}}^+(Y)$ can be represented by a bounded below complex of *quasi-coherent* sheaves. We will describe (2.3.5) on such a complex $\mathscr{G}^\bullet$. It is essential that all $\mathscr{G}^r$'s are quasi-coherent, or else the following description will not work. Let $I^{\bullet\bullet}$ be the canonical truncation in rows $\leq n$ of a Cartan-Eilenberg resolution of $f^\sharp\mathscr{G}^\bullet$ in the category of quasi-coherent $\mathcal{O}_X$-modules. In particular, all $I^{p,q}$'s are f_*-acyclic, since f_* has cohomological dimension $\leq n$ on the category of *quasi-coherent* $\mathcal{O}_X$-modules. Thus, $f_*(\mathrm{Tot}^\oplus I^{\bullet\bullet})$ represents $\mathbf{R}f_*(f^\sharp\mathscr{G}^\bullet)$. The natural map (which involves no intervention of signs)

(2.3.6)
$$f_*(\mathrm{Tot}^\oplus I^{\bullet\bullet}) \longrightarrow f_*(I^{\bullet-n,n})/\mathrm{im}(f_*I^{\bullet-n,n-1}) =\!\!= R^n f_*(f^\sharp \mathscr{G}^\bullet)[-n]$$

$$\parallel$$

$$R^n f_*(\omega \otimes f^* \mathscr{G}^\bullet)$$

is a map of *complexes* (recall the sign issues in the first part of Lemma 2.1.1).
Moreover, since all $\mathscr{G}^r$'s are quasi-coherent, the natural projection formula map

(2.3.7) $$R^n f_*(\omega_{\mathbf{P}_Y^n/Y}) \otimes \mathscr{G}^r \to R^n f_*(\omega_{\mathbf{P}_Y^n/Y} \otimes f^* \mathscr{G}^r)$$

is an isomorphism. Recall that by definition, (2.3.7) is equal to

$$R^n f_*(\omega_{\mathbf{P}_Y^n/Y}) \otimes \mathscr{G}^r \to R^n f_*(\omega_{\mathbf{P}_Y^n/Y}) \otimes f_* f^* \mathscr{G}^r \xrightarrow{\cup} R^n f_*(\omega_{\mathbf{P}_Y^n/Y} \otimes f^* \mathscr{G}^r),$$

so the map (2.3.7) makes sense when $\mathscr{G}^r$ is replaced by an arbitrary $\mathcal{O}_Y$-module.
If we combine (2.3.6), (2.3.7), and (2.3.1), we arrive at a map of complexes

$$f_*(\mathrm{Tot}^\oplus I^{\bullet\bullet}) \to \mathscr{G}^\bullet$$

which, *by definition*, represents (2.3.5) on $\mathscr{G}^\bullet$. By [**RD**, III, 4.3], this is a quasi-isomorphism, so the first map in (2.3.6) is a quasi-isomorphism.

We now come to a rather subtle point, analogous to Corollary 2.1.2. When
$\mathscr{G}^\bullet = \mathcal{O}_Y[0]$, we claim that (2.3.5) on H^0's recovers (2.3.1). Such an assertion is
ill-posed unless we remove sign ambiguity in the definition of the isomorphism

(2.3.8) $$\mathbf{R} f_*(\omega_{\mathbf{P}_Y^n/Y}[n]) = \mathrm{H}^0(\mathbf{R} f_*(\omega_{\mathbf{P}_Y^n/Y}[n]))[0] \simeq R^n f_*(\omega_{\mathbf{P}_Y^n/Y})[0],$$

where the first equality is a special case of the general derived category isomorphism

$$C^\bullet \xleftarrow{\simeq} \tau_{\leq m}(C^\bullet) \xrightarrow{\simeq} \mathrm{H}^m(C^\bullet)[-m]$$

for a complex $C^\bullet$ with $\mathrm{H}^i(C^\bullet) = 0$ for $i \neq m$. We refer the reader to (1.3.4)
and the subsequent discussion there for an explanation of why the second map
in (2.3.8) is of ambiguous nature (up to a sign of $(-1)^n$); this ambiguity must
be (uniquely) eliminated in such a way that (2.3.5) on H^0's recovers (2.3.1).

Choose an f_*-acyclic resolution $\omega_{\mathbf{P}_Y^n/Y} \to \mathscr{I}^\bullet$ to be used to compute the ordinary derived functors $R^\bullet f_*(\omega_{\mathbf{P}_Y^n/Y})$, and regard the shifted complex $\mathscr{I}^{\bullet+n}$
(*without* changing the differentials by signs!) as an f_*-acyclic resolution of
$\omega_{\mathbf{P}_Y^n/Y}[n]$. Since $R^i f_*(\omega_{\mathbf{P}_Y^n/Y}) = 0$ for $i \neq n$, there is an obvious derived category
isomorphism

$$f_*(\mathscr{I}^{\bullet+n}) \simeq \mathrm{H}^n(f_*\mathscr{I}^\bullet)[0]$$

defined without the intervention of signs, and *by definition* this represents (2.3.8);
that is, we compute

$$\mathbf{R} f_*(\omega_{\mathbf{P}_Y^n/Y}[n]) = f_*(\mathscr{I}^{\bullet+n}), \quad R^n f_*(\omega_{\mathbf{P}_Y^n/Y}) = \mathrm{H}^n(f_*\mathscr{I}^\bullet).$$

The reason that this convention for defining (2.3.8) makes (2.3.5) on H^0's re-
cover (2.3.1) is that when $\mathscr{I}^\bullet$ is an injective resolution, then placing $\tau_{\leq n}(\mathscr{I}^\bullet)$
in the $-n$th column gives the truncation $I^{\bullet\bullet}$ of a Cartan-Eilenberg resolution

of $\omega_{\mathbf{P}_Y^n/Y}[n]$ as required in the definition of Trp. Thus, the $-n$th column of this double complex computes the derived functors $\mathrm{R}^j f_*(\omega_{\mathbf{P}_Y^n/Y})$ while the associated total complex quasi-isomorphism is $\omega_{\mathbf{P}_Y^n/Y}[n] \to \mathscr{I}^{\bullet+n}$! An equivalent formulation of the assertion that (2.3.5) on H^0's recovers (2.3.1) is that the *composite* of (2.3.8) and (2.3.1) is $\mathrm{Trp}(\mathcal{O}_Y)$.

Note that the above definition of (2.3.8) *violates* our general convention (1.3.4), and if we had used (1.3.4) to define (2.3.8) then (2.3.8) would change by a sign of $(-1)^n$. The desire to have analogues of "recovering (2.3.1) from (2.3.5) on H^0's" in the general proper smooth cases later on will force a similar non-standard sign convention in the explication of the derived category trace on the level of higher direct image sheaves (there will be no such sign interference on the level of the derived category). This convention plays a role in both Theorem B.2.2 and Theorem B.4.1, which provide two natural explications of Grothendieck's trace map in the case of proper smooth curves. If the reader prefers the alternative definition of (2.3.8) which is off from our chosen definition by a sign of $(-1)^n$, then this same sign must be introduced in several later explications of derived category results. For example, the sign in Theorem B.2.2 would be eliminated and a sign would have to be introduced in Theorem B.4.1. For various reasons, it seems to the author that having no sign in Theorem B.4.1 is more pleasing than having no sign in Theorem B.2.2. In [D, Appendice (b)] Deligne appears to adopt the other point of view, using the alternative definition of (2.3.8) that is compatible with (1.3.4). This is consistent with the fact that the final sentence in [D, Appendice(c)] is off from Theorem B.2.2 by a factor of -1. For the benefit of the reader who prefers Deligne's convention, we will keep track of this subtle sign as it arises later on.

We must emphasize that this sign issue is *irrelevant* in the definition of (2.3.5), but paying attention to it is *crucial* when proving derived category results by reduction to calculations with ordinary derived functors (e.g., as in Lemma 2.8.2). One reason not to get too worried about this issue (which will arise in other contexts later on) is that the exponent in this kind of sign will always involve relative dimensions or codimensions, so in the context of $\mathscr{E}xt$ sheaves and higher direct image sheaves, assertions such as compatibility with respect to scheme morphisms will always turn out to be unaffected by such sign ambiguities (since $(-1)^n(-1)^m = (-1)^{n+m}$). Also, some explicit compatibility assertions will have the same sign factor appearing an even number of times (and therefore cancelling out).

We now return to the study of (2.3.5). By considering bounded below complexes $\mathscr{G}^\bullet$ of quasi-coherent sheaves, it follows from an analysis of a Cartan-Eilenberg resolution of $f^\sharp(\mathscr{G}^\bullet)$ that (2.3.5) is compatible with flat base change and respects translations, where the translation-compatibility of $\mathbf{R}f_*$ is defined without the intervention of signs and the translation-compatibility of the functor $f^\sharp$ is defined with a sign of $(-1)^{nm}$ intervening in every degree (see the *definitions* of (1.3.6) and (2.2.7), as well as the discussion following (2.2.7)). The *proof* of the translation compatibility of (2.3.5) is essentially identical to the verification that (2.1.1) commutes up to a sign of $(-1)^{nm}$, except that this sign is now (in some sense) cancelled by the sign in the translation compatibility for $f^\sharp$. More precisely, if we carry over the study of (2.1.1), we arrive at the problem of verifying the commutativity (in the derived category) of

(2.3.9)

$$\mathbf{R}f_*(f^\sharp\mathcal{G}^\bullet)[m] \xoverline{\quad\quad\quad} \mathbf{R}f_*((f^\sharp\mathcal{G}^\bullet)[m]) \xrightarrow{(-1)^{nm}} \mathbf{R}f_*(f^\sharp(\mathcal{G}^\bullet[m]))$$

down with $\simeq$ vertical maps,

$$\mathbf{R}^n f_*(f^\sharp\mathcal{G}^\bullet)[-n][m] \xrightarrow{(-1)^{nm}} \mathbf{R}^n f_*((f^\sharp\mathcal{G}^\bullet)[m])[-n] \xrightarrow{(-1)^{nm}} \mathbf{R}^n f_*(f^\sharp(\mathcal{G}^\bullet[m]))[-n]$$

no sign (left), no sign (right),

$$\mathbf{R}^n f_*(\omega \otimes f^*\mathcal{G}^\bullet)[m] \xoverline{\quad\quad \text{no sign} \quad\quad} \mathbf{R}^n f_*(\omega \otimes f^*\mathcal{G}^\bullet[m])$$

$\simeq$ vertical maps,

$$\mathcal{G}^\bullet[m] \xoverline{\quad\quad\quad\quad\quad} \mathcal{G}^\bullet[m]$$

where $\mathcal{G}^\bullet$ is a bounded below complex of quasi-coherent sheaves and the composites down the left and right are $\mathrm{Trp}_f(\mathcal{G}^\bullet)[m]$ and $\mathrm{Trp}_f(\mathcal{G}^\bullet[m])$ respectively. The upper left corner in (2.3.9) commutes by the analysis of (2.1.1), the upper right corner commutes by functoriality, the bottom part obviously commutes, and the middle part commutes because $(-1)^{nm}(-1)^{nm} = 1$ (this is the 'cancellation' of signs referred to above).

There are two important compatibility properties satisfied by the projective trace map (2.3.5). First, the diagram for Trp given in [**RD**, III, 4.4] is only sign-commutative in general and the proof in [**RD**] is erroneous, since the commutativity of a diagram in the derived category is not a local statement (moreover, even in the quasi-coherent setup in that proof, the result does not seem to "follow easily from the definitions"). We give the corrected diagram in the form of a theorem.

THEOREM 2.3.2. [**RD**, III, 4.4] *Let Y be a noetherian scheme with finite Krull dimension. Denote by $f : X = \mathbf{P}_Y^n \to Y$ the projection and choose $\mathcal{F}^\bullet, \mathcal{G}^\bullet \in \mathbf{D}_{\mathrm{qc}}^b(Y)$ with at least one of these having finite Tor-dimension. The diagram in $\mathbf{D}(Y)$*

(2.3.10)

$$
\begin{array}{ccc}
\mathbf{R}f_* f^\sharp(\mathcal{F}^\bullet) \overset{\mathbf{L}}{\otimes} \mathcal{G}^\bullet & \xrightarrow{\;\simeq\;} & \mathbf{R}f_*(f^\sharp(\mathcal{F}^\bullet) \overset{\mathbf{L}}{\otimes} f^*\mathcal{G}^\bullet) \\
{\scriptstyle \mathrm{Trp}(\mathcal{F}^\bullet)\overset{\mathbf{L}}{\otimes}1}\Big\downarrow & & \Big\uparrow{\scriptstyle \simeq} \\
\mathcal{F}^\bullet \overset{\mathbf{L}}{\otimes} \mathcal{G}^\bullet & \xleftarrow[\mathrm{Trp}(\mathcal{F}^\bullet\overset{\mathbf{L}}{\otimes}\mathcal{G}^\bullet)]{} & \mathbf{R}f_* f^\sharp(\mathcal{F}^\bullet \overset{\mathbf{L}}{\otimes} \mathcal{G}^\bullet)
\end{array}
$$

*commutes, where the top map is the projection formula (2.1.10) and the right side is $\mathbf{R}f_*$ applied to the obvious canonical map [**RD**, III, 2.4(a)] defined by replacing $\mathcal{F}^\bullet$ or $\mathcal{G}^\bullet$ with a bounded complex of flats.*

The finite Tor-dimension condition ensures $\mathscr{F}^\bullet \overset{\mathbf{L}}{\otimes} \mathscr{G}^\bullet$ is bounded (below), so $\mathrm{Trp}(\mathscr{F}^\bullet \overset{\mathbf{L}}{\otimes} \mathscr{G}^\bullet)$ makes sense. Also, if (2.3.10) were known to be commutative up to a universal sign ϵ_n only depending on n, then the case $\mathscr{F}^\bullet = \mathscr{G}^\bullet = \mathscr{O}_Y[0]$ forces $\epsilon_n = 1$. Finally, a stronger version of the projection formula (2.1.10) should yield the same theorem by essentially the same proof without finite Krull dimension hypotheses, but the above formulation suffices for our purposes.

By the compatibility (2.1.11) of the projection formula with respect to iterated $\overset{\mathbf{L}}{\otimes}$'s, one can reduce (2.3.10) to the special case where $\mathscr{F}^\bullet = \mathscr{O}_Y[0]$. In other words, the real claim is that when $\mathscr{F}^\bullet = \mathscr{O}_Y[0]$, the long way around (2.3.10) from lower right to lower left provides an 'explicit' definition of (2.3.5) in terms of the projective trace map (2.3.1) for $\omega_{\mathbf{P}_Y^n/Y}$ (using (2.3.8)). Even the special case $\mathscr{F}^\bullet = \mathscr{O}_Y[0]$ is rather difficult, so we will give its proof (as well as the easy reduction to the case $\mathscr{F}^\bullet = \mathscr{O}_Y[0]$) in the next section. It should be noted that the construction of (2.3.5) is useful in other settings (e.g., the proof of the important compatibility theorem [**RD**, III, 10.1] relating traces in the case of projective space and the case of finite maps; cf. (2.8.5)), so using Theorem 2.3.2 with $\mathscr{F}^\bullet = \mathscr{O}_Y[0]$ to define $\mathrm{Trp}(\mathscr{G}^\bullet)$ for $\mathscr{G}^\bullet \in \mathbf{D}_{qc}^b(Y)$ would *not* save any effort. The special case $\mathscr{F}^\bullet = \mathscr{O}_Y[0]$ in Theorem 2.3.2 illustrates the fact (which underlies our analysis of base change for traces of proper Cohen-Macaulay morphisms in Chapter 4) that the trace map for the dualizing sheaf determines the general theory of the trace in 'nice' situations.

A more difficult compatibility is Theorem 2.3.3 below. This theorem is not mentioned in [**RD**], but it is essential in the proof of [**RD**, VII, 3.4(TRA3)] (see Lemma 3.4.3 below), which is what enables one to show (by reduction to the case of the canonical projection $\mathbf{P}_Y^n \to Y$) the fundamental result of duality theory: duality morphisms arising from (suitable) proper maps are isomorphisms.

THEOREM 2.3.3. *Let Y be a separated noetherian scheme with finite Krull dimension. Denote by $f : X = \mathbf{P}_Y^n \to Y$ the projection and choose $\mathscr{F}^\bullet \in \mathbf{D}_c^-(Y)$, $\mathscr{G}^\bullet \in \mathbf{D}_{qc}^+(Y)$. The following diagram of isomorphisms in $\mathbf{D}(Y)$ commutes:*

$$
(2.3.11) \qquad
\begin{array}{ccc}
\mathbf{R}f_* \mathbf{R}\mathscr{H}om_X^\bullet(f^*\mathscr{F}^\bullet, f^\sharp\mathscr{G}^\bullet) & \longleftarrow & \mathbf{R}f_* f^\sharp \mathbf{R}\mathscr{H}om_Y^\bullet(\mathscr{F}^\bullet, \mathscr{G}^\bullet) \\
\downarrow & & \downarrow {\scriptstyle \mathrm{Trp}} \\
\mathbf{R}\mathscr{H}om_Y^\bullet(\mathscr{F}^\bullet, \mathbf{R}f_* f^\sharp \mathscr{G}^\bullet) & \underset{\mathrm{Trp}}{\longrightarrow} & \mathbf{R}\mathscr{H}om_Y^\bullet(\mathscr{F}^\bullet, \mathscr{G}^\bullet)
\end{array}
$$

*The left side is [**RD**, II, 5.10], defined by replacing $f^\sharp\mathscr{G}^\bullet$ with a bounded below complex of injectives, and the top is $\mathbf{R}f_*$ applied to [**RD**, III, 2.4(b)], which is defined by replacing $\mathscr{G}^\bullet$ with a bounded below complex of injectives.*

This will be proven in the next section. The analogous assertion for finite maps [**RD**, III, 6.9(b)] (valid for any locally noetherian Y) is much easier to prove. We note that the separatedness hypothesis on Y in Theorem 2.3.3 is purely an artifact of our method of proof (we use that for open affines U in Y, the open immersions $j : U \hookrightarrow Y$ are affine maps, so j_* is exact on quasi-coherent sheaves), and this condition should not be necessary for the truth of the theorem. Theorem 2.3.3 is adequate for the construction of Grothendieck duality later on.

2.4. Proofs of Properties of the Projective Trace

We want to prove Theorems 2.3.2 and 2.3.3. Both proofs require intricate manipulations with various resolutions. The reader is strongly advised to skip this section on a first reading (similar arguments will be used to prove Theorem 2.5.2 below, whose proof the reader is also advised to skip on a first reading).

PROOF. (of Theorem 2.3.2) If we set $\mathscr{F}^\bullet = \mathcal{O}_Y[0]$, then the assertion of the theorem is that for any $\mathscr{G}^\bullet \in \mathbf{D}^b_{qc}(Y)$, the composite

(2.4.1)

$$\mathbf{R}f_*f^\sharp(\mathscr{G}^\bullet) \xLongequal{} \mathbf{R}f_*(\omega_{X/Y}[n] \overset{\mathbf{L}}{\otimes} f^*\mathscr{G}^\bullet) \xleftarrow{\;\simeq\;} \mathbf{R}f_*(\omega_{X/Y}[n]) \overset{\mathbf{L}}{\otimes} \mathscr{G}^\bullet \xrightarrow{\;\simeq\;} \mathscr{G}^\bullet$$

is $\mathrm{Trp}(\mathscr{G}^\bullet)$, where the middle map in (2.4.1) is the projection formula quasi-isomorphism (2.1.10) and the last map in (2.4.1) is $\mathrm{Trp}(\mathcal{O}_Y) \overset{\mathbf{L}}{\otimes} 1_{\mathscr{G}^\bullet}$, where $\mathrm{Trp}(\mathcal{O}_Y)$ is the composite of (2.3.8) and (2.3.1), as we explained below (2.3.8). *Assuming* the special case (2.4.1), and using (2.1.11), it follows easily that the following diagram of isomorphisms commutes in $\mathbf{D}(Y)$ and the outer edge is exactly Theorem 2.3.2:

$$\mathscr{F}^\bullet \overset{\mathbf{L}}{\otimes} \mathscr{G}^\bullet$$

$$\mathbf{R}f_*(f^\sharp(\mathcal{O}_Y) \overset{\mathbf{L}}{\otimes} f^*\mathscr{F}^\bullet) \overset{\mathbf{L}}{\otimes} \mathscr{G}^\bullet \xleftarrow{\;\simeq\;} (\mathbf{R}f_*(f^\sharp\mathcal{O}_Y) \overset{\mathbf{L}}{\otimes} \mathscr{F}^\bullet) \overset{\mathbf{L}}{\otimes} \mathscr{G}^\bullet$$

$$\downarrow{\simeq} \qquad\qquad\qquad\qquad \downarrow{\simeq}$$

$$\mathbf{R}f_*((f^\sharp\mathcal{O}_Y \overset{\mathbf{L}}{\otimes} f^*\mathscr{F}^\bullet) \overset{\mathbf{L}}{\otimes} f^*\mathscr{G}^\bullet) \qquad \mathbf{R}f_*(f^\sharp\mathcal{O}_Y) \overset{\mathbf{L}}{\otimes} (\mathscr{F}^\bullet \overset{\mathbf{L}}{\otimes} \mathscr{G}^\bullet)$$

$$\downarrow{\simeq} \qquad\qquad\qquad\qquad \downarrow{\simeq}$$

$$\mathbf{R}f_*(f^\sharp\mathcal{O}_Y \overset{\mathbf{L}}{\otimes} (f^*\mathscr{F}^\bullet \overset{\mathbf{L}}{\otimes} f^*\mathscr{G}^\bullet)) \xrightarrow{\;\simeq\;} \mathbf{R}f_*(f^\sharp\mathcal{O}_Y \overset{\mathbf{L}}{\otimes} f^*(\mathscr{F}^\bullet \overset{\mathbf{L}}{\otimes} \mathscr{G}^\bullet))$$

It therefore suffices to prove that (2.4.1) is equal to $\mathrm{Trp}(\mathscr{G}^\bullet)$. The reason this is somewhat non-trivial to prove is that the projection formula in the middle of (2.4.1) is described in terms of a bounded above complex of flats which is quasi-isomorphic to $\mathscr{G}^\bullet$, while the definition of $\mathrm{Trp}(\mathscr{G}^\bullet)$ is described in terms of a bounded below complex of quasi-coherent sheaves which is quasi-isomorphic to $\mathscr{G}^\bullet$. These two types of complexes are quite different from each other, so relating them is a somewhat delicate matter.

For simplicity, we now write ω for $\omega_{X/Y}$. We can choose $\mathscr{G}^\bullet$ to be a bounded above complex of flats on Y, with bounded quasi-coherent cohomology (the $\mathscr{G}^r$'s are not quasi-coherent in general). By Lemma 2.1.6, pick a quasi-isomorphism $\varphi: \mathscr{G}^\bullet \to \mathscr{G}^\bullet_{qc}$, where $\mathscr{G}^\bullet_{qc}$ is a bounded below complex of *quasi-coherent* (injective) sheaves. Let $\mathscr{I}^{\bullet\bullet}$ (resp. $\mathscr{I}^{\bullet\bullet}_{\mathrm{qc}}$) be a Cartan-Eilenberg resolution of $f^{\sharp}\mathscr{G}^{\bullet}$ (resp. $f^{\sharp}\mathscr{G}^{\bullet}_{\mathrm{qc}}$) in the category of $\mathcal{O}_X$-modules (resp. quasi-coherent $\mathcal{O}_X$-modules), and let $\mathscr{I}'^{\bullet\bullet}$ (resp. $\mathscr{I}'^{\bullet\bullet}_{\mathrm{qc}}$) be the canonical truncation of $\mathscr{I}^{\bullet\bullet}$ (resp. $\mathscr{I}^{\bullet\bullet}_{\mathrm{qc}}$) in rows $\leq n$. Choose a double complex map $\Phi : \mathscr{I}^{\bullet\bullet} \to \mathscr{I}^{\bullet\bullet}_{\mathrm{qc}}$ over φ and let Φ' denote the induced map on canonical truncations in rows $\leq n$. From the commutative diagram (where the vertical maps are the canonical ones)

$$
\begin{array}{ccc}
f^{\sharp}\mathscr{G}^{\bullet} & \xrightarrow{\;f^{\sharp}(\varphi)\;} & f^{\sharp}\mathscr{G}^{\bullet}_{\mathrm{qc}} \\
{\scriptstyle \epsilon}\downarrow & & \downarrow \\
\mathrm{Tot}^{\oplus}(\mathscr{I}'^{\bullet\bullet}) & \xrightarrow[\;\mathrm{Tot}^{\oplus}\Phi'\;]{} & \mathrm{Tot}^{\oplus}(\mathscr{I}'^{\bullet\bullet}_{\mathrm{qc}})
\end{array}
$$

we conclude that $\mathrm{Tot}^{\oplus}\Phi'$ is a quasi-isomorphism since all other sides are quasi-isomorphisms. Let

$$
\alpha_{\mathrm{qc}} : \mathrm{Tot}^{\oplus}(\mathscr{I}'^{\bullet\bullet}_{\mathrm{qc}}) \to \mathrm{Tot}^{\oplus}(\mathscr{I}^{\bullet\bullet}_{\mathrm{qc}})
$$

be the canonical map, which is a quasi-isomorphism to a bounded below complex of *quasi-coherent* injectives. Define $\alpha = \alpha_{\mathrm{qc}} \circ \mathrm{Tot}^{\oplus}\Phi'$, so $\alpha \circ \epsilon : f^{\sharp}\mathscr{G}^{\bullet} \to \mathrm{Tot}^{\oplus}(\mathscr{I}^{\bullet\bullet}_{\mathrm{qc}})$ is a quasi-isomorphism.

Let $\omega \to \mathscr{J}^{\bullet}$ be an injective resolution of ω (supported in degrees ≥ 0), which we shall use to compute derived functors of ω. We view $\mathscr{J}^{\bullet+n}$ as an injective resolution of $\omega[n]$, supported in degrees $\geq -n$. Letting $\mathscr{J}'^{\bullet}$ denote the canonical truncation of $\mathscr{J}^{\bullet}$ in degrees $\leq n$, $\mathscr{J}'^{\bullet+n}$ is an f_*-acyclic resolution of $\omega[n]$ since $\mathrm{R}^i f_*(\omega) = 0$ for $i > n$. Thus, since $\mathscr{J}'^{\bullet} \to \mathscr{J}^{\bullet}$ a quasi-isomorphism, $f_*\mathscr{J}'^{\bullet} \to f_*\mathscr{J}^{\bullet}$ is also a quasi-isomorphism. Because $\mathscr{G}^{\bullet}$ is a bounded above complex of flats, the naturally induced map of complexes

$$
f^{\sharp}\mathscr{G}^{\bullet} \to \mathscr{J}'^{\bullet+n} \otimes f^*\mathscr{G}^{\bullet}
$$

is a quasi-isomorphism. By Lemma 2.1.3 and Lemma 2.1.4, there is a unique way, up to homotopy, to fill in the following diagram (where the diagonal map is a quasi-isomorphism) so that it commutes up to homotopy:

$$
\begin{array}{ccc}
f^{\sharp}\mathscr{G}^{\bullet} & \xrightarrow{\;\alpha\circ\epsilon\;} & \mathrm{Tot}^{\oplus}(\mathscr{I}^{\bullet\bullet}_{\mathrm{qc}}) \\
& \searrow & \Big\uparrow{\scriptstyle \gamma} \\
& & \mathscr{J}'^{\bullet+n} \otimes f^*\mathscr{G}^{\bullet}
\end{array}
$$

Fix a choice of such a γ. Note that γ is a quasi-isomorphism.

The natural composite map

$$
(2.4.2) \qquad \beta : f_*\mathscr{J}'^{\bullet+n} \otimes \mathscr{G}^{\bullet} \to f_*(\mathscr{J}'^{\bullet+n} \otimes f^*\mathscr{G}^{\bullet}) \xrightarrow{\;f_*(\gamma)\;} f_*\,\mathrm{Tot}^{\oplus}(\mathscr{I}^{\bullet\bullet}_{\mathrm{qc}})
$$

represents the derived category projection formula map $\mathrm{R}f_*(\omega[n]) \overset{\mathrm{L}}{\otimes} \mathscr{G}^{\bullet} \to \mathrm{R}f_*(\omega[n] \overset{\mathrm{L}}{\otimes} f^*\mathscr{G}^{\bullet})$, which is an isomorphism by [**RD**, II, 5.6], so β is a quasi-isomorphism. Thus, it makes sense to consider the commutativity in $\mathbf{D}(Y)$ of the following diagram of complexes of $\mathcal{O}_Y$-modules, in which all maps are quasi-isomorphisms except possibly the two 'factors' of β:

(2.4.3)

$$
\begin{array}{ccc}
f_* \operatorname{Tot}^{\oplus} \mathscr{I}'^{\bullet\bullet}_{\mathrm{qc}} & \longrightarrow & f_* \mathscr{I}'^{\bullet-n,n}_{\mathrm{qc}}/\operatorname{im}(f_* \mathscr{I}'^{\bullet-n,n-1}_{\mathrm{qc}}) \\
\downarrow {\scriptstyle f_*(\alpha_{\mathrm{qc}})} & & \| \\
f_* \operatorname{Tot}^{\oplus}(\mathscr{I}^{\bullet\bullet}_{\mathrm{qc}}) & & \mathrm{R}^n f_*(\omega \otimes f^* \mathscr{G}^{\bullet}_{\mathrm{qc}}) \\
\uparrow & & \simeq \uparrow \cup \\
\beta \left(f_*(\mathscr{I}'^{\bullet+n} \otimes f^* \mathscr{G}^{\bullet}) \right. & & \mathrm{R}^n f_*(\omega)[0] \otimes \mathscr{G}^{\bullet}_{\mathrm{qc}} \\
\uparrow & & \simeq \uparrow {\scriptstyle 1\otimes\varphi} \\
f_* \mathscr{I}'^{\bullet+n} \otimes \mathscr{G}^{\bullet} & \longrightarrow & \mathrm{R}^n f_*(\omega)[0] \otimes \mathscr{G}^{\bullet}
\end{array}
$$

Since γ is well-defined up to homotopy, the diagram (2.4.3) in $\mathbf{D}(Y)$ is independent of the choice of γ in the definition of (2.4.2). Let us describe some parts of this diagram a little more carefully in order to remove any possibility of confusion. The top row in (2.4.3) makes sense because f_* commutes with the formation of direct sums (as Y is locally noetherian) and this top row is a quasi-isomorphism because in $\mathbf{D}(Y)$ its composite with the right column is the isomorphism $\operatorname{Trp}(\mathscr{G}^{\bullet})$ in $\mathbf{D}(Y)$, using (2.3.1). The equality

$$f_* \mathscr{I}'^{\bullet-n,n}_{\mathrm{qc}}/\operatorname{im}(f_* \mathscr{I}'^{\bullet-n,n-1}_{\mathrm{qc}}) = \mathrm{R}^n f_*(\omega \otimes f^* \mathscr{G}^{\bullet}_{\mathrm{qc}})$$

in the right column of (2.4.3) is an 'equality' of complexes without the intervention of signs, because the isomorphism $(\omega[n] \otimes f^*\mathscr{G}^{\bullet}_{\mathrm{qc}})[-n] \simeq \omega \otimes f^*\mathscr{G}^{\bullet}_{\mathrm{qc}}$ involves no intervention of signs. Also, the bottom row in (2.4.3) uses the identification $\mathrm{H}^0(\mathscr{I}'^{\bullet+n}) \simeq \mathrm{R}^n f_*(\omega)$. Using the 'explicit' description of $\operatorname{Trp}(\mathcal{O}_Y)$ in terms of (2.3.8), it is not difficult to check that commutativity of (2.4.3) in $\mathbf{D}(Y)$ implies that (2.4.1) is the general projective space trace map, as desired.

We now check that (2.4.3) commutes in $\mathbf{D}(Y)$. Let $I'^{\bullet\bullet}$ be the upper-half plane double complex associated (in the sense defined at the end of §1.2) to an injective resolution of $f^{\sharp}\mathscr{G}^{\bullet}$ in the category of complexes of $\mathcal{O}_X$-modules, and similarly let $I^{\bullet\bullet}_{\mathrm{qc}}$ be the double complex associated to an injective resolution of $f^{\sharp}\mathscr{G}^{\bullet}_{\mathrm{qc}}$ in the category of complexes of quasi-coherent $\mathcal{O}_X$-modules. Let $I'^{\bullet\bullet}$ and $I'^{\bullet\bullet}_{\mathrm{qc}}$ denote the respective canonical truncations in rows $\leq n$. By the theory of injective resolutions in abelian categories, we can choose a map of double complexes $\rho_1 : I^{\bullet\bullet} \to I^{\bullet\bullet}_{\mathrm{qc}}$ over $f^{\sharp}\mathscr{G}^{\bullet} \to f^{\sharp}\mathscr{G}^{\bullet}_{\mathrm{qc}}$ and a map of double complexes $\rho_2 : \mathscr{I}^{\bullet\bullet}_{\mathrm{qc}} \to I^{\bullet\bullet}_{\mathrm{qc}}$ over $f^{\sharp}\mathscr{G}^{\bullet}_{\mathrm{qc}}$. Let ρ'_1 and ρ'_2 denote the induced maps on the canonical truncations in rows $\leq n$.

Since f_* has cohomological dimension $\leq n$ on quasi-coherent sheaves on X and $\operatorname{Tot}^{\oplus}(\rho'_2)$ is a map between complexes which are compatibly quasi-isomorphic to $f^{\sharp}\mathscr{G}^{\bullet}_{\mathrm{qc}}$ and consist of f_*-acyclics, applying f_* to $\operatorname{Tot}^{\oplus}(\rho'_2)$ yields a quasi-isomorphism. Beware that applying f_* to $\operatorname{Tot}^{\oplus}\rho'_1$ is probably not a quasi-isomorphism. The quasi-isomorphism $f_*(\operatorname{Tot}^{\oplus}\rho'_2)$ fits into the bottom row of the commutative diagram of complexes

$$(2.4.4) \qquad\qquad \begin{array}{ccc} f_* \operatorname{Tot}^{\oplus} \mathscr{I}_{\mathrm{qc}}^{\bullet\bullet} & \xrightarrow{\ f_*(\rho_2)\ } & f_* \operatorname{Tot}^{\oplus} I_{\mathrm{qc}}^{\bullet\bullet} \\ {\scriptstyle f_*(\alpha_{\mathrm{qc}})} \Big\uparrow & & \Big\uparrow \\ f_* \operatorname{Tot}^{\oplus} \mathscr{I}'^{\bullet\bullet}_{\mathrm{qc}} & \xrightarrow[\ f_*(\rho_2')\ ]{} & f_* \operatorname{Tot}^{\oplus} I'^{\bullet\bullet}_{\mathrm{qc}} \end{array}$$

Note that $\operatorname{Tot}^{\oplus} I_{\mathrm{qc}}^{\bullet\bullet}$ and $\operatorname{Tot}^{\oplus} I'^{\bullet\bullet}_{\mathrm{qc}}$ are *not* generally bounded below. We claim that all of the maps in (2.4.4) are quasi-isomorphisms. Since cohomology commutes with direct sums on a noetherian topological space, so a direct sum of f_*-acyclic $\mathscr{O}_X$-modules is f_*-acyclic, all total complexes in (2.4.4) consist of f_*-acyclics. These total complexes are all compatibly quasi-isomorphic to $\mathscr{G}_{\mathrm{qc}}^{\bullet}$, so the assertion that all maps in (2.4.4) are quasi-isomorphisms follows from the fact that f_* has *finite* cohomological dimension on the category of quasi-coherent $\mathscr{O}_X$-modules. A reflection of the diagram (2.4.4) across its main diagonal fits naturally 'on the left' of the commutative diagram of complexes

(2.4.5)

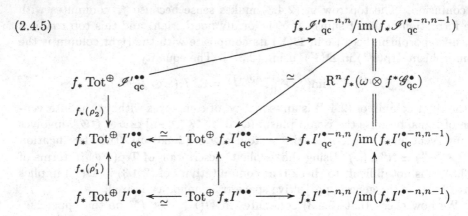

By using (2.4.4) and (2.4.5), as well as the naturality of cup products and the commutativity of

$$\begin{array}{ccc} f_* I'^{\bullet-n,n}_{\mathrm{qc}}/\mathrm{im}(f_* I'^{\bullet-n,n-1}_{\mathrm{qc}}) & =\!\!=\!\!= & \mathrm{R}^n f_*(\omega \otimes f^* \mathscr{G}_{\mathrm{qc}}^{\bullet}) \\ \Big\uparrow & & \Big\uparrow \\ f_* I'^{\bullet-n,n}/\mathrm{im}(f_* I'^{\bullet-n,n-1}) & =\!\!=\!\!= & \mathrm{R}^n f_*(\omega \otimes f^* \mathscr{G}^{\bullet}) \end{array}$$

it is easy to check via a diagram chase starting at $f_* \mathscr{J}'^{\bullet} \otimes \mathscr{G}^{\bullet}$ that the commutativity of (2.4.3) in $\mathbf{D}(Y)$ will follow if we can construct a map of complexes

$$(2.4.6) \qquad\qquad \psi \colon \mathscr{J}'^{\bullet+n} \otimes f^* \mathscr{G}^{\bullet} \to \operatorname{Tot}^{\oplus} I'^{\bullet\bullet}$$

which makes the diagram of complexes

(2.4.7)

$$f_* \mathscr{J}'^{\bullet+n} \otimes \mathscr{G}^\bullet \longrightarrow f_*(\mathscr{J}'^{\bullet+n} \otimes f^*\mathscr{G}^\bullet) \xrightarrow{f_*(\psi)} \mathrm{Tot}^\oplus f_* I'^{\bullet\bullet}$$

$$\mathrm{R}^n f_*(\omega)[0] \otimes \mathscr{G}^\bullet \xrightarrow[\cup]{} \mathrm{R}^n f_*(\omega \otimes f^*\mathscr{G}^\bullet) = \!\!= f_* I'^{\bullet-n,n}/\mathrm{im}(f_* I'^{\bullet-n,n-1})$$

commute and which makes the diagram of complexes

(2.4.8)

$$\mathscr{J}'^{\bullet+n} \otimes f^*\mathscr{G}^\bullet \xrightarrow{\psi} \mathrm{Tot}^\oplus I'^{\bullet\bullet} \xrightarrow{\rho_1'} \mathrm{Tot}^\oplus I_{\mathrm{qc}}'^{\bullet\bullet}$$
$$\gamma \downarrow$$
$$\mathrm{Tot}^\oplus \mathscr{I}_{\mathrm{qc}}^{\bullet\bullet} \xrightarrow{\rho_2} \mathrm{Tot}^\oplus I_{\mathrm{qc}}^{\bullet\bullet}$$

yield a commutative diagram in $\mathbf{D}(Y)$ after applying f_*.

We will construct (2.4.6) so that it respects the natural quasi-isomorphism from $f^\sharp \mathscr{G}^\bullet$ to each side (this forces ψ to be a quasi-isomorphism). Let us first see that this compatibility property is enough to force $f_*((2.4.8))$ to commute in $\mathbf{D}(Y)$. Since $\mathscr{G}^\bullet \to \mathscr{G}_{\mathrm{qc}}^\bullet$ is a quasi-isomorphism, it follows that all maps in (2.4.8) are quasi-isomorphisms. The map $f_*(\rho_2)$ along the bottom of $f_*((2.4.8))$ is a quasi-isomorphism, as noted in (2.4.4). Since the lower left corner $\mathrm{Tot}^\oplus \mathscr{I}_{\mathrm{qc}}^{\bullet\bullet}$ in (2.4.8) is a bounded below complex of injectives, it follows from [**W**, Lemma 10.4.6] that there is a map of double complexes

$$s : \mathrm{Tot}^\oplus I_{\mathrm{qc}}^{\bullet\bullet} \to \mathrm{Tot}^\oplus \mathscr{I}_{\mathrm{qc}}^{\bullet\bullet}$$

for which $s \circ \rho_2$ is homotopic to 1, so $f_*(s)$ is an inverse to the isomorphism $f_*(\rho_2)$ in $\mathbf{D}(Y)$. Thus, it is enough to prove that both composites around (2.4.8) yield homotopic maps when composed with s. By Lemma 2.1.4, such commutativity up to homotopy is a consequence of the commutativity of (2.4.8) in $\mathbf{D}(X)$. However, everything in (2.4.8) is compatible with $f^\sharp \mathscr{G}^\bullet \to f^\sharp \mathscr{G}_{\mathrm{qc}}^\bullet$ via the natural quasi-isomorphism from $f^\sharp \mathscr{G}^\bullet$ or $f^\sharp \mathscr{G}_{\mathrm{qc}}^\bullet$ to each complex in (2.4.8), so the commutativity of (2.4.8) in $\mathbf{D}(X)$ is clear.

It is now enough to concentrate on (2.4.7), provided (2.4.6) is constructed to be compatible with the natural map from $f^\sharp \mathscr{G}^\bullet$ to each side. In particular, we have eliminated any need to work with $\mathscr{G}_{\mathrm{qc}}^\bullet$ and its associated quasi-coherent injective and Cartan-Eilenberg resolutions. We have reduced ourselves to a problem that makes sense for any bounded above complex of flat sheaves $\mathscr{G}^\bullet$ on Y, without any hypotheses on its cohomology, and we now work in such generality (with the notation $\mathscr{J}^\bullet$, $\mathscr{J}'^\bullet$, $I^{\bullet\bullet}$, $I'^{\bullet\bullet}$ as above). Let

$$C^{p,q} = \mathscr{J}^q \otimes f^*\mathscr{G}^{n+p}$$

with vertical and horizontal differentials

$$d_v^{p,q} = d_{\mathscr{J}^\bullet}^q \otimes 1, \quad d_h^{p,q} = (-1)^{n+q}(1 \otimes d_{f^*\mathscr{G}^\bullet}^{n+p}),$$

so $C^{\bullet\bullet}$ is an upper half-plane double complex with *exact* columns (as all $\mathscr{G}^r$'s are flat). There is an augmentation map $f^\sharp \mathscr{G}^\bullet \to C^{\bullet,0}$ to the bottom row of $C^{\bullet\bullet}$, inducing the canonical map

$$f^\sharp \mathscr{G}^\bullet \to \operatorname{Tot}^\oplus C^{\bullet\bullet} = \mathscr{J}^{\bullet+n} \otimes f^* \mathscr{G}^\bullet$$

arising from $\omega[n] \to \mathscr{J}^{\bullet+n}$.

By the theory of injective resolutions in abelian categories, there is a map of double complexes $C^{\bullet\bullet} \to I^{\bullet\bullet}$ over $f^\sharp \mathscr{G}^\bullet$. Forming the canonical truncations $C'^{\bullet\bullet}$, $I'^{\bullet\bullet}$ in rows $\leq n$ and passing to the associated total complexes, we arrive at a map

$$\psi : \mathscr{J}'^{\bullet+n} \otimes f^* \mathscr{G}^\bullet = \operatorname{Tot}^\oplus C'^{\bullet\bullet} \to \operatorname{Tot}^\oplus I'^{\bullet\bullet}$$

over $f^\sharp \mathscr{G}^\bullet$. It remains to check that this ψ makes (2.4.7) commute.

If we consider how ψ was constructed out of maps between double complexes, and we observe that the commutativity of (2.4.7) can be checked in each degree separately, we can easily reduce ourselves to an analogous problem where $\mathscr{G}^\bullet$ is replaced by a single flat sheaf on Y. More precisely, let $f : X \to Y$ be a map of ringed spaces, $\mathscr{G}$ a flat sheaf on Y, $\mathscr{F}$ an $\mathscr{O}_X$-module (such as ω above), and $\mathscr{F} \otimes f^* \mathscr{G} \to I^\bullet$ an injective resolution. Fix $n \in \mathbf{Z}$ and let $\mathscr{F} \to J^\bullet$ be an injective resolution (such as $\omega \to \mathscr{J}^\bullet$ above). By flatness, $\mathscr{F} \otimes f^* \mathscr{G} \to J^\bullet \otimes f^* \mathscr{G}$ is a resolution. Thus, there is a map

$$\psi : J^\bullet \otimes f^* \mathscr{G} \to I^\bullet$$

as resolutions of $\mathscr{F} \otimes f^* \mathscr{G}$. For any $n \in \mathbf{Z}$, we claim that the diagram of sheaves

$$(2.4.9) \qquad \begin{array}{ccccc} \mathrm{H}^n(f_* J^\bullet \otimes \mathscr{G}) & \longrightarrow & \mathrm{H}^n(f_*(J^\bullet \otimes f^* \mathscr{G})) & \xrightarrow{\mathrm{H}^n(f_*(\psi))} & \mathrm{H}^n(f_* I^\bullet) \\ \| & & & & \| \\ \mathrm{H}^n(f_* J^\bullet) \otimes \mathscr{G} & = \!\!=\!\!= & \mathrm{R}^n f_*(\mathscr{F}) \otimes \mathscr{G} & \xrightarrow[\cup]{} & \mathrm{R}^n f_*(\mathscr{F} \otimes f^* \mathscr{G}) \end{array}$$

commutes. It is not hard to see that this implies the desired commutativity of (2.4.7). Since (2.4.9) is 'independent' of the choices of $I^\bullet$, $J^\bullet$, ψ, and thus is functorial in $\mathscr{G}$ and compatible with localizing on Y, by chasing sections we immediately reduce to the trivial case $\mathscr{G} = \mathscr{O}_Y$.   ■

PROOF. (of Theorem 2.3.3) We may take $\mathscr{G}^\bullet$ to be a bounded below complex of quasi-coherent injectives and $\mathscr{F}^\bullet$ to be a bounded above complex with coherent cohomology sheaves. Also, as in the previous proof, we let $\omega = \omega_{\mathbf{P}_Y^n/Y}$. It is important to note that we do not require the $\mathscr{F}^r$'s to be quasi-coherent (in fact, later on we will want to suppose that all $\mathscr{F}^r$'s are flat on Y). Although the complex $\mathscr{H}om_Y^\bullet(\mathscr{F}^\bullet, \mathscr{G}^\bullet)$ is bounded below, has quasi-coherent cohomology sheaves, and its terms involve only finite products, these terms are usually not quasi-coherent since the $\mathscr{F}^r$'s cannot generally be assumed to be coherent. Thus, we will need to be a bit careful about using the 'explicit' description of (2.3.5), which is only valid on a bounded below complex of *quasi-coherent* sheaves. By unwinding definitions, we will ultimately reduce ourselves to the analysis of several 'explicit' diagrams of complexes which we will show either commute on the level of complexes or else at least commute up to homotopy. Some of our manipulations are motivated by the proof of Theorem 2.3.2, via relations between $\mathscr{H}om$ and $\otimes$, but the calculations are more complicated.

As a first step, we want to describe projective trace morphisms such as $\mathrm{Trp}(\mathscr{H}om^{\bullet}(\mathscr{F}^{\bullet}, \mathscr{G}^{\bullet}))$ and $\mathrm{Trp}(\mathscr{G}^{\bullet})$ in terms of maps between complexes. Choose a quasi-isomorphism $\mathscr{H}om_Y^{\bullet}(\mathscr{F}^{\bullet}, \mathscr{G}^{\bullet}) \to \mathscr{A}^{\bullet}$ to a bounded below complex of quasi-coherent injectives, and let $I_{\mathrm{qc}}^{\bullet\bullet}$ be a Cartan-Eilenberg resolution of $f^{\sharp}\mathscr{A}^{\bullet}$ in the category of quasi-coherent sheaves on X. Denote by $I'^{\bullet\bullet}_{\mathrm{qc}}$ its canonical truncation in rows $\leq n$. Similarly, let $I^{\bullet\bullet}$ denote a Cartan-Eilenberg resolution of $f^{\sharp}\mathscr{H}om_Y^{\bullet}(\mathscr{F}^{\bullet}, \mathscr{G}^{\bullet})$ in the category of $\mathscr{O}_X$-modules and let $I'^{\bullet\bullet}$ denote its canonical truncation in rows $\leq n$. Choose a map of double complexes $\varphi : I^{\bullet\bullet} \to I_{\mathrm{qc}}^{\bullet\bullet}$ over $f^{\sharp}\mathscr{H}om_Y^{\bullet}(\mathscr{F}^{\bullet}, \mathscr{G}^{\bullet}) \to f^{\sharp}\mathscr{A}^{\bullet}$, so we get an induced quasi-isomorphism

$$\mathrm{Tot}^{\oplus}(\varphi') : \mathrm{Tot}^{\oplus} I'^{\bullet\bullet} \to \mathrm{Tot}^{\oplus} I'^{\bullet\bullet}_{\mathrm{qc}}$$

between bounded below total complexes. Applying f_*, this fits into the top row of the following commutative diagram of complexes:

(2.4.10)

$$
\begin{array}{ccc}
f_* \mathrm{Tot}^{\oplus} I'^{\bullet\bullet} & \xrightarrow{\quad f_* \mathrm{Tot}^{\oplus} \varphi' \quad} & f_* \mathrm{Tot}^{\oplus} I'^{\bullet\bullet}_{\mathrm{qc}} \\
\downarrow & & \downarrow \\
f_* I'^{\bullet -n,n}/\mathrm{im}(f_* I'^{\bullet -n,n-1}) & \longrightarrow & f_* I'^{\bullet -n,n}_{\mathrm{qc}}/\mathrm{im}(f_* I'^{\bullet -n,n-1}_{\mathrm{qc}}) \\
\| & & \| \\
\mathrm{R}^n f_*(\omega \otimes f^* \mathscr{H}om_Y^{\bullet}(\mathscr{F}^{\bullet}, \mathscr{G}^{\bullet})) & \longrightarrow & \mathrm{R}^n f_*(\omega \otimes f^* \mathscr{A}^{\bullet}) \\
\uparrow & & \uparrow {\scriptstyle \simeq} \\
\mathrm{R}^n f_*(\omega) \otimes \mathscr{H}om_Y^{\bullet}(\mathscr{F}^{\bullet}, \mathscr{G}^{\bullet}) & \xrightarrow[\mathrm{qism}]{} & \mathrm{R}^n f_*(\omega) \otimes \mathscr{A}^{\bullet}
\end{array}
$$

The map in the bottom row is a quasi-isomorphism, and the composite *in* $\mathbf{D}(Y)$

$$f_* \mathrm{Tot}^{\oplus} I'^{\bullet\bullet}_{\mathrm{qc}} \to \mathrm{R}^n f_*(\omega) \otimes \mathscr{H}om_Y^{\bullet}(\mathscr{F}^{\bullet}, \mathscr{G}^{\bullet}) \simeq \mathscr{H}om_Y^{\bullet}(\mathscr{F}^{\bullet}, \mathscr{G}^{\bullet})$$

going through the right column of (2.4.10) and using (2.3.1) is the isomorphism $\mathrm{Trp}(\mathscr{H}om_Y^{\bullet}(\mathscr{F}^{\bullet}, \mathscr{G}^{\bullet}))$. In particular, the top map in the right column of (2.4.10) is a quasi-isomorphism.

Next, we want to give an 'explicit' diagram which describes $\mathrm{Trp}(\mathscr{G}^{\bullet})$ in terms of maps between complexes. Let $\mathscr{I}^{\bullet\bullet}$ denote a Cartan-Eilenberg resolution of $f^{\sharp}\mathscr{G}^{\bullet}$ in the category of *quasi-coherent* sheaves on X, and let $\mathscr{I}'^{\bullet\bullet}$ denote its canonical truncation in rows $\leq n$. Since $\mathrm{Tot}^{\oplus} \mathscr{I}'^{\bullet\bullet} \to \mathrm{Tot}^{\oplus} \mathscr{I}^{\bullet\bullet}$ is a quasi-isomorphism between bounded below complexes of f_*-acyclic sheaves, applying f_* to this yields a quasi-isomorphism. Choose a quasi-isomorphism $f_* \mathrm{Tot}^{\oplus} \mathscr{I}^{\bullet\bullet} \to \mathscr{J}^{\bullet}$ to a bounded below complex of quasi-coherent injectives. By Lemma 2.1.4 and the fact that $\mathscr{G}^{\bullet}$ is a bounded below complex of quasi-coherent injectives, there exists a map $\alpha : \mathscr{J}^{\bullet} \to \mathscr{G}^{\bullet}$, unique up to homotopy, such that the diagram

(2.4.11)

$$f_* \operatorname{Tot}^\oplus \mathscr{I}'^{\bullet\bullet} \longrightarrow f_* \mathscr{I}'^{\bullet-n,n}/\mathrm{im}(f_* \mathscr{I}'^{\bullet-n,n-1}) =\!=\!= R^n f_*(\omega \otimes f^* \mathscr{G}^\bullet)$$

$$\mathscr{I}^\bullet \xrightarrow{\quad \alpha \quad} \mathscr{G}^\bullet \xleftarrow{\quad \simeq \quad} R^n f_*(\omega) \otimes \mathscr{G}^\bullet$$

with the left vertical map labeled "qism" (quasi-isomorphism) and the right vertical map labeled "$\simeq$".

is homotopy-commutative. The left column in (2.4.11) is a quasi-isomorphism and the composite map across the top and right side of (2.4.11) represents (via $R^n f_*(\omega) \simeq \mathscr{O}_Y$) the isomorphism $\mathrm{Trp}(\mathscr{G}^\bullet)$ in $\mathbf{D}(Y)$, so α is a *quasi-isomorphism*.

Now suppose without loss of generality that $\mathscr{F}^\bullet$ is a bounded above complex of flats. In order to combine (2.4.11) and (2.4.10), we need to introduce another map β. By our finite Krull dimension hypothesis on Y, f_* has finite cohomological dimension on $\mathscr{O}_X$-modules. Thus, we may choose a quasi-isomorphism $\gamma : f^* \mathscr{F}^\bullet \to F^\bullet$ where $F^\bullet$ is a bounded above complex of f_*-acyclics, so the isomorphism $\rho_f(\mathscr{F}^\bullet) : \mathscr{F}^\bullet \to \mathbf{R}f_* \mathbf{L}f^* \mathscr{F}^\bullet$ from [**RD**, II, 5.10] is represented by the map of complexes

$$\mathscr{F}^\bullet \to f_* f^* \mathscr{F}^\bullet \xrightarrow{f_*(\gamma)} f_* F^\bullet.$$

Since $\mathscr{H}om_X^\bullet(F^\bullet, \operatorname{Tot}^\oplus \mathscr{I}^{\bullet\bullet})$ represents $\mathbf{R}\mathscr{H}om_X^\bullet(\mathbf{L}f^* \mathscr{F}^\bullet, f^\sharp \mathscr{G}^\bullet) \in \mathbf{D}_{\mathrm{qc}}^+(X)$, choose a quasi-isomorphism

$$\gamma : \mathscr{H}om_X^\bullet(F^\bullet, \operatorname{Tot}^\oplus \mathscr{I}^{\bullet\bullet}) \to \mathscr{K}^\bullet$$

to a bounded below complex of injective $\mathscr{O}_X$-modules. By Lemma 2.1.4, up to homotopy there is a unique map of complexes $\beta : \operatorname{Tot}^\oplus I'^{\bullet\bullet}_{\mathrm{qc}} \to \mathscr{K}^\bullet$ which makes the diagram

(2.4.12)

$$\mathscr{H}om_X^\bullet(f^* \mathscr{F}^\bullet, f^\sharp \mathscr{G}^\bullet) \longrightarrow \mathscr{H}om_X^\bullet(f^* \mathscr{F}^\bullet, \operatorname{Tot}^\oplus \mathscr{I}^{\bullet\bullet})$$

with left vertical map labeled ψ and right vertical map labeled λ,

$$f^\sharp \mathscr{H}om_Y^\bullet(\mathscr{F}^\bullet, \mathscr{G}^\bullet) \qquad\qquad \mathscr{H}om_X^\bullet(F^\bullet, \operatorname{Tot}^\oplus \mathscr{I}^{\bullet\bullet})$$

with right vertical map labeled γ,

$$\operatorname{Tot}^\oplus I'^{\bullet\bullet} \qquad\qquad\qquad \mathscr{K}^\bullet$$

with diagonal map labeled β,

$$\operatorname{Tot}^\oplus I'^{\bullet\bullet}_{\mathrm{qc}}$$

commute in $\mathbf{D}(X)$. The vertical maps in (2.4.12) are all quasi-isomorphisms, with the top map in the left column the canonical map arising from (2.1.9), which does not involve the intervention of signs (this is essentially a modified version of [**RD**, II, 5.8]). Also, observe that the map λ in the right column of (2.4.12) is a quasi-isomorphism between complexes of finite products of flasque sheaves (which are therefore f_*-acyclic), so $f_*(\lambda)$ is a quasi-isomorphism.

Recall that for any ringed space Z, any flat $\mathcal{O}_Z$-module $\mathcal{H}$, and any injective $\mathcal{O}_Z$-module $\mathcal{I}$, the isomorphism

$$(2.4.13) \qquad \mathrm{Hom}_Z(\cdot, \mathcal{H}om_Z(\mathcal{H}, \mathcal{I})) \simeq \mathrm{Hom}_Z((\cdot) \otimes \mathcal{H}, \mathcal{I})$$

implies that the sheaf $\mathcal{H}om_Z(\mathcal{H}, \mathcal{I})$ is injective. Thus,

$$\mathcal{H}om_X^\bullet(f^*\mathcal{F}^\bullet, \mathrm{Tot}^\oplus \mathcal{I}^{\bullet\bullet})$$

is a bounded below complex of *injectives*, as is $\mathcal{H}om_Y^\bullet(\mathcal{F}^\bullet, \mathcal{G}^\bullet)$. In particular, the quasi-isomorphism

$$\mathcal{H}om_Y^\bullet(\mathcal{F}^\bullet, \mathcal{G}^\bullet) \to \mathcal{A}^\bullet$$

has a homotopy inverse and we can use

$$\lambda : \mathcal{H}om_X^\bullet(F^\bullet, \mathrm{Tot}^\oplus \mathcal{I}^{\bullet\bullet}) \to \mathcal{H}om_X^\bullet(f^*\mathcal{F}^\bullet, \mathrm{Tot}^\oplus \mathcal{I}^{\bullet\bullet})$$

as the quasi-isomorphism γ in the right column of (2.4.12). We can therefore regard β as a map

$$\beta : \mathrm{Tot}^\oplus I'^{\bullet\bullet}_{\mathrm{qc}} \to \mathcal{H}om_X^\bullet(f^*\mathcal{F}^\bullet, \mathrm{Tot}^\oplus \mathcal{I}^{\bullet\bullet}).$$

Putting together (2.4.10)–(2.4.12), it is not difficult to check that our task is equivalent to proving the commutativity in $\mathbf{D}(Y)$ of the following diagram of complexes:

$$(2.4.14)$$

$$
\begin{array}{ccc}
f_* \mathrm{Tot}^\oplus I'^{\bullet\bullet}_{\mathrm{qc}} & \longrightarrow & f_*(I'^{\bullet-n,n}_{\mathrm{qc}})/\mathrm{im}(f_* I'^{\bullet-n,n-1}_{\mathrm{qc}}) \\
{\scriptstyle f_*(\beta)} \big\downarrow & & \big\downarrow {\scriptstyle \simeq} \\
f_* \mathcal{H}om_X^\bullet(f^*\mathcal{F}^\bullet, \mathrm{Tot}^\oplus \mathcal{I}^{\bullet\bullet}) & & R^n f_*(\omega \otimes f^*\mathcal{A}^\bullet) \\
\big\downarrow & & \big\uparrow \\
\mathcal{H}om_Y^\bullet(f_* f^*\mathcal{F}^\bullet, f_* \mathrm{Tot}^\oplus \mathcal{I}^{\bullet\bullet}) & & R^n f_*(\omega \otimes f^*\mathcal{H}om_Y^\bullet(\mathcal{F}^\bullet, \mathcal{G}^\bullet)) \\
\big\downarrow & & \big\uparrow \\
\mathcal{H}om_Y^\bullet(\mathcal{F}^\bullet, \mathcal{J}^\bullet) & \xrightarrow{\;\;\alpha\;\;} & \mathcal{H}om_Y^\bullet(\mathcal{F}^\bullet, \mathcal{G}^\bullet)
\end{array}
$$

In (2.4.14), the bottom map in the right column is obtained from the canonical map $\mathcal{H} \simeq R^n f_*(\omega) \otimes \mathcal{H} \to R^n f_*(\mathcal{H} \otimes f^*\mathcal{H})$ for any $\mathcal{O}_Y$-module $\mathcal{H}$. Note that the commutativity of (2.4.14) in $\mathbf{D}(Y)$ is equivalent to its homotopy-commutativity as a diagram of complexes, since the complex $R^n f_*(\omega \otimes f^*\mathcal{A}^\bullet)$ is canonically isomorphic to $R^n f_*(\omega) \otimes \mathcal{A}^\bullet \simeq \mathcal{A}^\bullet$, which is a bounded below complex of injective sheaves. As in the proof of Theorem 2.3.2, the idea behind our analysis of (2.4.14) is to try to eliminate all references to $I^{\bullet\bullet}_{\mathrm{qc}}$ and to reduce to a commutativity assertion involving flat sheaves, which we will then check locally.

In order to carry out such a plan, we need to replace (2.4.14) by a more tractable diagram. This will require the introduction of some more maps and an auxiliary general lemma which will also be useful in the proof of Theorem 2.5.2. We noted above that the quasi-isomorphism $\mathcal{H}om_Y^\bullet(\mathcal{F}^\bullet, \mathcal{G}^\bullet) \to \mathcal{A}^\bullet$ has a homotopy inverse, so applying $f^\sharp$ to this shows that there is a double

complex map $\varphi^{-1} : I_{qc}^{\bullet\bullet} \to I^{\bullet\bullet}$ between Cartan-Eilenberg resolutions which is
a double complex homotopy inverse to φ. From the *construction* of φ^{-1} as in
[**CE**, XVII, Prop 1.2], the *homotopies* between $\varphi^{-1} \circ \varphi$ and 1, as well as between
$\varphi \circ \varphi^{-1}$ and 1, can be chosen to be compatible with respect to vertical canonical
truncation. Thus, the induced map $\varphi' : I'^{\bullet\bullet} \to I'^{\bullet\bullet}_{qc}$ on canonical truncations in
rows $\leq n$ is a double complex homotopy equivalence, so

$$(2.4.15) \qquad f_*(\mathrm{Tot}^{\oplus}\varphi') : f_*(\mathrm{Tot}^{\oplus} I'^{\bullet\bullet}) \to f_*(\mathrm{Tot}^{\oplus} I'^{\bullet\bullet}_{qc})$$

is a homotopy equivalence, hence a quasi-isomorphism. In order to check the
commutativity of (2.4.14) in $\mathbf{D}(Y)$, it is enough to check after composing the
two composites around the diagram, beginning in the upper left, with the *quasi-isomorphism* (2.4.15). By using this trick, we want to relate $f_*(\mathrm{Tot}^{\oplus}\varphi')$ and
$f_*(\beta)$ in order to replace (2.4.14) with a simpler diagram of complexes.

Since $\mathscr{F}^\bullet$ is bounded above, we can find a quasi-isomorphism to $\mathscr{F}^\bullet$ from
a bounded above complex each of whose terms is a direct sum of (the Y-flat)
sheaves of the form $j_!(\mathcal{O}_U)$ for various open subschemes $j : U \hookrightarrow Y$. The general
isomorphism

$$(2.4.16) \qquad \mathscr{H}om_Y(j_!\mathcal{O}_U, \cdot) \simeq j_*((\cdot)|_U)$$

shows that this is an *exact* functor on quasi-coherent sheaves if j is an affine map.
But recall that we assumed Y is separated, so by taking the U's to be open affines
(which we can do without loss of generality) we may suppose all j's are indeed
affine maps. We may (and do now) assume that the $\mathscr{F}^r$'s are direct sums of
sheaves of the type $j_!\mathcal{O}_U$ (for affine opens U), since $\mathscr{F}^\bullet$ only matters up to quasi-isomorphism. It then follows that each $f^*\mathscr{F}^r$ is an analogous such direct sum
on X, so $\mathscr{H}om_X(f^*\mathscr{F}^r, \cdot)$ is an exact functor on *quasi-coherent* $\mathcal{O}_X$-modules
(since an arbitrary product of exact sequences of quasi-coherent sheaves is an
exact sequence of sheaves). Using the Lemma on Way-Out Functors [**RD**, I, 7.1],
it follows that the functor $\mathscr{H}om_X^\bullet(f^*\mathscr{F}^\bullet, \cdot)$ from bounded below complexes of
quasi-coherent $\mathcal{O}_X$-modules to bounded below complexes of $\mathcal{O}_X$-modules takes
quasi-isomorphisms to quasi-isomorphisms. In particular,

$$(2.4.17) \qquad \mathscr{H}om_X^\bullet(f^*\mathscr{F}^\bullet, f^\sharp\mathscr{G}^\bullet) \to \mathscr{H}om_X^\bullet(f^*\mathscr{F}^\bullet, \mathrm{Tot}^\oplus \mathscr{I}^{\bullet\bullet})$$

is a quasi-isomorphism. The composite of (2.4.17) with the canonical map

$$(2.4.18) \qquad \psi : f^\sharp \mathscr{H}om_Y^\bullet(\mathscr{F}^\bullet, \mathscr{G}^\bullet) \longrightarrow \mathscr{H}om_X^\bullet(f^*\mathscr{F}^\bullet, f^\sharp\mathscr{G}^\bullet)$$

represents the derived category isomorphism [**RD**, II, 2.4(b)], so ψ is a quasi-isomorphism.

We want to construct a map (over ψ) from the Cartan-Eilenberg resolution
$I^{\bullet\bullet}$ of $f^\sharp \mathscr{H}om_Y^\bullet(\mathscr{F}^\bullet, \mathscr{G}^\bullet)$ to the double complex associated to a well-chosen resolution of $\mathscr{H}om_X^\bullet(f^*\mathscr{F}^\bullet, f^\sharp\mathscr{G}^\bullet)$ by injective complexes of $\mathcal{O}_X$-modules. Before
we can say what we mean by 'well-chosen', we need to state a lemma whose
proof is an easy exercise in chasing signs (and which will again be useful in the
proof of Theorem 2.5.2).

LEMMA 2.4.1. *Let Z be a ringed space, $\mathscr{F}^\bullet$ a bounded above complex of
$\mathcal{O}_Z$-modules, $\mathscr{G}^{\bullet\bullet}$ an upper-half plane double complex of $\mathcal{O}_Z$-modules. Define*

$C^{p,q} = \mathcal{H}om^p_Z(\mathcal{F}^\bullet, \mathcal{G}^{\bullet,q})$, so $C^{p,q} = 0$ for $q < 0$. Define the differentials for $C^{\bullet\bullet}$ as follows: the vertical differential $d^{p,q}_v : C^{p,q} \to C^{p,q+1}$ is induced by the vertical differential $\mathcal{G}^{\bullet,q} \to \mathcal{G}^{\bullet,q+1}$ and the horizontal differential $d^{p,q}_h : C^{p,q} \to C^{p+1,q}$ is $d_{h,\mathcal{G}^{\bullet\bullet}} + (-1)^{p+q+1}d_{\mathcal{F}^\bullet}$.

With these definitions, $C^{\bullet\bullet}$ is a double complex and the canonical inclusion of sheaves

$$\bigoplus_{r\in\mathbf{Z}} C^{r,n-r} \hookrightarrow \mathcal{H}om^n_Z(\mathcal{F}^\bullet, \mathrm{Tot}^\oplus \mathcal{G}^{\bullet\bullet})$$

identifies $\mathrm{Tot}^\oplus C^{\bullet\bullet}$ with a subcomplex of $\mathcal{H}om^\bullet_Z(\mathcal{F}^\bullet, \mathrm{Tot}^\oplus \mathcal{G}^{\bullet\bullet})$. This inclusion is an equality when $\mathrm{Tot}^\oplus \mathcal{G}^{\bullet\bullet}$ is bounded below with each term a finite direct sum (i.e., $\mathcal{G}^{n-q,q} = 0$ for $n \ll 0$ independent of q and for $q \geq q(n)$). In particular, if each row of $\mathcal{G}^{\bullet\bullet}$ is bounded below and $C'^{\bullet\bullet}$ (resp. $\mathcal{G}'^{\bullet\bullet}$) denotes the canonical truncation in rows $\leq n$, then $\mathrm{Tot}^\oplus C'^{\bullet\bullet}$ is equal to $\mathcal{H}om^\bullet_Z(\mathcal{F}^\bullet, \mathrm{Tot}^\oplus \mathcal{G}'^{\bullet\bullet})$.

To apply Lemma 2.4.1, let $K^{\bullet\bullet}$ be the upper half-plane double complex associated to a resolution of $f^\sharp\mathcal{G}^\bullet$ by injective complexes of quasi-coherent sheaves, with each (necessarily exact) row $K^{\bullet,q}$ bounded below. Let $K'^{\bullet\bullet}$ be the canonical truncation in rows $\leq n$. By Lemma 2.4.1, we get an upper half-plane double complex $C^{\bullet\bullet}$ with

$$C^{p,q} = \mathcal{H}om^p_X(f^*\mathcal{F}^\bullet, K^{\bullet,q})$$

and the canonical truncation in rows $\leq n$ has associated total complex

$$\mathcal{H}om^\bullet_X(f^*\mathcal{F}^\bullet, \mathrm{Tot}^\oplus K'^{\bullet\bullet}).$$

Moreover, from the construction, the map $C^{\bullet,q} \to C^{\bullet,q+1}$ (which is not a map of complexes) is induced by applying $\mathcal{H}om^\bullet_X(f^*\mathcal{F}^\bullet, \cdot)$ to $K^{\bullet,q} \to K^{\bullet,q+1}$ (which is not a map of complexes). Due to the special type of Y-flat $\mathcal{F}^r$'s we are considering and the fact that $K^{\bullet,q}$ is a quasi-coherent injective resolution of $f^\sharp\mathcal{G}^q$, each $C^{p,q}$ is an injective $\mathcal{O}_X$-module and the pth column $C^{p,\bullet}$ is an injective resolution of $\mathcal{H}om^p_X(f^*\mathcal{F}^\bullet, f^\sharp\mathcal{G}^\bullet)$. Also, each row $C^{\bullet,q}$ is an exact complex, because for any bounded below exact complex $\mathcal{I}^\bullet$ of injective $\mathcal{O}_X$-modules, the complex $\mathcal{H}om^\bullet_X(f^*\mathcal{F}^\bullet, \mathcal{I}^\bullet)$ is exact.

We conclude that $C^{\bullet\bullet}$ is the double complex associated to an injective resolution of $\mathcal{H}om^\bullet_X(f^*\mathcal{F}^\bullet, f^\sharp\mathcal{G}^\bullet)$ in the abelian category of complexes of $\mathcal{O}_X$-modules, so there is a map

$$I^{\bullet\bullet} \to C^{\bullet\bullet}$$

of double complexes over the quasi-isomorphism (2.4.18). The induced maps of sheaves

$$(2.4.19) \qquad I'^{p,q} \to C'^{p,q} \overset{\mathrm{def}}{=} \mathcal{H}om^p_X(f^*\mathcal{F}^\bullet, K'^{\bullet,q})$$

give rise to a double complex map $I'^{\bullet\bullet} \to C'^{\bullet\bullet}$ of canonical truncations in rows $\leq n$, and hence a map of complexes

$$\mathrm{Tot}^\oplus I'^{\bullet\bullet} \to \mathcal{H}om^\bullet_X(f^*\mathcal{F}^\bullet, \mathrm{Tot}^\oplus K'^{\bullet\bullet})$$

over (2.4.18). Applying f_* to (2.4.19) and using the adjointness of f_* and f^*, we get maps

$$f_* I'^{p,q} \longrightarrow f_* \mathcal{H}om_X^p(f^*\mathcal{F}^\bullet, K'^{\bullet,q})$$

$$\Big\downarrow \simeq$$

$$\mathcal{H}om_Y^p(\mathcal{F}^\bullet, f_*K'^{\bullet,q}/\mathrm{im}(f_*K'^{\bullet,q-1})) \longleftarrow \mathcal{H}om_Y^p(\mathcal{F}^\bullet, f_*K'^{\bullet,q})$$

which induce maps

(2.4.20) $\quad f_* I'^{p-n,n}/\mathrm{im}(f_*I'^{p-n,n-1}) \longrightarrow \mathcal{H}om_Y^{p-n}(\mathcal{F}^\bullet, f_*K'^{\bullet,n}/f_*K'^{\bullet,n-1})$

$$\Big\| $$

$$\mathcal{H}om_Y^p(\mathcal{F}^\bullet, \mathrm{H}_v^n(f_*K'^{\bullet-n,\bullet}))$$

where H_v^n denotes the complex of nth homologies with respect to vertical differentials. Since $(-1)^{p-n+n+1} = (-1)^{p+1}$, the composite map (2.4.20) is a map of complexes (indexed by p), and can be rephrased more conceptually as a map of complexes

(2.4.21) $\quad \mathrm{R}^n f_*(\omega \otimes f^*\mathcal{H}om_Y^\bullet(\mathcal{F}^\bullet, \mathcal{G}^\bullet)) \to \mathcal{H}om_Y^\bullet(\mathcal{F}^\bullet, \mathrm{R}^n f_*(\omega \otimes f^*\mathcal{G}^\bullet)).$

Recalling the definitions of α and β and using the *homotopy equivalence* (2.4.15), a straightfoward manipulation of all of our maps establishes that the commutativity of (2.4.14) in $\mathbf{D}(Y)$ is equivalent to the commutativity of the outside edge of

(2.4.22)

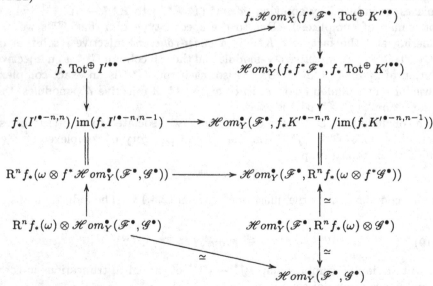

in $\mathbf{D}(Y)$, where the two middle horizontal maps are (2.4.20) and (2.4.21). In fact, we claim that (2.4.22) even commutes on the level of complexes of $\mathcal{O}_Y$-modules.

The commutativity of the top and middle parts follows from the definitions. For the bottom part, we can chase sections of the complex $\mathscr{H}om_Y^\bullet(\mathscr{F}^\bullet, \mathscr{G}^\bullet)$ in each separate degree. Since each term of this complex is a finite product and the construction of (2.4.20) does not depend upon the fact that $\mathscr{F}^\bullet$ has coherent cohomology sheaves, we can use functoriality with respect to $\mathscr{F}^\bullet$ and $\mathscr{G}^\bullet$ to reduce to the following general situation.

Let $f : X \to Y$ be a map of ringed spaces, $\mathscr{F}$ a flat $\mathcal{O}_Y$-module and $\mathscr{G}$ an $\mathcal{O}_Y$-module. Let ω denote an $\mathcal{O}_X$-module, and choose respective f_*-acyclic and injective resolutions

$$\omega \otimes f^* \mathscr{H}om_Y(\mathscr{F}, \mathscr{G}) \to I^\bullet, \quad \omega \otimes f^* \mathscr{G} \to K^\bullet.$$

Since $\mathscr{H}om_X(f^*\mathscr{F}, K^\bullet)$ is a complex of *injectives* (as $\mathscr{F}$ is Y-flat) with kernel $\mathscr{H}om_X(f^*\mathscr{F}, \omega \otimes f^*\mathscr{G})$ in degree 0, we can choose a map of complexes (unique up to homotopy)

$$\mu : I^\bullet \to \mathscr{H}om_X(f^*\mathscr{F}, K^\bullet)$$

over the canonical map $\omega \otimes f^* \mathscr{H}om_Y(\mathscr{F}, \mathscr{G}) \to \mathscr{H}om_X(f^*\mathscr{F}, \omega \otimes f^*\mathscr{G})$. By adjointness of f_* and f^*, μ gives rise to a map of complexes

$$f_* I^\bullet \to f_* \mathscr{H}om_X(f^*\mathscr{F}, K^\bullet) \simeq \mathscr{H}om_Y(\mathscr{F}, f_* K^\bullet),$$

which induces maps

(2.4.23) $\mu_n : \mathrm{H}^n(f_* I^\bullet) \to \mathrm{H}^n(\mathscr{H}om_Y(\mathscr{F}, f_* K^\bullet)) \to \mathscr{H}om_Y(\mathscr{F}, \mathrm{H}^n(f_* K^\bullet))$

for all $n \in \mathbf{Z}$. For our purposes above, it suffices to prove that the diagram of sheaves

(2.4.24)

$$\begin{array}{ccc}
\mathrm{H}^n(f_* I^\bullet) & \xrightarrow{\mu_n} & \mathscr{H}om_Y(\mathscr{F}, \mathrm{H}^n(f_* K^\bullet)) \\
\| & & \| \\
\mathrm{R}^n f_*(\omega \otimes f^* \mathscr{H}om_Y(\mathscr{F}, \mathscr{G})) & & \mathscr{H}om_Y(\mathscr{F}, \mathrm{R}^n f_*(\omega \otimes f^*\mathscr{G})) \\
\uparrow & & \uparrow \\
\mathrm{R}^n f_*(\omega) \otimes \mathscr{H}om_Y(\mathscr{F}, \mathscr{G}) & \longrightarrow & \mathscr{H}om_Y(\mathscr{F}, \mathrm{R}^n f_*(\omega) \otimes \mathscr{G})
\end{array}$$

commutes. Note that this construction is local on Y, functorial in $\mathscr{F}$ and $\mathscr{G}$, and changing μ up to homotopy or changing $I^\bullet$, $K^\bullet$ does not affect whether or not (2.4.24) commutes.

To check the commutativity of (2.4.24), we can use functoriality in $\mathscr{F}$ and the left exactness of $\mathscr{H}om_Y(\cdot, \cdot)$ in the first variable to reduce to the case where $\mathscr{F}$ is a direct sum of sheaves of the form $j_! \mathcal{O}_U$ for various open immersions $j : U \hookrightarrow Y$. By the universal mapping property of direct sums, we can assume $\mathscr{F} = j_! \mathcal{O}_U$ for a single open immersion $j : U \hookrightarrow Y$. Since two maps from an $\mathcal{O}_Y$-module to

$$\mathscr{H}om_Y(j_! \mathcal{O}_U, \mathscr{H}) \simeq j_*(\mathscr{H}|_U)$$

coincide if and only if their restrictions to U coincide, we can reduce to the case $Y = U$, so $\mathscr{F} = \mathcal{O}_Y$. This case is trivial. ∎

2.5. The Fundamental Local Isomorphism

The non-trivial relations between the functors $(\cdot)^{\sharp}$ and $(\cdot)^{\flat}$ are given in [**RD**, III, §7, §8]. The basic tool underlying this is the so-called 'fundamental local isomorphism,' which enjoys several key properties. In this section, we give the basic definitions and formulate how this isomorphism is compatible with respect to suitable composites of scheme morphisms [**RD**, III, 7.4(a)]. This crucial compatibility property is not proven in [**RD**], but the proof is non-trivial (as we shall see).

Let $i : Y \hookrightarrow X$ be an lci map with pure codimension n, $\mathscr{F}$ an *arbitrary* $\mathcal{O}_X$-module (it is technically important to have such generality later on). The 'fundamental local isomorphism' [**RD**, III, 7.2] (whose construction will be reviewed shortly) is an isomorphism of $\mathcal{O}_Y$-modules

$$(2.5.1) \qquad \eta_{Y/X} = \eta : \mathscr{E}xt^n_X(i_*\mathcal{O}_Y, \mathscr{F}) \simeq \omega_{Y/X} \otimes i^*(\mathscr{F}),$$

compatible with Zariski localization on X. There is (for now) no risk of confusion with the 'double duality' map (1.3.16), which [**RD**] also denotes by η.

To motivate the definition of (2.5.1), we first consider a commutative algebra analogue. For a ring A, an A-module M, and a regular sequence $\{f_1, \ldots, f_n\}$ generating an ideal J of A, the map

$$Y = \operatorname{Spec}(A/J) \hookrightarrow \operatorname{Spec}(A) = X$$

is an lci map with pure codimension n. We can therefore δ-functorially compute

$$\operatorname{Ext}^i_A(A/J, M) = \operatorname{H}^i(\operatorname{Hom}^{\bullet}_A(K_{\bullet}(\mathbf{f}), M)),$$

and for an arbitrary $\mathcal{O}_X$-module $\mathscr{F}$ we can δ-functorially compute

$$\mathscr{E}xt^i_X(i_*\mathcal{O}_Y, \mathscr{F}) = \operatorname{H}^i(\mathscr{H}om^{\bullet}_X(\widetilde{K_{\bullet}(\mathbf{f})}, \mathscr{F})),$$

where $\widetilde{K_{\bullet}(\mathbf{f}, \mathscr{F})}$ is the complex of quasi-coherent sheaves on $X = \operatorname{Spec}(A)$ attached to $K_{\bullet}(\mathbf{f})$. Using the isomorphism $\psi_{\mathbf{f},M}$ from (1.3.28) and the isomorphism $A/J \simeq \bigwedge^n(J/J^2)$ determined by $1 \mapsto f_1 \wedge \cdots \wedge f_n$, we get a composite isomorphism

$$(2.5.2) \quad \varphi_{\mathbf{f},M} : \operatorname{Ext}^n_A(A/J, M) \xrightarrow{\psi_{\mathbf{f},M}} M/JM \xrightarrow{\simeq} \bigwedge^n(J/J^2)^{\vee} \otimes_{A/J} M/JM$$

which is independent of the choice of regular sequence of n elements generating J, so (2.5.2) can instead be denoted $\varphi_{J,M}$. Recall that the definition of $\psi_{\mathbf{f},M}$ implicitly involves a sign of $(-1)^{n(n+1)/2}$. The *method* of construction of (2.5.2) is independent of the choice of regular sequence $f_1, \ldots, f_n$, so it globalizes to an arbitrary local complete intersection $i : Y \hookrightarrow X$ with pure codimension n and an *arbitrary* $\mathcal{O}_X$-module $\mathscr{F}$, defining the 'fundamental local isomorphism' (2.5.1) for $n > 0$. When $n = 0$, so $i : Y \hookrightarrow X$ is an isomorphism onto a closed and open subscheme, (2.5.1) is defined to be the canonical 'identity' (without the intervention of signs). With a little care, it is not hard to show that (2.5.1) is compatible with flat base change over X for arbitrary $\mathscr{F}$ [**RD**, III, 7.4(b)] and with *arbitrary* base change over X preserving the 'lci of pure codimension n' property of i for *quasi-coherent* X-flat $\mathscr{F}$ (using the base change theory for

$\mathscr{E}xt$'s in [**AK2**, §1] in the latter case, in which case the proof reduces to the easy analogous base change assertion for (2.5.2)).

Now that the isomorphism (2.5.1) is defined, the second part of [**RD**, III, 7.2] ensures the 'erasability' hypothesis in Lemma 2.1.1, so (2.5.1) yields a functorial isomorphism in $\mathbf{D}(Y)$, for variable $\mathscr{F}^\bullet$ in $\mathbf{D}(X)$:

$$(2.5.3) \qquad \eta_i : i^\flat(\mathscr{F}^\bullet) \simeq \omega_{Y/X}[-n] \overset{\mathbf{L}}{\otimes} \mathbf{L}i^*(\mathscr{F}^\bullet)$$

Due to the sign in (2.1.1), we see that by defining the translation compatibility of the right side of (2.5.3) via (1.3.6), the isomorphism (2.5.3) *is* compatible with translations! This is crucial in the proof of Lemma A.2.1. Beware that if we compose (2.5.3) with the isomorphism

$$(2.5.4) \qquad \omega_{Y/X}[-n] \overset{\mathbf{L}}{\otimes} \mathbf{L}i^*(\mathscr{F}^\bullet) \simeq (\omega_{Y/X} \overset{\mathbf{L}}{\otimes} \mathbf{L}i^*(\mathscr{F}^\bullet))[-n],$$

(which does not involve the intervention of signs) then via the conventions in §1.3 this modification of (2.5.3) is *not* compatible with translations in general! For this reason, we will always express the right side of (2.5.3) in the form given in (2.5.3), rather than in the form on the right side of (2.5.4). When $n = 0$, so $i : Y \hookrightarrow X$ is an isomorphism onto an open and closed subscheme, (2.5.3) is the canonical 'identity' map. By Corollary 2.1.2, when $\mathscr{F}^\bullet = \mathscr{F}[0]$ is concentrated in degree 0, the induced map on H^n's in (2.5.3) is *exactly* (2.5.1), without the intervention of signs in the identification. Beware that the statement of the isomorphism (2.5.3) in [**RD**, III, 7.3] has $\mathbf{L}i^*(\mathscr{F}^\bullet) \overset{\mathbf{L}}{\otimes} \omega_{Y/X}[-n]$ on the right side.

Compatibility of (2.5.3) with respect to composites in i [**RD**, III, 7.4(a)] is rather important. It is not proven in [**RD**], so we give a proof in Theorem 2.5.1. The proof depends on our definition of (1.3.28) via Koszul complexes, as this was used to define (2.5.1)! If we had omitted the sign in the definition of (1.3.28), then (2.5.3) would change by a sign of $(-1)^{n(n+1)/2}$. Later on, we will occasionally keep track of the consequences of omitting this mysterious sign (in order to keep clear where it is crucial and where it is irrelevant).

The study of the map η_i in (2.5.3) requires us to give an 'explicit' description, as in (2.1.2). Since i^* has finite homological dimension $\leq n$, every complex in $\mathbf{D}(X)$ is quasi-isomorphic to a complex of i^*-acyclics, so we just consider complexes $\mathscr{F}^\bullet$ with each $\mathscr{F}^r$ an i^*-acyclic object. It suffices to describe η_i on such complexes. Let $C'^{\bullet\bullet}$ be the canonical truncation in rows $\leq n$ of a Cartan-Eilenberg resolution of $\mathscr{F}^\bullet$, so $C'^{p\bullet}$ is the injective resolution used to compute

$\mathscr{E}xt_X^{\bullet}(\cdot, \mathscr{F}^p)$ and for each j there is a natural map

(2.5.5) $$\mathscr{H}om_X^j(i_*\mathscr{O}_Y, \mathrm{Tot}^{\oplus}(C'^{\bullet\bullet}))$$

$$\downarrow$$

$$\mathscr{H}om_X(i_*\mathscr{O}_Y, C'^{j-n,n})/\mathrm{im}(\mathscr{H}om_X(i_*\mathscr{O}_Y, C'^{j-n,n-1}))$$

$$\|$$

$$\mathscr{E}xt_X^n(i_*\mathscr{O}_Y, \mathscr{F}^{j-n})$$

$$\simeq\downarrow$$

$$\omega_{Y/X} \otimes i^*(\mathscr{F}^{j-n})$$

(using (2.5.1) at the end). By viewing (2.5.5) as a map of $\mathscr{O}_Y$-modules and watching out for the sign error in [**RD**, I, 7.4] as discussed below Lemma 2.1.1, we obtain (*without* the intervention of signs) a map in $\mathbf{D}(Y)$

(2.5.6) $$i^{\flat}(\mathscr{F}^{\bullet}) \to (\omega_{Y/X} \otimes i^*\mathscr{F}^{\bullet})[-n] \simeq \omega_{Y/X}[-n] \overset{\mathbf{L}}{\otimes} \mathbf{L}i^*(\mathscr{F}^{\bullet})$$

which is exactly η_i (this is just a special case of the general discussion of (2.1.2)). We emphasize that this description *only* applies when $\mathscr{F}^{\bullet}$ is a complex of i^*-*acyclic* sheaves. We will later need to know (e.g., for the proof of Theorem 2.7.2(1)) that we can replace the truncated Cartan-Eilenberg resolution $C'^{\bullet\bullet}$ above by the canonical truncation $K'^{\bullet\bullet}$ in rows $\le n$ of *any* double complex $K^{\bullet\bullet}$ of $\mathscr{H}om_X(i_*\mathscr{O}_Y, \cdot)$-acyclics for which there is an augmentation map $\mathscr{F}^{\bullet} \to K^{\bullet,0}$ that induces a resolution $\mathscr{F}^p \to K^{p,\bullet}$ for all p. The proof is easy, by comparing everything to the double complex associated to a resolution of $\mathscr{F}^{\bullet}$ by injective complexes (in the sense defined near the end of §1.2).

From the explicit description of η_i, we readily deduce that if $\mathscr{E}$ is a locally free sheaf with finite rank on X and $\mathscr{F}^{\bullet} \in \mathbf{D}(X)$, then the diagram

(2.5.7)

$$i^{\flat}(\mathscr{E} \otimes \mathscr{F}^{\bullet}) \overset{\simeq}{\longrightarrow} i^*\mathscr{E} \otimes i^{\flat}\mathscr{F}^{\bullet}$$

$$\eta_i(\mathscr{E}\otimes\mathscr{F}^{\bullet})\downarrow \qquad\qquad\qquad\qquad \downarrow\eta_i(\mathscr{F}^{\bullet})$$

$$\omega_{Y/X}[-n] \overset{\mathbf{L}}{\otimes} \mathbf{L}i^*(\mathscr{E} \otimes \mathscr{F}^{\bullet}) \qquad i^*\mathscr{E}[0] \overset{\mathbf{L}}{\otimes} (\omega_{Y/X}[-n] \overset{\mathbf{L}}{\otimes} \mathbf{L}i^*\mathscr{F}^{\bullet})$$

$$\simeq\downarrow \qquad\qquad\qquad\qquad\qquad \downarrow\simeq$$

$$(\omega_{Y/X}[-n] \overset{\mathbf{L}}{\otimes} i^*\mathscr{E}[0]) \overset{\mathbf{L}}{\otimes} \mathbf{L}i^*\mathscr{F}^{\bullet} \overset{\simeq}{\longrightarrow} (i^*\mathscr{E}[0] \overset{\mathbf{L}}{\otimes} \omega_{Y/X}[-n]) \overset{\mathbf{L}}{\otimes} \mathbf{L}i^*\mathscr{F}^{\bullet}$$

commutes in $\mathbf{D}(Y)$. This commutativity is important later on when dealing with the ambiguity of dualizing complexes up to tensoring with an invertible sheaf. Also, the explicit description makes it clear that η_i is of formation compatible with *flat* base change on X (as we have already noted this for (2.5.1)); this is [**RD**, III, 7.4(b)].

We conclude this section by discussing non-trivial compatibility properties of η_i (proofs will be given in the next section). There are three which we will prove. The first of these properties (Theorem 2.5.1) is the compatibility of η_i with respect to composites in i [RD, III, 7.4(a)] and is stated without proof in [RD]. The proof is *hard*. The other two compatibilities (Theorem 2.5.2) relate η_i to compatibilities for $(\cdot)^\flat$ and are needed to verify some basic properties in the global duality theory. These two extra compatibilities of η_i are not explicitly stated in [RD], and their proof involves a mixture of the methods used to prove Theorem 2.3.3 above and Theorem 2.5.1 below.

THEOREM 2.5.1. [RD, III, 7.4(a)] *Let* $Z \xrightarrow{j} Y \xrightarrow{i} X$ *be two local complete intersection morphisms of schemes, and assume that* j *(resp.* i*) has pure codimension* m *(resp.* n*), so* ij *is a local complete intersection morphism with pure codimension* $n + m$*. For* $\mathscr{F}^\bullet \in \mathbf{D}(X)$*, the diagram in* $\mathbf{D}(Z)$

(2.5.8)

$$
\begin{array}{ccc}
(ij)^\flat(\mathscr{F}^\bullet) & \xrightarrow{\ \eta_{ij}\ } & \omega_{Z/X}[-m-n] \overset{\mathbf{L}}{\otimes} \mathbf{L}(ij)^*(\mathscr{F}^\bullet) \\
\simeq \downarrow & & \downarrow \simeq \\
j^\flat i^\flat(\mathscr{F}^\bullet) & & \omega_{Z/X}[-m-n] \overset{\mathbf{L}}{\otimes} \mathbf{L}j^* \mathbf{L}i^*(\mathscr{F}^\bullet) \\
j^\flat(\eta_i) \downarrow & & \downarrow \zeta'_{i,j} \\
j^\flat(\omega_{Y/X}[-n] \overset{\mathbf{L}}{\otimes} \mathbf{L}i^* \mathscr{F}^\bullet) & \xrightarrow{\ \eta_j\ } & \omega_{Z/Y}[-m] \overset{\mathbf{L}}{\otimes} \mathbf{L}j^*(\omega_{Y/X}[-n] \overset{\mathbf{L}}{\otimes} \mathbf{L}i^*(\mathscr{F}^\bullet))
\end{array}
$$

commutes.

If the implicit sign in the definition of (1.3.28) were left out (so the definition of the fundamental local isomorphism would change by a sign), then the diagram (2.5.8) would only commute up to a sign of

$$(-1)^{m(m+1)/2}(-1)^{n(n+1)/2}(-1)^{(n+m)(n+m+1)/2} = (-1)^{nm}.$$

THEOREM 2.5.2. *Let* $p : X \to Y$ *be a separated smooth map with pure relative dimension* n *and let* i *be a section of* p*, so the closed immersion* i *is an lci map with pure codimension* n*. Assume that* Y *is locally noetherian.*

1. *For* $\mathscr{F}^\bullet \in \mathbf{D}_{qc}^+(X)$ *and* $\mathscr{G}^\bullet \in \mathbf{D}_{qc}^b(X)_{\mathrm{fTd}}$*, the diagram in* $\mathbf{D}(Y)$

(2.5.9)

$$
\begin{array}{ccc}
\omega_{Y/X}[-n] \overset{\mathbf{L}}{\otimes} \mathbf{L}i^*\mathscr{F}^\bullet \overset{\mathbf{L}}{\otimes} \mathbf{L}i^*\mathscr{G}^\bullet & \xrightarrow{\ \simeq\ } & \omega_{Y/X}[-n] \overset{\mathbf{L}}{\otimes} \mathbf{L}i^*(\mathscr{F}^\bullet \overset{\mathbf{L}}{\otimes} \mathscr{G}^\bullet) \\
\eta_i \otimes 1 \uparrow & & \uparrow \eta_i \\
i^\flat \mathscr{F}^\bullet \overset{\mathbf{L}}{\otimes} \mathbf{L}i^*\mathscr{G}^\bullet & \xrightarrow{\ \simeq\ } & i^\flat(\mathscr{F}^\bullet \overset{\mathbf{L}}{\otimes} \mathscr{G}^\bullet)
\end{array}
$$

commutes, where the bottom map is [RD, III, 6.9(a)]*, defined by replacing* $\mathscr{G}^\bullet$ *with a bounded complex of flats.*

2. *Assume moreover that Y is separated. For $\mathcal{F}^{\bullet} \in \mathbf{D}_c^-(Y)$ and $\mathcal{G}^{\bullet} \in \mathbf{D}_{qc}^+(Y)$, the diagram in $\mathbf{D}(Y)$*

(2.5.10)

$$
\begin{array}{ccc}
\omega_{Y/X}[-n] \overset{\mathbf{L}}{\otimes} \mathbf{L}i^*p^*\mathbf{R}\mathscr{H}om_Y^{\bullet}(\mathcal{F}^{\bullet},\mathcal{G}^{\bullet}) & \xleftarrow{\;\eta_i\;} & i^{\flat}p^*\mathbf{R}\mathscr{H}om_Y^{\bullet}(\mathcal{F}^{\bullet},\mathcal{G}^{\bullet}) \\
\simeq \big\uparrow & & \big\downarrow \simeq \\
\mathbf{R}\mathscr{H}om_Y^{\bullet}(\mathcal{F}^{\bullet},\omega_{Y/X}[-n] \overset{\mathbf{L}}{\otimes} \mathbf{L}i^*p^*\mathcal{G}^{\bullet}) & & i^{\flat}\mathbf{R}\mathscr{H}om_X^{\bullet}(p^*\mathcal{F}^{\bullet},p^*\mathcal{G}^{\bullet}) \\
\eta_i \big\uparrow & \nearrow \simeq & \\
\mathbf{R}\mathscr{H}om_Y^{\bullet}(\mathcal{F}^{\bullet},i^{\flat}p^*\mathcal{G}^{\bullet}) & &
\end{array}
$$

*commutes, where the bottom and lower right maps are [**RD**, III, 6.9(b)] (defined by replacing $p^*\mathcal{G}^{\bullet}$ with a bounded below complex of injectives) and [**RD**, II, 5.8] (defined by replacing $\mathcal{G}^{\bullet}$ with a bounded below complex of injectives) respectively.*

We note that in Theorem 2.5.2, the only reason for assuming that Y is locally noetherian in the first part and that (in addition) Y is separated in the second part is because it is needed for the proof we give (cf. Theorem 2.3.3). These conditions should not be needed, but they suffice for the construction of Grothendieck duality (*without* separatedness conditions on the base).

2.6. Proofs of Properties of the Fundamental Local Isomorphism

In this section, we give proofs of Theorem 2.5.1 and Theorem 2.5.2. We remind the reader that the sign in the definition of (1.3.28), which is used in the definition of (2.5.2), is essential in the proof of Theorem 2.5.1.

PROOF. (of Theorem 2.5.1) The first part of the proof is concerned with the abstract nonsense that reduces us to the crucial special case $\mathcal{F}^{\bullet} = \mathcal{F}[0]$ and the commutativity in degree $n + m$ cohomology (cf. Lemma 2.6.2). The study of this special case will reveal the important role of the implicit sign in (1.3.28).

Without loss of generality, n and m are positive (or else everything is trivial). Since i^* and $(ij)^*$ have finite homological dimension, we may also assume that $\mathcal{F}^{\bullet}$ is a complex of sheaves which are simultaneously i^*-acyclic and $(ij)^*$-acyclic. Thus, $i^*\mathcal{F}^{\bullet}$ is a complex of j^*-acyclic sheaves, so

$$
\omega_{Y/X}[-n] \otimes i^*\mathcal{F}^{\bullet} \simeq \omega_{Y/X}[-n] \overset{\mathbf{L}}{\otimes} \mathbf{L}i^*\mathcal{F}^{\bullet}
$$

is also a complex of j^*-acyclic sheaves. Let $\mathscr{I}^{\bullet\bullet}$ be the upper half-plane double complex associated to a resolution of $\mathcal{F}^{\bullet}$ by injective complexes of $\mathcal{O}_X$-modules. Let $I^{\bullet\bullet}$ and $J^{\bullet\bullet}$ denote the canonical truncations of $\mathscr{I}^{\bullet\bullet}$ in rows $\leq n$ and $\leq n + m$ respectively, so $I^{\bullet\bullet}$ is a double subcomplex of $J^{\bullet\bullet}$, which in turn is a double subcomplex of $\mathscr{I}^{\bullet\bullet}$. In terms of these resolutions, we want to describe all of the maps in (2.5.8). With such descriptions, we will be able to use some general claims about the behavior of differentials in certain spectral sequences in order to reduce to the special case in Lemma 2.6.2 below.

First, we describe all sides of the square (2.5.8). By construction, the top map η_{ij} is represented by the composite

$$(2.6.1) \qquad \mathcal{H}om_X(\mathcal{O}_Z, \mathrm{Tot}^\oplus J^{\bullet\bullet}) \longrightarrow H_v^{n+m}(\mathcal{H}om_X(\mathcal{O}_Z, J^{\bullet-n-m,n+m}))$$

$$\Big\|$$

$$\omega_{Z/X}[-n-m] \otimes (ij)^* \mathcal{F}^\bullet \xleftarrow{\ \simeq\ } \mathcal{E}xt_X^{n+m}(\mathcal{O}_Z, \mathcal{F}^\bullet)[-n-m]$$

where the bottom map in (2.6.1) is induced by (2.5.1). Term-by-term, (2.6.1) is just a restatement of (2.5.5). Since the canonical map

$$(2.6.2)$$
$$\mathcal{H}om_Y(\mathcal{O}_Z, \mathcal{H}om_X(\mathcal{O}_Y, \mathrm{Tot}^\oplus I^{\bullet\bullet})) \to \mathcal{H}om_Y(\mathcal{O}_Z, \mathcal{H}om_X(\mathcal{O}_Y, \mathrm{Tot}^\oplus J^{\bullet\bullet}))$$

is a quasi-isomorphism (argue as in the proof of [**RD**, III, 6.2]), the derived category composite in the left column of (2.5.8) is represented by the diagram of complexes of $\mathcal{O}_Z$-modules

$$(2.6.3) \qquad\qquad \mathcal{H}om_X(\mathcal{O}_Z, \mathrm{Tot}^\oplus J^{\bullet\bullet})$$

$$\simeq\Big\downarrow$$

$$\mathcal{H}om_Y(\mathcal{O}_Z, \mathcal{H}om_X(\mathcal{O}_Y, \mathrm{Tot}^\oplus J^{\bullet\bullet}))$$

$$\Big\uparrow{\scriptstyle\text{qism}}$$

$$\mathcal{H}om_Y(\mathcal{O}_Z, \mathcal{H}om_X(\mathcal{O}_Y, \mathrm{Tot}^\oplus I^{\bullet\bullet}))$$

$$\Big\downarrow$$

$$\mathcal{H}om_Y(\mathcal{O}_Z, \mathcal{H}om_X(\mathcal{O}_Y, I^{\bullet-n,n})/\mathrm{im}(\mathcal{H}om_X(\mathcal{O}_Y, I^{\bullet-n,n-1})))$$

$$\Big\|$$

$$\mathcal{H}om_Y(\mathcal{O}_Z, \mathcal{E}xt_X^n(\mathcal{O}_Y, \mathcal{F}^\bullet)[-n])$$

$$\simeq\Big\downarrow$$

$$\mathcal{H}om_Y(\mathcal{O}_Z, \omega_{Y/X}[-n] \otimes i^* \mathcal{F}^\bullet)$$

$$\Big\downarrow$$

$$j^b(\omega_{Y/X}[-n] \otimes i^* \mathcal{F}^\bullet)$$

where the second-from-bottom map in (2.6.3) is induced by (2.5.1), the bottom map in (2.6.3) is the canonical map from a left exact functor to its total derived functor, and the map labelled 'qism' is the quasi-isomorphism (2.6.2). The right column composite in (2.5.8) is represented by the composite

$$(2.6.4) \quad \omega_{Z/X}[-n-m] \otimes (ij)^* \mathscr{F}^\bullet \xrightarrow[\simeq]{\zeta'_{i,j}} \omega_{Z/Y}[-m] \otimes j^* \omega_{Y/X}[-n] \otimes j^* i^* \mathscr{F}^\bullet$$

$$\|$$

$$\omega_{Z/Y}[-m] \otimes j^*(\omega_{Y/X}[-n] \otimes i^* \mathscr{F}^\bullet)$$

The 'computation' of the remaining map η_j in the bottom row of (2.5.8) is a bit more delicate. Since $\omega_{Y/X}[-n] \otimes i^* \mathscr{F}^\bullet$ is a complex of j^*-acyclics and (as is implicit in (2.6.3)) we have an isomorphism of complexes

(2.6.5)

$$\omega_{Y/X}[-n] \otimes i^* \mathscr{F}^\bullet \xrightarrow{\quad \simeq \quad} \mathscr{E}xt_X^n(\mathscr{O}_Y, \mathscr{F}^\bullet)[-n]$$

$$\simeq \Big\downarrow$$

$$\mathscr{H}om_X(\mathscr{O}_Y, I^{\bullet-n,n})/\mathrm{im}(\mathscr{H}om_X(\mathscr{O}_Y, I^{\bullet-n,n-1})),$$

what we want to do is use $J^{\bullet\bullet}$ to compute the canonical truncation in rows $\leq m$ of a resolution of

$$\mathscr{H}om_X(\mathscr{O}_Y, I^{\bullet-n,n})/\mathrm{im}(\mathscr{H}om_X(\mathscr{O}_Y, I^{\bullet-n,n-1}))$$

by complexes of $\mathscr{H}om_Y(\mathscr{O}_Z, \cdot)$-acyclics. Since $m \geq 1$, we have $J^{\bullet,n} = \mathscr{I}^{\bullet,n}$ (and of course $J^{\bullet,n-1} = I^{\bullet,n-1} = \mathscr{I}^{\bullet,n-1}$). Defining the double complex $\mathscr{J}^{\bullet\bullet}$ by

$$\mathscr{J}^{\bullet,q} = \begin{cases} \mathscr{I}^{\bullet,n+q} & \text{if } q > 0 \\ 0 & \text{if } q \leq 0 \end{cases},$$

we get an upper-half plane double complex

$$(2.6.6) \qquad\qquad \mathscr{H}om_X(\mathscr{O}_Y, \mathscr{J}^{\bullet\bullet})$$

$$\uparrow$$

$$\mathscr{H}om_X(\mathscr{O}_Y, J^{\bullet,n})/\mathrm{im}(\mathscr{H}om_X(\mathscr{O}_Y, J^{\bullet,n-1}))$$

There is also a canonical map of complexes

$$(2.6.7) \qquad\qquad \mathscr{H}om_X(\mathscr{O}_Y, I^{\bullet,n})/\mathrm{im}(\mathscr{H}om_X(\mathscr{O}_Y, I^{\bullet,n-1}))$$

$$\Big\downarrow$$

$$\mathscr{H}om_X(\mathscr{O}_Y, J^{\bullet,n})/\mathrm{im}(\mathscr{H}om_X(\mathscr{O}_Y, J^{\bullet,n-1}))$$

to the bottom row of the upper-half plane double complex (2.6.6).

The functor $\mathscr{H}om_X(\mathscr{O}_Y, \cdot)$ has cohomological dimension $\leq n$, so via (2.6.7) the pth column of (2.6.6) is a resolution of

$$\mathscr{H}om_X(\mathscr{O}_Y, I^{p,n})/\mathrm{im}(\mathscr{H}om_X(\mathscr{O}_Y, I^{p,n-1})).$$

By (2.6.5), it follows that $\eta_j(\omega_{Y/X}[-n] \overset{\mathbf{L}}{\otimes} \mathbf{L}i^* \mathscr{F}^\bullet)$ can be 'calculated' in terms of the canonical truncation in rows $\leq m$ of (2.6.6), once we verify that all terms in (2.6.6) are $\mathscr{H}om_Y(\mathscr{O}_Z, \cdot)$-acyclic. But $\mathscr{H}om_X(\mathscr{O}_Y, \cdot)$ takes injective $\mathscr{O}_X$-modules to injective $\mathscr{O}_Y$-modules, so we just have to check the $\mathscr{H}om_Y(\mathscr{O}_Z, \cdot)$-acyclicity of the terms along the bottom row of (2.6.6). In other words, we want to prove

$$(2.6.8) \qquad \mathscr{E}xt_Y^r(\mathscr{O}_Z, \mathscr{H}om_X(\mathscr{O}_Y, \mathscr{I}^{\bullet,n})/\mathrm{im}(\mathscr{H}om_X(\mathscr{O}_Y, \mathscr{I}^{\bullet,n-1}))) = 0$$

for all $r \geq 1$ (recall $\mathscr{I}^{\bullet,n} = J^{\bullet,n}$ and $\mathscr{I}^{\bullet,n-1} = J^{\bullet,n-1}$).

Since all $\mathscr{H}om_X(\mathscr{O}_Y, \mathscr{I}^{\bullet\bullet})$'s are injective $\mathscr{O}_Y$-modules, for $r \geq 1$ we have

$$\mathscr{E}xt_{\mathscr{O}_Y}^r(\mathscr{O}_Z, \mathscr{H}om_X(\mathscr{O}_Y, \mathscr{I}^{\bullet,n})/\mathrm{im}(\mathscr{H}om_X(\mathscr{O}_Y, \mathscr{I}^{\bullet,n-1})))$$

$$\cong \downarrow$$

$$\mathscr{E}xt_Y^{r+1}(\mathscr{O}_Z, \mathrm{im}(\mathscr{H}om_X(\mathscr{O}_Y, \mathscr{I}^{\bullet,n-1}))).$$

However, the i^*-acyclicity of the $\mathscr{F}^p$'s implies $\mathscr{E}xt_X^t(\mathscr{O}_Y, \mathscr{F}^p) = 0$ for all $t \neq n$ and all p [**RD**, III, 7.2]. Since $n \geq 1$, we conclude that for a fixed value of $\bullet$, the complex

$$0 = \mathscr{H}om_X(\mathscr{O}_Y, \mathscr{F}^{\bullet}) \longrightarrow \mathscr{H}om_X(\mathscr{O}_Y, \mathscr{I}^{\bullet,0})$$

$$\downarrow$$

$$\cdots$$

$$\downarrow$$

$$\mathscr{H}om_X(\mathscr{O}_Y, \mathscr{I}^{\bullet,n-1})$$

$$\downarrow$$

$$0 \longleftarrow \mathrm{im}(\mathscr{H}om_X(\mathscr{O}_Y, \mathscr{I}^{\bullet,n-1}))$$

is exact. Thus, $\mathrm{im}(\mathscr{H}om_X(\mathscr{O}_Y, \mathscr{I}^{\bullet,n-1}))$ is an injective $\mathscr{O}_Y$-module, which implies the desired vanishing result (2.6.8).

We can now compute the map η_j on the bottom row of (2.5.8), by using the canonical truncation in rows $\leq m$ of the double complex (2.6.6). More precisely, define $K^{\bullet,q} = \mathscr{H}om_X(\mathscr{O}_Y, J^{\bullet,q+n})$ for $q \geq 1$, so $K^{\bullet,q} = 0$ for $q > m$, and define

$$K^{\bullet,0} = \mathscr{H}om_X(\mathscr{O}_Y, J^{\bullet,n})/\mathrm{im}(\mathscr{H}om_X(\mathscr{O}_Y, J^{\bullet,n-1})).$$

The canonical truncation of (2.6.6) in rows $\leq m$ is exactly the double complex $K^{\bullet\bullet}$, and we have a commutative diagram of complexes

(2.6.9)

$$\mathscr{H}om_X(\mathscr{O}_Y, \mathrm{Tot}^{\oplus} J^{\bullet\bullet}) \longrightarrow \mathrm{Tot}^{\oplus} K^{\bullet-n,\bullet}$$

$$\uparrow \qquad\qquad\qquad\qquad\qquad\qquad\qquad \uparrow$$

$$\mathscr{H}om_X(\mathscr{O}_Y, \mathrm{Tot}^{\oplus} I^{\bullet\bullet}) \longrightarrow \mathscr{H}om_X(\mathscr{O}_Y, I^{\bullet-n,n})/\mathrm{im}(\mathscr{H}om_X(\mathscr{O}_Y, I^{\bullet-n,n-1}))$$

in which the right column is a quasi-isomorphism to a complex of $\mathscr{H}om_Y(\mathscr{O}_Z, \cdot)$-acyclics (and hence by (2.6.5) is suitable for computing the term

$$j^{\flat}(\omega_{Y/X}[-n] \otimes i^*\mathscr{F}^{\bullet})$$

at the bottom of (2.6.3)). Thus, recalling (2.6.3), the composite of the left column and bottom side of (2.5.8) is represented by the composite

(2.6.10)
$$\mathscr{H}om_X(\mathscr{O}_Z, \mathrm{Tot}^{\oplus} J^{\bullet\bullet})$$
$$\downarrow \simeq$$
$$\mathscr{H}om_Y(\mathscr{O}_Z, \mathscr{H}om_X(\mathscr{O}_Y, \mathrm{Tot}^{\oplus} J^{\bullet\bullet}))$$
$$\xi \qquad \downarrow$$
$$\mathscr{H}om_Y(\mathscr{O}_Z, \mathrm{Tot}^{\oplus} K^{\bullet-n,\bullet})$$
$$\downarrow$$
$$\mathscr{H}om_Y(\mathscr{O}_Z, K^{\bullet-n-m,m})/\mathrm{im}(\mathscr{H}om_Y(\mathscr{O}_Z, K^{\bullet-n-m,m-1}))$$
$$\|$$
$$\mathscr{E}xt_Y^m(\mathscr{O}_Z, \mathscr{E}xt_X^n(\mathscr{O}_Y, \mathscr{F}^{\bullet})[-n])[-m]$$
$$\simeq \downarrow$$
$$\mathscr{E}xt_Y^m(\mathscr{O}_Z, (\omega_{Y/X} \otimes i^*\mathscr{F}^{\bullet})[-n])[-m]$$
$$\simeq \downarrow$$
$$(\omega_{Z/Y} \otimes j^*(\omega_{Y/X} \otimes i^*\mathscr{F}^{\bullet})[-n])[-m]$$
$$\|$$
$$\omega_{Z/Y}[-m] \otimes j^*\omega_{Y/X}[-n] \otimes (ij)^*\mathscr{F}^{\bullet}$$

where the commutativity of (2.6.9) ensures that the top two maps in the long column represent (2.6.3), and we use (2.6.5), Lemma 2.1.1 (as used in the definition of (2.5.3)), and several applications of (2.5.1). Note also that the equality at the bottom of (2.6.10) involves no intervention of signs. In addition, the 'top half' is commutative, where ξ is defined to be the composite of the top map in (2.6.1) and the equality

$$\mathscr{H}om_Y(\mathscr{O}_Z, J^{\bullet-n-m,n+m})/\mathrm{im}(\mathscr{H}om_Y(\mathscr{O}_Z, J^{\bullet-n-m,n+m-1}))$$
$$\|$$
$$\mathscr{H}om_Y(\mathscr{O}_Z, K^{\bullet-n-m,m})/\mathrm{im}(\mathscr{H}om_Y(\mathscr{O}_Z, K^{\bullet-n-m,m-1}))$$

Combining (2.6.10) with (2.6.1) and (2.6.4), our problem is to prove the commutativity in $\mathbf{D}(Z)$ of the following diagram of complexes:

$$H_v^m(\mathcal{H}om_X(\mathcal{O}_Z, K^{\bullet-n-m,\bullet}))$$

$$\mathcal{E}xt_Y^m(\mathcal{O}_Z, \mathcal{E}xt_X^n(\mathcal{O}_Y, \mathcal{F}^\bullet)[-n])[-m] \qquad H_v^{n+m}(\mathcal{H}om_X(\mathcal{O}_Z, J^{\bullet-n-m,\bullet}))$$

$$\simeq \downarrow \qquad\qquad\qquad \|$$

$$\mathcal{E}xt_Y^m(\mathcal{O}_Z, (\omega_{Y/X} \otimes i^*\mathcal{F}^\bullet)[-n])[-m] \qquad \mathcal{E}xt_X^{n+m}(\mathcal{O}_Z, \mathcal{F}^\bullet)[-n-m]$$

$$\simeq \downarrow \qquad\qquad\qquad \simeq \downarrow$$

$$(\omega_{Z/Y} \otimes (j^*\omega_{Y/X} \otimes (ij)^*\mathcal{F}^\bullet)[-n])[-m] \qquad (\omega_{Z/X} \otimes (ij)^*\mathcal{F}^\bullet)[-n-m]$$

$$\text{no signs! } \| \qquad\qquad\qquad \|$$

$$\omega_{Z/Y}[-m] \otimes j^*\omega_{Y/X}[-n] \otimes (ij)^*\mathcal{F}^\bullet \xleftarrow{\quad\zeta'_{i,j}\otimes 1\quad} \omega_{Z/X}[-n-m] \otimes (ij)^*\mathcal{F}^\bullet$$

We make the stronger claim that this diagram commutes on the level of complexes. This can be checked separately in each degree. In order to do this, consider the following problem.

Let $\mathcal{F}$ be an $\mathcal{O}_X$-module which is i^* and $(ij)^*$-acyclic, and let

$$J^0 \to \cdots \to J^{n+m} \to 0$$

be the canonical truncation in degrees $\leq n+m$ of an injective resolution of $\mathcal{F}$. The example we have in mind is any of the $\mathcal{F}^r$'s above and the complex $J^{r,\bullet}$. From the above arguments (see the analysis of (2.6.8)), we know that

$$\mathcal{H}om_X(\mathcal{O}_Y, J^n)/\operatorname{im}(\mathcal{H}om_X(\mathcal{O}_Y, J^{n-1})) \longrightarrow \mathcal{H}om_X(\mathcal{O}_Y, J^{n+1})$$

$$\cdots$$

$$0 \longleftarrow \mathcal{H}om_X(\mathcal{O}_Y, J^{n+m})$$

is a $\mathcal{H}om_Y(\mathcal{O}_Z, \cdot)$-acyclic resolution of $\mathcal{E}xt_X^n(\mathcal{O}_Y, \mathcal{F})$. By Lemma 2.6.2 below, we just need to show the commutativity of the resulting diagram (with surjective columns)

$$(2.6.11) \qquad \mathcal{H}om_X(\mathcal{O}_Z, J^{n+m}) \xrightarrow{\simeq} \mathcal{H}om_Y(\mathcal{O}_Z, \mathcal{H}om_X(\mathcal{O}_Y, J^{n+m}))$$

$$\downarrow \qquad\qquad\qquad\qquad \downarrow$$

$$\mathcal{E}xt_X^{n+m}(\mathcal{O}_Z, \mathcal{F}) \longrightarrow \mathcal{E}xt_Y^m(\mathcal{O}_Z, \mathcal{E}xt_X^n(\mathcal{O}_Y, \mathcal{F}))$$

where the bottom map arises from the Grothendieck spectral sequence for the functors $\mathscr{H}om_X(\mathcal{O}_Y, \cdot)$ and $\mathscr{H}om_Y(\mathcal{O}_Z, \cdot)$, with cohomological dimensions $\leq n$ and $\leq m$ respectively (and with composite functor $\mathscr{H}om_X(\mathcal{O}_Z, \cdot)$ having cohomological dimension $\leq n + m$).

It is convenient to prove a commutativity claim more general than (2.6.11), in which we remove the annoying condition that $\mathscr{F}$ be i^*-acyclic and $(ij)^*$-acyclic. Let

$$\mathscr{A} \xrightarrow{F} \mathscr{B} \xrightarrow{G} \mathscr{C}$$

be left-exact functors between abelian categories, where $\mathscr{A}$ and $\mathscr{B}$ have enough injectives and F takes injectives to G-acyclics (e.g., $F = \mathscr{H}om_X(\mathcal{O}_Y, \cdot)$, $G = \mathscr{H}om_Y(\mathcal{O}_Z, \cdot)$). Assume that G has cohomological dimension $\leq m$ and that X is an object in $\mathscr{A}$ with $R^j F(X) = 0$ for all $j > n$ (e.g., F has cohomological dimension $\leq n$), where $n, m \geq 1$. Since $R^i G \circ R^j F(X) = 0$ for $i > m$ or $j > n$, the Grothendieck spectral sequence yields a map (even an isomorphism)

(2.6.12) $R^{n+m}(GF)(X) \to R^m G(R^n F(X))$.

Meanwhile, if

$$I^0 \to \cdots \to I^{n+m} \to 0$$

is the canonical truncation in degrees $\leq n + m$ of an injective resolution of X, the complex

$$0 \to R^n F(X) \to F(I^n)/\mathrm{im}(F(I^{n-1})) \to F(I^{n+1}) \to \cdots \to F(I^{n+m}) \to 0$$

is exact. This resolution of $R^n F(X)$ admits a unique-up-to-homotopy map to an injective resolution of $R^n F(X)$, so we get a well-defined map

(2.6.13) $G(F(I^{n+m})) \to R^m G(R^n F(X))$.

Combining (2.6.12) and (2.6.13) with the canonical surjective map $GF(I^{n+m}) \to R^{n+m}(GF)(X)$, we are led to make the following assertion, which implies the commutativity of (2.6.11) as a special case:

LEMMA 2.6.1. *In the above general situation, the diagram*

$$GF(I^{n+m}) \longrightarrow R^m G(R^n F(X))$$
$$\downarrow \qquad \nearrow \simeq$$
$$R^{n+m}(GF)(X)$$

commutes.

The analogue of this lemma when $m = 0$ or $n = 0$ is trivial. In addition to proving Lemma 2.6.1, the only other thing which remains to be proven is the following lemma, which was invoked above (2.6.11) and is even a special case of Theorem 2.5.1 with $\mathscr{F}^\bullet = \mathscr{F}[0]$, since we have noted via Corollary 2.1.2 that applying H^n to (2.5.3) recovers (2.5.1) without the intervention of signs.

LEMMA 2.6.2. *For any $\mathcal{O}_X$-module $\mathcal{F}$, the diagram*

$$
\begin{array}{ccc}
\mathcal{E}xt_X^{n+m}(\mathcal{O}_Z,\mathcal{F}) & \longrightarrow & \mathcal{E}xt_Y^m(\mathcal{O}_Z,\mathcal{E}xt_X^n(\mathcal{O}_Y,\mathcal{F})) \\
\eta_{Z/X}\downarrow\simeq & & \simeq\downarrow\eta_{Y/X} \\
\omega_{Z/X}\otimes(ij)^*\mathcal{F} & & \mathcal{E}xt_Y^m(\mathcal{O}_Z,\omega_{Y/X}\otimes i^*\mathcal{F}) \\
\zeta'_{i,j}\otimes 1\downarrow\simeq & & \simeq\downarrow\eta_{Z/Y} \\
\omega_{Z/Y}\otimes j^*\omega_{Y/X}\otimes(ij)^*\mathcal{F} & =\!=\!= & \omega_{Z/Y}\otimes j^*(\omega_{Y/X}\otimes i^*\mathcal{F})
\end{array}
$$

commutes, where the top row is the map arising from the Grothendieck spectral sequence. In particular, the top row is an isomorphism.

To prove Lemma 2.6.2, we can check on stalks, which converts all $\mathcal{E}xt_X$'s into $\operatorname{Ext}_{\mathcal{O}_{X,x}}$'s, since $\mathcal{O}_Z$ and $\mathcal{O}_Y$ locally have finite free resolutions on X, etc. (one also needs to observe that formation of the Grothendieck spectral sequence is compatible with passage to stalks, which is easy to check by using its construction). Thus, Lemma 2.6.2 follows from the following purely algebraic assertion (in which we consider the module $\mathcal{F}_x$ over the ring $\mathcal{O}_{X,x}$ in the application above).

LEMMA 2.6.3. *Let A be a ring, M an A-module. Let $\{f_1,\dots,f_n\}$ and $\{f_1,\dots,f_n,g_1,\dots,g_m\}$ be regular sequences in A, generating ideals J and K respectively. The diagram of A/K-modules*

(2.6.14)

$$
\begin{array}{ccc}
\operatorname{Ext}_A^{n+m}(A/K,M) & \longrightarrow & \operatorname{Ext}_{A/J}^m(A/K,\operatorname{Ext}_A^n(A/J,M)) \\
\varphi_{K,M}\downarrow\simeq & & \simeq\downarrow(\varphi_{K/J,\wedge^n(J/J^2)^\vee\otimes M})\circ\varphi_{J,M} \\
\bigwedge^{n+m}(K/K^2)^\vee\otimes_A M & \xrightarrow[\simeq]{} & \bigwedge^m((K/J)/(K/J)^2)^\vee\otimes_{A/J}\bigwedge^n(J/J^2)^\vee\otimes_A M
\end{array}
$$

commutes, where the bottom map is determined by $(g^\vee\wedge f^\vee)\otimes\mu\mapsto g^\vee\otimes f^\vee\otimes\mu$ and the top map arises from the Grothendieck spectral sequence.

The truth of Lemma 2.6.3 depends heavily on our definition of (2.5.2), which uses (1.3.28) (in which there is an implicit sign). If the sign were removed from the definition of (1.3.28), then (2.6.14) would only commute up to $(-1)^{nm}$. It should also be noted that what makes Lemma 2.6.3 somewhat non-trivial is the fact that the vertical maps in (2.6.14) use the realization of Ext as a derived functor in the first variable, while the top horizontal map in (2.6.14) uses the realization of Ext as a derived functor in the second variable. The proofs of Lemma 2.6.3 and Lemma 2.6.1 require a careful analysis of the construction of the Grothendieck spectral sequence. We prove Lemma 2.6.1 first, because it is easier and is needed in the proof of Lemma 2.6.3! These two results, proven below, complete the proof of Theorem 2.5.1.   ∎

PROOF. (of Lemma 2.6.1) Consider a Cartan-Eilenberg resolution $J^{\bullet\bullet}$ of the bounded complex $F(I^{\bullet})$. Note that $F(I^{\bullet})$ is supported in degrees from 0 to $n+m$. Since F is left-exact and G has cohomological dimension $\leq m \leq m+n$, it follows that $F(I^{n+m})$ is G-acyclic and $J^{\bullet\bullet}$ is the canonical truncation in *columns* $\leq m + n$ of a Cartan-Eilenberg resolution of $F(\widehat{I}^{\bullet})$, where $\widehat{I}^{\bullet}$ is an injective resolution of X whose canonical truncation in degrees $\leq m+n$ is $I^{\bullet}$. Thus, $J^{\bullet\bullet}$ can be used to compute the Grothendieck spectral sequence of interest. Since $R^{i}F(X) = 0$ for $i > n$, the complex $F(I^{\bullet})$ is exact in degrees $\geq n+1$. Thus, all rows $J^{\bullet,q}$ are exact in degrees $\geq n+1$. Of course, each column $J^{p,\bullet}$ is an injective resolution of $F(I^{p})$. Since all $F(I^{p})$'s are G-acyclic, it follows that $G(J^{\bullet\bullet})$ has exact columns and has rows exact in degrees $\geq n+1$: the exactness of the rows in degrees $\geq n+1$ is a consequence of the fact that

$$J^{n,q} \to J^{n+1,q} \to \cdots \to J^{n+m,q} \to 0$$

is an exact sequence of injectives with all kernels and cokernels injective (so exactness is preserved after applying G).

Thus, the following first quadrant double complex $\mathscr{J}^{\bullet\bullet}$ consists of injectives, has exact rows and columns, and all columns (resp. all rows) stay exact (resp. stay exact in degrees $\geq n+1$) after applying G:

(2.6.15)

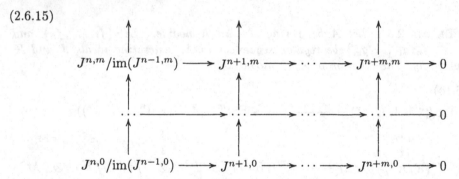

The complex $\mathscr{I}^{\bullet}$ of 'vertical direction kernels' along the bottom of (2.6.15) is

$$(2.6.16) \qquad F(I^{n})/\mathrm{im}(F(I^{n-1})) \to F(I^{n+1}) \to \cdots \to F(I^{n+m}) \to 0,$$

which is a resolution of $R^{n}F(X)$, as we saw in the discussion below (2.6.12). Also, the kernel complex along the left side of (2.6.15) is the complex

$$H_{h}^{n}(J^{\bullet,0}) \to \cdots \to H_{h}^{n}(J^{\bullet,m}) \to \cdots$$

of (horizontal) cohomologies, which is an injective resolution of $R^{n}F(X)$. Thus, the augmentation

$$G(H_{h}^{n}(J^{\bullet\bullet})) \to G(\mathrm{Tot}^{\oplus} \mathscr{J}^{\bullet\bullet})$$

is a quasi-isomorphism, while the other augmentation gives a map of complexes $G(\mathscr{I}^{\bullet}) \to G(\mathrm{Tot}^{\oplus} \mathscr{J}^{\bullet\bullet})$. Passing to cohomology, we therefore get a map

$$H^{m}(G(\mathscr{I}^{\bullet})) \to H^{m}(G(\mathrm{Tot}^{\oplus} \mathscr{J}^{\bullet\bullet})) \xleftarrow{\cong} H^{m}(G(H_{h}^{n}(J^{\bullet\bullet}))) = R^{m}G(R^{n}F(X)).$$

Composing this with the canonical surjective map $GF(I^{n+m}) \to \mathrm{H}^m(G(\mathscr{I}^\bullet))$, we arrive at a map

(2.6.17) $$GF(I^{n+m}) \to \mathrm{R}^m G(\mathrm{R}^n F(X)).$$

In terms of chasing 'members' in the sense of MacLane [**Mac**, Ch VIII, §4], (2.6.17) is obtained by mapping a member of $GF(I^{n+m})$ into $G(J^{n+m,0})$ and working 'backwards up the staircase' in $G(\mathscr{I}^{\bullet\bullet})$ to $G(\mathrm{H}^n_h(J^{\bullet,m}))$ and then passing to the image in the cohomology object $\mathrm{R}^m G(\mathrm{R}^n F(X))$.

Since G has cohomological dimension $\leq m$, $G(J^{\bullet\bullet})$ is exact in degrees $\geq m + 1$ along all columns, as well as along all columns of 'horizontal' kernels, cokernels, and cohomologies of row maps. Combining this with an analysis of the filtration used to define the Grothendieck spectral sequence, we see that (2.6.17) is equal to the composite

$$GF(I^{n+m}) \to \mathrm{R}^{n+m}(GF)(X) \to \mathrm{R}^m G(\mathrm{R}^n F(X))$$

obtained with the maps as in the statement of Lemma 2.6.1.

Thus, we are reduced to the following general claim. Suppose $0 \to Y \to C^\bullet$ is an exact sequence in an abelian category $\mathscr{B}$ with enough injectives (such as the resolution (2.6.16) of $\mathrm{R}^n F(X)$), and $J^{\bullet\bullet}$ is a (first quadrant) Cartan-Eilenberg resolution of $C^\bullet$. Let $I^q = \ker(J^{0,q} \to J^{1,q})$, so $I^\bullet$ is an injective resolution of Y. Let $G : \mathscr{B} \to \mathscr{C}$ be a left-exact functor between abelian categories, so the double complex $G(J^{\bullet\bullet})$ has exact rows and therefore the augmentation

$$\alpha : G(I^\bullet) \to G(\mathrm{Tot}^\oplus J^{\bullet\bullet})$$

is a quasi-isomorphism. Let

$$\beta : G(C^\bullet) \to G(\mathrm{Tot}^\oplus J^{\bullet\bullet})$$

denote the other augmentation. In this way, we get induced maps on cohomology

$$\mathrm{H}^i(\alpha)^{-1} \circ \mathrm{H}^i(\beta) : \mathrm{H}^i(G(C^\bullet)) \to \mathrm{H}^i(G(I^\bullet)) = \mathrm{R}^i G(Y).$$

We claim that this is equal to the map induced by the unique-up-to-homotopy map of resolutions $C^\bullet \to I^\bullet$ over Y.

Indeed, the augmentation $\epsilon_1 : I^\bullet \to \mathrm{Tot}^\oplus J^{\bullet\bullet}$ is a map between injective resolutions of Y, and so is a homotopy equivalence. Choose a homotopy inverse ϵ_1^{-1}. If $\epsilon_2 : C^\bullet \to \mathrm{Tot}^\oplus J^{\bullet\bullet}$ is the other augmentation, then $\epsilon_1^{-1} \circ \epsilon_2$ is a map $C^\bullet \to I^\bullet$ of resolutions of Y. Applying G, the induced map on the ith cohomology is clearly $\mathrm{H}^i(\alpha)^{-1} \circ \mathrm{H}^i(\beta)$, as desired.   ∎

PROOF. (of Lemma 2.6.3) Without loss of generality, $n, m \geq 1$. Also, since A/J and A/K have finite free resolutions over A, everything is easily of formation compatible with direct limits in M and the functors of M are covariant and *right exact*. Thus, we immediately reduce to the case $M = A$. For any A-module N, we let $N[J]$ denote the 'J-torsion' submodule $\mathrm{Hom}_A(A/J, N) \subseteq N$; similar shorthand will be used for any module over any ring (with respect to a chosen ideal). There is no risk of confusion with our notation "[r]" for translating complexes. We let $\overline{A} = A/J$, $\overline{g}_i = g_i \bmod J$, $\overline{K} = K \bmod J$. We also choose an injective A-module resolution $I^\bullet$ of A (concentrated in degrees ≥ 0, as usual).

Recall that (2.2.3) corresponds to the isomorphism of A/K-modules

$$(K/K^2)^\vee \simeq (\overline{K}/\overline{K}^2)^\vee \otimes_{A/K} (A/K \otimes_{A/J} (J/J^2)^\vee)$$

determined by

$$g^\vee \wedge f^\vee \mapsto g^\vee \otimes (1 \otimes f^\vee),$$

where $g^\vee = g_1^\vee \wedge \cdots \wedge g_m^\vee$ and $f^\vee = f_1^\vee \cdots \wedge f_n^\vee$. Since $g^\vee \wedge f^\vee = (-1)^{nm} f^\vee \wedge g^\vee$ but we don't know whether $\{g_1, \ldots, g_m, f_1, \ldots, f_n\}$ is a regular sequence, we need to verify the commutativity of the diagram

$$(2.6.18) \qquad \operatorname{Ext}_A^{n+m}(A/K, A) \longrightarrow \operatorname{Ext}_{A/J}^m(A/K, \operatorname{Ext}_A^n(A/J, A))$$

$$\psi_{\mathbf{f},\mathbf{g};A} \Big\downarrow \simeq \qquad\qquad\qquad \simeq \Big\downarrow \psi_{\overline{\mathbf{g}},\overline{A}} \circ \psi_{\mathbf{f},A}$$

$$A/K =\!\!=\!\!=\!\!=\!\!=\!\!=\!\!=\!\!=\!\!=\!\!=\!\!=\!\!=\!\!=\!\!=\!\!=\!\!= \overline{A}/\overline{K}$$
$$\varepsilon$$

where the bottom row is multiplication by $\varepsilon = (-1)^{nm}$ and the top row is the map from the Grothendieck spectral sequence. If we knew that

$$\{g_1, \ldots, g_m, f_1, \ldots, f_n\}$$

were a regular sequence, then we could have used $\psi_{\mathbf{g},\mathbf{f};A}$ in the left column and would prove commutativity with the identity map on the bottom row. Thus, the presence of the sign ε is rather artificial, so we introduce the notation ε in order to distinguish this from the sign $(-1)^{nm}$ which arises below from Lemma 2.6.4 (even if $\{\mathbf{g}, \mathbf{f}\}$ were a regular sequence and we had used $\psi_{\mathbf{g},\mathbf{f};A}$).

The key point is to describe all sides of (2.6.18) in 'derived category' terms. We begin with the vertical sides, and then consider the bottom and top respectively. Recall that by the *definition* of (1.3.28), $(-1)^{n(n+1)/2}\psi_{\mathbf{f},A}$ is the map induced on degree n cohomology by the diagram of augmentation map quasi-isomorphisms

$$\operatorname{Hom}_A^\bullet(A/J, I^\bullet) \overset{\simeq}{\longrightarrow} \operatorname{Hom}_A^\bullet(K_\bullet(\mathbf{f}), I^\bullet) \overset{\simeq}{\longleftarrow} \operatorname{Hom}_A^\bullet(K_\bullet(\mathbf{f}), A)$$

We have similar descriptions of

$$(-1)^{m(m+1)/2}\psi_{\overline{\mathbf{g}},\overline{A}}, \quad (-1)^{(n+m)(n+m+1)/2}\psi_{\mathbf{f},\mathbf{g};A}.$$

The complexes $K_\bullet(\mathbf{f}, \mathbf{g})$ and $K_\bullet(\mathbf{f}) \otimes K_\bullet(\mathbf{g})$ of finite free A-modules are both resolutions of A/K via canonical augmentation maps in degree 0 (without the intervention of signs), with the standard isomorphism of projective resolutions

$$K_\bullet(\mathbf{f}) \otimes K_\bullet(\mathbf{g}) \simeq K_\bullet(\mathbf{f}, \mathbf{g})$$

(corresponding to the identity map on $(A/K)[0]$ in $\mathbf{D}(A)$) defined without the intervention of signs, since our definition of the complex $K_\bullet(\mathbf{f})$ coincides with $K_\bullet(f_1) \otimes \cdots \otimes K_\bullet(f_n)$. Thus, we see that the standard isomorphism of complexes

$$(2.6.19) \qquad\qquad K_\bullet(\mathbf{g}) \otimes K_\bullet(\mathbf{f}) \simeq K_\bullet(\mathbf{f}, \mathbf{g})$$

does involve an intervention of signs; in particular, a sign of $\varepsilon = (-1)^{nm}$ intervenes in the degree $-n - m$ terms $A \otimes A \simeq A$ and *no* sign intervenes in degree

0. We conclude that the map on the bottom row of (2.6.18) is the map induced in degree $n + m$ cohomology by the composite map of complexes

$$\operatorname{Hom}_A^\bullet(K_\bullet(\mathbf{f}, \mathbf{g}), A) \xrightarrow{\ \simeq\ } \operatorname{Hom}_A^\bullet(K_\bullet(\mathbf{g}) \otimes K_\bullet(\mathbf{f}), A)$$

$$\operatorname{Hom}_{\overline{A}}^\bullet(K_\bullet(\overline{\mathbf{g}}), \overline{A}[-n]) \longleftarrow \operatorname{Hom}_A^\bullet(K_\bullet(\mathbf{g}), \operatorname{Hom}_A^\bullet(K_\bullet(\mathbf{f}), A))$$

where the first step involves an intervention of signs (as just indicated) but the other steps are defined without the intervention of signs. This complicated 'derived category' description of the bottom row of (2.6.18) will be useful shortly.

By Lemma 2.6.1, the top row of (2.6.18) is the map induced in degree $n + m$ cohomology by

$$\operatorname{Hom}_A^\bullet(A/K, I^\bullet) = \operatorname{Hom}_{\overline{A}}^\bullet(\overline{A}/\overline{K}, I^\bullet[J]) \to \operatorname{Hom}_{\overline{A}}^\bullet(\overline{A}/\overline{K}, \tau_{\geq n}(I^\bullet[J]))$$

since

$$(2.6.20) \qquad 0 \to \mathrm{H}^n(I^\bullet[J]) \to \tau_{\geq 0}(I^{\bullet + n}[J])$$

is an injective $\overline{A}$-module resolution of

$$\mathrm{H}^n(I^\bullet[J]) \simeq \operatorname{Ext}_A^n(A/J, M).$$

One point we need to verify is that $I^n[J]/\operatorname{im}(I^{n-1}[J]) = \tau_{\geq 0}(I^{\bullet + n}[J])^0$ *is* an injective $\overline{A}$-module. Since $\operatorname{Ext}_A^i(A/J, A) \simeq \mathrm{H}^i(\mathbf{f}, A) = 0$ for $i < n$ (here is where it is important to have reduced to the case $M = A$, or at least the case of A-flat M), the sequence

$$0 \to I^0[J] \to \cdots \to I^{n-1}[J] \to \operatorname{im}(I^{n-1}[J]) \to 0$$

is exact and consequently is split (as each $I^\bullet[J]$ is an injective $\overline{A}$-module), so $\operatorname{im}(I^{n-1}[J])$ is an injective $\overline{A}$-module. Thus, the injection

$$\operatorname{im}(I^{n-1}[J]) \hookrightarrow I^n[J]$$

must be split and so has injective cokernel (as $I^n[J]$ is an injective $\overline{A}$-module). This completes the verification that (2.6.20) is an injective $\overline{A}$-module resolution.

Recall the augmentation maps $K_\bullet(\mathbf{f}) \to (A/J)[0]$ and $K_\bullet(\overline{\mathbf{g}}) \to (\overline{A}/\overline{K})[0]$. Using these, it makes sense to consider the following diagram of complexes, in which all maps are the evident ones and the maps labelled 'qism' are quasi-isomorphisms:

(2.6.21)

$$\mathrm{Hom}^{\bullet}_{\overline{A}}(K_{\bullet}(\overline{\mathbf{g}}), \mathrm{H}^n(\mathrm{Hom}^{\bullet}_A(K_{\bullet}(\mathbf{f}), I^{\bullet}))[-n])$$

$$\alpha_2 \Big\uparrow \simeq$$

$$\mathrm{Hom}^{\bullet}_{\overline{A}}(K_{\bullet}(\overline{\mathbf{g}}), \mathrm{H}^n(I^{\bullet}[J])[-n])$$

$$\alpha_3 \Big\downarrow \text{qism}$$

$$\mathrm{Hom}^{\bullet}_{\overline{A}}(\overline{A}/\overline{K}, \tau_{\geq n}(I^{\bullet}[J])) \longrightarrow \mathrm{Hom}^{\bullet}_{\overline{A}}(K_{\bullet}(\overline{\mathbf{g}}), \tau_{\geq n}(I^{\bullet}[J]))$$

$$\alpha \Big\uparrow \qquad\qquad \alpha_4 \Big\uparrow$$

$$\mathrm{Hom}^{\bullet}_A(A/K, I^{\bullet}) \qquad\qquad \mathrm{Hom}^{\bullet}_A(K_{\bullet}(\mathbf{g}), I^{\bullet}[J]) \qquad\qquad \alpha_1$$

$$\text{qism} \Big\downarrow \qquad\qquad \text{qism} \Big\downarrow$$

$$\mathrm{Hom}^{\bullet}_A(K_{\bullet}(\mathbf{f}, \mathbf{g}), I^{\bullet}) \overset{\text{signs!}}{=\!=\!=\!=\!=} \mathrm{Hom}^{\bullet}_A(K_{\bullet}(\mathbf{g}), \mathrm{Hom}^{\bullet}_A(K_{\bullet}(\mathbf{f}), I^{\bullet}))$$

$$\text{qism} \Big\uparrow \qquad\qquad \text{qism} \Big\downarrow$$

$$\mathrm{Hom}^{\bullet}_A(K_{\bullet}(\mathbf{f}, \mathbf{g}), A) \underset{\text{signs!}}{=\!=\!=\!=\!=} \mathrm{Hom}^{\bullet}_A(K_{\bullet}(\mathbf{g}), \mathrm{Hom}^{\bullet}_A(K_{\bullet}(\mathbf{f}), A))$$

$$\mathrm{Hom}^{\bullet}_{\overline{A}}(K_{\bullet}(\overline{\mathbf{g}}), \mathrm{H}^n(\mathrm{Hom}^{\bullet}_A(K_{\bullet}(\mathbf{f}), A))[-n])$$

All maps are defined without the intervention of signs, except for the two lower horizontal equalities (for which we've already noted the presence of signs, due to 'flipping' the order of a tensor product of complexes as in (2.6.19)). In degree $n + m$ cohomology, the map α (respectively the bottom two maps) in (2.6.21) induces the top (respectively bottom) row of (2.6.18), the left column maps labelled 'qism' in (2.6.21) induce $(-1)^{(n+m)(n+m+1)/2}$ times the left column of (2.6.18), and the composite $\alpha_1^{-1} \circ \alpha_2 \circ \alpha_3^{-1} \circ \alpha_4$ in (2.6.21) induces $(-1)^{m(m+1)/2}(-1)^{n(n+1)/2}(-1)^{nm}$ times the right column of (2.6.18), where the sign $(-1)^{nm}$ is the universal sign arising in Lemma 2.6.4 below (this has nothing to do with the sign ε which was introduced earlier).

Since the signs we've mentioned along the vertical sides satisfy the relation

$$(-1)^{(n+m)(n+m+1)/2} = (-1)^{m(m+1)/2}(-1)^{n(n+1)/2}(-1)^{nm},$$

the commutativity of (2.6.18) is equivalent to the commutativity in degree $n+m$ cohomology of the outer part of (2.6.21). But the lower square and right part of (2.6.21) commute in degree $n + m$ cohomology. Indeed, for the lower square this follows by functoriality with respect to $A[0] \to I^{\bullet}$. Meanwhile, for the right part, it is enough to check commutativity in $\mathbf{D}(A)$. This right part can be identified (as a diagram of complexes of A-*modules*) with the result of applying

$\operatorname{Hom}_A^\bullet(K_\bullet(\mathbf{g}), \cdot)$ to the following diagram, in which *all* vertical maps are quasi-isomorphisms:

(2.6.22)

$$
\begin{array}{ccccc}
I^\bullet[J] & \longrightarrow & \tau_{\geq n}(I^\bullet[J]) & \xleftarrow{\text{qism}} & H^n(I^\bullet[J])[-n] \\
\downarrow & & \downarrow & & \downarrow \\
\operatorname{Hom}_A^\bullet(K_\bullet(\mathbf{f}), I^\bullet) & \longrightarrow & \tau_{\geq n}(\operatorname{Hom}_A^\bullet(K_\bullet(\mathbf{f}), I^\bullet)) & \longleftarrow & H^n(\operatorname{Hom}_A^\bullet(K_\bullet(\mathbf{f}), I^\bullet))[-n] \\
\uparrow & & \uparrow & & \uparrow \\
\operatorname{Hom}^\bullet(K_\bullet(\mathbf{f}), A) & \longrightarrow & \tau_{\geq n}(\operatorname{Hom}^\bullet(K_\bullet(\mathbf{f}), A)) & \overset{\text{qism}}{=\!=\!=} & H^n(\operatorname{Hom}^\bullet(K_\bullet(\mathbf{f}), A))[-n]
\end{array}
$$

The operation

$$
\operatorname{Hom}_A^\bullet(K_\bullet(\mathbf{g}), \cdot)
$$

on complexes of A-modules takes quasi-isomorphisms to quasi-isomorphisms, and consequently takes commutative diagrams in $\mathbf{D}(A)$ to commutative diagrams in $\mathbf{D}(A)$, so we just have to check that (2.6.22) commutes in $\mathbf{D}(A)$. But this follows from the functoriality of the derived category diagram

$$
C^\bullet \to \tau_{\geq n}(C^\bullet) \leftarrow H^n(C^\bullet)[-n]
$$

with respect to the left column of quasi-isomorphisms in (2.6.22).

Thus, we conclude that the lower square and right part of (2.6.21) commute in degree $n + m$ cohomology, so it suffices to prove that the upper square in (2.6.21) induces a diagram in degree $n + m$ cohomology which is commutative. This upper square in (2.6.21) is the outside edge of the diagram of complexes

(2.6.23)

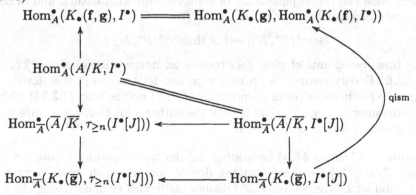

where all maps are defined without the intervention of signs, except for the top horizontal map (in which many signs intervene). It suffices to prove this commutes on the level of complexes. The commutativity of the lower triangle and square in (2.6.23) is trivial to check. Meanwhile, the commutativity of the 'upper' part of (2.6.23) follows by a direct verification in each separate degree, the point being that the isomorphism (2.6.19), which is implicit in the definition of the bottom map in the left column in (2.6.23), involves *no* intervention of signs in degree 0.

∎

The crucial general nonsense lemma used in the preceding argument is

LEMMA 2.6.4. *Let R be a ring, N an R-module, $P^\bullet \to N$ a projective resolution and $N \to I^\bullet$ an injective resolution (concentrated in respective degrees ≤ 0 and ≥ 0). Let $m, n \geq 0$ be integers. Let $\mathrm{Hom}^\bullet = \mathrm{Hom}_R^\bullet$. Consider the diagram*

$$(2.6.24) \qquad \begin{array}{ccc} \mathrm{H}^m(\mathrm{Hom}^\bullet(N, I^\bullet)) & =\!=\!=\!=\!= & \mathrm{H}^{n+m}(\mathrm{Hom}^\bullet(N, I^{\bullet - n})) \\ \cong \downarrow & & \downarrow \cong \\ \mathrm{H}^m(\mathrm{Hom}^\bullet(P^\bullet, I^\bullet)) & & \mathrm{H}^{n+m}(\mathrm{Hom}^\bullet(P^\bullet, I^{\bullet - n})) \\ \cong \uparrow & & \uparrow \cong \\ \mathrm{H}^m(\mathrm{Hom}^\bullet(P^\bullet, N)) & =\!=\!=\!=\!= & \mathrm{H}^{n+m}(\mathrm{Hom}^\bullet(P^\bullet, N[-n])) \end{array}$$

where the vertical maps are induced by quasi-isomorphism augmentation maps and the horizontal maps are defined without *the intervention of signs. This diagram commutes up to a sign of $(-1)^{nm}$.*

Before giving the proof, we make some comments about the horizontal sides of (2.6.24). The canonical isomorphism of complexes $\mathrm{Hom}^\bullet(N, I^{\bullet - n}) = \mathrm{Hom}^{\bullet - n}(N, I^\bullet)$ can be defined without the intervention of signs, so the top side of (2.6.24) is natural. Meanwhile, $\mathrm{Hom}^\bullet(P^\bullet, N[-n]) \simeq \mathrm{Hom}^\bullet(P^\bullet, N)[-n]$ is defined without the intervention of signs, so the bottom side of (2.6.24) is natural. However, these two isomorphisms are of somewhat different nature, and in fact the natural isomorphism

$$\mathrm{Hom}^\bullet(P^\bullet, N)[-n] \simeq \mathrm{Hom}^{\bullet - n}(P^\bullet, N)$$

relating these two points of view *does* require an intervention of signs. This is "why" (2.6.24) only commutes up to a sign and it is this sign that forces us to define the fundamental local isomorphism (2.5.3) via the map (1.3.28) which involves the intervention of a sign; it took the author a long time to realize that (2.6.24) does not commute in general.

PROOF. (of Lemma 2.6.4) For conceptual clarity, we consider a more general problem. Let $C^{\bullet\bullet}$ be a first quadrant double complex with exact rows and columns, and let $L^\bullet$ (respectively $B^\bullet$) denote the 'kernel complex' along the left column (respectively bottom row). That is,

$$L^\bullet = \ker(C^{0,\bullet} \to C^{1,\bullet}), \quad B^\bullet = \ker(C^{\bullet,0} \to C^{\bullet,1}).$$

Let $C'^{\bullet\bullet}$ denote the 'same' double complex, except the *horizontal* differentials $d_h^{p,q} : C^{p,q} \to C^{p+1,q}$ from the pth column to the $(p+1)$th column are multiplied by a sign e_p for all $p \geq 0$ (so $C'^{\bullet\bullet}$ is still a double complex). Define $L'^\bullet$, $B'^\bullet$

analogously to $L^\bullet$, $B^\bullet$, so $L'^\bullet = L^\bullet$ but $B'^\bullet$ is obtained from $B^\bullet$ by multiplying the pth differential by e_p. Fix $m \geq 0$ and consider the diagram

$$(2.6.25) \qquad \begin{array}{ccccc} H^m(B^\bullet) & \xrightarrow{\;\simeq\;} & H^m(\mathrm{Tot}^\oplus(C^{\bullet\bullet})) & \xleftarrow{\;\simeq\;} & H^m(L^\bullet) \\ \| & & & & \| \\ H^m(B'^\bullet) & \xrightarrow{\;\simeq\;} & H^m(\mathrm{Tot}^\oplus(C'^{\bullet\bullet})) & \xleftarrow{\;\simeq\;} & H^m(L'^\bullet) \end{array}$$

Here, the vertical maps are defined *without* the intervention of signs. We claim that this diagram commutes up to a sign of $e_0 \ldots e_{m-1}$ (which means 1 if $m = 0$). This is readily checked by 'walking backwards up the staircase' in $C^{\bullet\bullet}$ and $C'^{\bullet\bullet}$ from B^m (resp. B'^m) to L^m (resp. L'^m).

Now consider the special case where $C^{p,q} = \mathrm{Hom}(P^{-p}, I^q)$ with the usual vertical and horizontal differentials as in the definition of the total complex construction $\mathrm{Hom}^\bullet$. Using $e_p = (-1)^n$ for all $p \geq 0$ then makes (2.6.25) into exactly (2.6.24), so we get the desired commutativity of (2.6.24) up to a sign of $(-1)^{nm}$. ∎

With Theorem 2.5.1 completely proven, it remains to prove Theorem 2.5.2. This is simpler, but the argument is still non-trivial. We advise the reader to skip this proof on a first reading.

PROOF. (of Theorem 2.5.2)

We begin by proving the first part, which is easier. Without loss of generality, $\mathcal{G}^\bullet$ is a bounded complex of flats on X and $\mathcal{F}^\bullet$ is a bounded below complex of i^*-acyclics. Let $I^{\bullet\bullet}$ be the canonical truncation in rows $\leq n$ of a Cartan-Eilenberg resolution of $\mathcal{F}^\bullet$, so $\mathcal{F}^\bullet \to \mathrm{Tot}^\oplus I^{\bullet\bullet}$ is a quasi-isomorphism to a bounded below complex of $\mathcal{H}om_X(\mathcal{O}_Y, \cdot)$-acyclics. Since $\mathcal{G}^\bullet$ is a bounded complex of flats, the natural map of complexes $\varphi : \mathcal{F}^\bullet \otimes \mathcal{G}^\bullet \to (\mathrm{Tot}^\oplus I^{\bullet\bullet}) \otimes \mathcal{G}^\bullet$ is a quasi-isomorphism. Let $\Phi : J^{\bullet\bullet} \to K^{\bullet\bullet}$ be a map of double complexes over φ, where $J^{\bullet\bullet}$ (resp. $K^{\bullet\bullet}$) is the canonical truncation in rows $\leq n$ of a Cartan-Eilenberg resolution of $\mathcal{F}^\bullet \otimes \mathcal{G}^\bullet$ (resp. $(\mathrm{Tot}^\oplus I^{\bullet\bullet}) \otimes \mathcal{G}^\bullet$). In particular, $\mathrm{Tot}^\oplus \Phi$ is a quasi-isomorphism between complexes of $\mathcal{H}om_X(\mathcal{O}_Y, \cdot)$-acyclics, so applying $\mathcal{H}om_X(\mathcal{O}_Y, \cdot)$ to $\mathrm{Tot}^\oplus \Phi$ yields a quasi-isomorphism.

Since $\mathcal{F}^\bullet$ is a bounded below complex of i^*-acyclics, so is $\mathcal{F}^\bullet \otimes \mathcal{G}^\bullet$. Thus, we can describe η_i on $\mathcal{F}^\bullet$ (resp. $\mathcal{F}^\bullet \otimes \mathcal{G}^\bullet$) in terms of the truncated Cartan-Eilenberg resolution $I^{\bullet\bullet}$ (resp. $J^{\bullet\bullet}$), as in (2.1.2). Combining this with functoriality of (2.5.1) for *arbitrary* $\mathcal{O}_X$-modules (such as the $\mathcal{F}^r$'s), we conclude that (2.5.9) is represented by the following diagram of complexes of $i_*\mathcal{O}_Y$-modules, which we want to commute in $\mathbf{D}(Y)$:

$$(2.6.26) \quad \mathscr{H}om_X(\mathscr{O}_Y, \mathrm{Tot}^{\oplus} I^{\bullet\bullet}) \otimes \mathscr{G}^{\bullet} \xrightarrow{\ \mathrm{qism}\ } \mathrm{H}_v^n(\mathscr{H}om_X(\mathscr{O}_Y, I^{\bullet-n,\bullet})) \otimes \mathscr{G}^{\bullet}$$

$$\mathscr{H}om_X(\mathscr{O}_Y, (\mathrm{Tot}^{\oplus} I^{\bullet\bullet}) \otimes \mathscr{G}^{\bullet}) \qquad\qquad \mathscr{E}xt_X^n(\mathscr{O}_Y, \mathscr{F}^{\bullet})[-n] \otimes \mathscr{G}^{\bullet}$$

$$\mathscr{H}om_X(\mathscr{O}_Y, \mathrm{Tot}^{\oplus} K^{\bullet\bullet}) \qquad\qquad \mathscr{E}xt_X^n(\mathscr{O}_Y, \mathscr{F}^{\bullet} \otimes \mathscr{G}^{\bullet})[-n]$$

$$\Phi \uparrow \mathrm{qism}$$

$$\mathscr{H}om_X(\mathscr{O}_Y, \mathrm{Tot}^{\oplus} J^{\bullet\bullet}) \xrightarrow[\mathrm{qism}]{} \mathrm{H}_v^n(\mathscr{H}om_X(\mathscr{O}_Y, J^{\bullet-n,\bullet}))$$

For typographical reasons, we have essentially rotated (2.5.9) clockwise 90 degrees. We want to improve some of the choices of resolutions so as to replace (2.6.26) by a diagram of complexes which is in a form that we will be able to prove commutes on the level of complexes (and not just in the derived category).

The main point is to define a double complex $\mathscr{J}^{\bullet\bullet}$ which will replace $J^{\bullet\bullet}$ but which is not quite a truncated Cartan-Eilenberg resolution of $\mathscr{F}^{\bullet} \otimes \mathscr{G}^{\bullet}$. We use a construction analogous to Lemma 2.4.1, with $\otimes$ replacing $\mathscr{H}om^{\bullet}$. Define

$$\mathscr{J}^{p,q} = \bigoplus_{r\in\mathbf{Z}} I^{r,q} \otimes \mathscr{G}^{p-r}$$

with differentials

$$\mathrm{d}_h^{p,q} : \mathscr{J}^{p,q} \to \mathscr{J}^{p+1,q}, \quad \mathrm{d}_v^{p,q} : \mathscr{J}^{p,q} \to \mathscr{J}^{p,q+1}$$

given by

$$\mathrm{d}_h^{p,q}(x_{r,q} \otimes y_{p-r}) = \mathrm{d}_{h,I^{\bullet\bullet}}^{r,q} x_{r,q} \otimes y_{p-r} + (-1)^{q+r} x_{r,q} \otimes \mathrm{d}_{\mathscr{G}^{\bullet}}^{p-r} y_{p-r},$$

$$\mathrm{d}_v^{p,q}(x_{r,q} \otimes y_{p-r}) = \mathrm{d}_{v,I^{\bullet\bullet}}^{r,q} x_{r,q} \otimes y_{p-r}$$

respectively. It is easy to see that each $\mathscr{J}^{p,q}$ is a finite direct sum and $\mathscr{J}^{\bullet\bullet}$ forms a double complex, with the natural map $\mathscr{F}^{\bullet} \to I^{\bullet,0}$ inducing a map of complexes

$$(2.6.27) \qquad\qquad \mathscr{F}^{\bullet} \otimes \mathscr{G}^{\bullet} \to \mathrm{Tot}^{\oplus} \mathscr{J}^{\bullet\bullet}$$

such that the right side of (2.6.27) is naturally identified with $(\mathrm{Tot}^{\oplus} I^{\bullet\bullet}) \otimes \mathscr{G}^{\bullet}$ (this is similar to the conclusion of Lemma 2.4.1). Since $\mathscr{G}^{\bullet}$ is a bounded complex of flats, the pth column of $\mathscr{J}^{\bullet\bullet}$ is a *resolution* of $(\mathscr{F}^{\bullet} \otimes \mathscr{G}^{\bullet})^p$, via an augmentation map inducing (2.6.27) on the level of total complexes. Although none of the the $\mathscr{J}^{p,q}$'s are injectives in general, so we can't take $\mathscr{J}^{\bullet\bullet}$ in the role of the truncated Cartan-Eilenberg resolution $J^{\bullet\bullet}$ above, $\mathscr{J}^{\bullet\bullet}$ is 'almost as good' for what we need.

The point is that $\mathscr{J}^{\bullet\bullet}$ has columns which are resolutions of length n and all terms are $\mathscr{H}om_X(\mathcal{O}_Y,\cdot)$-acyclic. To see this acyclicity, it suffices to show that if I is a $\mathscr{H}om_X(\mathcal{O}_Y,\cdot)$-acyclic sheaf on X and $\mathscr{G}$ is flat on X, then $I\otimes\mathscr{G}$ is $\mathscr{H}om_X(\mathcal{O}_Y,\cdot)$-acyclic. More generally, if $I^{\bullet}\in\mathbf{D}^+(X)$ and $\mathscr{G}^{\bullet}\in\mathbf{D}^b(X)_{\mathrm{fTd}}$, then $i^b(I^{\bullet}\overset{L}{\otimes}\mathscr{G}^{\bullet})$ is isomorphic to $i^b(I^{\bullet})\overset{L}{\otimes}Li^*\mathscr{G}^{\bullet}$, by [**RD**, III, 6.9(a)]. Thus, we may use $\mathscr{J}^{\bullet\bullet}$ to compute $\eta_i(\mathscr{F}^{\bullet}\otimes\mathscr{G}^{\bullet})$. This enables us to replace (2.6.26) by the following diagram of complexes:

(2.6.28)

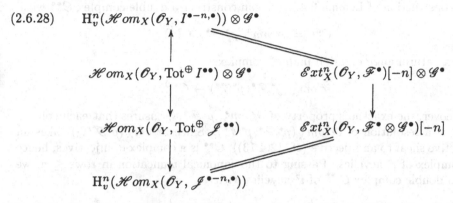

We claim that this diagram of complexes commutes (not just in the derived category). Since

$$\mathscr{J}^{r-n,n}=\bigoplus_{p+q=r} I^{p-n,n}\otimes\mathscr{G}^q,$$

we can define (without the intervention of any signs) an obvious map ψ of complexes from the upper left to lower left of (2.6.28), where ψ is compatible with the left column of (2.6.28). This map ψ is also compatible with the right column of (2.6.28) because of how the natural map

$$\mathscr{E}xt_X^{\bullet}(\mathscr{H},\mathscr{F})\otimes\mathscr{G}\to\mathscr{E}xt_X^{\bullet}(\mathscr{H},\mathscr{F}\otimes\mathscr{G})$$

can be defined for $\mathcal{O}_X$-flat $\mathscr{G}$ by using a $\mathscr{H}om_X(\mathscr{H},\cdot)$-acyclic resolution of $\mathscr{F}$. This completes the proof of the first part of Theorem 2.5.2.

Now we prove the second part of Theorem 2.5.2. This will be a little bit harder. We will use the construction in Lemma 2.4.1, and the hypotheses for equality of complexes in Lemma 2.4.1 will always hold in our setting. Without loss of generality, $\mathscr{G}^{\bullet}$ is a bounded below complex of quasi-coherent injectives on Y and $\mathscr{F}^{\bullet}$ is a bounded above complex on Y with coherent cohomology sheaves and all $\mathscr{F}^r$'s equal to a direct sum of sheaves of type $j_!\mathcal{O}_U$ for various open affine subschemes $j:U\hookrightarrow Y$. Our hypothesis that Y is separated ensures that all such maps j are affine maps, so j_* is exact on quasi-coherent sheaves. This will be essential in our method of proof, and is the reason for the separatedness assumption on Y.

As an example of the usefulness of open immersions $U\hookrightarrow Y$ being affine maps when U is affine, recall (as we saw in the proof of Theorem 2.3.3) that $\mathscr{H}om_Y^{\bullet}(\mathscr{F}^{\bullet},\cdot)$ and $\mathscr{H}om_X^{\bullet}(p^*\mathscr{F}^{\bullet},\cdot)$ take bounded below complexes to bounded below complexes, with quasi-isomorphisms going over to quasi-isomorphisms. In particular, the natural map

$$\mathcal{H}om_X^{\bullet}(p^*\mathcal{F}^{\bullet}, p^*\mathcal{G}^{\bullet}) \to \mathbf{R}\mathcal{H}om_X^{\bullet}(p^*\mathcal{F}^{\bullet}, p^*\mathcal{G}^{\bullet})$$

in $\mathbf{D}(X)$ is an isomorphism (cf. (2.4.17)).

Let $\mathcal{K}^{\bullet\bullet}$ be a Cartan-Eilenberg resolution of $p^*(\mathcal{G}^{\bullet})$ using *quasi-coherent* sheaves, and let $\mathcal{K}'^{\bullet\bullet}$ denote the canonical truncation of $\mathcal{K}^{\bullet\bullet}$ in rows $\leq n$. By the construction of Lemma 2.4.1, we can construct a double complex $C^{\bullet\bullet}$ with

$$C^{a,b} = \mathcal{H}om_X^a(p^*\mathcal{F}^{\bullet}, \mathcal{K}^{\bullet,b})$$

and a natural augmentation map of complexes

$$\mathcal{H}om_X^{\bullet}(p^*\mathcal{F}^{\bullet}, p^*\mathcal{G}^{\bullet}) \to C^{\bullet,0}.$$

Moreover, the exactness property of $\mathcal{H}om_X^{\bullet}(p^*\mathcal{F}^{\bullet}, \cdot)$ ensures that each column $C^{a,\bullet}$ is a resolution of $\mathcal{H}om_X^a(p^*\mathcal{F}^{\bullet}, p^*\mathcal{G}^{\bullet})$. Since $\mathcal{H}om_X(p^*\mathcal{F}^r, \cdot)$ takes an injective sheaf to an injective (cf. (2.4.13)), $C^{\bullet\bullet}$ is a complex of injectives, hence a complex of i^b-acyclics. Passing to the canonical truncation in rows $\leq n$, we get a double complex $C'^{\bullet\bullet}$ of i^b-acyclics with

(2.6.29) $$C'^{a,b} = \mathcal{H}om_X^a(p^*\mathcal{F}^{\bullet}, \mathcal{K}'^{\bullet,b}).$$

Since there is a natural augmentation map

$$\mathcal{H}om_X^{\bullet}(p^*\mathcal{F}^{\bullet}, p^*\mathcal{G}^{\bullet}) \to C'^{\bullet,0},$$

we can use $C'^{\bullet\bullet}$ to compute $\eta_i(\mathbf{R}\mathcal{H}om_X^{\bullet}(p^*\mathcal{F}^{\bullet}, p^*\mathcal{G}^{\bullet}))$, provided we verify that $\mathcal{H}om_X^{\bullet}(p^*\mathcal{F}^{\bullet}, p^*\mathcal{G}^{\bullet})$ is a complex of i^*-acyclics.

Since $p^*(\cdot)$ is i^*-acyclic and $\mathcal{G}^{\bullet}$ is a complex of quasi-coherent sheaves, in order to show that all sheaves $\mathcal{H}om_X^r(p^*\mathcal{F}^{\bullet}, p^*\mathcal{G}^{\bullet})$ are i^*-acyclic, one is easily reduced (using (2.4.16)) to the assertion that an arbitrary product of quasi-coherent i^*-acyclic sheaves is i^*-acyclic. By working locally and using the finite Koszul resolution of $i_*\mathcal{O}_Y$ over $\mathcal{O}_X$, it remains to observe that a product of an exact sequence of quasi-coherent sheaves is an exact sequence of sheaves. This enables us to use $\mathcal{K}'^{\bullet\bullet}$ to describe $\eta_i(\mathbf{R}\mathcal{H}om_X^{\bullet}(p^*\mathcal{F}^{\bullet}, p^*\mathcal{G}^{\bullet}))$, via $C'^{\bullet\bullet}$.

In order to relate this description to the map $\eta_i(p^*\mathcal{G}^{\bullet})$, let $J^{\bullet\bullet}$ denote the canonical truncation in rows $\leq n$ of a Cartan-Eilenberg resolution of $\mathcal{G}^{\bullet}$ on Y. Since p^* is exact, we may choose (by the construction of Cartan-Eilenberg resolutions) a map $p^*J^{\bullet\bullet} \to \mathcal{K}'^{\bullet\bullet}$ of double complexes over $p^*\mathcal{G}^{\bullet}$. Also, let $I^{\bullet\bullet}$ (resp. $\mathcal{I}^{\bullet\bullet}$) denote the canonical truncation in degrees $\leq n$ of a Cartan-Eilenberg resolution of $p^*\mathcal{H}om_Y^{\bullet}(\mathcal{F}^{\bullet}, \mathcal{G}^{\bullet})$ (resp. $\mathcal{H}om_X^{\bullet}(p^*\mathcal{F}^{\bullet}, p^*\mathcal{G}^{\bullet})$), and choose a map of double complexes $I^{\bullet\bullet} \to \mathcal{I}^{\bullet\bullet}$ over the quasi-isomorphism

$$p^*\mathcal{H}om_Y^{\bullet}(\mathcal{F}^{\bullet}, \mathcal{G}^{\bullet}) \to \mathcal{H}om_X^{\bullet}(p^*\mathcal{F}^{\bullet}, p^*\mathcal{G}^{\bullet})$$

(recall, as was explained above, that $\mathcal{H}om_X^{\bullet}(p^*\mathcal{F}^{\bullet}, \cdot)$ takes quasi-isomorphisms to quasi-isomorphisms on bounded below complexes of quasi-coherent sheaves).

It follows that (2.5.10) is represented by the following diagram of complexes:
(2.6.30)

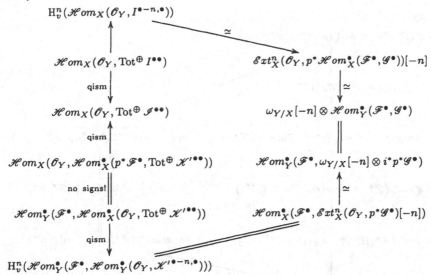

where, for typographical reasons, we have rotated (2.5.10), with the complex

$$\mathscr{H}om_X(\mathcal{O}_Y, \mathrm{Tot}^{\oplus} I^{\bullet\bullet})$$

corresponding to $i^{\flat}p^*\mathbf{R}\mathscr{H}om_X^{\bullet}(\mathscr{F}^{\bullet}, \mathscr{G}^{\bullet})$.

To simplify this mess, we make three observations. First of all, since the composite map of complexes

$$(2.6.31) \quad i^*p^*\mathscr{H}om_Y^{\bullet}(\mathscr{F}^{\bullet}, \mathscr{G}^{\bullet}) \to i^*\mathscr{H}om_X^{\bullet}(p^*\mathscr{F}^{\bullet}, p^*\mathscr{G}^{\bullet}) \to \mathscr{H}om_Y^{\bullet}(\mathscr{F}^{\bullet}, \mathscr{G}^{\bullet})$$

is the canonical isomorphism and the first map in (2.6.31) is a quasi-isomorphism (it is obtained by applying i^* to a quasi-isomorphism between complexes of i^*-acyclic sheaves), we see that the two maps in (2.6.31) are inverses in $\mathbf{D}(Y)$ (using the isomorphism $i^*p^* = 1$). Second, we have a commutative diagram of complexes
(2.6.32)

$$\begin{array}{ccc}
\mathscr{H}om_X(\mathcal{O}_Y, \mathrm{Tot}^{\oplus} I^{\bullet\bullet}) & \longrightarrow \cdots \longrightarrow & \omega_{Y/X}[-n] \otimes i^*p^*\mathscr{H}om_Y^{\bullet}(\mathscr{F}^{\bullet}, \mathscr{G}^{\bullet}) \\
\downarrow{\scriptstyle\text{qism}} & & \downarrow{\scriptstyle\text{qism}} \\
\mathscr{H}om_X(\mathcal{O}_Y, \mathrm{Tot}^{\oplus} \mathscr{I}^{\bullet\bullet}) & \longrightarrow \cdots \longrightarrow & \omega_{Y/X}[-n] \otimes i^*\mathscr{H}om_X^{\bullet}(p^*\mathscr{F}^{\bullet}, p^*\mathscr{G}^{\bullet})
\end{array}$$

in which the horizontal maps are the 'explicit' descriptions (2.5.5) of η_i on the complexes of i^* acyclics

$$p^*\mathscr{H}om_Y^{\bullet}(\mathscr{F}^{\bullet}, \mathscr{G}^{\bullet}), \quad \mathscr{H}om_X^{\bullet}(p^*\mathscr{F}^{\bullet}, p^*\mathscr{G}^{\bullet})$$

Third, the columns of the double complexes $\mathscr{I}^{\bullet\bullet}$ and $C'^{\bullet\bullet}$ give $i^{\flat}$-acyclic resolutions of the terms of $\mathscr{H}om_X^{\bullet}(p^*\mathscr{F}^{\bullet}, p^*\mathscr{G}^{\bullet})$. We remind the reader that $C'^{\bullet\bullet}$ is the natural double complex whose terms were defined in (2.6.29).

Putting these observations together, it is enough to prove the commutativity in $\mathbf{D}(Y)$ of the diagram of complexes

(2.6.33)

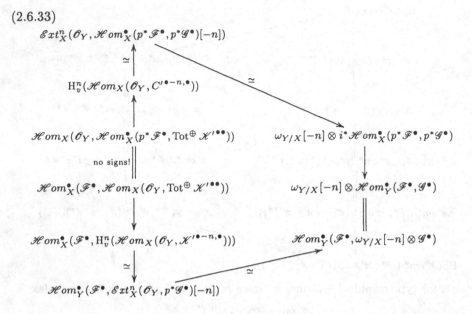

We make the stronger claim that (2.6.33) even commutes on the level of complexes. Note that (2.6.33) makes perfectly good sense without any assumption that the cohomology sheaves of $\mathscr{F}^\bullet$ are coherent (or even quasi-coherent), so we drop this assumption on the cohomology of $\mathscr{F}^\bullet$. Since everything in sight is compatible with respect to translations in $\mathscr{F}^\bullet$ (note this has no effect on the signs of the differentials in the construction of Lemma 2.4.1) and is functorial in $\mathscr{F}^\bullet$, we easily reduce to the case where $\mathscr{F}^\bullet$ is concentrated in degree 0. Arguing as in the end of the proof of Theorem 2.3.3, we may assume $\mathscr{F}^\bullet = (j_! \mathscr{O}_U)[0]$ for a single open immersion $j : U \hookrightarrow Y$, and then adjointness between j_* and j^* reduces us to the trivial case $Y = U$. ∎

2.7. Compatibilities between $(\cdot)^\sharp$ and $(\cdot)^\flat$

In order to use (2.5.3) as in [**RD**, III, §8] to 'glue' the functors $(\cdot)^\sharp$ and $(\cdot)^\flat$, it is necessary to study some compatibilities between $(\cdot)^\sharp$ and $(\cdot)^\flat$. We begin by recalling the basic construction from which everything else follows. If $\pi : X \to Y$ is a smooth *separated* map with pure relative dimension n and i is a section, then [**RD**, III, 8.1] gives a natural isomorphism

(2.7.1) $\psi_{i,\pi} : \mathscr{G}^\bullet \simeq i^\flat \pi^\sharp \mathscr{G}^\bullet$

for $\mathscr{G}^{\bullet} \in \mathbf{D}(Y)$. Since the construction in [**RD**] has several sign problems, we quickly give the definition in a form adapted to our sign conventions:

(2.7.2)
$$\mathscr{G}^{\bullet} \xrightarrow{\;\zeta'_{i,\pi}\otimes 1_{\mathscr{G}^{\bullet}}\;} (\omega_{Y/X}[-n] \otimes i^{*}\omega_{X/Y}[n]) \otimes \mathscr{G}^{\bullet}$$

$$\downarrow \simeq$$

$$\omega_{Y/X}[-n] \overset{\mathbf{L}}{\otimes} \mathbf{L}i^{*}(\omega_{X/Y}[n] \otimes \pi^{*}\mathscr{G}^{\bullet})$$

$$\simeq \uparrow \eta_i$$

$$i^{\flat}(\omega_{X/Y}[n] \otimes \pi^{*}\mathscr{G}^{\bullet})$$

$$\|$$

$$i^{\flat}\pi^{\sharp}\mathscr{G}^{\bullet}$$

Recall that the isomorphism $\zeta'_{i,\pi}$ in (2.7.2) is locally given by

$$1 \mapsto (x_1^{\vee} \wedge \cdots \wedge x_n^{\vee}) \otimes i^{*}(dx_n \wedge \cdots \wedge dx_1),$$

where the x_j's cut out i and $\omega_{X/Y}[n] \otimes \pi^{*}\mathscr{G}^{\bullet}$ *is a complex of i^{*}-acyclics* (as $\pi \circ i = 1$ and π are flat maps); this is why the second map in (2.7.2) is an isomorphism. The isomorphism $\psi_{i,\pi}$ in (2.7.1) commutes with flat base change (as one sees by using [**RD**, III, 2.1, 6.3, 7.4(b)]) and commutes with *translations* since η_i does. Omitting the sign in the definition of (1.3.28) would change (2.7.2) by $(-1)^{n(n+1)/2}$ (since (2.5.3) would be changed by this sign).

Let us make (2.7.2) explicit in a special case that will arise several times later on. Let $\mathscr{G}^{\bullet} = \mathscr{G}[m]$ for an integer m and an $\mathcal{O}_Y$-module $\mathscr{G}$. Let $\mathscr{I}^{\bullet}$ be an injective resolution of $\omega_{X/Y} \otimes \pi^{*}\mathscr{G}$, and view $\mathscr{I}^{\bullet+m+n}$ as an injective resolution of $\pi^{\sharp}(\mathscr{G}[m]) = (\omega_{X/Y} \otimes \pi^{*}\mathscr{G})[m+n]$ (this amounts to viewing $\mathscr{I}^{\bullet}$ placed in the $-(m+n)$th column as a Cartan-Eilenberg resolution of $\pi^{\sharp}\mathscr{G}^{\bullet}$). Since the sheaf $\omega_{X/Y} \otimes \pi^{*}\mathscr{G}$ is i^{*}-acyclic, following (2.5.5) we see that for 'Cartan-Eilenberg' reasons much like the ones discussed in the definition in (2.3.8), the map induced on H^{-m}'s by

$$\psi_{i,\pi} : \mathscr{G}[m] \simeq i^{\flat}((\omega_{X/Y} \otimes \pi^{*}\mathscr{G})[m+n]) \overset{\mathrm{def}}{=} i^{\flat}(\mathscr{I}^{\bullet+m+n})$$

is the composite

(2.7.3)
$$\mathscr{G} \xrightarrow[\simeq]{\;\zeta'_{i,\pi}\;} \omega_{Y/X} \otimes i^{*}\omega_{X/Y} \otimes \mathscr{G} \xrightarrow[\simeq]{\;\eta_i\;} \mathscr{E}\!xt_X^n(\mathcal{O}_Y, \omega_{X/Y} \otimes \pi^{*}\mathscr{G}),$$

where the $\mathscr{E}\!xt^n$ on the right is computed with the injective resolution $\mathscr{I}^{\bullet}$ specified above. We emphasize that if we computed $i^{\flat}(\pi^{\sharp}\mathscr{G}^{\bullet})$ by using $\mathscr{I}^{\bullet}[m+n]$ but computed $\mathscr{E}\!xt_X^n(\mathcal{O}_Y, \omega_{X/Y} \otimes \pi^{*}\mathscr{G})$ using $\mathscr{I}^{\bullet}$, then the above explication on H^{-m}'s would change by a sign of $(-1)^{m(m+n)}$ (which is harmless if $m = \pm n$ or $m = 0$, for example).

The isomorphism $\psi_{i,\pi}$ is generalized in [**RD**] as follows. Let $f : X \to Y$ and $g : Y \to Z$ be two maps between locally noetherian schemes. Assume that

g is separated, smooth with pure relative dimension n and f is finite. When gf is finite (resp. smooth and separated with pure relative dimension m), [**RD**, III, 8.2, 8.4] uses (2.7.1) to define respective isomorphisms

(2.7.4) $$\psi_{f,g} : (gf)^\flat \simeq f^\flat g^\sharp$$

and

(2.7.5) $$\psi_{f,g} : (gf)^\sharp \simeq f^\flat g^\sharp$$

as functors on $\mathbf{D}^+_{qc}(Z)$.

Briefly, the definitions of (2.7.4) and (2.7.5) go as follows. When gf is finite, we consider the diagram

(2.7.6)

$$
\begin{array}{ccc}
X \times_Z Y & \xrightarrow{\;p_2\;} & Y \\
{\scriptstyle i}\Big\uparrow\Big\downarrow{\scriptstyle p_1} \;\;\nearrow{\scriptstyle f} & & \Big\downarrow{\scriptstyle g} \\
X & \xrightarrow[\;gf\;]{} & Z
\end{array}
$$

and define (2.7.4) to be the composite

(2.7.7) $$\psi_{f,g} : (gf)^\flat \xrightarrow[\;\simeq\;]{\psi_{i,p_1}} i^\flat p_1^\sharp (gf)^\flat \xrightarrow{\;\simeq\;} i^\flat p_2^\flat g^\sharp =\!=\!= f^\flat g^\sharp$$

using the easy Lemma 2.7.3 in the middle map and the compatibility of $(\cdot)^\flat$ with respect to composition at the end. When gf is smooth and separated with pure relative dimension m, we use (2.7.6) and the diagram (2.7.11) below to define (2.7.5) as follows:

(2.7.8)

$$
\begin{array}{ccc}
(gf)^\sharp & & \\
\;\;\simeq\Big\downarrow{\scriptstyle \psi_{i,p_1}} & & \\
i^\flat p_1^\sharp (gf)^\sharp & =\!=\!=\!=\!=\!=\!=\!=\!= & \Delta^\flat (1 \times f)^\flat p_2^\sharp g^\sharp \\
& & \Big\uparrow{\scriptstyle \simeq} \\
f^\flat g^\sharp & \xleftarrow{\;\psi_{\Delta, q_2}(f^\flat g^\sharp)\;} & \Delta^\flat q_2^\sharp f^\flat g^\sharp \\
{\scriptstyle \zeta'_{\Delta, q_2}}\Big\downarrow{\scriptstyle \simeq} & & \simeq\Big\downarrow{\scriptstyle \eta_\Delta} \\
\omega_{X/X\times_Z X}[-m] \otimes \Delta^* \omega_{X\times_Z X/X}[m] \overset{\mathbf{L}}{\otimes} f^\flat g^\sharp & \xleftarrow[\;\simeq\;]{} & \omega_{X/X\times_Z X}[-m] \overset{\mathbf{L}}{\otimes} \mathbf{L}\Delta^* q_2^\sharp f^\flat g^\sharp
\end{array}
$$

where we have used the compatibility of $(\cdot)^\flat$, $(\cdot)^\sharp$ with respect to compositions, as well as the easy Lemma 2.7.3 again (and the bottom square is commutative, by the *definition* of (2.7.4)).

It is useful to note that the definitions of (2.7.4), (2.7.5) just given would *not* change if we replace $X \times_Z Y$ with $Y \times_Z X$ (and use $f \times 1$ instead of $1 \times f$); this is an exercise in going through the definitions and using (2.2.4) mostly in the trivial cases where some of the relevant scheme morphisms are isomorphisms. Omitting the sign in (1.3.28) would change (2.7.4) and (2.7.5) by respective signs

$$(-1)^{n(n+1)/2}, \quad (-1)^{n(n+1)/2}(-1)^{m(m+1)/2}.$$

When f is a section to g, (2.7.4) coincides with (2.7.1). By construction, one sees readily that (2.7.4) and (2.7.5) are compatible with translations and flat base change. In the case when the Z-morphism $f : X \hookrightarrow Y$ is a closed immersion between separated smooth Z-schemes with pure relative dimensions m and n respectively, (2.7.5) evaluated on $\mathscr{O}_Z[0]$ gives an isomorphism

$$\omega_{X/Z} \simeq \mathscr{E}xt_Y^{n-m}(\mathscr{O}_X, \omega_{Y/Z})$$

This isomorphism plays a crucial role in the base change theory for dualizing sheaves, and it will be made explicit in Lemma 3.5.3.

There are a number of compatibilities that (2.7.4) and (2.7.5) must satisfy in order to use them to 'glue' the δ-functors $(\cdot)^{\sharp}$ and $(\cdot)^{\flat}$ into a single δ-functor $(\cdot)^!$ for suitable morphisms (such as projective morphisms) which factor into a finite map followed by a separated smooth map. First of all, for a finite smooth (i.e., finite étale) map $f : X \to Y$ between locally noetherian schemes, there are two δ-functors $f^{\flat}$ and $f^{\sharp}$ which must be related. Using (2.7.4), we get an isomorphism

$$(2.7.9) \qquad\qquad \psi_{f,1} : f^{\flat} \simeq f^{\sharp} = f^*$$

between δ-functors on $\mathbf{D}_{qc}^+(Y)$, and this is insensitive to the sign in (1.3.28). In concrete terms (as noted in [**RD**, III, p.187]), it is easy to check that the *inverse* of (2.7.9) is obtained from the map of $f_*\mathscr{O}_X$-modules

$$f_*\mathscr{O}_X \otimes \mathscr{G} \to \mathscr{H}om_Y(f_*\mathscr{O}_X, \mathscr{G})$$

defined by

$$(2.7.10) \qquad\qquad a \otimes s \mapsto (a' \mapsto \mathrm{Tr}(aa')s),$$

where $\mathrm{Tr} : f_*\mathscr{O}_X \to \mathscr{O}_Y$ is the classical trace.

It is quite important in the global theory that (2.7.4) and (2.7.5) are compatible with (2.7.9). More precisely, using (2.7.9), one wants to know:

LEMMA 2.7.1. *For maps $f : X \to Y$ and $g : Y \to Z$ between locally noetherian schemes, with f finite and g separated smooth with pure relative dimension,*

- *when f is smooth, hence finite étale, (2.7.5) is equal to $(gf)^{\sharp} \simeq f^{\sharp}g^{\sharp}$,*
- *when g is finite, hence finite étale, (2.7.4) is equal to $(gf)^{\flat} \simeq f^{\flat}g^{\flat}$,*
- *when gf is finite étale, (2.7.4) is equal to (2.7.5).*

As a 'safety check', we observe that the truth of the three assertions in Lemma 2.7.1 is insensitive to the sign in (1.3.28).

PROOF. Using our concrete descriptions, such as (2.7.10) and the fact that η_i is the 'identity' when i is an open and closed immersion, the last two cases are easy to prove: one unwinds the definitions and reduces to assertions involving commutative diagrams of quasi-coherent sheaves which are easily checked after suitable étale surjective base changes (to make finite étale maps totally split).

As an example, the proof of the second part boils down to the fact that if $A \to B$ is finite étale with a section $B \twoheadrightarrow A$, then the idempotent e of B corresponding to the factor ring A of this section satisfies $\mathrm{Tr}_{B/A}(e) = 1$. To prove this, we may assume A is strictly henselian local, so B is a finite product of copies of A and the trace calculation is then trivial.

The first case in the lemma is somewhat more complicated, so we now give the proof. The relevant scheme diagram is

$$(2.7.11)$$

$$
\begin{array}{ccc}
X \times_Z X & \xrightarrow{\ 1 \times f\ } & X \times_Z Y \\
\end{array}
$$

with f finite étale and g separated smooth with pure relative dimension n. After carefully unwinding *all* steps in the definition of (2.7.5) evaluated on $\mathscr{G}^\bullet \in \mathbf{D}^+_{\mathrm{qc}}(Z)$, as well as all steps in the definition of (2.7.9), the functoriality of η_i, η_Δ, and the isomorphism $(1 \times f)^\flat \simeq (1 \times f)^\sharp$ reduce us to a compatibility of η_i and η_Δ with respect to the isomorphism

$$(2.7.12) \qquad i^\flat \simeq \Delta^\flat \circ (1 \times f)^\flat \simeq \Delta^\flat \circ (1 \times f)^\sharp = \Delta^\flat \circ (1 \times f)^*.$$

To be precise, upon winding all of the definitions we get a large diagram in which everything is expressed in terms of $\mathscr{F}^\bullet = \omega_{Y/Z}[n] \otimes g^* \mathscr{G}^\bullet \in \mathbf{D}^+_{\mathrm{qc}}(Y)$, but since

- $p_2^\sharp$ takes any $\mathscr{O}_Y$-module to an i^*-acyclic sheaf (as p_2 and $p_2 \circ i = f$ are flat),
- $(1 \times f)^* \circ p_2^\sharp$ takes any $\mathscr{O}_Y$-module to a Δ^*-acyclic sheaf,
- applying (2.7.12) to $p_2^\sharp(\cdot)$ coincides with the composite

$$
i^\flat p_2^\sharp \,=\!=\, \Delta^\flat (1 \times f)^\flat p_2^\sharp \xleftarrow{\ \simeq\ } \Delta^\flat q_2^\sharp f^\flat \xrightarrow{\ \psi_{f,1}\ } \Delta^\flat q_2^\sharp f^\sharp \,=\!=\, \Delta^\flat (1 \times f)^\sharp p_2^\sharp
$$

(as is easily checked using (2.7.6)),

we are reduced (with the help of (2.7.6)) to showing that for any $\mathscr{F}^{\bullet} \in \mathbf{D}_{\mathrm{qc}}^{+}(Y)$, the diagram of isomorphisms in $\mathbf{D}(X)$

(2.7.13)

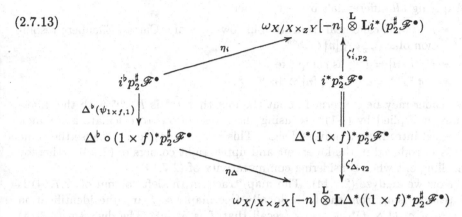

commutes.

Since

$$\Delta^{\flat}(1 \times f)^* \simeq \Delta^{\flat}(1 \times f)^{\sharp} \simeq \Delta^{\flat}(1 \times f)^{\flat} \simeq i^{\flat},$$

it follows that $(1 \times f)^*$ takes injectives to $\Delta^{\flat}$-acyclics. Thus, if

$$\mathscr{H}^{\bullet} \in \mathbf{D}_{\mathrm{qc}}^{+}(X \times_Z Y)$$

is a bounded below complex and $I^{\bullet\bullet}$ is the canonical truncation in rows $\leq n$ of a Cartan-Eilenberg resolution of $\mathscr{H}^{\bullet}$, then the exact functor $(1 \times f)^*$ takes $I^{\bullet\bullet}$ to an augmented double complex whose columns are $\Delta^{\flat}$-acyclic resolutions. Applying this with $\mathscr{H}^{\bullet} = p_2^* \mathscr{F}^{\bullet}$, the commutativity of (2.7.13) is reduced to the more general claim that if $\mathscr{H}^{\bullet} \in \mathbf{D}_{\mathrm{qc}}^{+}(X \times_Z Y)$ is a bounded below complex of i^*-*acyclics* (so descriptions such as (2.5.5), (2.5.6) can be used) and $I^{\bullet\bullet}$ is the canonical truncation in rows $\leq n$ of a Cartan-Eilenberg resolution of $\mathscr{H}^{\bullet}$, the following diagram of complexes commutes in $\mathbf{D}(X)$:

(2.7.14)

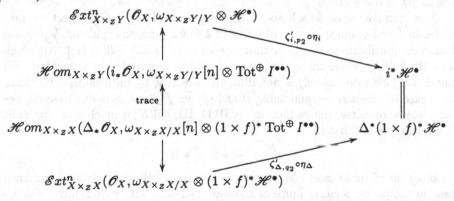

We now make some remarks concerning how to check that (2.7.14) really does 'compute' (2.7.13), as just claimed. The key points are that (2.5.5) computes

(2.5.3) without the intervention of signs, and that if $\widetilde{I}^{\bullet\bullet}$ is the double complex obtained from $\omega_{X\times_Z Y/Y}\otimes I^{\bullet\bullet}$ by shifting columns n units to the left and multiplying *all* differentials by $(-1)^n$, then

- $\widetilde{I}^{\bullet\bullet}$ is the canonical truncation in rows $\leq n$ of a Cartan-Eilenberg resolution of $\omega_{X\times_Z Y/Y}[n]\otimes\mathcal{H}^{\bullet}$,
- $\widetilde{I}^{\bullet,n}[-n] = I^{\bullet,n}$ as complexes,
- $\mathrm{Tot}^{\oplus}\widetilde{I}^{\bullet\bullet} = \omega_{X\times_Z Y/Y}[n]\otimes\mathrm{Tot}^{\oplus}I^{\bullet\bullet}$.

The reader may be concerned about the fact that $\widetilde{I}^{p\bullet}$ is $I^{p+n,\bullet}$ with the differentials multiplied by $(-1)^n$, so using these resolutions to calculate a common $\mathcal{E}xt$-sheaf introduces a universal sign. This is not a problem, because the same sign is introduced in the lower left and upper right corners of (2.7.14), thereby cancelling out when considering commutativity of (2.7.14).

Now we analyze (2.7.14). The map 'trace' in the left column of (2.7.14) is the trace arising from the finite locally free map $1\times f$ and the identification $\omega_{X\times_Z X/X}\simeq (1\times f)^*\omega_{X\times_Z Y/Y}$ (recall that f is étale). The diagram (2.7.14) even commutes as a diagram of complexes of sheaves. Indeed, it is easy to check that flat base change does not affect the acyclicity properties of the (possibly non-quasi-coherent) sheaves I^{pq} with respect to the functors

$$\mathcal{H}om_{X\times_Z Y}(i_*\mathcal{O}_X,\cdot),\quad \mathcal{H}om_{X\times_Z X}(\Delta_*\mathcal{O}_X,\cdot),$$

whose derived functors can be computed locally via Koszul resolutions in the first variable. Thus, by suitable étale base change we immediately reduce to the case where the finite étale map $f:X\to Y$ is totally split, and then the identity map (so $i=\Delta$), in which case the commutativity of (2.7.14) is clear.   ∎

For convenience, when f and g are *both* separated smooth with pure relative dimension (resp. both finite), we define $\psi_{f,g}$ to be the canonical isomorphism $(gf)^{\sharp}\simeq f^{\sharp}g^{\sharp}$ (resp. $(gf)^{\flat}\simeq f^{\flat}g^{\flat}$). By Lemma 2.7.1, it is not difficult to check that via (2.7.9), these isomorphisms are compatible with (2.7.4) and (2.7.5) when f, g, or gf is finite étale.

Now it makes sense to ask about compatibility of $\psi_{f,g}$ with respect to composites in f and g, much like the identity (2.2.4) for the isomorphisms $\zeta'_{f,g}$. A list of such compatibilities is given without proof in [**RD**, III, 8.6(*b*),(*c*)]. Although [**RD**, III, 8.6(*c*)] (where the smooth map should be required to be separated with bounded relative dimension) is not difficult to verify, by unraveling definitions, being careful about signs, and using that $(gf)^{\flat}\simeq f^{\flat}g^{\flat}$ respects flat base change and triple composites, the verification of [**RD**, III, 8.6(*b*)] is much more difficult. This result asserts that if

$$X\xrightarrow{\ f\ }Y\xrightarrow{\ g\ }Z\xrightarrow{\ h\ }W$$

is a diagram of three morphisms between locally noetherian schemes and each possible composite is either finite or separated smooth with pure relative dimension, and we write $(\cdot)^{!}$ in place of $(\cdot)^{\sharp}$ and $(\cdot)^{\flat}$ in the smooth and finite cases respectively (making use of (2.7.9) and Lemma 2.7.1 to avoid any confusion), then the diagram of δ-functors on $\mathbf{D}_{\mathrm{qc}}^{+}(W)$

$$(2.7.15) \qquad \begin{array}{ccc} (hgf)^{!} & \xrightarrow{\psi_{f,hg}} & f^{!}(hg)^{!} \\ \psi_{gf,h} \downarrow & & \downarrow f^{!}(\psi_{g,h}) \\ (gf)^{!}h^{!} & \xrightarrow{\psi_{f,g}} & f^{!}g^{!}h^{!} \end{array}$$

commutes in $\mathbf{D}(X)$. The two trivial cases are when f, g, h are all finite or all separated smooth maps with pure relative dimension, since then (2.7.15) just asserts the triple composite compatibility of $(\cdot)^{\flat}$ and $(\cdot)^{\sharp}$ respectively.

There are many more cases to be considered, with a bewildering array of complexity. It seems almost beyond human ability to check all of these directly, and presumably there is a small number of such cases from which all others can be formally deduced. Fortunately, aside from the two trivial cases, there are only *three* other cases of (2.7.15) which are needed in the subsequent development of Grothendieck's duality theory:

- f, g finite, h smooth, hg finite (so gf and hgf are finite),
- f finite, g smooth, gf finite, h, hg smooth, hgf finite,
- same as the second case, except hgf is smooth.

To be explicit, [**RD**, III, 8.7] and [**RD**, VI, §2, p.316, (v)] require the first two cases, the development of duality theory for proper Cohen-Macaulay morphisms (as in Chapter 4) and the proof of [**RD**, III, §9, R2] (cf. (R2) in Appendix A) require the second case, and a useful generalization of [**RD**, VI, p.331, VAR5] (cf. Theorem 3.2.2) requires the third case. We therefore believe that one should be able to formally deduce all cases of (2.7.15) from the two trivial cases and the three non-trivial cases we have just mentioned, but we ignore this point, as it is not needed (however, see Theorem 2.8.1).

The above few cases of (2.7.15) are the only ones needed in the development of the theory (though the theory of the residue symbol [**RD**, III, §9] also requires Theorem 2.8.1, as we explain in Appendix A), but it is still an interesting question to determine an efficient way to prove all possible cases of (2.7.15) in general, based on the direct verification of a small number of cases. Let us briefly explain how we can at least deduce the commutativity of (2.7.15) with $\mathbf{D}_{c}^{+}$ in place of $\mathbf{D}_{\mathrm{qc}}^{+}$ in most cases of interest. Nearly all applications of Grothendieck's duality theory are in the case of maps between schemes of finite type over a regular noetherian scheme (such as finite type schemes over $\mathbf{Z}$, a field, or a complete local noetherian ring). Schemes of this type admit dualizing complexes (discussed in §3.1), and for any finite type map $f : X \to Y$ between noetherian schemes admitting a dualizing complex, it is shown in [**RD**, VII, 3.4(a)] (and explained in more detail in §3.3) that there is a theory of a δ-functor $f^{!} : \mathbf{D}_{c}^{+}(Y) \to \mathbf{D}_{c}^{+}(X)$ and for finite (resp. smooth separated) f, this is compatible with the theory of $f^{\flat}$ (resp. $f^{\sharp}$); moreover, there is a δ-functorial isomorphism $(gf)^{!} \simeq f^{!}g^{!}$ which is 'associative' with respect to triple composites and is compatible with the $\psi_{f,g}$ isomorphisms when each of f, g, gf is either finite or separated smooth with pure relative dimension. Thus, we obtain (2.7.15) for finite type maps between noetherian schemes admitting a dualizing complex, at least on the derived categories $\mathbf{D}_c^+$ in place of $\mathbf{D}_{qc}^+$.

The important point is that the *construction* of a good theory of $(\cdot)^!$ by means of dualizing complexes requires proving a few non-trivial cases of [**RD**, III, 8.6(b)] in advance, namely the three non-trivial cases mentioned above. It may be possible to deduce the commutativity of (2.7.15) in complete generality by means of comparisons with Lipman's alternate development of duality theory in [**LLT**], but it might be difficult to relate the $\psi_{f,g}$ isomorphisms above with isomorphisms in Lipman's theory.

Of the three non-trivial cases of (2.7.15) mentioned above, the third case can be easily reduced to two special instances of the second case (by using functoriality, the *definition* of (2.7.5), and the special case

$$\Gamma_{gf}^b \simeq \Gamma_f^b \circ (1_X \times_W g)^\sharp$$

of (2.7.4), where Γ_{gf}, Γ_f are graph maps). However, the proof of the second non-trivial case in (2.7.15) depends in an essential way on the signs in the definition of (2.2.3). We now present proofs of the first two cases of (2.7.15).

THEOREM 2.7.2. *Let* $X \xrightarrow{f} Y \xrightarrow{g} Z \xrightarrow{h} W$ *be three maps between locally noetherian schemes such that each of the composites is either finite or smooth and separated with pure relative dimension.*

1. *Assume that* f, g *are finite,* h *is smooth, and* hg *is finite (so* gf *and* hgf *are finite). Then the diagram of* δ*-functors on* $\mathbf{D}_{qc}^+(W)$

(2.7.16)
$$\begin{array}{ccc}
(hgf)^b & \xrightarrow{\psi_{gf,h}} & (gf)^b h^\sharp \\
{\scriptstyle \simeq}\downarrow & & \downarrow{\scriptstyle \simeq} \\
f^b(hg)^b & \xrightarrow[f^b(\psi_{g,h})]{} & f^b g^b h^\sharp
\end{array}$$

commutes in $\mathbf{D}(X)$.

2. *Assume that* f *is finite,* g *is smooth,* gf *is finite,* h, hg *are smooth, and* hgf *is finite. Then the diagram of* δ*-functors on* $\mathbf{D}_{qc}^+(W)$

(2.7.17)
$$\begin{array}{ccc}
(hgf)^b & \xrightarrow{\psi_{gf,h}} & (gf)^b h^\sharp \\
{\scriptstyle \psi_{f,hg}}\downarrow & & \downarrow{\scriptstyle \psi_{f,g}} \\
f^b(hg)^\sharp & \xrightarrow[\simeq]{} & f^b g^\sharp h^\sharp
\end{array}$$

commutes in $\mathbf{D}(X)$.

The proof of the commutativity of (2.7.17) relies crucially on the fact that the translation compatibility of η_i is defined with respect to the translation compatibility of the functor $\omega_i[-n] \overset{\mathbf{L}}{\otimes} \mathbf{L}i^*(\cdot)$, whose translation compatibility involves an intervention of signs, as in (1.3.6).

PROOF. We begin by proving the commutativity of (2.7.16). If we go back to the definition of (2.7.4) and use [**RD**, III, 8.6(c)], we wind up with a diagram which is a special case of (2.7.16) with $Y = W$ and $hg = 1$. Thus, we may suppose that h is separated smooth with pure relative dimension n and $W = Y$ with $hg = 1$, so g is an lci map with pure codimension n. Then the relevant scheme diagram is

$$
\begin{array}{ccc}
X \times_Y Z & \xrightarrow{\ p_2\ } & Z \\
\Gamma_{gf} \upharpoonleft\downharpoonright \; p_1 & \nearrow_{gf} \; g \upharpoonleft\downharpoonright h & \\
X & \xrightarrow{\ f\ } & Y
\end{array}
$$

(with Γ_{gf} an lci map of pure codimension n, as g has this property), and the commutativity of (2.7.16) is equivalent to the commutativity of

(2.7.18)
$$
\begin{array}{ccc}
f^\flat & \xrightarrow{\ \psi_{gf,h}\ } & (gf)^\flat h^\sharp \\
{\scriptstyle f^\flat(\psi_{g,h})} \downarrow & \swarrow_{\simeq} & \\
f^\flat g^\flat h^\sharp & &
\end{array}
$$

in $\mathbf{D}(X)$. By unwinding the definitions of (2.7.1) and (2.7.4), and using functoriality with respect to $p_2^\flat h^* \simeq p_1^* f^\flat$, the commutativity of (2.7.18) reduces to the commutativity in $\mathbf{D}(X)$ of the diagram of functors on $\mathbf{D}_{qc}^+(Y)$

(2.7.19)
$$
\begin{array}{ccc}
f^\flat(\omega_{Y/Z}[-n] \overset{\mathbf{L}}{\otimes} (\mathbf{L}g^* \circ h^*)) & \xrightarrow{\ \simeq\ } & \omega_{X/X\times_Y Z}[-n] \overset{\mathbf{L}}{\otimes} f^\flat(\mathbf{L}g^* \circ h^*) \\
{\scriptstyle \eta_g} \uparrow & & \downarrow {\scriptstyle \simeq} \\
f^\flat g^\flat h^* & & \omega_{X/X\times_Y Z}[-n] \overset{\mathbf{L}}{\otimes} (\mathbf{L}\Gamma_{gf}^* \circ p_1^* f^\flat) \\
\| & & \downarrow {\scriptstyle \simeq} \\
\Gamma_{gf}^\flat p_2^\flat h^* & \xrightarrow[\ \eta_{\Gamma_{gf}}\ ]{} & \omega_{X/X\times_Y Z}[-n] \overset{\mathbf{L}}{\otimes} (\mathbf{L}\Gamma_{gf}^* \circ p_2^\flat h^*)
\end{array}
$$

(note that $\mathbf{L}g^* \circ h^*$ and $\mathbf{L}\Gamma_{gf}^* \circ p_1^*$ are the identity functors, since $hg = 1$, $p_1\Gamma_{gf} = 1$).

Pick a bounded below complex of quasi-coherent *injectives* $\mathscr{G}^\bullet \in \mathbf{D}_{qc}^+(Y)$, and let $I^{\bullet\bullet}$ be the canonical truncation in rows $\leq n$ of a Cartan-Eilenberg resolution $\overline{I}^{\bullet\bullet}$ of $h^*\mathscr{G}^\bullet$ with each $\overline{I}^{p,q}$ quasi-coherent. Since $p_2^\flat h^* \simeq p_1^* f^\flat$ has vanishing higher cohomology when evaluated on an injective $\mathcal{O}_Y$-module, the $h^*\mathscr{G}^r$'s are $p_2^\flat$-acyclic. Also, $\mathcal{H}om_Z(p_{2*}\mathcal{O}_{X\times_Y Z}, \cdot)$ takes injectives to injectives, as this can be checked locally (by the remark following Lemma 2.1.5) and we can locally factor the finite map p_2 into a closed immersion followed by a finite flat map (for each of which the analogous assertion is easy to check). Thus,

$$
\mathscr{I}^{\bullet\bullet} \overset{\mathrm{def}}{=} \mathcal{H}om_Z(p_{2*}\mathcal{O}_{X\times_Y Z}, I^{\bullet\bullet})
$$

is the canonical truncation in rows $\leq n$ of a double complex of (quasi-coherent) injectives

$$\overline{\mathscr{I}}^{\bullet\bullet} \stackrel{\text{def}}{=} \mathscr{H}om_Z(p_{2*}\mathcal{O}_{X\times_Y Z}, \overline{I}^{\bullet\bullet})$$

with exact columns and a canonical 'kernel augmentation' from

$$G^{\bullet} \stackrel{\text{def}}{=} p_1^*\mathscr{H}om_Y(f_*\mathcal{O}_X, \mathscr{G}^{\bullet}) \simeq \mathscr{H}om_Z(p_{2*}\mathcal{O}_{X\times_Y Z}, h^*\mathscr{G}^{\bullet}) \simeq p_2^{\flat}(h^*\mathscr{G}^{\bullet})$$

along the bottom row.

Note also that the terms in $G^{\bullet}$ and $\mathscr{I}^{\bullet\bullet}$ are $\Gamma^{\flat}_{gf}$-acyclics. This allows us to explicate the 'evaluation' of (2.7.19) on $\mathscr{G}^{\bullet} \in \mathbf{D}^+_{\mathrm{qc}}(Y)$ in terms of $I^{\bullet\bullet}$, since $G^{\bullet} = p_1^*(\cdot)$ and $h^*\mathscr{G}^{\bullet}$ are Γ^*_{gf}-acyclic and g^*-acyclic respectively (as $p_1\Gamma_{gf} = 1$, $hg = 1$ with p_1 and h flat). Indeed, by the comments following (2.5.6), it follows that the description (2.5.5) of $\eta_{\Gamma_{gf}}(G^{\bullet})$ can be given with the truncation $\mathscr{I}^{\bullet\bullet}$ of $\overline{\mathscr{I}}^{\bullet\bullet}$, even though $\overline{\mathscr{I}}^{\bullet\bullet}$ is generally not a Cartan-Eilenberg resolution of $G^{\bullet}$. But $\mathscr{G}^{\bullet}$ is a complex of injectives, so

$$\mathscr{H}om_Z(\mathcal{O}_Y, I^{\bullet-n,n})/\mathrm{im}(\mathscr{H}om_Z(\mathcal{O}_Y, I^{\bullet-n,n-1})) \longrightarrow \mathscr{E}xt^n_Z(\mathcal{O}_Y, h^*\mathscr{G}^{\bullet})[-n]$$
$$\downarrow \simeq$$
$$\omega_{Y/Z}[-n] \otimes \mathscr{G}^{\bullet}$$

is a diagram of complexes of $f^{\flat}$-acyclics. Thus, if we also use $I^{\bullet\bullet}$ in (2.5.5) to describe $\eta_g(h^*\mathscr{G}^{\bullet})$, then keeping track of the 'Tot' part of (2.5.5) reduces the commutativity of (2.7.19) to the commutativity of the following diagram, which we will prove even commutes on the level of complexes:

$$\begin{array}{ccc}
\mathscr{H}om_Y(f_*\mathcal{O}_X, \omega_{Y/Z}[-n] \otimes \mathscr{G}^{\bullet}) & \xrightarrow{\simeq} & \omega_{X/X\times_Y Z}[-n] \otimes \mathscr{H}om_Y(f_*\mathcal{O}_X, \mathscr{G}^{\bullet}) \\
\big\uparrow {\scriptstyle \eta_g} & & \big\| \\
\mathscr{H}om_Y(f_*\mathcal{O}_X, \mathscr{E}xt^n_Z(\mathcal{O}_Y, h^*\mathscr{G}^{\bullet})[-n]) & & \omega_{X/X\times_Y Z}[-n] \otimes \Gamma^*_{gf}G^{\bullet} \\
\big\| & & \big\uparrow {\scriptstyle \simeq \, \eta_{\Gamma_{gf}}} \\
\mathrm{H}^n_v(\mathscr{H}om_Y(f_*\mathcal{O}_X, \mathscr{H}om_Z(\mathcal{O}_Y, I^{\bullet-n,\bullet}))) & & \mathscr{E}xt^n_{X\times_Y Z}(\mathcal{O}_X, G^{\bullet}[-n]) \\
\big\uparrow & & \\
\mathrm{H}^n_v(\mathscr{H}om_{X\times_Y Z}(\mathcal{O}_X, \mathscr{I}^{\bullet-n,\bullet})) & &
\end{array}$$

By studying each degree separately, it is enough to treat the following problem. Let $\mathscr{G}$ be a quasi-coherent $\mathcal{O}_Y$-module, $h^*\mathscr{G} \to I^{\bullet}$ an injective resolution. Let $J^{\bullet}$ be an injective resolution of the $\mathcal{O}_{X\times_Y Z}$-module

$$p_1^*\mathscr{H}om_Y(f_*\mathcal{O}_X, \mathscr{G}) \simeq \mathscr{H}om_Z(p_{2*}\mathcal{O}_{X\times_Y Z}, h^*\mathscr{G}).$$

There is a map (unique up to homotopy)

(2.7.20) $J^{\bullet} \to \mathscr{H}om_Z(p_{2*}\mathcal{O}_{X\times_Y Z}, I^{\bullet})$

over $\mathscr{H}om_X(p_{2*}\mathcal{O}_{X\times_Y Z}, h^*\mathscr{G})$, since the right side of (2.7.20) is an augmented complex of injectives. It is enough to prove the commutativity of the following diagram of sheaves:

(2.7.21)

$$\mathscr{H}om_Y(f_*\mathcal{O}_X, \mathscr{E}xt_Z^n(\mathcal{O}_Y, h^*\mathscr{G})) \xrightarrow[\simeq]{\eta_g} \mathscr{H}om_Y(f_*\mathcal{O}_X, \omega_{Y/Z}\otimes\mathscr{G})$$

$$\mathscr{H}om_Y(f_*\mathcal{O}_X, \mathrm{H}^n(\mathscr{H}om_Z(\mathcal{O}_Y, I^\bullet))) \qquad \omega_{X/X\times_Y Z}\otimes\mathscr{H}om_Y(f_*\mathcal{O}_X, \mathscr{G})$$

$$\mathrm{H}^n(\mathscr{H}om_{X\times_Y Z}(\Gamma_{gf_*}\mathcal{O}_X, J^\bullet)) =\!\!=\!\!=\!\!= \mathscr{E}xt_{X\times_Y Z}^n(\mathcal{O}_X, \mathscr{H}om_Z(p_{2*}\mathcal{O}_{X\times_Y Z}, h^*\mathscr{G}))$$

(with maps labelled $\simeq$ on the right, $\eta\mathsf{r}_{gf}$)

This commutativity is clearly independent of the choices of $I^\bullet$, $J^\bullet$, and (2.7.20).

We will reduce the commutativity of (2.7.21) to a problem about modules and Koszul resolutions. Working locally on Y, we may factorize the finite map $f : X \to Y$ into a closed immersion followed by a finite locally free map, and it is easy to see that when $f = f_1 \circ f_2$ for finite f_1, f_2, it is enough to prove (2.7.21) in general for the triples of maps (h, g, f_1) and (h', g', f_2) where h', g' are the base changes of h, g by f_1. Thus, we may assume that f is either a closed immersion or finite locally free. We will explain the case of finite locally free f; the case where f is a closed immersion is handled by a similar method, except the use of (2.7.22) below is replaced by the use of the unique δ-functorial isomorphism

$$\mathscr{E}xt_{X\times_Y Z}^\bullet(\mathcal{O}_X, \cdot) \simeq \mathscr{E}xt_Z^\bullet(g_*\mathcal{O}_Y, p_{2*}(\cdot))$$

(over $\mathcal{O}_Y \twoheadrightarrow \mathcal{O}_X$) which is the canonical map in degree 0.

Now consider finite locally free f. In this case we may take (2.7.20) to be an equality and the lower left map in (2.7.21) is an isomorphism. There is a unique δ-functorial isomorphism

(2.7.22) $\quad \mathscr{E}xt_{X\times_Y Z}^\bullet(\mathcal{O}_X, \mathscr{H}om_Z(p_{2*}\mathcal{O}_{X\times_Y Z}, \cdot)) \simeq \mathscr{H}om_Y(f_*\mathcal{O}_X, \mathscr{E}xt_Z^\bullet(\mathcal{O}_Y, \cdot))$

on quasi-coherent $\mathcal{O}_Z$-modules which is the expected map in degree 0. The obvious explicit description of this map in terms of injective resolutions makes it clear that with $\bullet = n$, the evaluation of (2.7.22) on $h^*\mathscr{G}$ fits into (2.7.21) to make the bottom part commute. Thus, our task is to show that for quasi-coherent $\mathscr{G}$ on Y, the diagram

(2.7.23)

$$\mathscr{E}xt_{X\times_Y Z}^n(\mathcal{O}_X, \mathscr{H}om_Z(p_{2*}\mathcal{O}_{X\times_Y Z}, h^*\mathscr{G})) \xrightarrow{\simeq} \mathscr{H}om_Y(f_*\mathcal{O}_X, \mathscr{E}xt_Z^n(\mathcal{O}_Y, h^*\mathscr{G}))$$

$$\eta_g \downarrow \simeq \qquad\qquad\qquad \simeq \downarrow \eta\mathsf{r}_{gf}$$

$$\omega_{X/X\times_Y Z}\otimes\mathscr{H}om_Y(f_*\mathcal{O}_X, \mathscr{G}) =\!\!=\!\!=\!\!= \mathscr{H}om_Y(f_*\mathcal{O}_X, \omega_{Y/Z}\otimes\mathscr{G})$$

commutes, where the top horizontal map is the map in degree n induced by (2.7.22).

We may assume X and Y are affine, and we want to reduce to the case where Z is affine. If Y is local, then we may replace Z by any open affine around the image of the closed point of Y under g. Thus, to reduce to the case where all three schemes are affine, it suffices to check that (2.7.23) is compatible with flat base change over Y, such as $\operatorname{Spec} \mathcal{O}_{Y,y} \to Y$ for $y \in Y$. The only non-trivial flat base change compatibility is for the top map in (2.7.23), and it suffices to check this locally over Z and $X \times_Y Z$, and hence in the case of affine Z. We will explicitly calculate (2.7.22) below in the affine setting, and this will make it clear that (2.7.23) respects base change. Thus, we now can suppose that $X = \operatorname{Spec}(B)$, $Y = \operatorname{Spec}(A)$, $Z = \operatorname{Spec}(A')$. Let $B' = B \otimes_A A'$.

The map $A \to A'$ is flat with an lci section $A' \twoheadrightarrow A$ of pure codimension n and $A \to B$ is finite locally free. Let I be the kernel of $A' \twoheadrightarrow A$, so $J = B \otimes_A I$ is the the kernel of $B' \twoheadrightarrow B$. We will construct an explicit B-linear δ-functorial isomorphism

$$(2.7.24) \qquad \operatorname{Ext}^\bullet_{B'}(B, \operatorname{Hom}_{A'}(B', M)) \simeq \operatorname{Hom}_A(B, \operatorname{Ext}^\bullet_{A'}(A, M))$$

(for variable A'-modules M) giving the expected map in degree 0, so this must be the 'same' as (2.7.22) by uniqueness. We will then check for this explicit construction that (2.7.24) respects flat base change and the diagram

(2.7.25)

$$\operatorname{Ext}^n_{B'}(B, \operatorname{Hom}_{A'}(B', A' \otimes_A N)) \xrightarrow{\;\simeq\;} \operatorname{Hom}_A(B, \operatorname{Ext}^n_{A'}(A, A' \otimes_A N))$$

$$\Big\downarrow{\simeq} \qquad\qquad\qquad\qquad\qquad\qquad\qquad\qquad \Big\downarrow{\simeq}$$

$$\textstyle\bigwedge^n (J/J^2)^\vee \otimes_B \operatorname{Hom}_A(B, N) \xrightarrow{\;\simeq\;} \operatorname{Hom}_A(B, \textstyle\bigwedge^n (I/I^2)^\vee \otimes_A N)$$

(analogous to (2.7.23)) commutes for any A-module N. This will complete the proof of the first part of the theorem.

To construct (2.7.24), choose a projective resolution $P^\bullet$ of A as an A'-module, so $B \otimes_A P^\bullet = B' \otimes_{A'} P^\bullet$ is a projective resolution of B as a B'-module. We then have a natural B'-linear composite isomorphism of complexes (functorial in variable A'-modules M)

$$\operatorname{Hom}^\bullet_{B'}(B' \otimes_{A'} P^\bullet, \operatorname{Hom}_{A'}(B', M)) \xrightarrow{\;\simeq\;} \operatorname{Hom}^\bullet_{A'}(P^\bullet, \operatorname{Hom}_A(B, M))$$

$$\Big\downarrow{\simeq}$$

$$\operatorname{Hom}^\bullet_A(B, \operatorname{Hom}_{A'}(P^\bullet, M)) \xleftarrow{\;\simeq\;} \operatorname{Hom}^\bullet_{A'}(P^\bullet \otimes_A B, M)$$

defined without the intervention of signs. Since the δ-functorial structure of Ext in the second variable can be recovered by taking projective resolutions in the first variable (cf. [**Tohoku**, II, 2.3, p.144] and our discussion in §1.3), we obtain the δ-functorial isomorphism (2.7.24) as expected in degree 0 (and by uniqueness, the choice of $P^\bullet$ does not matter). Since we can use any $P^\bullet$, such as a complex of *finite* A'-modules, we readily see that (2.7.24) respects *flat* base change over A. Thus, it suffices to check the commutativity of (2.7.25) when A is local. But for local A, we can take $P^\bullet$ to be a Koszul complex resolution, and then the commutativity of (2.7.25) is clear by the definition of the vertical maps!

Now we prove the second part of Theorem 2.7.2. We first recall the following easy base change lemma, which will be used frequently:

LEMMA 2.7.3. [**RD**, III, 6.4] *Let* $f : X \to Y$ *be a finite map of locally noetherian schemes,* $u : Y' \to Y$ *smooth, and let* $u' : X' = X \times_Y Y' \to X$, $f' : X' \to Y'$ *be the projections. There is a canonical isomorphism* $u'^{\sharp} f^{\flat} \simeq f'^{\flat} u^{\sharp}$ *as* δ-*functors on* $\mathbf{D}^{+}(Y)$, *and this isomorphism respects composites in* u *and in* f.

For what we are about to do, the relevant scheme diagrams are the following, which have various cartesian squares:

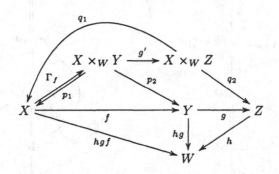

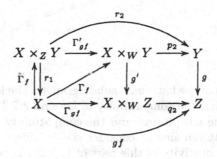

If one writes out the definition of (2.7.17) in terms of (2.5.3), the commutativity of (2.7.17) is equivalent to the commutativity of the "upper long rectangle" (with *vertical* sides and *horizontal* base) in the following diagram of isomorphisms of functors (with m, n, $m+n$ the respective pure relative dimensions of h, g, and hg)

(2.7.26)

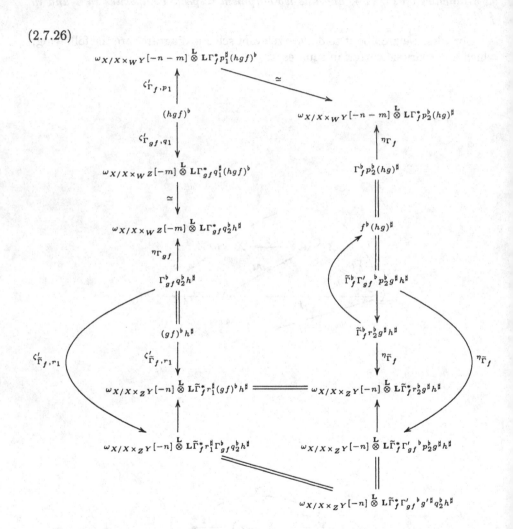

The commutativity of the four small subdiagrams in the lower half of (2.7.26) follows from functoriality, the compatibility of Lemma 2.7.3 below with respect to composites of scheme morphisms, and the compatibility of $(\cdot)^\flat$ with respect to triple composites. It remains to consider the outside edge of (2.7.26). By functoriality, the commutativity of this part of (2.7.26) is equivalent to that of the diagram of isomorphisms

(2.7.27)

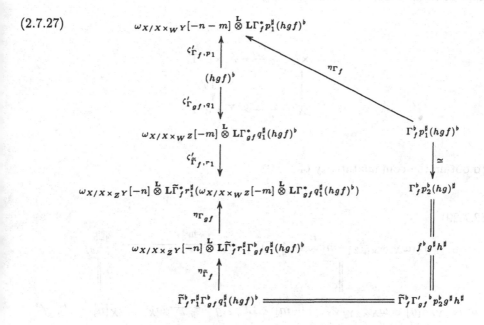

where the bottom row consists of three applications of Lemma 2.7.3.

In order to simplify (2.7.27), we make three observations. First, since Theorem 2.5.1 can be applied to the commutative scheme diagram

$$X \xrightarrow{\widetilde{\Gamma}_f} X \times_Z Y \xrightarrow{\Gamma'_{gf}} X \times_W Y$$
$$X \xrightarrow{\quad\quad\Gamma_f\quad\quad} X \times_W Y$$

(in which all maps are lci's), we obtain the commutativity of

(2.7.28)

$$\Gamma_f^{\flat} \xrightarrow{\eta_{\Gamma_f}} \omega_{X/X\times_W Y}[-n-m] \overset{\mathbf{L}}{\otimes} \mathbf{L}\Gamma_f^{*}$$

$$\widetilde{\Gamma}_f^{\flat}{\Gamma'_{gf}}^{\flat} \quad \omega_{X/X\times_Z Y}[-n] \overset{\mathbf{L}}{\otimes} \mathbf{L}\widetilde{\Gamma}_f^{*}(\omega_{X\times_Z Y/X\times_W Y}[-m] \overset{\mathbf{L}}{\otimes} \mathbf{L}\Gamma'_{gf}^{*}(\cdot))$$

$$\widetilde{\Gamma}_f^{\flat}(\omega_{X\times_Z Y/X\times_W Y}[-m] \overset{\mathbf{L}}{\otimes} \mathbf{L}\Gamma'_{gf}^{*}(\cdot))$$

Second, the flat base change compatibility of (2.5.3) [**RD**, III, 7.4(*b*)] can be applied to the cartesian square (with g' flat)

$$
\begin{array}{ccc}
X \times_Z Y & \xrightarrow{\ \Gamma'_{gf}\ } & X \times_W Y \\
{\scriptstyle r_1}\Big\downarrow & & \Big\downarrow{\scriptstyle g'} \\
X & \xrightarrow{\ \Gamma_{gf}\ } & X \times_W Z
\end{array}
$$

to obtain the commuatativity of

(2.7.29)

$$
r_1^{\sharp}(\omega_{X/X\times_W Z}[-m] \overset{\mathbf{L}}{\otimes} \mathbf{L}\Gamma^*_{gf}(\cdot)) \xleftarrow{\quad\eta_{\Gamma_{gf}}\quad} r_1^{\sharp}\Gamma^{\flat}_{gf}
$$

$$
\Big\| \qquad\qquad\qquad\qquad\qquad\qquad\qquad\qquad\qquad\qquad\quad \Big\|
$$

$$
\omega_{X\times_Z Y/X}[n] \overset{\mathbf{L}}{\otimes} \omega_{X\times_Z Y/X\times_W Y}[-m] \overset{\mathbf{L}}{\otimes} \mathbf{L}\Gamma'^*_{gf} g'^* \xleftarrow{\ \eta_{\Gamma'_{gf}}\ } \omega_{X\times_Z Y/X}[n] \overset{\mathbf{L}}{\otimes} \Gamma'^{\flat}_{gf} g'^*
$$

Finally, since $\Gamma'^*_{gf}\omega_{X\times_W Y/X\times_W Z} \simeq \omega_{X\times_Z Y/X}$ and (2.5.3) is compatible both with tensoring by an invertible sheaf (see (2.5.7)) and with *translations* (using (1.3.6)!), the diagram

(2.7.30)

$$
\omega_{X\times_Z Y/X}[n] \overset{\mathbf{L}}{\otimes} \omega_{X\times_Z Y/X\times_W Y}[-m] \overset{\mathbf{L}}{\otimes} \mathbf{L}\Gamma'^*_{gf} g'^* \xleftarrow{\ \eta_{\Gamma'_{gf}}\ } \omega_{X\times_Z Y/X}[n] \overset{\mathbf{L}}{\otimes} \Gamma'^{\flat}_{gf} g'^*
$$

$$
{\scriptstyle signs!}\Big\| \qquad\qquad\qquad\qquad\qquad\qquad\qquad\qquad\qquad\qquad\quad \Big\downarrow{\scriptstyle \simeq}
$$

$$
\omega_{X\times_Z Y/X\times_W Y}[-m] \overset{\mathbf{L}}{\otimes} \mathbf{L}\Gamma'^*_{gf} g'^{\sharp} \xleftarrow{\quad\eta_{\Gamma'_{gf}}\quad} \Gamma'^{\flat}_{gf} g'^{\sharp}
$$

commutes, where the top row is defined using the isomorphism

(2.7.31)

$$
\omega_{X\times_Z Y/X}[n] \otimes \omega_{X\times_Z Y/X\times_W Y}[-m] \xrightarrow[\simeq]{(-1)^{nm}} \omega_{X\times_Z Y/X\times_W Y}[-m] \otimes \omega_{X\times_Z Y/X}[n].
$$

Putting together (2.7.28), (2.7.29), and (2.7.30) with the help of functoriality, we may replace (2.7.27) with the diagram

(2.7.32)

$$\omega_{X/X\times_W Y}[-n-m] \overset{\mathbf{L}}{\otimes} \mathbf{L}\Gamma_f^* p_1^*(hgf)^{\flat}$$

$$\Big\uparrow {\scriptstyle \zeta'_{\Gamma_f, p_1}}$$

$$(hgf)^{\flat} \qquad\qquad\qquad \overset{\eta\Gamma_f}{\Big\nwarrow}$$

$$\Big\downarrow {\scriptstyle \zeta'_{\Gamma_{gf}, q_1}}$$

$$\omega_{X/X\times_W Z}[-m] \overset{\mathbf{L}}{\otimes} \mathbf{L}\Gamma_{gf}^* q_1^{\sharp}(hgf)^{\flat} \qquad\qquad\qquad \Gamma_f^{\flat} p_1^{\sharp}(hgf)^{\flat}$$

$$\Big\downarrow {\scriptstyle \zeta'_{\widetilde{\Gamma}_f, r_1}} \qquad\qquad\qquad\qquad\qquad \Big\Vert$$

$$\omega_{X/X\times_Z Y}[-n] \overset{\mathbf{L}}{\otimes} \mathbf{L}\widetilde{\Gamma}_f^* r_1^{\sharp}(\omega_{X/X\times_W Z}[-m] \overset{\mathbf{L}}{\otimes} \mathbf{L}\Gamma_{gf}^* q_1^{\sharp}(hgf)^{\flat}) \qquad \Gamma_f^{\flat} p_2^{\flat}(hg)^{\sharp}$$

$$\Big\Vert {\scriptstyle \text{sign!}} \qquad\qquad\qquad\qquad\qquad\qquad \Big\Vert$$

$$\omega_{X/X\times_Z Y}[-n] \overset{\mathbf{L}}{\otimes} \mathbf{L}\widetilde{\Gamma}_f^*(\omega_{X\times_Z Y/X\times_W Y}[-m] \overset{\mathbf{L}}{\otimes} \mathbf{L}\Gamma'_{gf}^* g'^{\sharp} q_1^{\sharp}(hgf)^{\flat}) \qquad f^{\flat}(hg)^{\sharp}$$

$$\Big\uparrow {\scriptstyle \zeta'_{\widetilde{\Gamma}_f, \Gamma'_{gf}}} \qquad\qquad\qquad\qquad\qquad \Big\Vert$$

$$\omega_{X/X\times_W Y}[-n-m] \overset{\mathbf{L}}{\otimes} \mathbf{L}\Gamma_f^* g'^{\sharp} q_1^{\sharp}(hgf)^{\flat} \qquad\qquad\qquad \Gamma_f^{\flat} p_2^{\flat} g^{\sharp} h^{\sharp}$$

$$\Big\uparrow {\scriptstyle \eta\Gamma_f} \qquad\qquad\qquad\qquad\qquad\qquad \Big\uparrow {\scriptstyle \simeq}$$

$$\Gamma_f^{\flat} g'^{\sharp} q_1^{\sharp}(hgf)^{\flat} \xrightarrow{\;\;\simeq\;\;} \Gamma_f^{\flat} g'^{\sharp} q_2^{\flat} h^{\sharp}$$

where the map labelled 'qism' in the left column is defined using the top row of (2.7.30).

We now prove the commutativity of (2.7.32), by reducing it to two simpler diagrams. First, we want the commutativity of the outside edge of

(2.7.33)

$$\begin{array}{ccc}
\Gamma_f^{\flat} p_1^{\sharp}(hgf)^{\flat} & \xrightarrow{\;\simeq\;} \Gamma_f^{\flat} p_2^{\flat}(hg)^{\sharp} = & f^{\flat}(hg)^{\sharp} \\[4pt]
\Big\Vert & \Big\Vert & \Big\Vert \\[4pt]
\Gamma_f^{\flat} g'^{\sharp} q_1^{\sharp}(hgf)^{\flat} & \Gamma_f^{\flat} p_2^{\flat} g^{\sharp} h^{\sharp} = & f^{\flat} g^{\sharp} h^{\sharp} \\[4pt]
\Big\downarrow {\scriptstyle \simeq} & \nearrow {\scriptstyle \simeq} & \\[4pt]
\Gamma_f^{\flat} g'^{\sharp} q_2^{\flat} h^{\sharp} & &
\end{array}$$

The right half of (2.7.33) trivially commutes, we so only need to consider the left half. Once this is verified, then by the *functoriality* of $\eta\Gamma_f$, the commutativity of (2.7.32) will follow from that of

(2.7.34)

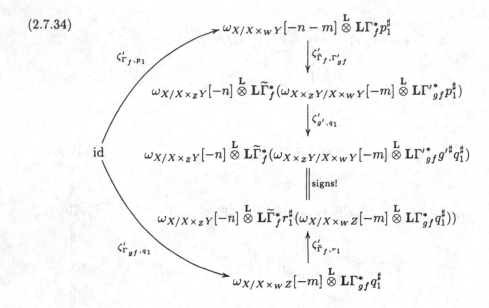

where the isomorphism labelled 'signs!' involves (2.7.31) after writing out the definitions of $g'^{\sharp}$ and $r_1^{\sharp}$. Note that (2.7.33) and (2.7.34) do *not* involve the subtle isomorphisms (2.5.1), (2.5.3).

For the left half of (2.7.33), it is enough to prove commutativity in after 'cancelling' the $\Gamma_f^{\flat}$ throughout. By using the scheme diagram with cartesian squares

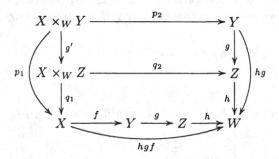

we can appeal to the fact that Lemma 2.7.3 respects composites of scheme morphisms. The proof of the commutativity of (2.7.34) requires a little more work, as follows. It is easy to check that the commutativity of (2.7.34) is equivalent to the assertion that the composite of sheaf isomorphisms

$$\mathcal{O}_X$$

$$\downarrow \zeta'_{\tilde{\Gamma}_f, r_1} \otimes \zeta'_{\Gamma_{gf}, q_1}$$

$$\omega_{X/X\times_Z Y} \otimes \tilde{\Gamma}_f^* \omega_{X\times_Z Y/X} \otimes \omega_{X/X\times_W Z} \otimes \Gamma_{gf}^* \omega_{X\times_W Z/X}$$

$$\| (-1)^{nm}$$

$$\omega_{X/X\times_Z Y} \otimes \omega_{X/X\times_W Z} \otimes \tilde{\Gamma}_f^* \omega_{X\times_Z Y/X} \otimes \Gamma_{gf}^* \omega_{X\times_W Z/X}$$

$$\|$$

$$\omega_{X/X\times_Z Y} \otimes \tilde{\Gamma}_f^* \omega_{X\times_Z Y/X\times_W Y} \otimes \tilde{\Gamma}_f^* \Gamma_{gf}'^* \omega_{X\times_W Y/X\times_W Z} \otimes \Gamma_f^* g'^* \omega_{X\times_W Z/X}$$

$$\simeq \uparrow \zeta'_{\tilde{\Gamma}_f, \Gamma'_{gf}} \otimes \Gamma_f^*(\zeta'_{g', q_1})$$

$$\omega_{X/X\times_W Y} \otimes \Gamma_f^* \omega_{X\times_W Y/X}$$

is equal to ζ'_{Γ_f, p_1}.

More conceptually and more generally, if

$$
\begin{array}{ccc}
X' & \xrightarrow{f'} & Y' \\
{\scriptstyle g'}\downarrow & \searrow^{h} & \downarrow{\scriptstyle g} \\
X & \xrightarrow{f} & Y
\end{array}
$$

is a cartesian commutative diagram of smooth separated scheme maps, where g has pure relative dimension n and f has pure relative dimension m, and

$$
\begin{array}{ccc}
X' & \xleftarrow{j'} & Y' \\
{\scriptstyle i'}\uparrow & \nwarrow^{s} & \uparrow{\scriptstyle i} \\
X & \xleftarrow{j} & Y
\end{array}
$$

is a cartesian commutative diagram of sections, then we want the composite isomorphism of sheaves

$$(2.7.35) \qquad \mathcal{O}_Y \xrightarrow{\zeta'_{i,g} \otimes \zeta'_{j,f}} \omega_{Y/Y'} \otimes i^* \omega_{Y'/Y} \otimes \omega_{Y/X} \otimes j^* \omega_{X/Y}$$

$$\Big\| (-1)^{nm}$$

$$\omega_{Y/Y'} \otimes \omega_{Y/X} \otimes i^* \omega_{Y'/Y} \otimes j^* \omega_{X/Y}$$

$$\Big\|$$

$$\omega_{Y/Y'} \otimes i^* \omega_{Y'/X'} \otimes i^*(j'^* \omega_{X'/X}) \otimes s^* g'^* \omega_{X/Y}$$

$$\simeq \Big\uparrow \zeta'_{i,j'} \otimes s^*(\zeta'_{g',f})$$

$$\omega_{Y/X'} \otimes s^* \omega_{X'/Y}$$

to equal $\zeta'_{s,h}$. This is clearly just a matter of signs and is an easy calculation in local étale coordinates.

Briefly, it goes as follows. Let $x_1, \ldots, x_m$ be local coordinates along $j(Y)$ for $X \to Y$ and let $y_1, \ldots, y_n$ be local coordinates along $i(Y)$ for $Y' \to Y$, and define the sections

$$x^\vee = x_1^\vee \wedge \cdots \wedge x_m^\vee, \quad y^\vee = y_1^\vee \wedge \cdots \wedge y_n^\vee$$

of $\omega_{Y/X}$, $\omega_{Y/Y'}$ respectively and the sections

$$\mathrm{d}x = \mathrm{d}x_m \wedge \cdots \wedge \mathrm{d}x_1, \quad \mathrm{d}y = \mathrm{d}y_n \wedge \cdots \wedge \mathrm{d}y_1$$

of $\omega_{X/Y}$, $\omega_{Y'/Y}$ respectively (note the distinction in the orderings in these definitions). Finally, define $x'^\vee = g'^* x^\vee$, $y'^\vee = f'^* y^\vee$, $\mathrm{d}x' = g'^*(\mathrm{d}x)$, $\mathrm{d}y' = f'^*(\mathrm{d}y)$. The first two steps in (2.7.35) are determined by

$$1 \mapsto y^\vee \otimes i^* \mathrm{d}y \otimes x^\vee \otimes j^* \mathrm{d}x \mapsto (-1)^{nm} y^\vee \otimes x^\vee \otimes i^* \mathrm{d}y \otimes j^* \mathrm{d}x,$$

and the third step maps this element to

$$(-1)^{nm} y^\vee \otimes i^* x'^\vee \otimes i^* j'^* \mathrm{d}y' \otimes s^* \mathrm{d}x'.$$

This final expression is equal to

$$(-1)^{nm} \zeta'_{i,j'}(y^\vee \wedge x'^\vee) \otimes s^* \zeta'_{g',f}(\mathrm{d}y' \wedge \mathrm{d}x').$$

Using the sign to interchange $y^\vee$ and $x'^\vee$ gives us exactly

$$\zeta'_{s,h}((x^\vee \wedge y^\vee) \otimes (\mathrm{d}y' \wedge \mathrm{d}x')),$$

as desired.   ∎

One application of the $\psi_{f,g}$ isomorphisms which should not go unmentioned here is in the definition of a trace map on top differentials. If X and Y are separated smooth Z-schemes with pure relative dimension n, $f : X \to Y$ is a finite Z-map (hence *finite locally free*, so $f_* = \mathbf{R}f_*$), and $g : Y \to Z$ is the structure map, then when Z is locally noetherian we can use (2.7.5) to define an $\mathscr{O}_Y$-linear trace map

$$(2.7.36) \qquad \operatorname{Tr}_f : f_* \omega_{X/Z} \to \omega_{Y/Z}$$

to be the map induced on the $(-n)$th cohomology by

$$(2.7.37) \quad f_* \omega_{X/Z}[n] = \mathbf{R}f_*((gf)^\sharp \mathscr{O}_Z) \xrightarrow{\psi_{f,g}} \mathbf{R}f_* f^\flat g^\sharp \mathscr{O}_Z \xrightarrow{\operatorname{Trf}_f} g^\sharp \mathscr{O}_Z = \omega_{Y/Z}[n]$$

(recall that since f is finite locally free, Trf_f is just the 'evaluate at 1' map $\mathscr{H}om_Y(f_*\mathscr{O}_X, \cdot) \to (\cdot)$).

Due to lack of an adequate reference, we want to make this definition a little more explicit so that it makes sense without noetherian assumptions and so that we can prove that (2.7.36) is compatible with arbitrary base change over Z (this is useful in the theory of residues; cf. proof of (R10) in Appendix A). Without any noetherian assumptions, we will construct an isomorphism of quasi-coherent $\mathscr{O}_X$-modules

$$(2.7.38) \qquad \mathscr{H}om_Y(f_*\mathscr{O}_X, \omega_{Y/Z}) \simeq \omega_{X/Z}$$

which is compatible with arbitrary base change over Z and recovers $\psi_{f,g}^{-1}(\mathscr{O}_Z)$ in the locally noetherian case, so composing 'evaluation at 1' with the inverse of $f_*((2.7.38))$ will return the abstractly defined (2.7.36) in the locally noetherian case. This will then give a general definition of Tr_f on top differentials, compatible with arbitrary base change (for any scheme Z).

The relevant scheme diagram is

$$(2.7.39)$$

where $h = g \circ f$, $f' = f \times 1$ and Γ is the composite of the lci graph morphism Γ_f of f and the 'flip' isomorphism $X \times_Z Y \simeq Y \times_Z X$. We define the composite isomorphism (2.7.38) of quasi-coherent $\mathcal{O}_X$-modules as follows:

(2.7.40)
$$\mathscr{H}om_Y(f_*\mathcal{O}_X, \omega_{Y/Z})$$

$$\Big\downarrow \zeta'_{\Delta,p_1}$$

$$\omega_{X/X\times_Z X} \otimes \Delta^*(\omega_{X\times_Z X/X} \otimes p_1{}^* \mathscr{H}om_Y(f_*\mathcal{O}_X, \omega_{Y/Z}))$$

$$\Big\uparrow \eta_\Delta$$

$$\Delta^* \mathscr{E}xt^n_{X\times_Z X}(\Delta_*\mathcal{O}_X, \omega_{X\times_Z X/X} \otimes p_1{}^* \mathscr{H}om_Y(f_*\mathcal{O}_X, \omega_{Y/Z}))$$

$$\Big\| $$

$$\Delta^* \mathscr{E}xt^n_{X\times_Z X}(\Delta_*\mathcal{O}_X, \mathscr{H}om_{Y\times_Z X}(f'_*\mathcal{O}_{X\times_Z X}, \omega_{Y\times_Z X/Y} \otimes \pi^*\omega_{Y/Z}))$$

$$\Big\uparrow \zeta'_{\pi,g}$$

$$\Delta^* \mathscr{E}xt^n_{X\times_Z X}(\Delta_*\mathcal{O}_X, \mathscr{H}om_{Y\times_Z X}(f'_*\mathcal{O}_{X\times_Z X}, \omega_{Y\times_Z X/Z}))$$

$$\Big\| \alpha$$

$$\Gamma^* \mathscr{E}xt^n_{Y\times_Z X}(\Gamma_*\mathcal{O}_X, \omega_{Y\times_Z X/Z})$$

$$\Big\downarrow \eta_\Gamma$$

$$\omega_{X/Y\times_Z X} \otimes \Gamma^*\omega_{Y\times_Z X/Z}$$

$$\Big\uparrow \zeta'_{\Gamma, g\circ\pi}$$

$$\omega_{X/Z},$$

where the isomorphism α arises from universal δ-functor considerations. Note that this composite map uses the fundamental local isomorphisms η_Δ, η_Γ for the lci maps Δ, Γ with the *same* pure codimension, namely n. Thus, the implicit sign of $(-1)^{n(n+1)/2}$ in the definition of (2.5.1) (via (1.3.28)) 'cancels out' in the composite (2.7.40).

It is simply a matter of unwinding the definition of (2.7.5) and using functoriality and (2.7.3) to see that in the *locally noetherian* case, the inverse of (2.7.40) coincides with $\psi_{f,g}(\mathcal{O}_Z)$, so the inverse of $f_*((2.7.40))$ recovers (2.7.36) via 'evaluation at 1'. To verify this, note that the use of $Y\times_Z X$ in (2.7.40) does not cause problems with respect to the use of $X\times_Z Y$ in the definition of (2.7.5), because (as we noted earlier) the definition of (2.7.5) can be given in terms of $Y\times_Z X$ as well. The other input that one needs is that for the second projection $p_2 : Y\times_Z X \to X$, we have

$$(1 \otimes \Gamma^*(\zeta'_{p_2,h})) \circ \zeta'_{\Gamma, g\circ\pi} = \zeta'_{\Gamma,p_2} \otimes 1.$$

This follows from (2.2.4), since $\zeta'_{p_2\Gamma,h} = \zeta'_{1,h}$ is the identity.

Since (2.5.1) is compatible with any base change (preserving the 'lci of pure codimension n' property) when evaluated on a flat quasi-coherent sheaf, it is easy to check (by a consideration of stalks) that the composite (2.7.40) is compatible with *arbitrary* base change over Z. Thus, without any noetherian hypotheses, (2.7.36) is compatible with arbitrary base change over Z.

In order to 'explicitly compute' (2.7.36), we only consider the case where $Z = \mathrm{Spec}(A)$ with A a complete local noetherian ring, and we choose a section $s \in X(Z)$. The formal completion of X (resp. Y) along s (resp. $f(s)$) has the form $\mathrm{Spf}(B)$ (resp. $\mathrm{Spf}(C)$), with $B \simeq A[\![U_1, \ldots, U_n]\!]$ having the U-adic topology (resp. $C \simeq A[\![T_1, \ldots, T_n]\!]$ having the T-adic topology). Let $\Phi : C \to B$ be the corresponding natural map, so Φ makes B a finite free C-module (the finiteness rests on the fact that A is complete, not just henselian). The map (2.7.36) induces a C-linear map on formal completions

$$\widehat{\mathrm{Tr}}_f : B\, dU \to C\, dT,$$

where $dU = dU_1 \wedge \cdots \wedge dU_n$, $dT = dT_1 \wedge \cdots \wedge dT_n$. The element

$$d\Phi \overset{\mathrm{def}}{=} \det \left(\frac{\partial \Phi(T_i)}{\partial U_j} \right) dU \in B\, dU$$

makes sense, and the calculations of Tate in [**MR**, Appendix] imply (via (2.7.40)) that

(2.7.41) $$\widehat{\mathrm{Tr}}_f(b\, d\Phi) = \mathrm{Tr}_{B/C}(b)\, dT.$$

We will use this in the proof of (R7) in Appendix A, with A an artin local ring and $b = 1$.

2.8. Gluing $(\cdot)^{\sharp}$ and $(\cdot)^{\flat}$

Now that we have established the fundamental compatibility properties of the isomorphism $\psi_{f,g}$, we can uniquely 'glue' the δ-functors $(\cdot)^{\sharp}$ and $(\cdot)^{\flat}$, as is made precise in [**RD**, III, 8.7]. Namely, we fix a base scheme S and consider S-morphisms $f : X \to Y$ between locally noetherian S-schemes such that f is S-*embeddable* in the sense of [**RD**, III, p.189]; that is, f is a composite

(2.8.1) $$X \overset{i}{\longrightarrow} P \times_S Y \overset{\pi}{\longrightarrow} Y$$

where i is a finite S-map, with P smooth separated over S with pure relative dimension, and π is the projection (e.g., $S = \mathrm{Spec}(\mathbf{Z})$ and $P = \mathbf{P}_{\mathbf{Z}}^n$ for variable n); in particular, f is separated. We choose such a factorization of f and define $f_S^! \overset{\mathrm{def}}{=} i^{\flat}\pi^{\sharp} : \mathbf{D}_{\mathrm{qc}}^+(Y) \to \mathbf{D}_{\mathrm{qc}}^+(X)$. As is noted in the proof of [**RD**, III, 8.7], this definition is 'independent' of the factorization of f in the sense that the ψ isomorphisms allow one to easily construct an isomorphism of δ-functors

(2.8.2) $$i^{\flat}\pi^{\sharp} \simeq i'^{\flat}\pi'^{\sharp}$$

for any second such factorization $\pi' \circ i'$ of f; moreover, (2.8.2) is 'transitive' with respect to any third such factorization of f (this step requires using the second part of Theorem 2.7.2). For finite étale f, the canonical identifications $f_S^! \simeq f^{\flat}$ and $f_S^! \simeq f^{\sharp}$ are compatible with the isomorphism (2.7.9).

By using Theorem 2.7.2 to avoid well-definedness problems, an argument similar to the proof of [**RD**, VI, 3.1] yields a satisfactory theory [**RD**, III, 8.7] for the above δ-*functorial* notion of $(\cdot)_S^!$, including δ-functorial isomorphisms $(gf)_S^! \simeq f_S^! g_S^!$ which make the obvious analogue of (2.7.15) commute. This generalizes $(\cdot)^\flat$, $(\cdot)^\sharp$, and the relations between $(\cdot)^\flat$ and $(\cdot)^\sharp$. A corollary of [**RD**, III, 8.7] is the following result, which will only be needed in Appendix A:

THEOREM 2.8.1. [**RD**, III, 8.7] *For any scheme* S, *consider the category of locally noetherian* S-*schemes with* S-*embeddable morphisms. For any map* f *in this category, define* $f^! = f^\sharp$ *if* f *is separated smooth with pure relative dimension and define* $f^! = f^\flat$ *if* f *is finite (and use (2.7.9) to remove ambiguity in case* f *is finite étale). If* $f : X \to Y$ *and* $g : Y \to Z$ *are two* S-*embeddable maps such that each of* f, g, *and* gf *is either finite or separated smooth with pure relative dimension, then the diagram (2.7.15), using the various* ψ *isomorphisms, is commutative.*

Although the above notion of $(\cdot)_S^!$ in the case $S = \mathrm{Spec}(\mathbf{Z})$ includes any morphism $X \to Y$ of locally noetherian schemes which factors through a closed immersion into some $\mathbf{P}_Y^n$, a completely different method is needed to deal with proper maps that might not be projective (and the reader may wish to contemplate that the *proof* of canonical relative projectiveness of abstract stable proper curves of genus $g > 1$ requires the general theory of the relative dualizing sheaf, to be discussed in §3ff). However, the above notion of $(\cdot)_S^!$ is still an important preliminary step in the study of subsequent generalizations, since some proofs in the duality theory for proper maps in [**RD**] proceed by using Chow's Lemma to reduce to the projective case, where we have available a theory of $(\cdot)_S^!$ with various properties deduced from the finite and smooth cases. For example, as is noted in [**RD**, III, 8.8], for any S-embeddable $f : X \to Y$, if we factorize f as in (2.8.1) then we get functorial isomorphisms

$$(2.8.3) \qquad f_S^!(\mathscr{F}^\bullet) \overset{\mathbf{L}}{\otimes} \mathbf{L}f^*(\mathscr{G}^\bullet) \simeq f_S^!(\mathscr{F}^\bullet \overset{\mathbf{L}}{\otimes} \mathscr{G}^\bullet)$$

and

$$(2.8.4) \qquad \mathbf{R}\mathscr{H}om_X^\bullet(\mathbf{L}f^*\mathscr{F}^\bullet, f_S^!\mathscr{G}^\bullet) \simeq f_S^!(\mathbf{R}\mathscr{H}om_Y^\bullet(\mathscr{F}^\bullet, \mathscr{G}^\bullet)),$$

with $\mathscr{F}^\bullet \in \mathbf{D}_{\mathrm{qc}}^+(Y)$, $\mathscr{G}^\bullet \in \mathbf{D}_{\mathrm{qc}}^b(Y)_{\mathrm{fTd}}$ and $\mathscr{F}^\bullet \in \mathbf{D}_{\mathrm{c}}^-(Y)$, $\mathscr{G}^\bullet \in \mathbf{D}_{\mathrm{qc}}^+(Y)$ respectively, by using the analogues [**RD**, III, 2.4, 6.9] in the smooth and finite cases. By construction, (2.8.3) and (2.8.4) are compatible with translations in all variables.

It is important and non-trivial to check if the isomorphisms (2.8.3) and (2.8.4) are *independent* of the choice of factorization of f. Once this is known, the compatibility of these two isomorphisms with respect to general composites (using $(gf)_S^! \simeq f_S^! g_S^!$) follows from the compatibility of Lemma 2.7.3 with respect to the finite and smooth cases of (2.8.3) and (2.8.4); this latter compatibility is an easy consequence of writing out the definitions (upon replacing $\mathscr{F}^\bullet$, $\mathscr{G}^\bullet$ by suitable complexes of acyclics). By a routine argument going back to the definitions of various isomorphisms (such as the ψ's) in terms of (2.5.3), the fact that (2.8.3) and (2.8.4) are independent of the factorization of f can be reduced to the first and second parts of Theorem 2.5.2 respectively. This theorem was not easy to prove! Since we only proved the second part of Theorem 2.5.2 under a separatedness assumption, we have only proven the well-definedness of (2.8.4) when Y is *separated* (which forces X to be separated). This suffices for later applications (such as the proof of [**RD**, VII, 3.4(a)], as explained in the proof of Theorem 3.3.1(1)).

In §3.3, we will review the definition of $f^!$ for possibly non-projective maps $f : X \to Y$ (with some mild hypotheses on Y) by using residual complexes. An associated trace map can then be defined for such proper f by means of the theory of the trace for finite maps (cf. [**RD**, VI, 4.2] and §3.4). In order to prove anything about this 'abstract' trace in the proper case, one needs to check that in the case of the projection $f : \mathbf{P}_Y^n \to Y$, the 'abstract' trace equals the old projective space trace map (2.3.5) for the functor $f^{\sharp}$. The proof of this identification [**RD**, VII, 3.2] makes essential use of a basic fact [**RD**, III, 10.1] which relates the theories of trace for finite maps and projective space: for locally noetherian Y, $f : \mathbf{P}_Y^n \to Y$ the projection, and s a section of f, the composite map of functors on $\mathbf{D}_{\mathrm{qc}}^+(Y)$

$$(2.8.5) \qquad \mathrm{id} \simeq \mathbf{R}f_* \mathbf{R}s_* \xrightarrow{\psi_{s,\cdot}} \mathbf{R}f_* \mathbf{R}s_* s^{\flat} f^{\sharp} \xrightarrow{\mathrm{Trf}} \mathbf{R}f_* f^{\sharp} \xrightarrow{\mathrm{Trp}} \mathrm{id}$$

is the *identity*.

As we have remarked above Lemma 2.3.1, the proof of this claim in [**RD**, III, 10.1] is incomplete. Nevertheless, since all maps in (2.8.5) are *translation-compatible* and we have *already* proven via Lemma 2.3.1 that the definition of (2.3.1) is independent of the choice of projective coordinates, it is not too difficult to correct the argument in [**RD**, III, 10.1] to get to step 4 of that proof, which is the special case when (2.8.5) is evaluated on $\mathcal{O}_Y[0]$ and s is the section $[1, 0, \ldots, 0]$. Observe that the explication (2.7.3) of $\psi_{s,f}(\mathcal{O}_Y)$ and the explication (2.3.8) of $\mathrm{Trp}(\mathcal{O}_Y)$ *both* require computing (2.8.5) with the quasi-isomorphism $f^{\sharp} \mathcal{O}_Y \to \mathcal{I}^{\bullet + n}$, where $\omega_{X/Y} \to \mathcal{I}^{\bullet}$ is an injective resolution chosen to compute derived functors (such as $\mathcal{E}xt_X^{\bullet}(\mathcal{O}_Y, \cdot)$ and $\mathbf{R}^{\bullet}f_*(\cdot)$) of $\omega_{X/Y}$. Thus, the proof that (2.8.5) is the identity on $\mathcal{O}_Y[0]$ is exactly the following lemma:

LEMMA 2.8.2. *Let Y be a scheme, $f : X = \mathbf{P}_Y^n \to Y$ the projection, and s the section $[1, 0, \ldots, 0]$. The composite map of quasi-coherent $\mathcal{O}_Y$-modules*

(2.8.6)

$$\mathcal{O}_Y$$
$$\downarrow{\scriptstyle \zeta'_{\cdot,f}}$$
$$\omega_{Y/X} \otimes s^* \omega_{X/Y} \xleftarrow[\underset{\sim}{\eta}]{} \mathcal{E}xt_X^n(s_* \mathcal{O}_Y, \omega_{X/Y}) \xrightarrow{\beta} \mathbf{R}^n f_*(\omega_{X/Y}) \xrightarrow{\simeq} \mathcal{O}_Y$$

is the identity, where the last map is (2.3.1) and β is induced by the unique δ-functorial map of $\mathcal{O}_Y$-modules $\mathcal{E}xt_X^{\bullet}(s_* \mathcal{O}_Y, \cdot) \to \mathbf{R}^{\bullet} f_*(\cdot)$ which is the canonical map in degree 0.

It is at this step that we need to have made the correct choice of generator (2.3.3) of $\mathrm{H}^n(\mathbf{P}_{\mathbf{Z}}^n, \omega_{\mathbf{Z}})$, or more generally of $\mathrm{H}^n(\mathbf{P}_A^n, \omega_A)$, to define the isomorphism (2.3.1) used at the end of (2.8.6). Our Koszul complex sign conventions and our definition of the ζ'''s and η's will play a significant role in the calculation of (2.8.6) and thereby force our choice of sign in the definition of (2.3.3).

PROOF. Let $T_0, \ldots, T_n$ be the standard projective coordinates on $\mathbf{P}_Y^n$, $t_j = T_j/T_0$ for $1 \le j \le n$, and let U_j be the non-vanishing locus of T_j for $0 \le j \le n$. Clearly U_0 contains the section s and the isomorphism

$$\zeta'_{s,f} : \mathcal{O}_Y \simeq \omega_{Y/X} \otimes s^*\omega_{X/Y}$$

is determined by

$$1 \mapsto (t_1^\vee \wedge \cdots \wedge t_n^\vee) \otimes s^*(dt_n \wedge \cdots \wedge dt_1).$$

Without loss of generality, Y is affine and we can compute on the level of global sections. Using the explicit definition of (2.5.1) in terms of Koszul complexes (with an intervention of the sign $(-1)^{n(n+1)/2}$ as in the definition of (1.3.28)), $\eta^{-1}(\zeta'_{s,f}(1))$ is represented by the Koszul n-cocycle in $\mathrm{Hom}^n_{\mathcal{O}_Y(U_0)}(K_\bullet(\mathbf{t}), \omega(U_0))$ which is determined by

$$(2.8.7) \quad e_1 \wedge \cdots \wedge e_n \mapsto (-1)^{n(n+1)/2} dt_n \wedge \cdots \wedge dt_1 = (-1)^n dt_1 \wedge \cdots \wedge dt_n.$$

For any $\mathcal{O}_X$-module $\mathcal{F}$, define a map of *complexes*

$$(2.8.8) \quad \mathrm{Hom}^\bullet_{\mathcal{O}_Y(U_0)}(K_\bullet(\mathbf{t}), \mathcal{F}(U_0)) \to \check{C}^\bullet(\mathfrak{U}, \mathcal{F})$$

by sending the Koszul p-cochain c to the Čech p-cochain whose $(i_0, \ldots, i_p)$-coordinate (for $i_0 < \cdots < i_p$) is 0 if $i_0 > 0$ and is

$$\frac{\epsilon_p c(e_{i_1} \wedge \cdots \wedge e_{i_p})|_{U_{i_0} \cap \cdots \cap U_{i_p}}}{t_{i_1} \cdots t_{i_p}} \in \mathcal{F}(U_{i_0} \cap \cdots \cap U_{i_p})$$

if $i_0 = 0$ (this means $\epsilon_p c(1)$ if $p = 0$), where

$$\epsilon_{p+1} = (-1)^p \epsilon_p$$

for all $p \ge 0$, so $\epsilon_p = (-1)^{p(p-1)/2} \epsilon_0$. The recursion for the ϵ_p's says exactly that (2.8.8) is a map of complexes and so uses the calculation of the differential in the 'dual' Koszul complexes $\mathrm{Hom}^\bullet_A(K_\bullet(\mathbf{f}), M)$ (which amounts to replacing $(-1)^{j+1}$ with $(-1)^{j+p}$ in (1.3.26)). Since the image of s is *disjoint* from U_j for $j \ne 0$, the map (2.8.8) in degree 0 respects augmentation from the map

$$\mathrm{Hom}_X(s_*\mathcal{O}_Y, \mathcal{F}) \to \mathrm{H}^0(X, \mathcal{F})$$

precisely when $\epsilon_0 = 1$. Thus, we now define $\epsilon_p = (-1)^{p(p-1)/2}$ for all $p \ge 0$.

We claim that (2.8.8) with $\mathcal{F} = \omega_{X/Y}$ computes β over an affine base Y. Since the product of ϵ_n and the sign $(-1)^n$ in (2.8.7) is $(-1)^{n(n+1)/2}$, (2.8.8) sends the Koszul n-cocycle (2.8.7) to the Čech n-cocycle whose value on $e_1 \wedge \cdots \wedge e_n$ is (2.3.3). By the *definition* of (2.3.1), it follows that (2.8.6) is the identity map as long as (2.8.8) really computes β.

More generally, we prove that (2.8.8) computes the *unique* δ-functorial map $\mathrm{Ext}^\bullet_X(s_*\mathcal{O}_Y, \cdot) \to \mathrm{H}^n(X, \cdot)$ when evaluated on any *quasi-coherent* $\mathcal{O}_X$-module (where the δ-functorial map is required to be the canonical map in degree 0; this corresponds to the condition $\epsilon_0 = 1$ above). In the locally noetherian case (which is all we really need), this follows from Lemma 2.1.6 and a universal δ-functor argument. More generally, we need to introduce some double complexes, as follows.

Let $\mathscr{F} \to \mathscr{I}^\bullet$ be a resolution of an $\mathscr{O}_X$-module $\mathscr{F}$ by injective $\mathscr{O}_X$-modules. There is a first quadrant double complex $C^{\bullet\bullet} = \check{C}^\bullet(\mathfrak{U}, \mathscr{I}^\bullet)$ with

$$C^{p,q} = \check{C}^q(\mathfrak{U}, \mathscr{I}^p)$$

and a sign change of $(-1)^p$ along the pth column $\check{C}^\bullet(\mathfrak{U}, \mathscr{I}^p)$ in order to give a double complex rather than a commutative diagram (see the end of §1.2). It is well-known that the augmentation map $\Gamma(X, \mathscr{I}^\bullet) \to \mathrm{Tot}^\oplus(C^{\bullet\bullet})$ is a quasi-isomorphism, and in this way $\mathrm{Tot}^\oplus(C^{\bullet\bullet})$ computes derived functor cohomology as a δ-functor of the $\mathscr{O}_X$-module $\mathscr{F}$. Thus, by (1.3.24), the other augmentation map $\check{C}^\bullet(\mathfrak{U}, \mathscr{F}) \to \mathrm{Tot}^\oplus(C^{\bullet\bullet})$ induces the edge map from Čech cohomology to derived functor cohomology (which is an isomorphism when $\mathscr{F}$ is quasi-coherent).

Similarly, if we let $\mathscr{K}_\bullet(\mathbf{t})$ denote the 'sheafified' Koszul complex on U_0 (concentrated in degrees between $-n$ and 0) and we define the first quadrant double complex $K^{\bullet\bullet}$ by

$$K^{p,q} = \mathrm{Hom}_{\mathscr{O}_Y(U_0)}(K_{-q}(\mathbf{t}), \mathscr{I}^p(U_0))$$

with vertical differentials in the pth column $K^{p,\bullet}$ defined as $(-1)^p$ times the differentials along

$$\mathrm{Hom}_{\mathscr{O}_Y(U_0)}^\bullet(K_\bullet(\mathbf{t}), \mathscr{I}^p(U_0)[0]),$$

then thanks to $(-1)^{p+(q+1)} = (-1)^{(p+q)+1}$ we see that total complex of $K^{\bullet\bullet}$ is

$$\mathrm{Hom}_{U_0}^\bullet(\mathscr{K}_\bullet(\mathbf{t}), \mathscr{I}^\bullet|_{U_0}) = \mathbf{R}\,\mathrm{Hom}_{U_0}^\bullet(s_*\mathscr{O}_Y, \mathscr{F}).$$

Thus, the cohomology of $\mathrm{Tot}^\oplus(K^{\bullet\bullet})$ computes the δ-functor

$$\mathrm{Ext}_{U_0}^\bullet(s_*\mathscr{O}_Y, (\cdot)|_{U_0}) \simeq \mathrm{Ext}_X^\bullet(s_*\mathscr{O}_Y, \cdot).$$

The augmentation map $\mathrm{Hom}_{\mathscr{O}_Y(U_0)}^\bullet(K_\bullet(\mathbf{t}), \mathscr{F}(U_0)) \to \mathrm{Tot}^\oplus(K^{\bullet\bullet})$ is a quasi-isomorphism if $\mathscr{F}$ has vanishing higher cohomology on U_0. This includes any quasi-coherent $\mathscr{F}$, since U_0 is an *affine* scheme, and recovers the usual calculation of $\mathrm{Ext}_X^\bullet(s_*\mathscr{O}_Y, \cdot)$ on quasi-coherent $\mathscr{O}_X$-modules in terms of a Koszul resolution of $s_*\mathscr{O}_Y$ on U_0.

The map (2.8.8) induces a map of double complexes $K^{\bullet\bullet} \to C^{\bullet\bullet}$, and the induced map on the cohomology of the total complexes is a δ-functorial map

$$(2.8.9) \qquad \mathrm{Ext}_X^\bullet(s_*\mathscr{O}_Y, \cdot) \to \mathrm{H}^\bullet(X, \cdot)$$

which is as expected in degree 0. Moreover, in the case of a quasi-coherent $\mathscr{O}_X$-module, the above construction of (2.8.9) is visibly compatible with (2.8.8) via the computation of the left side (resp. right side) of (2.8.9) by means of $\mathrm{Hom}^\bullet(K_\bullet(\mathbf{t}), \cdot)$ (resp. $\check{C}^\bullet(\mathfrak{U}, \cdot)$). Thus, (2.8.8) with $\mathscr{F} = \omega_{X/Y}$ does compute β. ∎

Although [**RD**, VI, VII] constructs a trace map and duality theory for general proper morphisms over certain bases by building up from the case of finite

maps via residual complexes, as we will discuss in Chapter 3, a logically independent interlude [**RD**, III, 10.5, §11] claims to directly construct such a theory
for projective morphisms over a locally noetherian base. We end this section
by briefly explaining why this interlude [**RD**, III, 10.5, §11] seems to involve
some non-trivial problems of well-definedness. This has no affect on anything
in [**RD**, IVff], so it is not a serious concern to us. If $f : X \to Y$ is a map of
locally noetherian schemes which factors as $f = \pi \circ i$ with $i : X \hookrightarrow \mathbf{P}_Y^N$ a closed
immersion over Y and $\pi : \mathbf{P}_Y^N \to Y$ the projection, we want the trace morphism
of functors

$$(2.8.10) \qquad \mathbf{R}f_* f_Z^! \simeq \mathbf{R}\pi_* \mathbf{R}i_* i^b \pi^\sharp \simeq \mathbf{R}\pi_* \pi^\sharp \simeq \mathrm{id}$$

on $\mathbf{D}_{\mathrm{qc}}^+(Y)$ to be independent of the factorization of f, where we have used
the trace for the finite map i and the projective space map π. In order to
show that (2.8.10) is independent of the factorization of f, one easily reduces to
two special cases which we now formulate. Consider the cartesian diagram of
projective space maps

$$
\begin{array}{ccc}
\mathbf{P}_Y^m \times \mathbf{P}_Y^n & \xrightarrow{\ q'\ } & \mathbf{P}_Y^m \\
{\scriptstyle p'}\downarrow & \searrow{\scriptstyle h} & \downarrow{\scriptstyle p} \\
\mathbf{P}_Y^n & \xrightarrow{\ q\ } & Y
\end{array}
$$

Let $i : \mathbf{P}_Y^m \times \mathbf{P}_Y^n \hookrightarrow \mathbf{P}_Y^{(m+1)(n+1)-1}$ be the Segre embedding and let

$$\pi : \mathbf{P}_Y^{(m+1)(n+1)-1} \to Y$$

be the structure map. We need the following two compatibilities:

- Using the trace map (2.3.5) for projective space over a base, and regarding
  $\mathbf{P}_Y^m \times \mathbf{P}_Y^n$ as projective n-space (resp. m-space) over $\mathbf{P}_Y^m$ (resp. $\mathbf{P}_Y^n$), the
  following diagram of functors on $\mathbf{D}_{\mathrm{qc}}^+(Y)$ should commute:

$$(2.8.11)
\begin{array}{ccccc}
\mathbf{R}q_* \mathbf{R}p'_* p'^\sharp q^\sharp & \xleftarrow{\ \simeq\ } & \mathbf{R}h_* h^\sharp & \xrightarrow{\ \simeq\ } & \mathbf{R}p_* \mathbf{R}q'_* q'^\sharp p^\sharp \\
\downarrow & & & & \downarrow \\
\mathbf{R}q_* q^\sharp & \xrightarrow{\qquad} & \mathrm{id} & \xleftarrow{\qquad} & \mathbf{R}p_* p^\sharp
\end{array}
$$

- The common composite $\mathbf{R}h_* h^\sharp \to \mathrm{id}$ in (2.8.11) should be equal to

$$\mathbf{R}h_* h^\sharp \overset{\psi_{i,\pi}}{\simeq} \mathbf{R}\pi_* \mathbf{R}i_* i^b \pi^\sharp \to \mathbf{R}\pi_* \pi^\sharp \to \mathrm{id},$$

using the traces for the finite map i and the projective space map π.

Neither of these compatibilities seems clear (based on the methods in [**RD**])
and there is a serious error in the proof of them in [**RD**, III, 10.5]. The error
in the proof is that unlike properness and smoothness, the property of a map
being a 'projective space' (i.e., of the form $\mathbf{P}_Y^n \to Y$ for some n) is *not* preserved
under formation of products, so one cannot regard the product of two closed
immersions into projective spaces as a third such closed immersion.

CHAPTER 3

Duality Foundations

In this chapter, we discuss Grothendieck's notion of a residual complex. This concept allows one to construct a duality theory in the proper Cohen-Macaulay case without projectiveness assumptions (although some proofs ultimately reduce via Chow's Lemma to the analysis of projective space and finite maps, as treated in Chapter 2). The special role of CM maps are that these are exactly the morphisms for which one can define a relative dualizing *sheaf* (rather than a relative dualizing complex), generalizing the sheaf of top degree relative differential forms in the smooth case. The base change theory for dualizing sheaves is set up at the end of this chapter. This makes it possible to consider the base change compatibility of the trace map for proper CM morphisms, a problem we will address in Chapter 4.

3.1. Dualizing Complexes

We begin with a review of some facts from [**RD**, IV] concerning Cousin complexes. Let X be a locally noetherian scheme and let $Z^\bullet = \{Z^p\}$ be a filtration of X by subsets Z^p such that

- each Z^p is stable under specialization,
- $Z^p \subseteq Z^{p-1}$ for all p,
- $X = Z^p$ for some sufficiently negative p and $\cap Z^p = \emptyset$, so X is disjoint union of $Z^p - Z^{p+1}$ over $p \in \mathbf{Z}$,
- each $x \in Z^p - Z^{p+1}$ is not a specialization of any other point of Z^p.

If $Z^\bullet$ is such a filtration, we denote by $Z^\bullet[n]$ the filtration with $Z^\bullet[n]^p = Z^{p+n}$. The standard example of such a filtration is

$$Z^p = \{x \in X \mid \dim \mathscr{O}_{X,x} \geq p\};$$

we call this the *codimension filtration of* X. For any such $Z^\bullet$ as above, the additive category $\mathrm{Coz}(Z^\bullet, X)$ of *Cousin complexes of $\mathscr{O}_X$-modules on X with respect to $Z^\bullet$* is defined in [**RD**, IV, p.241]: these are complexes $\mathscr{F}^\bullet$ of $\mathscr{O}_X$-modules such that for all p,

$$(3.1.1) \qquad \mathscr{F}^p \simeq \bigoplus_{x \in Z^p - Z^{p+1}} i_{x*}(M_x),$$

where $i_x : \mathrm{Spec}(\mathscr{O}_{X,x}) \to X$ is the canonical map and M_x is a quasi-coherent sheaf supported at the closed point (i.e., consisting entirely of $\mathfrak{m}_x$-power torsion);

note that $i_{x*}(M_x)$ is necessarily quasi-coherent (as the separated i_x is quasi-compact, since X is locally noetherian) and is the constant sheaf on $\overline{\{x\}}$ attached to M_x. In particular, the terms of a Cousin complex are *quasi-coherent*. We summarize the condition (3.1.1) by saying that $\mathscr{F}^p$ is *supported in the Z^p/Z^{p+1}-skeleton*. Clearly $\mathscr{F}^\bullet[n] \in \mathrm{Coz}(Z^\bullet[n], X)$ if and only if $\mathscr{F}^\bullet \in \mathrm{Coz}(Z^\bullet, X)$. Note that if $Z^{p_0} = X$, then $\mathscr{F}^p = 0$ for $p < p_0$ for any $\mathscr{F}^\bullet \in \mathrm{Coz}(Z^\bullet, X)$, so Cousin complexes are automatically bounded below.

The most basic example occurs in the classical theory of the residue map on a smooth, connected curve over an algebraically closed field. In some sense, the entire theory of dualizing and residual complexes (and its relevance to duality on schemes) is just a vast generalization of this one example, and this example plays a crucial foundational role in the general theory. We will therefore return to this example (and variations on it) over and over, so here we describe the basic situation. Let k be an algebraically closed field and let $f : X \to \mathrm{Spec}(k)$ be a smooth, connected k-scheme, where X is 1-dimensional. Let ξ denote the unique generic point of X, with local ring $K = \mathcal{O}_{X,\xi}$ equal to the function field of X, and let $X^0 = X - \{\xi\}$ denote the set of closed points of X. For each point $x \in X$, we let $i_x : \mathrm{Spec}(\mathcal{O}_{X,x}) \to X$ denote the canonical map of schemes.

Let
$$\underline{\Omega}^1_{K/k} = i_{\xi*}i_\xi^*(\Omega^1_{X/k}) = i_{\xi*}\Omega^1_{X/k,\xi},$$
a quasi-coherent sheaf which is just the constant sheaf attached to $\Omega^1_{K/k}$. Since the injection
$$\Omega^1_{X/k} \to \underline{\Omega}^1_{K/k}$$
of quasi-coherent sheaves is an isomorphism at the generic point, we have an exact sequence of quasi-coherent sheaves

$$(3.1.2) \qquad 0 \to \Omega^1_{X/k} \to \underline{\Omega}^1_{K/k} \to \bigoplus_{x \in X^0} i_{x*}(\Omega^1_{X/k,\xi}/\Omega^1_{X/k,x}) \to 0$$

(where, for $x \in X^0$, the $\mathcal{O}_{X,x}$-module $\Omega^1_{X/k,\xi}/\Omega^1_{X/k,x}$ is supported at the closed point x of $\mathrm{Spec}(\mathcal{O}_{X,x})$). Since the two terms on the right are 'divisible' modules and X is Dedekind, this is an injective resolution of $\Omega^1_{X/k}$. In particular, we see that the complex $\Omega^1_{X/k}[1]$ is canonically quasi-isomorphic to the complex

$$(3.1.3) \qquad \cdots \to 0 \to i_{\xi*}\Omega^1_{X/k,\xi} \to \bigoplus_{x \in X^0} i_{x*}(\Omega^1_{X/k,\xi}/\Omega^1_{X/k,x}) \to 0 \dots,$$

where the non-zero terms are in respective degrees -1 and 0 and the induced map
$$\Omega^1_{X/k,\xi} \to \Omega^1_{X/k,\xi}/\Omega^1_{X/k,x}$$
on stalks at $x \in X^0$ is the the canonical projection. The complex (3.1.3) is a Cousin complex on X with respect to $Z^\bullet[1]$, where $Z^\bullet$ is the codimension filtration on X.

The complex (3.1.3) with its augmentation from $\Omega^1_{X/k}[1]$ plays a fundamental role in the classical construction of an isomorphism

$$\mathrm{H}^1(X, \Omega^1_{X/k}) \simeq k$$

via *residues* when X is *proper* over k. This isomorphism also determines Serre duality on such k-schemes. Grothendieck's theory vastly generalizes this construction. Before getting into the generalities, we highlight some features of the above example which illustrate the main themes in what follows. There are *two* complexes naturally lurking here: $\Omega^1_{X/k}[1]$ and (3.1.3). These are *canonically* quasi-isomorphic, but are of quite different nature. For example, (3.1.3) is a bounded below complex with *coherent* cohomology (see (3.1.2)) and its terms involve the quasi-coherent injective hulls at *all* $x \in X$, each appearing 'exactly once'. Meanwhile, $\Omega^1_{X/k}[1]$ has coherent cohomology and finite injective dimension (see (3.1.2)), with the natural map

$$\mathcal{O}_X \to \mathbf{R}\mathcal{H}om^{\bullet}_X(\mathbf{R}\mathcal{H}om^{\bullet}_X(\mathcal{O}_X, \Omega^1_{X/k}[1]), \Omega^1_{X/k}[1])$$

an isomorphism in $\mathbf{D}(X)$ (since $\Omega^1_{X/k}$ is invertible).

Of course, we could have carried out this construction with $\Omega^1_{X/k}[1]$ replaced by $\mathcal{L}[m]$ for any invertible sheaf $\mathcal{L}$ on X and any integer m, but the relation

$$\Omega^1_{X/k}[1] = f^{\sharp}(\mathcal{O}_{\mathrm{Spec}(k)})$$

is what makes $\mathcal{L} = \Omega^1_{X/k}$, $m = 1$ fit in well with the *relative* theory later on (since $\mathcal{O}_{\mathrm{Spec}(k)}[0]$ will be a "dualizing complex" on $\mathrm{Spec}(k)$, due to the freeness of vector spaces over a field). The complex (3.1.3) leads to the notion of *residual complex*, while the complex $\Omega^1_{X/k}[1]$ leads to the notion of *dualizing complex*. In the general setting later on, these two notions will be related by a construction like (3.1.2), to be made precise in Lemma 3.2.1.

We return to the generality of a locally noetherian scheme X with a filtration $Z^{\bullet}$ as above. For any $\mathcal{F}^{\bullet} \in \mathbf{D}^+(X)$ and any such $Z^{\bullet}$ as above, there is a naturally associated Cousin complex $E_{Z^{\bullet}}(\mathcal{F}^{\bullet})$ with respect to $Z^{\bullet}$, consisting of the $E_1^{p,0}$-terms of a certain spectral sequence, as in [**RD**, IV, p.241]. Before giving the definition, we need to introduce some terminology. Define $\underline{\Gamma}_{Z^p}$ to be the functor which assigns to any $\mathcal{O}_X$-module $\mathcal{F}$ the subsheaf of sections whose (closed) support has all points in Z^p (it suffices to consider generic points of the support). For example, if $Z^{\bullet}$ is the codimension filtration, then $\underline{\Gamma}_{Z^p}(\mathcal{F})$ consists of all sections of $\mathcal{F}$ supported in codimension at least p. The sheaf $\underline{\Gamma}_{Z^{p+1}}(\mathcal{F})$ is naturally a subsheaf of $\underline{\Gamma}_{Z^p}(\mathcal{F})$ and we denote the quotient sheaf by $\underline{\Gamma}_{Z^p/Z^{p+1}}(\mathcal{F})$. Beware that $\underline{\Gamma}_{Z^p/Z^{p+1}}$ is *not* generally a left-exact functor. Nevertheless, we can still use injective resolutions to form the hyperderived functor sheaves

$$\underline{H}^i_{Z^p/Z^{p+1}}(\mathcal{F}^{\bullet}) \overset{\mathrm{def}}{=} H^i(\mathbf{R}(\underline{\Gamma}_{Z^p/Z^{p+1}})(\mathcal{F}^{\bullet}))$$

for bounded below complexes $\mathcal{F}^{\bullet}$ of $\mathcal{O}_X$ modules.

Although there is a natural injection

$$\underline{\Gamma}_{Z^p/Z^{p+1}}(\mathcal{F}) \hookrightarrow \underline{H}^0_{Z^p/Z^{p+1}}(\mathcal{F})$$

for all $\mathcal{F}$, this is generally only an isomorphism for rather special sheaves $\mathcal{F}$ (e.g., the flasque ones, which are the only ones needed in the proof of [**RD**, p.226]). In general, one has a canonical δ-*functorial* description [**RD**, p.226]

$$(3.1.4) \qquad \underline{\mathrm{H}}^i_{Z^p/Z^{p+1}}(\mathscr{F}^\bullet) \simeq \bigoplus_{x \in Z^p - Z^{p+1}} i_{x*}(\mathrm{H}^i_x(\mathscr{F}^\bullet)),$$

where $i_x : \mathrm{Spec}(\mathscr{O}_{X,x}) \to X$ is the canonical map and $\mathrm{H}^i_x(\mathscr{F}^\bullet)$ denotes the usual $\mathfrak{m}_x$-power torsion ith local cohomology group (i.e., the composite of $\mathscr{F}^\bullet \rightsquigarrow \mathscr{F}^\bullet_x \in \mathbf{D}^+(\mathscr{O}_{X,x})$ and the ith derived functor of the *left exact* "elements supported at the closed point" functor Γ_x on $\mathscr{O}_{X,x}$-modules). The formula (3.1.4) implies that the sheaves $\underline{\mathrm{H}}^i_{Z^p/Z^{p+1}}(\mathscr{F}^\bullet)$ are supported in the Z^p/Z^{p+1}-skeleton. As an example, if $Z^{p+1} = \emptyset$ (so all points in Z^p are closed in X) and if $\mathscr{F}^\bullet = \mathscr{F}[0]$, then $\underline{\Gamma}_{Z^p/Z^{p+1}} = \underline{\Gamma}_{Z^p}$ is left-exact and (3.1.4) on variable $\mathscr{F}[0]$'s is the isomorphism of erasable δ-functors given by the canonical map for $i = 0$. For another example, if we drop the condition on Z^{p+1} but require $\mathscr{F}^\bullet = \mathscr{F}[0]$ for a *flasque* sheaf $\mathscr{F}$, then (3.1.4) is uniquely determined by the canonical isomorphism

$$\underline{\Gamma}_{Z^p/Z^{p+1}}(\mathscr{F})_x \simeq \Gamma_{\{x\}}(\mathscr{F}_x)$$

for $x \in Z^p - Z^{p+1}$.

If $u : X' \to X$ is a (not necessarily locally finite type) *flat* map of locally noetherian schemes and $Z'^\bullet$ is a filtration as above on X' such that $u^{-1}(Z^p) \subseteq Z'^p$ for all p, then there is a canonical map of δ-functors

$$(3.1.5) \qquad u^* \circ \underline{\mathrm{H}}^j_{Z^i/Z^{i+1}} \to \underline{\mathrm{H}}^j_{Z'^i/Z'^{i+1}} \circ u^*$$

on $\mathbf{D}^+(X)$. We insist that u be flat in (3.1.5) only because we do not see how to give a definition using $\mathbf{L}u^*$ more generally; the flat case is all we will need later on. Examples of such u are 'localization' maps $\mathrm{Spec}\,\mathscr{O}_{X,x} \to X$ and (strict) henselization maps $\mathrm{Spec}\,A^{\mathrm{sh}} \to \mathrm{Spec}\,A$ for local noetherian rings A (with $Z^\bullet$ and $Z'^\bullet$ the codimension filtrations up to a common shift).

For any bounded below $\mathscr{F}^\bullet$, the Cousin complex $E_{Z^\bullet}(\mathscr{F}^\bullet)$ is defined as follows. Choose a bounded below injective (or even just flasque) resolution $\mathscr{I}^\bullet$ of $\mathscr{F}^\bullet$, so the complex $\mathscr{I}^\bullet$ admits a decreasing exhaustive filtration by subcomplexes

$$\cdots \supseteq \underline{\Gamma}_{Z^p}(\mathscr{I}^\bullet) \supseteq \underline{\Gamma}_{Z^{p+1}}(\mathscr{I}^\bullet) \supseteq \cdots$$

This filtration is bounded above (since $X = Z^p$ for sufficiently negative p) and is stalkwise bounded below (since $\cap Z^p = \emptyset$ and the Z^p's are stable under specialization). Consider the spectral sequence $E_1^{p,q} \Rightarrow \mathrm{H}^{p+q}(\mathscr{I}^\bullet) = \mathrm{H}^{p+q}(\mathscr{F}^\bullet)$ for computing the cohomology of the filtered complex $\mathscr{I}^\bullet$. We have

$$E_0^{p,q} = \underline{\Gamma}_{Z^p}(\mathscr{I}^{p+q})/\underline{\Gamma}_{Z^{p+1}}(\mathscr{I}^{p+q}) = \underline{\Gamma}_{Z^p/Z^{p+1}}(\mathscr{I}^{p+q}),$$

so $E_0^{p,\bullet}$ is naturally a complex indexed by $p + \bullet$, and then

$$(3.1.6) \qquad E_1^{p,q} = \mathrm{H}^{p+q}(E_0^{p,\bullet}) = \underline{\mathrm{H}}^{p+q}_{Z^p/Z^{p+1}}(\mathscr{F}^\bullet).$$

The spectral sequence $E_1^{p,q} \Rightarrow \mathrm{H}^{p+q}(\mathscr{F}^\bullet)$ is independent of the choice of $\mathscr{I}^\bullet$, is stalkwise convergent (even globally convergent if X has finite Krull dimension and $Z^\bullet$ is the codimension filtration, up to a shift), and is of formation compatible with any flat base change u as considered in (3.1.5). The complex

$E_{Z^\bullet}(\mathscr{F}^\bullet)$ is defined to be the complex of terms $E_1^{p,0} = \underline{\mathrm{H}}_{Z^p/Z^{p+1}}^p(\mathscr{F}^\bullet)$ with horizontal differentials $\mathrm{d}_h^{p,0}$. By construction and (3.1.4), this is a *Cousin complex* with respect to $Z^\bullet$. The formation of $E_{Z^\bullet}(\mathscr{F}^\bullet)$ is local for the Zariski topology on X and is denoted $E(\mathscr{F}^\bullet)$ in [**RD**, p.241], but the dependence on $Z^\bullet$ is quite important. By looking at the definition, it is easy to construct a natural isomorphism of complexes

(3.1.7) $$E_{Z^\bullet}(\mathscr{F}^\bullet)[n] \simeq E_{Z^\bullet[n]}(\mathscr{F}^\bullet[n])$$

which is compatible with composite translations and involves no intervention of signs. More explicitly, if we use an injective resolution $\mathscr{I}^\bullet$ of $\mathscr{F}^\bullet$ to compute $E_{Z^\bullet}(\mathscr{F}^\bullet)$, (3.1.7) corresponds to using the injective resolution $\mathscr{I}^\bullet[n]$ of $\mathscr{F}^\bullet[n]$ to compute $E_{Z^\bullet[n]}(\mathscr{F}^\bullet[n])$; in degree p, (3.1.7) is the identity (defined without the intervention of signs)

$$\mathrm{H}^{p+n}(\underline{\Gamma}_{Z^{p+n}/Z^{p+n+1}}(\mathscr{I}^\bullet)) = \mathrm{H}^p(\underline{\Gamma}_{Z^{p+n}/Z^{p+n+1}}(\mathscr{I}^\bullet[n])).$$

It is likewise easy to formulate compatibility of $E_{Z^\bullet}$ with respect to tensoring by an invertible sheaf.

The functor $E_{Z^\bullet}$ is compatible with respect to suitable flat base change in certain cases. More precisely, let $u : X' \to X$ be a flat map between locally noetherian schemes with finite Krull dimension and assume that u has discrete fibers, so for all closed irreducible sets $Y \subseteq X$ of codimension p, $u^{-1}(Y)$ is of pure codimension p (for us, the most useful example of such maps are maps u which are *residually stable*, a notion to be defined later; the most interesting examples of such u are henselizations and strict henselizations, which are almost never locally of finite type). If $Z^\bullet$ (resp. $Z'^\bullet$) denotes the codimension filtration on X (resp. X'), up to some common shift, then (3.1.5) gives rise to a natural isomorphism of complexes

(3.1.8) $$u^* E_{Z^\bullet}(\mathscr{F}^\bullet) \simeq E_{Z'^\bullet}(u^* \mathscr{F}^\bullet)$$

which is compatible with translations via (3.1.7).

Due to the tremendous importance of the $E_{Z^\bullet}$ construction, as well as the prominent role of the example of smooth curves in the foundations of the general theory [**RD**, VII, §1], we want to make the construction of $E_{Z^\bullet}$ 'explicit' in the case of curves, or more precisely whenever the filtration $Z^\bullet$ on the locally noetherian scheme X satisfies

$$\emptyset = Z^2 \subseteq Z^1 \subseteq Z^0 = X.$$

Let $\mathscr{F}$ be a quasi-coherent sheaf on such an X, and let $\mathscr{I}^\bullet$ be an injective resolution of $\mathscr{F}$ in the usual sense (so we may view $\mathscr{I}^\bullet$ as an injective resolution of $\mathscr{F}[0]$). We have a short exact sequence of complexes

$$0 \to \underline{\Gamma}_{Z^1}(\mathscr{I}^\bullet) \to \underline{\Gamma}_{Z^0}(\mathscr{I}^\bullet) \to \underline{\Gamma}_{Z^0/Z^1}(\mathscr{I}^\bullet) \to 0.$$

The snake lemma then gives a coboundary map

$$(3.1.9) \qquad \underline{\mathrm{H}}^q_{Z^0/Z^1}(\mathscr{F}) =\!=\!= \mathrm{H}^q(\underline{\Gamma}_{Z^0/Z^1}(\mathscr{I}^{\bullet})) \longrightarrow \mathrm{H}^q(\underline{\Gamma}_{Z^1}(\mathscr{I}^{\bullet}))$$

$$\underline{\mathrm{H}}^{q+1}_{Z^1/Z^2}(\mathscr{F}) =\!=\!= \mathrm{H}^q(\underline{\Gamma}_{Z^1/Z^2}(\mathscr{I}^{\bullet}))$$

since $Z^2 = \emptyset$. When $q = 0$, this can be viewed via (3.1.6) as a map $E_1^{0,0} \to E_1^{1,0}$, and as such this is *exactly* the horizontal differential $\mathrm{d}_h^{0,0}$ in the spectral sequence $E_1^{p,q} \Rightarrow \mathrm{H}^{p+q}(\mathscr{F}[0])$. Thus, (3.1.9) gives the two-term complex $E_{Z^{\bullet}}(\mathscr{F}[0])$, supported in degree 0 and 1. If we *negate* this differential and relabel the degrees as -1 and 0, we get exactly $E_{Z^{\bullet}[1]}(\mathscr{F}[1])$, via (3.1.7).

The most interesting example of this setup is when X is a smooth curve over a field (or even over a local artin ring), with $Z^{\bullet}$ the codimension filtration and $\mathscr{F}$ the invertible sheaf of relative 1-forms. Before analyzing this example further, we introduce some convenient general terminology.

A complex $\mathscr{F}^{\bullet} \in \mathbf{D}^b(X)$ is said to be *Cohen-Macaulay with respect to* $Z^{\bullet}$ if there is an isomorphism $\mathscr{F}^{\bullet} \simeq E_{Z^{\bullet}}(\mathscr{F}^{\bullet})$ in $\mathbf{D}(X)$ [**RD**, IV, p.247]. If, in addition, all local hypercohomology $\mathscr{O}_{X,x}$-modules $\mathrm{H}^i_x(\mathscr{F}^{\bullet})$ are injective (e.g., $\mathrm{H}^i_x(\mathscr{F}^{\bullet}) = 0$), we say that $\mathscr{F}^{\bullet}$ is *Gorenstein with respect to* $Z^{\bullet}$ [**RD**, IV, p.248]. If $X = \mathrm{Spec}(A)$ is an affine noetherian scheme, $Z^{\bullet}$ is the codimension filtration of $\mathrm{Spec}(A)$, and $\mathscr{F}^{\bullet} = \widetilde{M}[0]$ with M a finite A-module having support $\mathrm{Supp}(M) = \mathrm{Spec}(A)$ (resp. $M = A$), we recover the usual notion of the A-module M (resp. the ring A) being Cohen-Macaulay (resp. Gorenstein), by [**RD**, IV, p.239-9, Prop 3.1] (resp. [**RD**, IV, p.249]).

Now we consider an example in the setting of smooth curves over a local artin ring. Let A be a local artin ring with *algebraically closed* residue field k and let X be a smooth, connected A-scheme with relative dimension 1. Let X^0 denote the set of closed points, ξ the generic point, $K = \mathscr{O}_{X,\xi}$. Since A is Cohen-Macaulay (but possibly not Gorenstein), the A-smooth X is certainly Cohen-Macaulay (but possibly not Gorenstein). Let $Z^{\bullet}$ be the codimension filtration on X and let $\mathscr{F}$ be a quasi-coherent sheaf on X (e.g., $\mathscr{F} = \Omega^1_{X/A}$). We have seen via (3.1.9) that the complex $E_{Z^{\bullet}}(\mathscr{F}[0])$ is a two-term complex concentrated in degrees 0 and 1, where it has the form

$$\underline{\mathrm{H}}^0_{Z^0/Z^1}(\mathscr{F}) \to \underline{\mathrm{H}}^1_{Z^1/Z^2}(\mathscr{F}).$$

Since

$$(3.1.10) \qquad \underline{\mathrm{H}}^0_{Z^0/Z^1} = i_{\xi *} i_\xi^*, \quad \underline{\mathrm{H}}^1_{Z^1/Z^2}(\cdot) = \bigoplus_{x \in X^0} i_{x *}(\mathrm{H}^1_x(\cdot))$$

by (3.1.4) and its explication for $\mathscr{F}^{\bullet} = \mathscr{F}[0]$, $Z^2 = \emptyset$, we conclude that the complex $E_{Z^{\bullet}}(\mathscr{F}[0])$ has its degree 0, 1 part given by some map of quasi-coherent sheaves

$$(3.1.11) \qquad i_{\xi *}(\mathscr{F}_\xi) \to \bigoplus_{x \in X^0} i_{x *}(\mathrm{H}^1_x(\mathscr{F})).$$

The natural map $\mathscr{F} \to i_{\xi_*}(\mathscr{F}_\xi)$ gives rise to a *natural* map of complexes $\mathscr{F}[0] \to E_{Z^\bullet}(\mathscr{F}[0])$, and if this is an isomorphism in $\mathbf{D}(X)$ then $\mathscr{F}[0]$ is certainly Cohen-Macaulay with respect to $Z^\bullet$ (by definition!). Before we give some interesting examples where this isomorphism condition holds in $\mathbf{D}(X)$, we want to make (3.1.10) and (3.1.11) more explicit.

Fix $x \in X^0$ and let $\mathscr{O}_x = \mathscr{O}_{X,x}$. With respect to the maximal-adic topology of $\mathscr{O}_x$, the set $\{J_i\}$ of ideals in $\mathscr{O}_x$ which are free of rank 1 constitute a base of opens. To see this, first note that any such ideal is trivially open (since $\mathscr{O}_x/J_i$ must have dimension 0). Next, we recall that X is A-smooth of relative dimension 1 and A is artinian with algebraically closed residue field k, so there exists a section in $X(A)$ lifting the k-rational point based at x on the (reduced) closed fiber. The ideal in $\mathscr{O}_x$ cutting out this section is free of rank 1 and a choice of generator t gives rise to an isomorphism of local A-algebras $\widehat{\mathscr{O}}_x \simeq A[\![t]\!]$ [**EGA**, IV$_4$, 17.12.2] Thus, the ideals $t^i \mathscr{O}_x$ are free of rank 1 and are a base of opens (since A is artinian). By applying the Weierstrass Preparation Theorem to $A[\![t]\!]$, we see that an element of $\mathscr{O}_x$ is not a zero divisor if and only if its image in $\widehat{\mathscr{O}}_{X,x} \simeq A[\![t]\!]$ has a unit coefficient somewhere, which says exactly that the element is non-zero in the local ring at x on the (reduced) fiber of $X \to \mathrm{Spec}(A)$. In other words (since X is connected), the regular sections of $\mathscr{O}_X$ are exactly the elements which are units at the generic point of X, so the artin ring $K = \mathscr{O}_{X,\xi}$ is canonically identified with the total ring of fractions of $\mathscr{O}_x$ for *any* $x \in X^0$.

Give the base of opens $\{J_i\}$ a partial ordering by reverse inclusion (i.e., $i' \geq i$ if $J_{i'} \subseteq J_i$). For $x \in X^0$, $\mathrm{H}_x^\bullet$ is the derived functor of "elements of $(\cdot)_x$ supported at x". Thus, by universal δ-functor nonsense there is a unique isomorphism of δ-functors from quasi-coherent $\mathscr{O}_X$-modules to $\mathscr{O}_x$-modules

$$(3.1.12) \qquad \mathrm{H}_x^\bullet \simeq \varinjlim \mathrm{Ext}_{\mathscr{O}_x}^\bullet(\mathscr{O}_x/J_i, (\cdot)_x)$$

giving the canonical isomorphism in degree 0, where the injective limit is taken with respect to the canonical projections $\mathscr{O}_x/J_{i'} \to \mathscr{O}_x/J_i$ for $i' \geq i$. Since each J_i is free of rank 1 over $\mathscr{O}_x$, (2.5.2) gives a canonical isomorphism

$$\mathrm{Ext}_{\mathscr{O}_x}^1(\mathscr{O}_x/J_i, M) \simeq \mathrm{Hom}_{\mathscr{O}_x/J_i}(J_i/J_i^2, \mathscr{O}_x/J_i) \otimes_{\mathscr{O}_x} M \simeq (J_i^{-1}/\mathscr{O}_x) \otimes_{\mathscr{O}_x} M$$

for any $\mathscr{O}_x$-module M (with $J_i^{-1} = \mathrm{Hom}_{\mathscr{O}_x}(J_i, \mathscr{O}_x)$) and the diagram

$$(3.1.13) \qquad \begin{array}{ccc} \mathrm{Ext}_{\mathscr{O}_x}^1(\mathscr{O}_x/J_i, M) & \xrightarrow{\ \simeq\ } & (J_i^{-1}/\mathscr{O}_x) \otimes_{\mathscr{O}_x} M \\ \downarrow & & \downarrow \\ \mathrm{Ext}_{\mathscr{O}_x}^1(\mathscr{O}_x/J_{i'}, M) & \xrightarrow{\ \simeq\ } & (J_{i'}^{-1}/\mathscr{O}_x) \otimes_{\mathscr{O}_x} M \end{array}$$

commutes, where the right column is the canonical map (go back to the *definition* of (2.5.2) to see the commutativity). Passing to the limit, from (3.1.12) we get a canonical isomorphism of $\mathscr{O}_x$-modules

$$(3.1.14) \qquad \mathrm{H}_x^1(\mathscr{F}) \simeq (K/\mathscr{O}_x) \otimes_{\mathscr{O}_x} \mathscr{F}_x$$

for any quasi-coherent $\mathscr{F}$ on X. This isomorphism involves $(-1)^{1(1+1)/2} = -1$ in the definition of (1.3.28), since the rows of (3.1.13) use (2.5.2). Our analysis of this example is concluded by:

LEMMA 3.1.1. *By means of* (3.1.11) *and* (3.1.14), $E_{Z\bullet}(\mathscr{F}[0])$ *is identified with a two-term complex*

$$i_{\xi_*}(\mathscr{F}_\xi) \to \bigoplus_{x \in X^0} i_{x*}((K/\mathscr{O}_x) \otimes_{\mathscr{O}_x} \mathscr{F}_x)$$

supported in degrees $0, 1$. *Localizing at* $x \in X^0$, *the resulting map*

$$(3.1.15) \qquad K \otimes_{\mathscr{O}_x} \mathscr{F}_x = \mathscr{F}_\xi \to (K/\mathscr{O}_x) \otimes_{\mathscr{O}_x} \mathscr{F}_x$$

is the negative of the canonical projection. In particular, this map is surjective with kernel equal to the image of $\mathscr{F}_x \to \mathscr{F}_\xi$, *so if* $\mathscr{F}_x \to \mathscr{F}_\xi$ *is injective for all* $x \in X^0$ *then* $\mathscr{F}[0]$ *is Cohen-Macaulay with respect to* $Z^\bullet$.

The maps $\mathscr{F}_x \to \mathscr{F}_\xi$ are always injective when $\mathscr{F} = f^*(M) \otimes \mathscr{L}$ for an A-module M and invertible $\mathscr{O}_X$-module $\mathscr{L}$ (which is the case that arises later on), since base change to any $\widehat{\mathscr{O}}_x$ transforms $\mathscr{F}_x \to \mathscr{F}_\xi$ into the canonical map

$$M \otimes_A A[\![t]\!] \to M \otimes_A A(\!(t)\!)$$

which is obviously injective. This injectivity condition ensures that $\Gamma(X, \mathscr{F}) = \cap \mathscr{F}_x$, the intersection taken inside of $\mathscr{F}_\xi$. This is why the CM assertion at the end of the lemma is a consequence of the rest of the lemma. Also, the sign in (3.1.15) is exactly the sign $(-1)^{1(1+1)/2} = -1$ in the definition of (2.5.2) for $n = 1$.

PROOF. Since quasi-coherent injective sheaves localize to injective modules, we may localize at $x \in X^0$ and thereby reduce ourselves to the following local problem. Let B be the local ring at a rational point on a smooth connected curve over a local artin ring A, and let K denote the total ring of fractions of B. Give $\mathrm{Spec}(B)$ the codimension filtration $Z^\bullet$. Let ξ and x be the generic and closed points of $\mathrm{Spec}(B)$. By identifying quasi-coherent sheaves on $\mathrm{Spec}(B)$ with B-modules, we want to show that (via (3.1.14)) for any B-module M, the spectral sequence map

$$d_h^{0,0} : K \otimes_B M = \underline{\mathrm{H}}^0_{Z^0/Z^1}(M) \to \underline{\mathrm{H}}^1_{Z^1/Z^2}(M) = \mathrm{H}^1_x(M) = (K/B) \otimes_B M$$

is the negative of the canonical projection.

In order to analyze $d_h^{0,0}$, we recall its definition via the snake lemma as in (3.1.9). Let $\{t_i\}$ be the set of regular elements in B, so

$$K = \varinjlim t_i^{-1} B,$$

where the limit is taken with respect to divisibility in B. Let $M \to I^\bullet$ be an injective resolution by B-modules. The canonical short exact sequence of complexes

$$0 \to \underline{\Gamma}_{Z^1}(I^\bullet) \to \underline{\Gamma}_{Z^0}(I^\bullet) \to \underline{\Gamma}_{Z^0/Z^1}(I^\bullet) \to 0$$

is exactly

$$0 \to \varinjlim \mathrm{Hom}_B(B/t_i, I^\bullet) \to I^\bullet \to K \otimes_B I^\bullet \to 0,$$

so $d_h^{0,0}$ is the coboundary map

$$d : K \otimes_B M = H^0(K \otimes_B I^\bullet) \longrightarrow H^1(\varinjlim \mathrm{Hom}_B(B/t_i, I^\bullet))$$

$$\|$$

$$(K/B) \otimes_B M \xleftarrow[\simeq]{} \varinjlim \mathrm{Ext}_B^1(B/t_i, M)$$

We want to prove that this map is the negative of the canonical projection.

A more precise assertion that holds "at finite level" is that for a fixed regular element $t \in B$, the exact sequence

$$0 \to \mathrm{Hom}_B(B/t, I^\bullet) \to I^\bullet \xrightarrow{t} I^\bullet \to 0$$

induces a coboundary map

(3.1.16) $M = H^0(I^\bullet) \to \mathrm{Ext}_B^1(B/t, M) \simeq M/t$

(where the isomorphism on the right is defined using (1.3.28)), and this coboundary map is the negative of the canonical projection. Once this is proven, passage to the limit over t yields that $d : K \otimes_B M \to (K/B) \otimes_B M$ is the negative of the canonical projection, as desired.

Applying the functor $\mathrm{Hom}_B(\cdot, M)$ to the short exact sequence

$$0 \to B \xrightarrow{t} B \to B/t \to 0,$$

gives a coboundary map $M \to \mathrm{Ext}_B^1(B/t, M)$ via δ-functoriality of Ext_B^1 in the second variable, and this is the coboundary map in (3.1.16). But as is noted near (1.3.14), we can compute this δ-functor structure using projective resolutions in the *first* variable, following the method of [**Tohoku**, II, 2.3, p.144], as follows. We use the commutative diagram of (vertical) projective resolutions:

(3.1.17)

$$
\begin{array}{ccccccccc}
0 & \longrightarrow & 0 & \longrightarrow & B & \xrightarrow{-1} & B & \longrightarrow & 0 \\
& & \downarrow & & \downarrow{\scriptstyle i_2} & & \downarrow{\scriptstyle t} & & \\
0 & \longrightarrow & B & \longrightarrow & B \oplus B & \longrightarrow & B & \longrightarrow & 0 \\
& & \| & & \downarrow{\scriptstyle \pi_1} & & \downarrow & & \\
0 & \longrightarrow & B & \xrightarrow{t} & B & \longrightarrow & B/t & \longrightarrow & 0
\end{array}
$$

where the maps in the middle row are $b \mapsto (tb, b)$ and $(b_1, b_2) \mapsto b_1 - tb_2$ respectively, and $i_2(b) = (0, b)$, $\pi_1(b_1, b_2) = b_1$. Restricting our attention to the top two rows in (3.1.17) and applying $\mathrm{Hom}_B^\bullet(\cdot, M)$ — not $\mathrm{Hom}_B(\cdot, M)$ — to the columns, we get the commutative diagram

(3.1.18)

$$
\begin{array}{ccccccccc}
0 & \longleftarrow & 0 & \longleftarrow & M & \xleftarrow{-1} & M & \longleftarrow & 0 \\
& & & & \uparrow{\scriptstyle -\pi_2} & & \uparrow{\scriptstyle -t} & & \\
0 & \longleftarrow & M & \longleftarrow & M \oplus M & \longleftarrow & M & \longleftarrow & 0
\end{array}
$$

where the maps in the bottom row are $m \mapsto (m, -tm)$ and $(m_1, m_2) \mapsto tm_1 + m_2$. The resulting 'snake lemma' coboundary map $M \to M/t$ is the canonical projection. Recall now that the right column in (3.1.18) calculates an isomorphism $\mathrm{Ext}^1_B(B/t, M) \simeq M/t$ that is off by a factor of $(-1)^{1(1+1)/2} = -1$ from (1.3.28), and so is the negative of the isomorphism (2.5.2) used in the definition of (3.1.14). ∎

Thanks to (3.1.7) and the remark following Lemma 3.1.1, keeping track of a shift [1] leads us to the following important result that will play a critical role in the general theory later on (with the invertible sheaf $\mathscr{L} = \Omega^1_{X/A}$):

COROLLARY 3.1.2. *Let* $f : X \to \mathrm{Spec}(A)$ *be a smooth, connected curve over a local artin ring* A *with algebraically closed residue field* k. *Let* $K = \mathscr{O}_{X,\xi}$ *be the local ring at the generic point and let* $Z^\bullet$ *be the codimension filtration. Then for any* A-*module* M *and any invertible sheaf* $\mathscr{L}$ *on* X, $f^*(M) \otimes \mathscr{L}[1]$ *is Cohen-Macaulay with respect to* $Z^\bullet[1]$.

The complex $E_{Z^\bullet[1]}(f^*(M) \otimes \mathscr{L}[1]) \simeq E_{Z^\bullet}(f^*(M) \otimes \mathscr{L})[1]$ *is a two-term complex concentrated in degrees* -1 *and* 0, *given by*

$$i_{\xi_*}(M \otimes_A \mathscr{L}_\xi) \to \bigoplus_{x \in X^0} i_{x_*}(\mathrm{H}^1_x(M \otimes_A \mathscr{L})),$$

and under the canonical isomorphism $\mathrm{H}^1_x(M \otimes_A \mathscr{L}) \simeq M \otimes_A (\mathscr{L}_\xi / \mathscr{L}_x)$ *from* (2.5.2) *and* (3.1.12), *the resulting stalk maps*

(3.1.19) $$M \otimes_A \mathscr{L}_\xi \to M \otimes_A (\mathscr{L}_\xi / \mathscr{L}_x)$$

at $x \in X^0$ *are the canonical projection maps.*

The above *computations* of local cohomology and Ext^1's use (2.5.2), so the fact that (3.1.19) is the canonical projection (rather than its negative) is due to the sign of $(-1)^{n(n+1)/2}$ in the definition of (1.3.28), with $n = 1$ in the setting of Corollary 3.1.2. The reader should not worry too much about this, because the fundamental local isomorphism η, whose definition uses (1.3.28), will also play another role in the subsequent theory of residues on curves, and the signs in these two applications of (1.3.28) will "cancel out," thereby giving results for the Grothendieck trace map on curves which are 'independent' of the sign in (1.3.28); see the discussion following (B.3.3).

Taking $A = k$ to be an algebraically closed field in Lemma 3.1.1, we see that for a smooth connected curve X over k, with codimension filtration $Z^\bullet$, the residual complex $E_{Z^\bullet}(\Omega^1_{X/k}[0])$ is the 2-term complex given in degrees 0 and 1 by

$$i_{\xi_*}(\Omega^1_{K/k}) \to \bigoplus_{x \in X^0} i_{x_*}(\Omega^1_{K/k} / \Omega^1_{\mathscr{O}_{X,x}/k}),$$

with the *negatives* of the canonical projection maps on the stalks at each $x \in X^0$. Since this is an *injective* resolution of $\Omega^1_{X/k}[0]$ via the canonical augmentation map, we conclude that $\Omega^1_{X/k}[0]$ is *Gorenstein* with respect to the codimension filtration $Z^\bullet$ on X (this is obvious, since $\Omega^1_{X/k}$ is invertible and X is regular, hence Gorenstein) and $E_{Z\bullet}(\Omega^1_{X/k}[0])$ is a complex of quasi-coherent injectives which has bounded cohomology (concentrated in degree 0).

This example is not anamolous. Back in the general situation, Gorenstein complexes with respect to $Z^\bullet$ form an additive category $\mathbf{D}^b(X)_{\mathrm{Gor}(Z^\bullet)}$ which is intimately related to the additive subcategory $\mathrm{Icz}(Z^\bullet, X) \subseteq \mathrm{Coz}(Z^\bullet, X)$ of Cousin complexes for which all (necessarily quasi-coherent) terms are *injective* and the cohomology is *bounded* (i.e., vanishes in sufficiently negative and positive degrees); note that [**RD**, IV, 3.4] accidentally omits the boundedness condition. If $Q(\cdot)$ denotes the additive functor from the category of complexes of $\mathcal{O}_X$-modules to the category $\mathbf{D}(X)$, then the key fact is:

THEOREM 3.1.3. [**RD**, IV, 3.4] *The functor $E_{Z\bullet}$ takes $\mathbf{D}^b(X)_{\mathrm{Gor}(Z^\bullet)}$ into* $\mathrm{Icz}(Z^\bullet, X)$, *the functor Q takes $\mathrm{Icz}(Z^\bullet, X)$ into $\mathbf{D}^b(X)_{\mathrm{Gor}(Z^\bullet)}$, and there are natural isomorphisms $\alpha : E_{Z\bullet} \circ Q \simeq 1$, $\beta : Q \circ E_{Z\bullet} \simeq 1$ such that*

$$E_{Z\bullet}(\beta) = \alpha(E_{Z\bullet}), \quad Q(\alpha) = \beta(Q).$$

These isomorphisms α, β respect translations (using (3.1.7)), Zariski localization on X, base change to $\mathrm{Spec}(\mathcal{O}_{X,x})$ for $x \in X$, and tensoring with an invertible sheaf.

Theorem 3.1.3 is important because it functorially transforms questions about derived category maps between Gorenstein complexes into questions about 'ordinary' maps between Cousin complexes. Unfortunately, the proof in [**RD**] that Q and $E_{Z\bullet}$ are quasi-inverses relies upon a false lemma [**RD**, IV, 3.2] which asserts that for any two maps $f_1^\bullet, f_2^\bullet : \mathscr{F}^\bullet \to \mathscr{G}^\bullet$ between $\mathscr{F}^\bullet, \mathscr{G}^\bullet \in \mathrm{Coz}(Z^\bullet, X)$ with $\mathrm{H}^i(f_1^\bullet) = \mathrm{H}^i(f_2^\bullet)$ for all i, the maps $f_1^\bullet$ and $f_2^\bullet$ are equal as maps of complexes (or, equivalently, if a map $f^\bullet : \mathscr{F}^\bullet \to \mathscr{G}^\bullet$ has all $\mathrm{H}^i(f^\bullet) = 0$, then $f^\bullet = 0$). For a counterexample, let $X = \mathbf{A}^1_k = \mathrm{Spec}(k[t])$ be the affine line over a field and let $Z^\bullet$ be the codimension filtration. Let $\mathscr{F}^\bullet = \mathscr{G}^\bullet$ be the same complex (of quasi-coherent sheaves), supported in degrees 0 and 1, given on the level of $k[t]$-modules by:

$$\cdots \to 0 \to k(t)[\epsilon] \xrightarrow{\epsilon} (k(t)/\mathcal{O}_{X,0})[\epsilon] \to 0 \to \cdots,$$

where $\epsilon^2 = 0$. This is trivially a Cousin complex on X with respect to $Z^\bullet$. The *non-zero* endomorphism $f^\bullet$ of $\mathscr{F}^\bullet$ given by $f^0 = 0$, $f^1 = \epsilon$ has $\mathrm{H}^i(f^\bullet) = 0$ for all i.

It is also important to note that the proof of [**RD**, IV, 3.4] constructs the δ-functorial isomorphisms α and β by an extremely unnatural procedure involving non-canonical choices of maps. It is hopeless under such a definition to have any kind of compatibility for α, β with respect to tensoring by an invertible sheaf or Zariski localization or base change to local rings on X (let alone with respect to the operation of residually stable base change, to be considered later). However, such compatibility is absolutely essential in the theory of residual complexes. Thus, the proof *must* produce canonical constructions for α and β (and this is also more aesthetically pleasing).

PROOF. (of Theorem 3.1.3) We first prove that $Q : \mathrm{Icz}(Z^\bullet, X) \to \mathbf{D}(X)$ is fully faithful, eliminating the need to appeal to the false [**RD**, IV, 3.2]. Since every object in $\mathrm{Icz}(Z^\bullet, X)$ is a bounded below complex of injectives, by Lemma 2.1.4 we just need to prove that if $\varphi : \mathscr{F}^\bullet \to \mathscr{G}^\bullet$ is a map between objects in $\mathrm{Icz}(Z^\bullet, X)$ such that φ is homotopic to 0, then $\varphi = 0$. Going back to the definition of 'homotopic', it suffices to show that any sheaf map $k^p : \mathscr{F}^p \to \mathscr{G}^{p-1}$ must vanish. Since $\mathscr{F}^p$ is in the Z^p/Z^{p+1}-skeleton and $\mathscr{G}^{p-1}$ is in the Z^{p-1}/Z^p-skeleton, this is clear.

Now by definition, if $\mathscr{F}^\bullet$ is in $\mathbf{D}^b(X)_{\mathrm{Gor}(Z^\bullet)}$ then $\mathscr{F}^\bullet$ is Cohen-Macaulay with respect to $Z^\bullet$, so there is an isomorphism $\mathscr{F}^\bullet \simeq E_{Z^\bullet}(\mathscr{F}^\bullet)$ in $\mathbf{D}(X)$ and the terms in the Cousin complex $E_{Z^\bullet}(\mathscr{F}^\bullet)$ are injective (by (3.1.4) and Lemma 2.1.5). Thus, $E_{Z^\bullet}(\mathscr{F}^\bullet)$ lies in $\mathrm{Icz}(Z^\bullet, X)$ and consequently Q is essentially surjective.

Conversely, if $\mathscr{G}^\bullet$ lies in $\mathrm{Icz}(Z^\bullet, X)$ then $\mathscr{G}^\bullet$ is a bounded below complex of *quasi-coherent* injectives and for fixed p there is an abstract isomorphism

$$(3.1.20) \qquad \mathscr{G}^p \simeq \bigoplus_{x \in Z^p - Z^{p+1}} i_{x*}(I_x)$$

for some $\mathcal{O}_{X,x}$-modules I_x. In particular,

$$(3.1.21) \qquad \mathscr{G}_x^r = 0 \quad \text{if} \quad x \in Z^p - Z^{p+1}, r > p.$$

To show that $Q(\mathscr{G}^\bullet) \in \mathbf{D}^b(X)$ is Gorenstein with respect to $Z^\bullet$, we need $\mathrm{H}_x^i(\mathscr{G}^\bullet)$ to be an injective $\mathcal{O}_{X,x}$-module for all i and

$$(3.1.22) \qquad \mathscr{G}^\bullet \simeq E_{Z^\bullet}(Q(\mathscr{G}^\bullet))$$

in $\mathbf{D}(X)$ (ignoring canonicalness of this latter isomorphism). An isomorphism of this latter type would give the desired injective module property for H_x^p with $x \in Z^p - Z^{p+1}$, by Lemma 2.1.5 and (3.1.4), and would force $\mathrm{H}_x^i(\mathscr{G}^\bullet) = 0$ otherwise (by (3.1.4) and [**RD**, IV, $3.1(iii) \Rightarrow (ii)$]). Thus, it is sufficient to produce an isomorphism as in (3.1.22). We will produce such an isomorphism α via a *canonical* construction, from which the desired compatibilities of α will be obvious.

By [**RD**, IV, 2.1], the condition (3.1.20) forces the natural maps

$$\mathscr{G}^p \leftarrow \underline{\Gamma}_{Z^p}(\mathscr{G}^p) \to \underline{\Gamma}_{Z^p/Z^{p+1}}(\mathscr{G}^p) = \underline{\mathrm{H}}^0_{Z^p/Z^{p+1}}(\mathscr{G}^p) = \bigoplus_{x \in Z^p - Z^{p+1}} i_{x*}(\Gamma_x(\mathscr{G}_x^p))$$

to be isomorphisms, where the two indicated equalities follow from the flasqueness of $\mathscr{G}^p$ and from (3.1.4) respectively. Since $\mathscr{G}^\bullet$ is a bounded below complex of flasques, the spectral sequence

$$E_1^{p,q} = \underline{\mathrm{H}}^{p+q}_{Z^p/Z^{p+1}}(\mathscr{G}^\bullet) \Rightarrow \mathrm{H}^{p+q}(\mathscr{G}^\bullet)$$

can be computed in terms of the decreasing filtration $\underline{\Gamma}_{Z^p}(\mathscr{G}^\bullet)$ on $\mathscr{G}^\bullet$.

Since

$$E_0^{p,q} = \underline{\Gamma}_{Z^p/Z^{p+1}}(\mathscr{G}^{p+q}) \simeq \underline{\mathrm{H}}^0_{Z^p/Z^{p+1}}(\mathscr{G}^{p+q}) \simeq \bigoplus_{x \in Z^p - Z^{p+1}} i_{x*}(\Gamma_x(\mathscr{G}_x^{p+q})),$$

we see by (3.1.21) that $E_0^{p,q} = 0$ if $q > 0$. If $q < 0$ and $x \in Z^p$, then

$$\mathscr{G}_x^{p+q} = \bigoplus_{y \in Z^{p+q} - Z^{p+q+1}} \left(i_{y_*} \Gamma_y(\mathscr{G}_y^{p+q}) \right)_x,$$

so $\Gamma_x(\mathscr{G}_x^{p+q}) = 0$ since $Z^p \cap (Z^{p+q} - Z^{p+q+1}) = \emptyset$. Thus, $d_0^{p,q} : E_0^{p,q} \to E_0^{p,q+1}$ vanishes for all q, so $E_1^{p,q} = 0$ for $q \neq 0$ and

$$E_1^{p,0} = E_0^{p,0} = \underline{\Gamma}_{Z^p/Z^{p+1}}(\mathscr{G}^p) \simeq \underline{\Gamma}_{Z^p}(\mathscr{G}^p),$$

with differentials fitting into the top row of a commutative diagram

$$
\begin{array}{ccc}
\underline{\Gamma}_{Z^p}(\mathscr{G}^p) \longrightarrow \underline{\Gamma}_{Z^p}(\mathscr{G}^{p+1}) \xleftarrow{\simeq} \underline{\Gamma}_{Z^{p+1}}(\mathscr{G}^{p+1}) \\
\simeq \downarrow \qquad\qquad \simeq \qquad\qquad \downarrow \simeq \\
\mathscr{G}^p \longrightarrow \mathscr{G}^{p+1}
\end{array}
$$

We therefore arrive at a *canonical* isomorphism of complexes

$$\alpha : E_{Z^\bullet}(Q(\mathscr{G}^\bullet)) \simeq \mathscr{G}^\bullet,$$

so Q takes $\mathrm{Icz}(Z^\bullet, X)$ to $\mathbf{D}^b(X)_{\mathrm{Gor}(Z^\bullet)}$ and there is a canonical isomorphism $\alpha : E_{Z^\bullet} \circ Q \simeq 1$, visibly compatible with tensoring by an invertible sheaf, Zariski localization, and base change to $\mathrm{Spec}(\mathscr{O}_{X,x})$. It is also clear that α respects translation (using (3.1.7)).

Since Q is fully faithful and essentially surjective, it follows that $E_{Z^\bullet}$ has the same properties and there is a unique isomorphism $\beta : Q \circ E_{Z^\bullet} \simeq 1$ such that $E_{Z^\bullet}(\beta) = \alpha(E_{Z^\bullet})$, so β consequently has the same compatibility properties as α. The identity $Q(\alpha) = \beta(Q)$ follows from applying $E_{Z^\bullet}$ to both sides and using the identity $E_{Z^\bullet} \circ Q(\alpha) = \alpha(E_{Z^\bullet} \circ Q)$, which is obvious from the *construction* of α. ∎

As an example of the preceding theorem, it follows from Corollary 3.1.2 and the *construction* of α in Theorem 3.1.3 that when X is a connected Dedekind scheme, $\mathscr{L}$ is an invertible sheaf on X, and $Z^\bullet$ is the codimension filtration, the canonical isomorphism in $\mathbf{D}(X)$

$$\beta : \mathscr{L}[1] \simeq Q \circ E_{Z^\bullet[1]}(\mathscr{L}[1])$$

is represented by the unique map of complexes $\beta_{\mathscr{L}} : \mathscr{L}[1] \to E_{Z^\bullet[1]}(\mathscr{L}[1])$ given in degree -1 by the canonical map

$$\mathscr{L} \to i_{\xi_*}(\mathscr{L}_\xi) = E_{Z^\bullet[1]}(\mathscr{L}[1])^{-1}.$$

To check this, one need only verify that $E_{Z^\bullet[1]}(Q(\beta_{\mathscr{L}})) = \alpha(E_{Z^\bullet[1]})$. This is an easy exercise in unwinding the construction of α in a particularly simple setting. For example, by Corollary 3.1.2 we see that Theorem 3.1.3 includes as a very special case the canonical quasi-isomorphism from $\Omega_{X/k}^1[1]$ to (3.1.3) when X is a smooth curve over an algebraically closed field k.

Probably inspired by this example, Grothendieck's method of construction of duality theory is to use residual complexes (which are analogous to objects in $\mathrm{Icz}(Z^\bullet, X)$ above) and pointwise dualizing complexes (which are analogous to objects in $\mathbf{D}^b(X)_{\mathrm{Gor}(Z\bullet)}$ above). We now review some basic facts concerning (pointwise) dualizing complexes, and in §3.2 we will discuss residual complexes.

A *dualizing complex* on a locally noetherian scheme X is a complex $\mathscr{R}^\bullet \in \mathbf{D}^b_c(X)$ which has *finite* injective dimension (i.e., $\mathscr{R}^\bullet$ is isomorphic in $\mathbf{D}(X)$ to a bounded complex of injectives, which can be assumed to be quasi-coherent) and for which the natural map (see (1.3.20)) of δ-functors on $\mathbf{D}_c(X)$

$$\eta_{\mathscr{R}\bullet} : \mathrm{id} \to \mathbf{R}\mathscr{H}om^\bullet(\mathbf{R}\mathscr{H}om^\bullet(\cdot, \mathscr{R}^\bullet), \mathscr{R}^\bullet)$$

is an isomorphism. It suffices to check that $\eta_{\mathscr{R}\bullet}(\mathscr{O}_X[0])$ is an isomorphism [**RD**, V, 2.1], and the proof of this fact depends in an essential way on the *global* hypothesis that $\mathscr{R}^\bullet$ has finite injective dimension.

For an example, if X is regular then $\mathscr{R}^\bullet = \mathscr{O}_X[0]$ is a dualizing complex provided X has finite Krull dimension [**RD**, V, Example 2.2]. Suppose in addition that X is connected and Dedekind with set of closed points X^0 and generic point ξ. Then $\mathscr{L}[0]$ is dualizing for any invertible sheaf $\mathscr{L}$ on X, where the canonical resolution

$$0 \to \mathscr{L} \to i_{\xi*}(\mathscr{L}_\xi) \to \bigoplus_{x \in X^0} i_{x*}(\mathscr{L}_\xi/\mathscr{L}_x) \to 0$$

by 'divisible' (hence injective) quasi-coherent sheaves makes explicit the *finiteness* of the injective dimension of $\mathscr{L}[0]$. If this Dedekind X is a smooth connected curve over a field k, the most interesting situation is when $\mathscr{L} = \Omega^1_{X/k}$. Another interesting example is $X = \mathrm{Spec}(\mathbf{Z})$ with $\mathscr{L} = \mathscr{O}_X$. At the other extreme, since regular local rings A are Gorenstein and therefore $\mathrm{injdim}(A) = \dim(A)$ for such rings, for a regular scheme X which does not have finite Krull dimension (cf. Nagata's example in [**AM**, Exer. 4, Ch 11]) the sheaf $\mathscr{O}_X$ does *not* have finite injective dimension. Thus, $\mathscr{O}_X[0]$ cannot be a dualizing complex for such X. In general, if a locally noetherian scheme admits a dualizing complex then it *must* be catenary and have finite Krull dimension [**RD**, V, 7.2].

If A is a noetherian ring, then one can likewise define the notion of a *dualizing complex* $R^\bullet$ in the derived category $\mathbf{D}(A)$ of A-modules. Namely, we require $R^\bullet \in \mathbf{D}_c(A)_{\mathrm{fid}}$ and that the natural map of functors on $\mathbf{D}_c(A)$

$$\eta_{R\bullet} : \mathrm{id} \to \mathbf{R}\mathrm{Hom}^\bullet(\mathbf{R}\mathrm{Hom}^\bullet(\cdot, R^\bullet), R^\bullet)$$

is an isomorphism. The proof of [**RD**, V, 2.1] carries over to this commutative algebra setting to show that it suffices to prove that $\eta_{R\bullet}(A)$ is an isomorphism. It is important for various reasons (e.g., the proof of many results in [**RD**, V], such as [**RD**, V, 3.4]) to know that dualizing complexes in $\mathbf{D}(\mathrm{Spec}(A))$ are closely related to dualizing complexes in $\mathbf{D}(A)$. From Lemma 2.1.6 we see that $\mathbf{D}^+_c(A)$ and $\mathbf{D}^+_c(\mathrm{Spec}(A))$ are 'essentially the same'.

Let $R^\bullet \in \mathbf{D}^+_c(A)$ and let $\mathscr{R}^\bullet \in \mathbf{D}^+_c(\mathrm{Spec}(A))$ be the associated object on $\mathrm{Spec}(A)$, so clearly $R^\bullet$ has finite injective dimension if and only if $\mathscr{R}^\bullet$ has finite injective dimension (by Lemma 2.1.3 and Lemma 2.1.6); this fact will be used without comment below when we pass between $\mathbf{D}^+_c(A)$ and $\mathbf{D}^+_c(\mathrm{Spec}(A))$ with A equal to the local ring at a point on a locally noetherian scheme. The connection between the algebraic geometry and the commutative algebra is the following easy lemma, which is used but not explicitly stated in [**RD**].

LEMMA 3.1.4. *For $R^\bullet$ and $\mathscr{R}^\bullet$ as above, $R^\bullet$ is a dualizing complex if and only if $\mathscr{R}^\bullet$ is a dualizing complex. In general, for a locally noetherian scheme X and an object $\mathscr{R}^\bullet \in \mathbf{D}_c^+(X)_{\mathrm{fid}}$ with finite injective dimension, $\mathscr{R}^\bullet$ is dualizing if and only if $\mathscr{R}_x^\bullet \in \mathbf{D}_c(\mathcal{O}_{X,x})_{\mathrm{fid}}$ is dualizing for all $x \in X$.*

PROOF. The key fact we need (which is invoked in the proof of [**RD**, V, 2.3]) is that the double duality map is compatible with passage to stalks under suitable boundedness conditions. More precisely, for any locally noetherian scheme X, any object $\mathscr{R}^\bullet \in \mathbf{D}_c^+(X)_{\mathrm{fid}}$, and any $\mathscr{F}^\bullet \in \mathbf{D}_c^b(X)$, clearly $\mathbf{R}\mathscr{H}om^\bullet(\mathscr{F}^\bullet, \mathscr{R}^\bullet) \in \mathbf{D}_c^b(X)$ and there is a canonical map

$$(3.1.23) \qquad \mathbf{R}\mathscr{H}om^\bullet(\mathscr{F}^\bullet, \mathscr{R}^\bullet)_x \to \mathbf{R}\mathrm{Hom}^\bullet_{\mathcal{O}_{X,x}}(\mathscr{F}_x^\bullet, \mathscr{R}_x^\bullet)$$

in $\mathbf{D}_c^b(\mathcal{O}_{X,x})$. We claim that this natural map is an isomorphism, so in particular the restriction to $\mathbf{D}_c^b(X)$ of the natural transformation $\eta_{\mathscr{R}^\bullet}$ on $\mathbf{D}_c(X)$ is naturally compatible with passage to stalks. To prove this, we immediately reduce to the case where $\mathscr{F}^\bullet = \mathscr{F}[0]$ for a coherent sheaf $\mathscr{F}$, and then (3.1.23) is represented by the map of complexes

$$\mathscr{H}om(\mathscr{F}, \mathscr{R}^\bullet)_x \to \mathrm{Hom}_{\mathcal{O}_{X,x}}(\mathscr{F}_x, \mathscr{R}_x^\bullet),$$

where we take $\mathscr{R}^\bullet$ to be a bounded below complex of quasi-coherent sheaves without loss of generality. This is clearly an isomorphism of complexes, since $\mathscr{F}$ is of finite presentation. The exact same argument applies with $\mathbf{D}(A)$ in place of $\mathbf{D}(X)$, using the 'local' derived categories $\mathbf{D}(A_\mathfrak{p})$ for primes $\mathfrak{p} \in \mathrm{Spec}(A)$.

As a consequence of this, we see that for any locally noetherian scheme X and any $\mathscr{R}^\bullet \in \mathbf{D}_c(X)_{\mathrm{fid}} \subseteq \mathbf{D}_c^b(X)$, the natural map

$$\eta_{\mathscr{R}^\bullet}(\mathcal{O}_X[0]) : \mathcal{O}_X[0] \to \mathbf{R}\mathscr{H}om^\bullet(\mathscr{R}^\bullet, \mathscr{R}^\bullet),$$

which is just (1.3.17) and involves no intervention of signs, is an isomorphism if and only if the map

$$\mathcal{O}_{X,x}[0] \to \mathbf{R}\,\mathrm{Hom}^\bullet_{\mathcal{O}_{X,x}}(\mathscr{R}_x^\bullet, \mathscr{R}_x^\bullet)$$

in $\mathbf{D}(\mathcal{O}_{X,x})$ is an isomorphism for all $x \in X$. Since passage to the stalk takes quasi-coherent injectives to injective modules over the local ring, we conclude that the given $\mathscr{R}^\bullet \in \mathbf{D}_c(X)_{\mathrm{fid}}$ is a dualizing complex on X if and only $\mathscr{R}_x^\bullet \in \mathbf{D}(\mathcal{O}_{X,x})$ is a dualizing complex for all $x \in X$; note that the complex $\mathcal{O}_X[0]$ for regular X of infinite Krull dimension gives a counterexample if we drop the *global* requirement that $\mathscr{R}^\bullet$ have finite injective dimension. The same arguments show that for any noetherian ring A and any $R^\bullet \in \mathbf{D}_c(A)_{\mathrm{fid}}$, $R^\bullet$ is a dualizing complex if and only if $R_\mathfrak{p}^\bullet \in \mathbf{D}(A_\mathfrak{p})$ is a dualizing complex for all $\mathfrak{p} \in \mathrm{Spec}(A)$.

Now consider $R^\bullet \in \mathbf{D}_c^+(A)$ and $\mathscr{R}^\bullet \in \mathbf{D}_c^+(\mathrm{Spec}(A))$ as in the first sentence of the lemma. Without loss of generality, both have finite injective dimension. For any $x = \mathfrak{p} \in X = \mathrm{Spec}(A)$, we have a canonical identification $R_\mathfrak{p}^\bullet \simeq \mathscr{R}_x^\bullet$ under the equivalence $\mathbf{D}(A_\mathfrak{p}) = \mathbf{D}(\mathcal{O}_{X,x})$. In view of the above discussion, the lemma follows.

■

To deal with the case of possibly infinite Krull dimension, [**RD**] introduces
the notion of a *pointwise dualizing complex*. The inspiration is two-fold: Krull's
fundamental theorem that the local schemes $\mathrm{Spec}(\mathcal{O}_{X,x})$ at points of a locally
noetherian scheme do have finite Krull dimension (even if X does not), and
the stalkwise criterion in Lemma 3.1.4. However, the *global* criterion of finite
injective dimension is essential in the proof of [**RD**, V, 2.1] (which reduces the
dualizing property to the case of $\eta_{\mathscr{F}^{\bullet}}(\mathcal{O}_X[0])$), so the validity of its 'pointwise'
analogue [**RD**, V, 8.1] is unclear (the problem is that the proof of [**RD**, V, 2.1]
does not appear to carry over, contrary to what is claimed in the proof of [**RD**,
V, 8.1]). Consequently, the notion of 'pointwise dualizing complex' as defined in
[**RD**, V, §8] seems problematic. In fact, there are *two* natural definitions of this
concept. I am grateful to Gabber for explaining the following points to me.

Let X be a locally noetherian scheme and choose $\mathscr{R}^{\bullet} \in \mathbf{D}_c^+(X)$. As in [**RD**],
we say that $\mathscr{R}^{\bullet}$ has *pointwise finite injective dimension* if $\mathscr{R}_x^{\bullet} \in \mathbf{D}_c^+(\mathcal{O}_{X,x})$ has
finite injective dimension for all $x \in X$ (and for $R^{\bullet} \in \mathbf{D}_c^+(A)$ with A a noetherian
ring, we say $R^{\bullet}$ has *pointwise finite injective dimension* if $R_{\mathfrak{p}}^{\bullet} \in \mathbf{D}_c^+(A_{\mathfrak{p}})$ has finite
injective dimension for all $\mathfrak{p} \in \mathrm{Spec}(A)$). Following Gabber, we say that $\mathscr{R}^{\bullet} \in$
$\mathbf{D}_c^+(X)$ is *weakly pointwise dualizing* if $\mathscr{R}_x^{\bullet} \in \mathbf{D}_c^+(\mathcal{O}_{X,x}) \simeq \mathbf{D}_c^+(\mathrm{Spec}(\mathcal{O}_{X,x}))$
is dualizing for all $x \in X$; note that this forces $\mathscr{R}^{\bullet}$ to have pointwise finite
injective dimension and Lemma 3.1.4 removes any ambiguity about working
with $\mathbf{D}_c^+(\mathcal{O}_{X,x})$ or $\mathbf{D}_c^+(\mathrm{Spec}(\mathcal{O}_{X,x}))$. If, in addition, $\mathscr{R}^{\bullet}$ has locally bounded
cohomology (denoted $\mathscr{R}^{\bullet} \in \mathbf{D}^{lb}(X)$) then we say that $\mathscr{R}^{\bullet}$ is a *strongly pointwise
dualizing complex*. Since strongly pointwise dualizing complexes are required to
be in $\mathbf{D}^+(X)$ — that is, to be *globally* bounded below — the 'local boundedness'
refers to local upper bounds on the cohomology of $\mathscr{R}^{\bullet}$.

Dualizing complexes are strongly pointwise dualizing, by Lemma 3.1.4, and
$\mathcal{O}_X[0]$ gives a counterexample to the converse when X is regular with infinite
Krull dimension. Due to the rather prominent role of schemes of finite type over
Z and over *local* noetherian rings later on, it is rather important that the notions
of dualizing and pointwise dualizing (in either sense) coincide if X is *noetherian*
with finite Krull dimension [**RD**, V, 8.2]. It is not immediately apparent that
either notion of pointwise dualizing (strong or weak) has anything to do with the
natural transformation $\eta_{\mathscr{F}^{\bullet}}$ on $\mathbf{D}_c(X)$ being an isomorphism, even if restricted
to $\mathbf{D}_c^b(X)$ (or just $\mathcal{O}_X[0]$). The essential problem (in view of the proof of Lemma
3.1.4) is that it is not clear if $\mathbf{R}\mathscr{H}om^{\bullet}(\cdot, \mathscr{R}^{\bullet})$ takes $\mathbf{D}_c^{lb}(X)$ back to itself. We
cannot expect this for weakly pointwise dualizing complexes, but for strongly
pointwise dualizing complexes this will be shown in Lemma 3.1.5 below.

The reason for introducing the notion of pointwise dualizing (in either the
weak or strong sense) is that Grothendieck's duality theory is constructed in
terms of *residual complexes* (to be defined later), but in order to *construct* a
good theory of residual complexes, the concept of a pointwise dualizing complex
is extremely useful. Even in the setting of noetherian schemes of finite Krull
dimension, where the notion of pointwise dualizing complex (in either sense)
coincides with the notion of dualizing complex, the local nature of the pointwise

dualizing definition is very convenient. The key fact one needs to show is that if $\mathscr{R}^\bullet$ is a pointwise dualizing complex (in either sense) on a locally noetherian scheme X and $f : X' \to X$ is a finite map (resp. a smooth map with bounded fiber dimension), then $f^\flat \mathscr{R}^\bullet$ (resp. $f^\sharp \mathscr{R}^\bullet$) is a pointwise dualizing complex on X' (in the same sense). The proofs of these facts in [**RD**, V, 2.4, 8.3] are only applicable with the weak sense of pointwise dualizing. For the notion of strongly pointwise dualizing, one can recover the same results by the same proofs, provided one knows the first part of the following lemma (which is needed with $\mathscr{F}^\bullet = \mathscr{F}[0]$ for a coherent $\mathscr{O}_X$-algebra $\mathscr{F}$); this lemma also ensures that even without a finite Krull dimension condition, there is a reasonable notion of 'double duality isomorphism' for strongly pointwise dualizing complexes (and so perhaps this is the more natural notion of pointwise dualizing?).

LEMMA 3.1.5. (Gabber) *Let X be a locally noetherian scheme, $\mathscr{R}^\bullet \in \mathbf{D}_c^{lb}(X)$ with pointwise finite injective dimension. Then for any $\mathscr{F}^\bullet \in \mathbf{D}_c^{lb}(X)$, the object $\mathbf{R}\mathscr{H}om^\bullet(\mathscr{F}^\bullet, \mathscr{R}^\bullet) \in \mathbf{D}_c(X)$ has locally bounded cohomology (and has bounded below cohomology if $\mathscr{F}^\bullet$ has bounded above cohomology and $\mathscr{R}^\bullet$ has bounded below cohomology).*

If $\mathscr{R}^\bullet \in \mathbf{D}_c^+(X)$ as well, then $\mathscr{R}^\bullet$ is strongly pointwise dualizing if and only the double duality map $\eta_{\mathscr{R}^\bullet}(\mathscr{O}_X[0])$ is an isomorphism, in which case $\eta_{\mathscr{R}^\bullet}(\mathscr{F}^\bullet)$ is an isomorphism for all $\mathscr{F}^\bullet \in \mathbf{D}_c^{lb}(X)$.

The proof below is taken from a letter from Gabber to the author. Note that the *local* boundedness conclusion in Lemma 3.1.5 is the most one can expect in general, even if $\mathscr{R}^\bullet, \mathscr{F}^\bullet \in \mathbf{D}_c^b(X)$. Indeed, for $\mathscr{R}^\bullet \in \mathbf{D}_c^b(X)$ with pointwise finite injective dimension and $\mathscr{F}^\bullet \in \mathbf{D}_c^b(X)$, $\mathbf{R}\mathscr{H}om^\bullet(\mathscr{F}^\bullet, \mathscr{R}^\bullet) \in \mathbf{D}_c^{lb}(X)$ does not generally lie in $\mathbf{D}_c^b(X)$ if X is not quasi-compact. For example, if

$$ X = \coprod_{n \geq 1} \mathbf{A}_k^n, \quad \mathscr{R}^\bullet = \mathscr{O}_X[0], \quad \mathscr{F}^\bullet = \mathscr{F}[0], $$

with $\mathscr{F}|_{\mathbf{A}_k^n}$ equal to the structure sheaf of the origin on $\mathbf{A}_k^n$, then the restriction of $\mathbf{R}\mathscr{H}om^\bullet(\mathscr{F}^\bullet, \mathscr{R}^\bullet)$ to the open $\mathbf{A}_k^n$ in X has cohomology concentrated in degree n, where it is a coherent sheaf supported at the origin, with (non-zero) stalk equal to

$$ \mathrm{Ext}^n_{\mathscr{O}_{\mathbf{A}_k^n,0}}(k, \mathscr{O}_{\mathbf{A}_k^n,0}) \simeq k. $$

PROOF. Once we know that $\mathbf{R}\mathscr{H}om^\bullet(\mathscr{F}^\bullet, \mathscr{R}^\bullet)$ has locally bounded cohomology for $\mathscr{R}^\bullet, \mathscr{F}^\bullet \in \mathbf{D}_c^{lb}(X)$ (obviously bounded below if $\mathscr{F}^\bullet$ has bounded above cohomology and $\mathscr{R}^\bullet$ has bounded below cohomology), then the proof of Lemma 3.1.4 can be used to show, via passage to stalks, that $\eta_{\mathscr{R}^\bullet}(\mathscr{F}^\bullet)$ is an isomorphism when $\mathscr{R}^\bullet$ is strongly pointwise dualizing and $\mathscr{F}^\bullet \in \mathbf{D}_c^{lb}(X)$. Since the dualizing property on the (finite Krull dimension!) stalks can be checked by looking at the structure sheaf, viewed as a complex concentrated in degree 0, all that remains is to prove the local boundedness of the cohomology of $\mathbf{R}\mathscr{H}om^\bullet(\mathscr{F}^\bullet, \mathscr{R}^\bullet)$ for $\mathscr{R}^\bullet, \mathscr{F}^\bullet \in \mathbf{D}_c^{lb}(X)$. Working locally on X, we may assume $\mathscr{R}^\bullet, \mathscr{F}^\bullet \in \mathbf{D}_c^b(X)$. Also, we immediately reduce to the case where $\mathscr{F}^\bullet = \mathscr{F}[0]$ for a coherent sheaf $\mathscr{F}$ on X.

We may assume $X = \text{Spec}(A)$ is (noetherian) affine, so we can work in the following commutative algebra setting. Let $R^\bullet \in \mathbf{D}_c^b(A)$ have pointwise finite injective dimension. We want to prove that $\mathbf{R}\,\text{Hom}_A(M, R^\bullet) \in \mathbf{D}_c^+(A)$ has bounded cohomology for every finite A-module M. In other words, we need to prove that the *finitely generated* A-module $\text{Ext}_A^i(M, R^\bullet)$ vanishes for large i (possibly depending on M). This will be proven by noetherian induction on the support $\text{Supp}(M)$, the case $M = 0$ being obvious. By the theory of associated primes, there is a finite filtration of M with successive quotients A/P_j for various primes P_j of A [**Mat**, Thm 6.4]. Thus, we may assume $M = A/P$ for a prime ideal P of A. If $f \notin P$, then the exact sequence

$$0 \to M \xrightarrow{f} M \to M/fM \to 0$$

gives rise to an A-linear injection

(3.1.24) $\text{Ext}_A^i(M, R^\bullet)/f\,\text{Ext}_A^i(M, R^\bullet) \hookrightarrow \text{Ext}_A^{i+1}(M/fM, R^\bullet)$

for all $i \geq 1$. Since M/fM has strictly smaller support than M, by induction there is some large N so that $\text{Ext}_A^j(M/fM, R^\bullet) = 0$ for $j > N$. The injection (3.1.24), together with Nakayama's lemma, implies that $\text{Ext}_A^i(M, R^\bullet)_Q = 0$ for $i > N - 1$ and primes Q of A containing f. It therefore suffices to show

$$\text{Ext}_A^i(M, R^\bullet)_f \simeq \text{Ext}_{A_f}^i(M_f, R_f^\bullet)$$

vanishes for large i. Note that $R_f^\bullet$ satisfies the same hypotheses relative to A_f that $R^\bullet$ does relative to A, but we do *not* try to replace A by A_f yet, as this might upset the noetherian induction argument.

Let $f_1, \ldots, f_n$ be an ordered set of generators of P, and choose $f \notin P$ such that the (co)homology modules of the Koszul complex $K_\bullet = K_\bullet(f_1, \ldots, f_n)$ are free $(A/P)_f$-modules (such f exists by standard direct limit arguments, since the $\text{H}^i(K_\bullet)$ are finite A/P-modules, all but finitely many of which are 0, and over the unique generic point of $\text{Spec}(A/P)$ the A/P-modules $\text{H}^i(K_\bullet)$ becomes vector spaces over the fraction field of A/P and hence are generically free of finite rank). By localizing throughout by this f, it suffices to prove the following general claim that has *nothing* to do with noetherian induction. Let A be an arbitrary noetherian ring, $R^\bullet \in \mathbf{D}_c(A)$ an object with pointwise finite injective dimension, and $P = (f_1, \ldots, f_n)$ a prime ideal in A such that the A/P-modules $\text{H}^i(K_\bullet)$ are all *free*. Let m be the largest integer such that $\text{H}^{-m}(K_\bullet) \neq 0$ and N_0 the largest integer such that $\text{H}^{N_0}(R^\bullet) \neq 0$ (with $N_0 = -\infty$ in case $R^\bullet = 0$); m makes sense since $\text{H}^0(K_\bullet) = A/P \neq 0$ and $K_\bullet$ is a bounded complex. Then we claim that $\text{Ext}_A^i(A/P, R^\bullet) = 0$ for all $i > N_0 + n - m$.

The formation of the A/P-module $\text{Ext}_A^i(A/P, R^\bullet)$ is compatible with localization at a prime of A containing P. Such localization preserves the hypotheses on $R^\bullet$, causes N_0 to at worst go down, and does not change the value of m (due to the *freeness* assumption on the $\text{H}^j(K_\bullet)$'s). Thus, we may assume A is local, so $R^\bullet$ has *finite injective dimension* and therefore $\text{Ext}_A^i(A/P, R^\bullet) = 0$ for large i. Let i_0 be the largest integer such that $\text{Ext}_A^{i_0}(A/P, R^\bullet) \neq 0$ (if no such i_0 exists, we are done). We want to show $i_0 \leq N_0 + n - m$.

Consider the first quadrant hypercohomology spectral sequence which computes the cohomology of $\mathbf{R}\operatorname{Hom}_A^\bullet(K_\bullet, R^\bullet)$:

$$E_2^{r,s} = \operatorname{Ext}_A^r(\mathrm{H}^{-s}(K_\bullet), R^\bullet) \Longrightarrow \operatorname{Ext}_A^{r+s}(K_\bullet, R^\bullet).$$

Since each $\mathrm{H}^{-s}(K_\bullet)$ is a finite direct sum of copies of A/P, we have $E_2^{r,s} = 0$ for $s > m$ and $E_2^{r,s} = 0$ for $r > i_0$. Thus,

$$0 \neq E_2^{i_0,m} = E_\infty^{i_0,m} = \operatorname{Ext}_A^{i_0+m}(K_\bullet, R^\bullet).$$

It remains to show that $\operatorname{Ext}_A^j(K_\bullet, R^\bullet) = 0$ for all $j > N_0 + n$. Without loss of generality, $R^\bullet$ is a bounded complex supported in degrees $\leq N_0$. Since $K_\bullet$ is a complex of projectives supported in degrees $-n$ to 0, it follows that $\operatorname{Ext}_A^j(K_\bullet, R^\bullet)$ is isomorphic to the jth cohomology of $\operatorname{Hom}_A^\bullet(K_\bullet, R^\bullet)$, which clearly vanishes for $j > N_0 + n$.   ∎

For any weakly pointwise dualizing complex $\mathscr{R}^\bullet$ on a locally noetherian scheme X and any $x \in X$, there is a unique integer $d = d_{\mathscr{R}^\bullet}(x)$ such that $\mathrm{H}^{-d}(\mathbf{R}\mathscr{H}om_{\operatorname{Spec}(\mathcal{O}_{X,x})}^\bullet(k(x), \mathscr{R}_x^\bullet))$ is non-zero [**RD**, V, 3.4]. By [**RD**, V, 7.1], this behaves like a (shifted) codimension function in the sense that $d_{\mathscr{R}^\bullet}(x_1) = d_{\mathscr{R}^\bullet}(x) + 1$ for any $x_1 \in \overline{\{x\}}$ of codimension 1. Moreover,

(3.1.25) $d_{f^! \mathscr{R}^\bullet}(x') = d_{\mathscr{R}^\bullet}(f(x')) - \operatorname{trdeg}(k(x')/k(f(x')))$

for any smooth map $f : X' \to X$ with bounded fiber dimension (by [**RD**, V, 8.4], which mistakenly has $+$ instead of $-$ in (3.1.25); the same error occurs in [**RD**, VI, 3.4]) and

(3.1.26) $d_{f^\flat \mathscr{R}^\bullet}(x') = d_{\mathscr{R}^\bullet}(f(x'))$

for any finite map $f : X' \to X$ (by duality for finite maps [**RD**, III, 6.7] and reduction to the case where $X = \operatorname{Spec}(A)$ for a complete local noetherian ring A). Clearly, $d_{\mathscr{R}^\bullet}$ differs from the codimension function by a constant on each irreducible component of X (or more generally on each connected union of irreducible components of X on which the (reduced) local rings are equidimensional). We call $d_{\mathscr{R}^\bullet}$ the *codimension function on X associated to $\mathscr{R}^\bullet$* and we define the *associated filtration $Z^\bullet(\mathscr{R}^\bullet)$ of X* by

(3.1.27) $Z^p(\mathscr{R}^\bullet) = \{x \in X \mid d_{\mathscr{R}^\bullet}(x) \geq p\}.$

In the Dedekind example above, with the dualizing complex $\mathscr{R}^\bullet = \mathscr{L}[1]$ for an invertible sheaf $\mathscr{L}$ on X, it is clear that $d_{\mathscr{R}^\bullet}$ is equal to -1 plus the usual codimension function on X, so $Z^\bullet(\mathscr{R}^\bullet)$ is the shift by [1] of the codimension filtration.

Since Grothendieck duality theory in [**RD**, VII] is defined in terms of choices of dualizing complexes, it is essential that dualizing complexes are almost unique. The form of this 'almost uniqueness' given in [**RD**, V, 3.1] is not quite strong enough, so we want to state the precise uniqueness assertion which is needed in the theory. By (1.3.11), if $\mathscr{R}^\bullet$ is a dualizing complex on X, $\mathscr{L}$ is an invertible sheaf, and n is a locally constant $\mathbf{Z}$-valued function on X, then $\mathscr{L}[n] \overset{\mathbf{L}}{\otimes} \mathscr{R}^\bullet$

is a dualizing complex on X. Clearly $Z^{\bullet}(\mathscr{L}[n] \overset{\mathbf{L}}{\otimes} \mathscr{R}^{\bullet}) = Z^{\bullet}(\mathscr{R}^{\bullet})[n]$ for any invertible sheaf $\mathscr{L}$ on X and any locally constant $\mathbf{Z}$-valued function n on X. In the converse direction, if $\mathscr{R}^{\bullet}$ and $\mathscr{R}'^{\bullet}$ are two dualizing complexes on X, then by [**RD**, V, 3.1] (or its proof) there is a unique locally constant $\mathbf{Z}$-valued function $n = n(\mathscr{R}^{\bullet}, \mathscr{R}'^{\bullet})$ on X for which the $\mathscr{O}_X$-module

$$\mathscr{L}(\mathscr{R}^{\bullet}, \mathscr{R}'^{\bullet}) = \mathrm{H}^{-n}(\mathbf{R}\mathscr{H}om^{\bullet}(\mathscr{R}^{\bullet}, \mathscr{R}'^{\bullet}))$$

is nowhere zero; moreover, $\mathscr{L}(\mathscr{R}^{\bullet}, \mathscr{R}'^{\bullet})$ is an invertible sheaf. For any three dualizing complexes $\mathscr{R}_1^{\bullet}$, $\mathscr{R}_2^{\bullet}$, $\mathscr{R}_3^{\bullet}$, there is a natural isomorphism

$$(3.1.28) \qquad \mathbf{R}\mathscr{H}om^{\bullet}(\mathscr{R}_2^{\bullet}, \mathscr{R}_3^{\bullet}) \overset{\mathbf{L}}{\otimes} \mathbf{R}\mathscr{H}om^{\bullet}(\mathscr{R}_1^{\bullet}, \mathscr{R}_2^{\bullet}) \simeq \mathbf{R}\mathscr{H}om^{\bullet}(\mathscr{R}_1^{\bullet}, \mathscr{R}_3^{\bullet})$$

(defined without the intervention of signs) which is suitably 'associative' with respect to a fourth dualizing complex, and this induces an isomorphism

$$(3.1.29)$$
$$\mathscr{L}(\mathscr{R}_2^{\bullet}, \mathscr{R}_3^{\bullet})[n(\mathscr{R}_2^{\bullet}, \mathscr{R}_3^{\bullet})] \otimes \mathscr{L}(\mathscr{R}_1^{\bullet}, \mathscr{R}_2^{\bullet})[n(\mathscr{R}_1^{\bullet}, \mathscr{R}_2^{\bullet})] \simeq \mathscr{L}(\mathscr{R}_1^{\bullet}, \mathscr{R}_3^{\bullet})[n(\mathscr{R}_1^{\bullet}, \mathscr{R}_3^{\bullet})]$$

with a similar 'associativity' property. The proof of [**RD**, V, 3.1] yields an isomorphism

$$(3.1.30) \qquad \beta_{\mathscr{R}^{\bullet}, \mathscr{R}'^{\bullet}} : \mathscr{L}(\mathscr{R}^{\bullet}, \mathscr{R}'^{\bullet})[n(\mathscr{R}^{\bullet}, \mathscr{R}'^{\bullet})] \overset{\mathbf{L}}{\otimes} \mathscr{R}^{\bullet} \simeq \mathscr{R}'^{\bullet}$$

(defined *without* the intervention of signs) and the strong form of 'almost uniqueness' for dualizing complexes is that the isomorphism (3.1.30) is 'transitive' with respect to a third dualizing complex, via the isomorphism (3.1.29). Because of (3.1.30), it is essential that we have kept track of compatibility of all constructions with respect to Zariski localization, translation, and tensoring with an invertible sheaf (and keeping track of behavior with respect to base change to $\mathrm{Spec}(\mathscr{O}_{X,x})$'s is required for Grothendieck's method of handling infinite Krull dimension cases by base change to local schemes, which have finite Krull dimension).

Once a dualizing complex $\mathscr{R}^{\bullet}$ is fixed on X (if one exists), we define $D = \mathbf{R}\mathscr{H}om^{\bullet}(\cdot, \mathscr{R}^{\bullet})$ as a functor on $\mathbf{D}_c(X)$. There are some convenient natural maps involving the duality functor D, but one must be careful about signs in the definitions. For example, for suitable $\mathscr{F}^{\bullet}$, $\mathscr{G}^{\bullet}$, there is an isomorphism [**RD**, V, 2.6(*b*)]

$$(3.1.31) \qquad \mathscr{F}^{\bullet} \overset{\mathbf{L}}{\otimes} D(\mathscr{G}^{\bullet}) \to D(\mathbf{R}\mathscr{H}om^{\bullet}(\mathscr{F}^{\bullet}, \mathscr{G}^{\bullet}))$$

which is translation-compatible in $\mathscr{F}^{\bullet}$, $\mathscr{G}^{\bullet}$ and is defined by multiplying the natural map

$$\mathscr{F}^p \otimes \mathscr{H}om_X(\mathscr{G}^q, \mathscr{R}^r) \to \mathscr{H}om_X(\mathscr{H}om_X(\mathscr{F}^p, \mathscr{G}^q), \mathscr{R}^r)$$

by $(-1)^{pr}(-1)^p = (-1)^{p(r+1)}$ in order to ensure that we get a map of complexes (so there is no intervention of signs if the complex $\mathscr{F}^{\bullet}$ is concentrated in degree 0). Note that the 'sign formalism' in §1.3 via (1.3.18), (1.3.19) does produce exactly this sign (thanks to the sign of $(-1)^{-p^2} = (-1)^{-p}$ in degree p in (1.3.16) when $\mathscr{G}^{\bullet} = \mathscr{O}_X[0]$ there). In a similar elementary way, many other isomorphisms arising from dualizing complexes in [**RD**, V, VI] require the intervention of signs and are compatible with translations and (3.1.30); the fact that $n(\mathscr{R}^{\bullet}, \mathscr{R}'^{\bullet})$ is merely *locally* constant does not cause any problems, since derived categories behave well with respect to a partitioning of the space into disjoint opens. This completes our review of dualizing complexes.

3.2. Residual Complexes

The theory of residual complexes [**RD**, VI] is a 'concrete' version of the theory of pointwise dualizing complexes (made precise by Lemma 3.2.1 below). For a motivating example, let X be a connected Dedekind scheme with generic point ξ and set of closed points X^0. Choose an invertible sheaf $\mathscr{L}$ on X (e.g., $\mathscr{L} = \Omega^1_{X/k}$ if X is a smooth connected curve over a field k, or $\mathscr{L} = \mathcal{O}_X$ if $X = \mathrm{Spec}(\mathbf{Z})$). We have seen that $\mathscr{L}[1] \in \mathbf{D}^b_c(X)$ is a dualizing complex on X. Consider the quasi-coherent injective resolution of $\mathscr{L}[1]$ given by the two-term complex $\mathscr{K}^{\bullet}$ supported in degrees -1, 0:

$$(3.2.1) \qquad i_{\xi*}(\mathscr{L}_{\xi}) \to \bigoplus_{x \in X^0} i_{x*}(\mathscr{L}_{\xi}/\mathscr{L}_x),$$

where localization at $x \in X^0$ yields the canonical projection map $\mathscr{L}_{\xi} \to \mathscr{L}_{\xi}/\mathscr{L}_x$. We regard this complex as a rather special isomorphism class representative (in $\mathbf{D}^b_c(X)$) for $\mathscr{L}[1]$, since up to isomorphism every injective hull at a point of X appears 'exactly once' in (3.2.1).

In general, a *residual complex* on a locally noetherian scheme X is a bounded below complex $\mathscr{K}^{\bullet}$ of quasi-coherent injectives in $\mathbf{D}_c(X)$ (note the assumed *coherence* of the cohomology) such that there is an isomorphism of $\mathcal{O}_X$-modules

$$\bigoplus_{p \in \mathbf{Z}} \mathscr{K}^p \simeq \bigoplus_{x \in X} \mathscr{J}(x),$$

where $\mathscr{J}(x) = i_{x*}J(x)$, with $i_x : \mathrm{Spec}(\mathcal{O}_{X,x}) \to X$ the canonical map and $J(x)$ the quasi-coherent sheaf on $\mathrm{Spec}(\mathcal{O}_{X,x})$ associated to an injective hull of $k(x)$ over $\mathcal{O}_{X,x}$ (so $J(x)$ is supported at $\{x\} \subseteq \mathrm{Spec}(\mathcal{O}_{X,x})$). Given a residual complex $\mathscr{K}^{\bullet}$ on X and a point $x \in X$, there is a unique integer $d = d_{\mathscr{K}^{\bullet}}(x)$ such that $\mathscr{J}(x)$ is a direct summand of $\mathscr{K}^d$ (see Lemma 2.1.5), so

$$(3.2.2) \qquad \mathscr{K}^p \simeq \bigoplus_{d_{\mathscr{K}^{\bullet}}(x)=p} \mathscr{J}(x).$$

For example, the complex (3.2.1) is a residual complex with $d_{\mathscr{K}^{\bullet}}$ equal to -1 plus the usual codimension function on X. Note also that in the general case, the *coherence* condition on the cohomology of $\mathscr{K}^{\bullet}$ imposes rather strong conditions (consider the consequence of the vanishing all differentials in $\mathscr{K}^{\bullet}$).

For any residual complex $\mathscr{K}^{\bullet}$ on X, the function $d_{\mathscr{K}^{\bullet}}$ jumps up by 1 under immediate specialization (this follows from [**RD**, IV, 1.1(a)], to be discussed shortly), so it is called the *codimension function on X associated to* $\mathscr{K}^{\bullet}$ and we define the associated filtration $Z^{\bullet}(\mathscr{K}^{\bullet})$ on X in the obvious way, analogous to (3.1.27). Since $d_{\mathscr{K}^{\bullet}}$ differs from the codimension function by a constant on each irreducible component of X, we see that for noetherian X with finite Krull dimension, residual complexes have bounded codimension functions and so are automatically bounded as complexes (which is stronger than having bounded cohomology).

In our Dedekind scheme example above, the functor $Q(\cdot)$ from complexes of $\mathscr{O}_X$-modules to $\mathbf{D}(X)$ takes the residual complex (3.2.1) to the dualizing complex $\mathscr{L}[1]$, preserving the associated codimension functions and filtrations. Conversely, if $Z^\bullet$ denotes the shift by [1] of the codimension filtration (i.e., $Z^\bullet(\mathscr{L}[1])$) and the Dedekind X is a smooth connected curve over an algebraically closed field, Lemma 3.1.1 implies that $E_{Z^\bullet}(\mathscr{L}[1])$ is exactly the residual complex (3.2.1).

More generally, in [**RD**, VI, 1.1(a),(b)] it is shown that for any locally noetherian scheme X, the functor Q from complexes of $\mathscr{O}_X$-modules to $\mathbf{D}(X)$ takes residual complexes to *weakly* pointwise dualizing complexes (preserving associated codimension functions and filtrations) and conversely, for any weakly pointwise dualizing complex $\mathscr{R}^\bullet$ on X, $E(\mathscr{R}^\bullet) \overset{\text{def}}{=} E_{Z^\bullet(\mathscr{R}^\bullet)}(\mathscr{R}^\bullet)$ is a residual complex with the same associated codimension function and filtration as $\mathscr{R}^\bullet$. Thus, we have a translation-compatible construction

$$E : \mathscr{R}^\bullet \rightsquigarrow E_{Z^\bullet(\mathscr{R}^\bullet)}(\mathscr{R}^\bullet)$$

from weakly pointwise dualizing complexes to residual complexes, and E preserves associated codimension functions and filtrations. In particular, a weakly pointwise dualizing complex $\mathscr{R}^\bullet$ is strongly pointwise dualizing (i.e., has locally bounded cohomology) if and only if the associated residual complex $E(\mathscr{R}^\bullet)$ has locally bounded cohomology. The sense in which E and Q are (sometimes) quasi-inverses will be discussed shortly.

The functorial nature of E is a little delicate. Consider a map $\varphi : \mathscr{R}_1^\bullet \to \mathscr{R}_2^\bullet$ between two weakly pointwise dualizing complexes, with $Z_i^\bullet = Z^\bullet(\mathscr{R}_i^\bullet)$. It only seems possible to define a functorial map

$$E(\varphi) : E_{Z_1^\bullet}(\mathscr{R}_1^\bullet) \to E_{Z_2^\bullet}(\mathscr{R}_2^\bullet)$$

when $Z_1^\bullet = Z_2^\bullet$. For an example with $Z_1^\bullet \neq Z_2^\bullet$, we can consider the case where $\mathscr{R}_2^\bullet = \mathscr{R}_1^\bullet[1]$ and $\varphi = \mathrm{d}_{\mathscr{R}_1^\bullet}$ is the map induced by the differential. Thus, E only makes sense as a *functor* when we fix a filtration $Z^\bullet$ of X which is a shift of the codimension filtration on each irreducible component (with codimension computed relative to that component) and we consider *only* those weakly pointwise dualizing complexes $\mathscr{R}^\bullet$ for which $Z^\bullet(\mathscr{R}^\bullet) = Z^\bullet$. Clearly it is reasonable to consider only those residual complexes with a *fixed* associated filtration (or, equivalently, a *fixed* associated codimension function), since everything is translation-compatible.

In [**RD**, VI, 1.1(c)], it is claimed that when X admits a residual complex with *bounded* cohomology, then E and Q are quasi-inverse functors between pointwise dualizing complexes and residual complexes. Even if we fix a choice of the weak or strong sense of pointwise dualizing, this does not make sense since E does not make sense as a functor on pointwise dualizing complexes $\mathscr{R}^\bullet$ unless we fix the filtration $Z^\bullet(\mathscr{R}^\bullet)$. Moreover, Q is generally *not faithful* if we do not fix this filtration. For example, consider the map of residual complexes

$d^\bullet_{\mathscr{K}^\bullet} : \mathscr{K}^\bullet \to \mathscr{K}^\bullet[1]$ induced by the differential. This map is trivially homotopic to 0, so $Q(d^\bullet_{\mathscr{K}^\bullet}) = 0$, but in general $d^\bullet_{\mathscr{K}^\bullet} \neq 0$.

The correct formulation of [**RD**, VI, $1.1(c)$] generalizing Theorem 3.1.3 must account for this filtration condition:

LEMMA 3.2.1. *Let X be a locally noetherian scheme which has a residual complex or weakly pointwise dualizing complex. Assume that there is such a complex with bounded cohomology (this holds if X is noetherian with finite Krull dimension, since pointwise dualizing complexes in either sense are then dualizing), and let $Z^\bullet$ be a filtration on X which is a shift of the codimension filtration on each irreducible component of X.*

The notions of weakly and strongly pointwise dualizing coincide on X and the functors E and Q are naturally quasi-inverse functors between pointwise dualizing complexes whose associated filtration is $Z^\bullet$ and residual complexes whose associated filtration is $Z^\bullet$. Moreover, the isomorphisms $\beta : QE \simeq 1$, $\alpha : EQ \simeq 1$ satisfy $E(\beta) = \alpha(E)$, $Q(\alpha) = \beta(Q)$ and are defined in a way which respects Zariski localization, base change to $\mathrm{Spec}(\mathscr{O}_{X,x})$, tensoring with an invertible sheaf, and translations on complexes and the filtration $Z^\bullet$.

The proof of Lemma 3.2.1 follows the method of proof of Theorem 3.1.3 above, thanks to Lemma 3.1.5. We refer the reader to [**RD**, VI, 1.1] for details, noting that the construction of a functorial $EQ \simeq 1$ for *local* noetherian schemes X (or more generally, noetherian X with finite Krull dimension) follows from Theorem 3.1.3 and the essential uniqueness of dualizing complexes for such X (see (3.1.30)). This allows one to use the localization method in [**RD**, VI, 1.2] to define functorially $EQ \simeq 1$ in general and to prove (full) faithfulness of Q in general (subject to the hypotheses in Lemma 3.2.1).

Since isomorphic residual complexes have the *same* associated filtration, Lemma 3.2.1 yields translation-compatible isomorphisms $EQ \simeq 1$, $QE \simeq 1$ which are in particular functorial with respect to *isomorphisms*. Beware that a residual complex with bounded cohomology does *not* need to be bounded as a complex (or equivalently, have a bounded codimension function). Indeed, by Lemma 3.2.1, it suffices to give an example of a weakly pointwise dualizing complex which has bounded cohomology but does not have finite injective dimension (i.e., which is not dualizing). As usual, $\mathscr{O}_X[0]$ is such a weakly pointwise dualizing complex for a regular noetherian scheme X with infinite Krull dimension. [**RD**, V, 8.2] ensures that this problem never occurs on a noetherian (i.e., quasi-compact) X with *finite* Krull dimension; i.e., residual complexes on such an X are automatically bounded as complexes.

Now that we have reviewed the basic facts relating residual and pointwise dualizing complexes, we end this section by proving a generalization of a result about residual complexes in [**RD**, VI] which we will need in §3.3. The proof makes essential use of the non-trivial Theorem 2.7.2. For the rest of this section, unless otherwise indicated, we only consider *locally finite type maps* between locally noetherian schemes, with *bounded* fiber dimension. For any such morphism $f : X \to Y$, we define (following [**RD**, VI, §2]) the translation-compatible functors between residual complexes

$$f^y = E \circ f^\flat \circ Q$$

if f is finite and

$$f^z = E \circ f^\sharp \circ Q$$

if f is separated and smooth. We use (2.7.9) to identify f^z and f^y when f is finite étale. This is all compatible with localization and tensoring by an invertible sheaf! Note that f^y and f^z are only functorial with respect to maps between residual complexes with the *same* associated filtrations, so in particular are functorial with respect to isomorphisms (which is all that we need). Also, when $f : X \to Y$ is an open immersion, f^z is canonically identified with the functor 'restrict to X'.

As an example, let A be a local artin ring with algebraically closed residue field k and let $f : X \to \mathrm{Spec}(A)$ be a smooth, connected curve, with generic point ξ and set of closed points X^0. Let $K = \mathscr{O}_{X,\xi}$, the total ring of fractions of any of the $\mathscr{O}_{X,x}$'s for $x \in X^0$, and let $Z^\bullet$ denote the codimension filtration on X. Choose an injective hull I for k over A, with corresponding quasi-coherent (even coherent) sheaf $\widetilde{I}$ on $\mathrm{Spec}(A)$. The complex $\widetilde{I}[0]$ is obviously a residual complex on $\mathrm{Spec}(A)$ which induces the usual codimension filtration on $\mathrm{Spec}(A)$, so (3.1.25) and Corollary 3.1.2 tell us that the residual complex $f^z(\widetilde{I})$ has corresponding filtration $Z^\bullet[1]$ and is exactly the two-term complex in degrees -1, 0 given by

$$i_{\xi*}(I \otimes_A \Omega^1_{K/A}) \to \bigoplus_{x \in X^0} i_{x*}(I \otimes_A (\Omega^1_{K/A}/\Omega^1_{\mathscr{O}_{X,x}/A})),$$

where the map on the stalk at $x \in X^0$ is the map

$$I \otimes_A \Omega^1_{K/A} \to I \otimes_A (\Omega^1_{K/A}/\Omega^1_{\mathscr{O}_{X,x}/A})$$

which is the canonical projection. For example, if $A = k$ then $\widetilde{k}[0]$ is a residual complex and dualizing complex on $\mathrm{Spec}(k)$, $f^\sharp(\widetilde{k}[0]) = \Omega^1_{X/k}[1]$ is a *dualizing* complex on X, and

$$f^z(\widetilde{k}[0]) = E_{Z^\bullet[1]}(\Omega^1_{X/k}[1])$$

is a *residual* complex on X, namely the example in (3.1.3).

Back in the general situation, since noetherian local rings have finite Krull dimension, by using localization and the functoriality of $QE \simeq 1$ with respect to *isomorphisms* in the noetherian finite Krull dimension case (such as the $\psi_{f,g}$'s introduced in §2.7), all of the compatibilities for $(\cdot)^\flat$ and $(\cdot)^\sharp$, such as (2.7.4) and (2.7.5), yield analogues for $(\cdot)^y$ and $(\cdot)^z$ [**RD**, VI, Lemma 1.2, §2]. In [**RD**, VI, 3.1], these analogues are used to 'glue' the constructions $(\cdot)^y$ and $(\cdot)^z$ into a construction denoted $(\cdot)^\Delta$ for locally finite type maps with bounded fiber dimension, and $(\cdot)^\Delta$ is equipped with the extra data consisting of translation-compatible natural isomorphisms $\psi_f : f^\Delta \simeq f^y$ for finite f, $\varphi_f : f^\Delta \simeq f^z$ for smooth separated f, and

(3.2.3) $c_{f,g} : (gf)^\Delta \simeq f^\Delta g^\Delta$

for general f, g, as well as a number of compatibilities with the $\psi_{f,g}$'s, tensoring with an invertible sheaf, localization, translations, etc. Everything here, including the f^Δ's, is only functorial with respect to maps between residual complexes with the *same* associated filtration, but since the role of residual complexes in the construction of Grothendieck duality theory is just as an intermediate tool that only matters up to isomorphism, translation, and tensoring with an invertible sheaf (due to (3.1.30) and Lemma 3.2.1), the filtration restriction on the functorial nature of $(\cdot)^\Delta$ will not be a problem. We note that by using (3.2.3), we obtain analogues of (3.2.2) and (3.1.26) for residual complexes. Namely, for any locally finite type $f : X \to Y$ with bounded fiber dimension and any residual complex $\mathscr{K}^\bullet$ on Y, we have

$$(3.2.4) \qquad d_{f^\Delta \mathscr{K}^\bullet}(x) = d_{\mathscr{K}^\bullet}(f(x)) - \operatorname{trdeg}(k(x)/k(f(x)))$$

for all $x \in X$.

The construction $f \rightsquigarrow f^\Delta$ is uniquely characterized by a list of properties [**RD**, VI, 3.1, VAR1–VAR5] and the proof of the existence and uniqueness of $(\cdot)^\Delta$ involves many unchecked commutative diagrams which are all essentially trivial to verify, based on what has come before. However, in Theorem 3.3.1 we will need a non-trivial generalized version of the property [**RD**, VI, 3.1, VAR5]. We want to explain this point in some detail.

First, we need to digress and briefly discuss one of the properties of $(\cdot)^y$ and $(\cdot)^z$ which is related to what we want to generalize. In [**RD**, VI, §2], it is shown that for a finite f with a factorization $f = p \circ i$ where i is a closed immersion and p is separated smooth, there is an isomorphism

$$(3.2.5) \qquad f^y \simeq i^y p^z$$

obtained from $QE \simeq 1$, (2.7.4), and localization to the noetherian finite Krull dimension case. The same argument shows that we have such an isomorphism when i is merely finite. Likewise, when i is finite but f is separated smooth, we get an isomorphism

$$(3.2.6) \qquad f^z \simeq i^y p^z$$

using (2.7.5) instead of (2.7.4). Both (3.2.5) with finite i and (3.2.6) are compatible with translations, tensoring with an invertible sheaf, and localization on the base.

Although the weaker version of (3.2.5) with i a closed immersion is sufficient for the "existence and uniqueness" characterization of $(\cdot)^\Delta$ for general maps in [**RD**, VI, 3.1, VAR1–VAR5], the case of (3.2.5) with finite i and the case of (3.2.6) with separated smooth f and finite i enables us to prove Theorem 3.2.2 below, which will be useful in §3.3. We note that [**RD**, VI, 3.1, VAR5] is the first case of Theorem 3.2.2 with i a closed immersion, and this special case is a triviality, since everything then follows from the *definition* of $c_{i,p}$ in (3.2.3).

THEOREM 3.2.2. *Consider a commutative diagram*

$$X \xrightarrow{\ i\ } P$$
$$f \searrow \quad \downarrow p$$
$$Y$$

where X, Y, P are locally noetherian schemes. Suppose that i is finite and p is separated and smooth with bounded fiber dimension.

1. *When f is finite, the diagram of isomorphisms*

$$
\begin{array}{ccc}
f^\Delta & \xrightarrow{\ c_{i,p}\ } & i^\Delta p^\Delta \\
\psi_f \downarrow & & \psi_i \downarrow \varphi_p \\
f^y & \xrightarrow{\ \simeq\ } & i^y p^z
\end{array}
$$

commutes, where the bottom row is (3.2.5).
2. *When f is smooth, the diagram of isomorphisms*

$$
\begin{array}{ccc}
f^\Delta & \xrightarrow{\ c_{i,p}\ } & i^\Delta p^\Delta \\
\varphi_f \downarrow & & \psi_i \downarrow \varphi_p \\
f^z & \xrightarrow{\ \simeq\ } & i^y p^z
\end{array}
$$

commutes, where the bottom row is (3.2.6).

PROOF. First suppose that i is a closed immersion. When f is finite, the definition of $c_{i,p}$ makes everything clear. When f is smooth, we just have to show that $f^z \simeq i^y p^z$ is a 'permissable' isomorphism in the sense of [**RD**, VI, p.322]. By using the scheme diagram

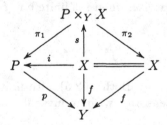

we are reduced to proving the commutativity of the diagram of isomorphisms

$$
\begin{array}{ccccc}
i^y p^z & \longrightarrow & s^y \pi_1^z p^z & \longleftarrow & s^y (p\pi_1)^z \\
\uparrow & & & & \| \\
f^z & \longrightarrow & s^y \pi_2^z f^z & \longleftarrow & s^y (f\pi_2)^z
\end{array}
$$

This follows from the second part of Theorem 2.7.2.

It remains to reduce to the case in which i is a closed immersion. This will be fairly easy when f is smooth, and less trivial when f is finite. Suppose f is smooth. Consider the diagram

(3.2.7)

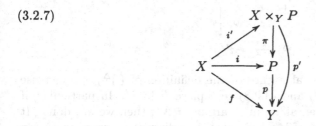

with π and p' smooth separated and i' a closed immersion. This yields a diagram of isomorphisms

(3.2.8)
$$
\begin{array}{ccccccc}
f^\Delta & \underset{c_{i,p}}{\overset{c_{i',p'}}{\rightleftarrows}} & i^\Delta p^\Delta & \underset{c_{i',\pi}}{\rightarrow} & i'^\Delta \pi^\Delta p^\Delta & \underset{c_{\pi,p}}{\leftarrow} & i'^\Delta p'^\Delta \\
\varphi_f \downarrow & & \psi_i \downarrow \; \varphi_p & & \psi_{i'} \downarrow \; \varphi_\pi \,\varphi_p & & \psi_{i'} \downarrow \; \varphi_{p'} \\
f^z & \underset{\gamma}{\rightarrow} & i^y p^z & \overset{\simeq}{\rightarrow} & i'^y \pi^z p^z & \overset{\simeq}{\leftarrow} & i'^y p'^z \\
& & & \gamma' & & &
\end{array}
$$

with the maps γ and γ' along the bottom equal to cases of (3.2.6). We want the left square in (3.2.8) to commute. The commutativity of the bottom part of (3.2.8) follows from the second part of Theorem 2.7.2, the commutativity of the top and right parts of (3.2.8) follows from [**RD**, VI, 3.1, VAR1, VAR3], and the commutativity of the middle square in (3.2.8) is the trivial case of the theorem with a finite map i factored into a closed immersion i' followed by a smooth separated map π. Thus, it suffices to prove the commutativity of the outer edges of (3.2.8), which reduces the case of smooth f to the known case when i is a closed immersion.

Finally, suppose that f is finite and i is finite. Since the composite

$$P \times_Y X \to P \to Y$$

is neither finite nor smooth in general, and i does not generally factor into a closed immersion followed by a smooth (separated) map to P, we cannot apply the second part of Theorem 2.7.2 so readily in this case. However, if X can be realized as a closed subscheme of a separated smooth P-scheme, then we can imitate the argument based on Theorem 2.7.2 in the 'f is smooth' case above. In particular, if P is affine then X is affine and there is a closed immersion $X \hookrightarrow \mathbf{A}_P^n$, so the case of affine P is settled. Thus, it is enough to reduce to the case of affine P.

Suppose that $j : U \hookrightarrow P$ is an open subscheme containing $i(X)$, so the induced map $i' : X \to U$ is finite. Since

$$
\begin{array}{ccc}
i^{\Delta} & \xrightarrow{c_{i',j}} & i'^{\Delta}j^{\Delta} \\
\psi_i \downarrow & & \psi_{i'} \downarrow \; \varphi_j \\
i^y & \xrightarrow[\;\simeq\;]{} & i'^y j^z
\end{array}
$$

trivially commutes (by the local nature of the definition of $(\cdot)^{\Delta}$), we can use diagrams analogous to (3.2.7) and (3.2.8) to replace P by U. In particular, if X is local, so there is an open affine in P around $i(X)$, then we are done. It remains to reduce to the case of local X. If X is a disjoint union of two open subschemes X_1 and X_2, it is clearly enough to study X_1 and X_2 separately in place of X. In particular, if Y is henselian local, so X is a finite disjoint union of (henselian) local schemes, we are done.

We can certainly assume without loss of generality that Y is local, so X is quasi-compact, and then we can replace P by a quasi-compact open subscheme around $i(X)$. Thus, without loss of generality, all of our schemes are noetherian with finite Krull dimension. This ensures that pointwise dualizing complexes are dualizing [**RD**, V, 8.2], so Lemma 3.2.1 applies, thereby simplifying the definitions of the isomorphisms relating $(\cdot)^y$, $(\cdot)^z$, and $(\cdot)^{\Delta}$ in our setting.

To reduce the case of local Y to the case of henselian local Y, we would like a base change theory for residual complexes with respect to henselization of the base. Note that if $A \to B$ is a flat local map of local noetherian rings, the functor $B \otimes_A (\cdot)$ from A-modules to B-modules usually destroys the property of being an injective hull, and hence does *not* generally take residual complexes to residual complexes. However, [**RD**, VI, §5] defines the notion of a *residually stable* map and proves that base change by such a map takes residual complexes to residual complexes and, via (3.1.8), there is a natural compatibility of $(\cdot)^{\Delta}$ with respect to residually stable base change [**RD**, VI, 5.5]. The definition of residual stability in [**RD**, VI, §5] is too restrictive. A more general definition, for which all proofs in [**RD**] go through, is the following: a (not necessarily locally finite type) morphism $f : X \to Y$ between locally noetherian schemes is said to be *residually stable* if

- f is flat,
- the fibers of f are discrete and for all $x \in X$, the extension $k(x)/k(f(x))$ is algebraic,
- the fibers of f are Gorenstein schemes

(in [**RD**], the second condition is replaced by the condition that f is an integral map, a property which is rarely satisfied by open immersions).

The commutativity of diagrams of quasi-coherent sheaves can be checked after faithfully flat base change, so in order to justify reduction to the case of henselian local Y it suffices to prove that the faithfully flat base change from a local noetherian ring A to its henselization is always residually stable. Let A be a local noetherian ring and $f : A \to A'$ a (strict) henselization. We claim that f is residually stable. By the construction of (strict) henselizations as a direct limit of étale maps, the only non-trivial issue is to check that the fibers of $\mathrm{Spec}(f)$ are discrete. Use [**EGA**, IV$_4$, 18.6.9(ii)] in the case of henselizations and [**EGA**, IV$_4$, 18.8.12(ii)] in the case of strict henselizations. This completes the proof of Theorem 3.2.2.

∎

All other unproven facts about residual and dualizing complexes in [**RD**, VI] amount to unravelling the definitions and using previously proven compatibilities (such as the results we have proved in Chapter 2). No serious difficulties arise. Since base change for residual complexes only makes sense for residually stable base change, the use of residual complexes to construct Grothendieck duality theory in [**RD**, VII] (and §3.4 below) makes the general base change compatibility of the trace morphism (1.1.1) quite non-obvious. This problem will be treated in Chapter 4.

3.3. The Functor $(\cdot)^!$ and Residual Complexes

In this section, unless otherwise specified, we only consider schemes which are noetherian and admit a dualizing complex (or equivalently, admitting a residual complex which is bounded as a complex [**RD**, V, 8.2; VI, 1.1]), but we will often remind the reader of this hypothesis. In particular, all schemes to be considered automatically have finite Krull dimension [**RD**, V, 7.2]. Any scheme of finite type over a regular ring with finite Krull dimension admits a dualizing complex [**RD**, p.299]. This includes finite type schemes over **Z**, a field, or a complete local noetherian ring. We want to review the ingredients that go into the 'residual complex' duality theorem [**RD**, VII, 3.4]. This will serve to fix the basic concepts which we will frequently use, and will provide what is needed for the construction of dualizing sheaves and duality theorems later on.

Let $f : X \to Y$ be a finite type morphism between noetherian schemes which admit a dualizing complex. There is a map of triangulated categories

$$(3.3.1) \qquad f^! : \mathbf{D}_c^+(Y) \to \mathbf{D}_c^+(X)$$

whose definition we will recall in (3.3.6)–(3.3.13) below and whose basic properties we will then review in the rest of this section. The duality theorem for f requires f to be proper and uses a trace map of δ-functors

$$(3.3.2) \qquad \mathrm{Tr}_f : \mathbf{R}f_* \circ f^! \to 1.$$

The idea is that Tr_f should make $f^!$ a right adjoint to $\mathbf{R}f_*$. In §3.4 we will define (3.3.2), review its basic properties, and then use (3.3.1) and (3.3.2) to give the definition of the duality morphism and the statement of the duality theorem (for noetherian schemes admitting a dualizing complex) in Theorem 3.4.4. Many definitions will be given in terms of *choices* of residual complexes, so it is important to check that everything we do is independent of such choices (in a suitable sense).

Throughout, we will need to use 'duality functors' which are defined in terms of residual complexes. Let $K^\bullet$ be a residual complex on Y, so $K^\bullet$ is a bounded complex of quasi-coherent injective $\mathscr{O}_Y$-modules and $K^\bullet$ has *coherent* cohomology sheaves. The contravariant 'duality' δ-functor

$$(3.3.3) \qquad D_{K^\bullet} = D_{Y,K^\bullet} : \mathbf{D}_c(Y) \to \mathbf{D}_c(Y)$$

is defined to be $\mathbf{R}\mathscr{H}om_Y^\bullet(\cdot, K^\bullet)$. If $K^\bullet$ is fixed, we will sometimes denote this by D_Y or even D. This δ-functor interchanges $\mathbf{D}_c^+(Y)$ and $\mathbf{D}_c^-(Y)$. Since residual complexes on Y are dualizing complexes in the derived category (by Lemma 3.2.1), there is a canonical isomorphism of δ-functors

$$(3.3.4) \qquad \eta = \eta_{K^\bullet} : 1 \simeq DD$$

which involves an intervention of signs as in (1.3.16); this is the notation used in [**RD**], and there seems to be only a small risk of confusion with the notation for the fundamental local isomorphism (2.5.3), which will only appear explicitly again in §3.5 and Appendix A, where the map (3.3.4) plays no explicit role. Note the important fact that the composite

$$(3.3.5) \qquad D \xrightarrow{\eta D} DDD \xrightarrow{D(\eta)} D$$

is the identity; this follows from (1.3.21).

Let $f : X \to Y$ be a finite type map between two noetherian schemes admitting a dualizing complex and let $K^\bullet$ be a residual complex on Y, so $f^\triangle K^\bullet$ is a residual complex on X. Since $D_{K^\bullet}$ and $D_{f^\triangle K^\bullet}$ induce contravariant 'autodualities' of $\mathbf{D}_c(Y)$ and $\mathbf{D}_c(X)$ respectively and $\mathbf{L}f^*$ is a left adjoint to $\mathbf{R}f_*$, the morphism of triangulated categories $f^! : \mathbf{D}_c^+(Y) \to \mathbf{D}_c^+(X)$ given by

$$(3.3.6) \qquad f_{K^\bullet}^! \overset{\text{def}}{=} D_{f^\triangle K^\bullet} \circ \mathbf{L}f^* \circ D_{K^\bullet}$$

is a right adjoint to $\mathbf{R}f_*$ on $\mathbf{D}_c^+(Y)$. We want to make precise the sense in which (3.3.6) is naturally independent of the choice of $K^\bullet$, thereby justifying the notation $f^!$.

Before we formulate the sense in which (3.3.6) is independent of the choice of $K^\bullet$, we need to recall the relation between different choices of residual complexes on Y and the relation between the corresponding duality functors. If $K'^\bullet$ is another residual complex on Y, it follows from [**RD**, V, 3.1], (3.1.28)–(3.1.30), and Lemma 3.2.1 that there is a locally constant $\mathbf{Z}$-valued function $n(K^\bullet, K'^\bullet)$ on Y and an invertible sheaf $\mathscr{L}(K^\bullet, K'^\bullet)$ on Y for which there is an isomorphism of complexes

$$(3.3.7) \qquad \beta_{K^\bullet,K'^\bullet} : K'^\bullet \simeq \mathscr{L}(K^\bullet, K'^\bullet)[n(K^\bullet, K'^\bullet)] \otimes K^\bullet$$

and if $K''^\bullet$ is a third residual complex on Y, there is an isomorphism of complexes

$$(3.3.8) \qquad \begin{array}{c} \mathscr{L}(K'^\bullet, K''^\bullet)[n(K'^\bullet, K''^\bullet)] \otimes \mathscr{L}(K^\bullet, K'^\bullet)[n(K^\bullet, K'^\bullet)] \\ \Big\downarrow \simeq \\ \mathscr{L}(K^\bullet, K''^\bullet)[n(K^\bullet, K''^\bullet)] \end{array}$$

compatible with Zarsiki localization, base change to $\mathrm{Spec}(\mathscr{O}_{Y,y})$, and with respect to a choice of a fourth residual complex on Y. Moreover, (3.3.7) is 'well-behaved' via (3.3.8) with respect to a third choice of residual complex on Y.

Let $\mathscr{L}$ be an invertible sheaf on Y and let n be a locally constant $\mathbf{Z}$-valued function on Y. By (3.1.7) and the local nature of the definition of $(\cdot)^\Delta$, there is a canonical isomorphism of complexes

$$(3.3.9) \qquad f^\Delta(\mathscr{L}[n] \otimes K^\bullet) \simeq (f^*\mathscr{L})[n] \otimes f^\Delta K^\bullet \simeq (f^*\mathscr{L} \otimes f^\Delta K^\bullet)[n]$$

which involves no intervention of signs. Also, for any two complexes of $\mathscr{O}_Y$-modules $\mathscr{F}^\bullet$, $\mathscr{G}^\bullet$ and any invertible $\mathscr{O}_Y$-module $\mathscr{L}$, there is a natural isomorphism of complexes

$$(3.3.10) \qquad \mathscr{H}om_Y^\bullet(\mathscr{L}[n] \otimes \mathscr{F}^\bullet, \mathscr{L}[n] \otimes \mathscr{G}^\bullet) \simeq \mathscr{H}om_Y^\bullet(\mathscr{F}^\bullet, \mathscr{G}^\bullet)$$

which involves a sign of $(-1)^{nm}$ in degree m, by (1.3.11). The isomorphism (3.3.10) therefore gives rise to a natural isomorphism

$$(3.3.11) \qquad D_{\mathscr{L}[n] \otimes K^\bullet}(\mathscr{L}[n] \otimes (\cdot)) \simeq D_{K^\bullet}$$

which involves an intervention of signs.

Using (3.3.7), (3.3.9), and (3.3.11), and letting $\mathscr{L} = \mathscr{L}(K^\bullet, K'^\bullet)$, $n = n(K^\bullet, K'^\bullet)$, we define the isomorphism of functors

$$(3.3.12) \qquad \phi_{f, K^\bullet, K'^\bullet} : f_{K'^\bullet}^! \simeq f_{K^\bullet}^!$$

by

$$f_{K'^\bullet}^! \xrightarrow{\ \simeq\ } D_{f^*\mathscr{L}[n] \otimes f^\Delta K^\bullet} \circ \mathbf{L}f^* \circ D_{\mathscr{L}[n] \otimes K^\bullet}$$

$$\Big\downarrow \simeq$$

$$D_{f^*\mathscr{L}[n] \otimes f^\Delta K^\bullet} \circ (f^*\mathscr{L}[n] \otimes \mathbf{L}f^* \circ D_{K^\bullet}(\cdot))$$

$$\Big\downarrow \simeq$$

$$D_{f^\Delta K^\bullet} \circ \mathbf{L}f^* \circ D_{K^\bullet}$$

$$\Big\| $$

$$f_{K^\bullet}^!$$

It is easy to check that $\phi_{f, K^\bullet, K^\bullet} = 1$ and for any three residual complexes $K_1^\bullet$, $K_2^\bullet$, $K_3^\bullet$ on Y,

$$(3.3.13) \qquad \phi_{f, K_2^\bullet, K_3^\bullet} \circ \phi_{f, K_1^\bullet, K_2^\bullet} = \phi_{f, K_1^\bullet, K_3^\bullet}$$

(this uses the compatibility of (1.3.11) with respect to "$[n_1 + n_2] = [n_1] \circ [n_2]$"). This is the strong sense in which we can define $f^! = f_{K^\bullet}^!$ and say that $f^!$ is 'independent' of the choice of $K^\bullet$. In many proofs and definitions, it will be essential that we can make certain convenient choices of residual complexes based on other choices of residual complexes, so the 'independence of the choice of $K^{\bullet}$' in the definition of $f^!$ is important.

If $X \xrightarrow{f} Y \xrightarrow{g} Z$ is a composite of finite type maps and $K^{\bullet}$ is a residual complex on Z, then $g^{\Delta} K^{\bullet}$ is a residual complex on Y and $f^{\Delta} g^{\Delta} K^{\bullet} \simeq (gf)^{\Delta} K^{\bullet}$ is a residual complex on X. Using these choices of residual complexes on X and Y which are 'compatible' with the choice on Z, we define the δ-functorial isomorphism

$$(3.3.14) \qquad\qquad c_{f,g} : (gf)^! \simeq f^! g^!$$

to be the composite

$$(3.3.15) \qquad (gf)^!_{K^{\bullet}} =\!\!=\!\!=\!\!=\!\!=\!\!=\!\!= D_{(gf)^{\Delta} K^{\bullet}} \mathbf{L}(gf)^* D_{K^{\bullet}}$$

$$\Big\downarrow \simeq$$

$$D_{f^{\Delta} g^{\Delta} K^{\bullet}} \mathbf{L} f^* \mathbf{L} g^* D_{K^{\bullet}}$$

$$\eta_{g^{\Delta} K^{\bullet}} \Big\uparrow \simeq$$

$$D_{f^{\Delta}(g^{\Delta} K^{\bullet})} \mathbf{L} f^* D_{g^{\Delta} K^{\bullet}} D_{g^{\Delta} K^{\bullet}} \mathbf{L} g^* D_{K^{\bullet}}$$

$$\Big\|$$

$$f^!_{g^{\Delta} K^{\bullet}} g^!_{K^{\bullet}}$$

It is a straightfoward matter to check that (3.3.14) is well-defined in the sense that (3.3.15) is independent of the choice of $K^{\bullet}$ via (3.3.12), and that (3.3.14) is 'associative' with respect to any third finite type map $h : Z \to W$. Lemma 3.3.2 below will remove any confusion between (3.2.3) and (3.3.14).

In addition to (3.3.14), three other important properties of $(\cdot)^!$ are compatibilities with $(\cdot)^{\flat}$ and $(\cdot)^{\sharp}$ for finite and separated smooth maps respectively, and behavior with respect to residually stable base change. These are given by (3.3.19), (3.3.21), and (3.3.24) below, as we now explain.

It follows from Lemma 3.2.1 and the definition of $(\cdot)^{\Delta}$ in [**RD**, VI, 3.1] that there are translation-compatible isomorphisms in the *derived category* (naturally compatible with respect to tensoring with an invertible sheaf)

$$(3.3.16) \qquad f^{\Delta}(K^{\bullet}) \simeq \begin{cases} f^{\flat}(K^{\bullet}), & \text{for finite } f \\ f^{\sharp}(K^{\bullet}), & \text{for separated smooth } f \end{cases}$$

for any residual complex $K^{\bullet}$ on Y (and when f is finite étale, (3.3.16) is compatible with the isomorphism $f^{\flat} \simeq f^{\sharp}$ from (2.7.9)). The isomorphism (3.3.16) is functorial with respect to maps $K_1^{\bullet} \to K_2^{\bullet}$ between residual complexes which have the *same* associated codimension function (or, equivalently, the *same* associated filtration), due to the intervention of Lemma 3.2.1 in the definition of (3.3.16).

A special case of (3.3.16) which will be important later is the case of a closed immersion $f = j : X \hookrightarrow Y$. In this case, since any residual complex $K^{\bullet}$ on Y is a bounded complex of (quasi-coherent) injectives which is a Cousin complex with respect to the associated filtration $Z^{\bullet}(K^{\bullet})$ on Y, we have *by definition* the equality

$$(3.3.17) \qquad j^{\Delta}(K^{\bullet}) = \mathscr{H}om_Y(j_*\mathscr{O}_X, K^{\bullet})$$

as complexes of quasi-coherent *injective* sheaves, and the isomorphism (3.3.16) for the map j is the derived category isomorphism

$$(3.3.18) \qquad j^{\Delta}(K^{\bullet}) = \mathscr{H}om_Y(j_*\mathscr{O}_X, K^{\bullet}) \simeq \mathbf{R}\mathscr{H}om_Y(j_*\mathscr{O}_X, K^{\bullet}) = j^b(K^{\bullet})$$

arising from the canonical map from any left exact functor to its total derived functor.

For a finite map $f : X \to Y$ and a residual complex $K^{\bullet}$ on Y, we use (3.3.16) to define the δ-functorial isomorphism $d_f : f^b \simeq f^!$ to be

$$(3.3.19) \qquad d_f : f^b \xrightarrow{\;\;\simeq\;\;} f^b \circ D_{Y,K^{\bullet}} \circ D_{Y,K^{\bullet}}$$

$$\|$$

$$f^b\mathbf{R}\mathscr{H}om_Y^{\bullet}(D_{Y,K^{\bullet}}(\cdot), K^{\bullet})$$

$$\text{qism} \Big\uparrow \simeq$$

$$\mathbf{R}\mathscr{H}om_X^{\bullet}(\mathbf{L}f^* \circ D_{Y,K^{\bullet}}(\cdot), f^b K^{\bullet} \simeq f^{\Delta}K^{\bullet})$$

$$\|$$

$$D_{X,f^{\Delta}K^{\bullet}} \circ \mathbf{L}f^* \circ D_{Y,K^{\bullet}},$$

where the isomorphism labelled 'qism' is a special case of the δ-bifunctorial isomorphism [**RD**, III, 6.9(b)]

$$(3.3.20) \qquad \mathbf{R}\mathscr{H}om_X^{\bullet}(\mathbf{L}f^*\mathscr{F}^{\bullet}, f^b\mathscr{G}^{\bullet}) \xrightarrow{\;\;\simeq\;\;} f^b\mathbf{R}\mathscr{H}om_Y^{\bullet}(\mathscr{F}^{\bullet}, \mathscr{G}^{\bullet})$$

for $\mathscr{F}^{\bullet} \in \mathbf{D}_c^-(Y)$, $\mathscr{G}^{\bullet} \in \mathbf{D}_{qc}^+(Y)$, defined by replacing $\mathscr{F}^{\bullet}$ with a bounded above complex of flats and $\mathscr{G}^{\bullet}$ with a bounded below complex of injectives. Meanwhile, for separated smooth f, we define the δ-functorial isomorphism $e_f : f^{\sharp} \simeq f^!$ to be

$$(3.3.21) \qquad e_f : f^{\sharp} \xrightarrow{\;\;\simeq\;\;} f^{\sharp} \circ D_{Y,K^{\bullet}} \circ D_{Y,K^{\bullet}}$$

$$\|$$

$$f^{\sharp}\mathbf{R}\mathscr{H}om_Y^{\bullet}(D_{Y,K^{\bullet}}(\cdot), K^{\bullet})$$

$$\text{qism} \Big\downarrow \sim$$

$$\mathbf{R}\mathscr{H}om_X^{\bullet}(\mathbf{L}f^* \circ D_{Y,K^{\bullet}}(\cdot), f^{\sharp} K^{\bullet} \simeq f^{\Delta}K^{\bullet})$$

$$\|$$

$$D_{X,f^{\Delta}K^{\bullet}} \circ \mathbf{L}f^* \circ D_{Y,K^{\bullet}},$$

where the isomorphism labelled 'qism' is a special case of the δ-bifunctorial isomorphism [**RD**, III, 2.4(b)]

$$(3.3.22) \qquad f^{\natural}\mathbf{R}\mathscr{H}om_Y^{\bullet}(\mathscr{F}^{\bullet},\mathscr{G}^{\bullet}) \xrightarrow{\;\simeq\;} \mathbf{R}\mathscr{H}om_X^{\bullet}(\mathbf{L}f^*\mathscr{F}^{\bullet}, f^{\natural}\mathscr{G}^{\bullet})$$

$$\downarrow{\simeq}$$

$$\mathbf{R}\mathscr{H}om_X^{\bullet}(f^*\mathscr{F}^{\bullet}, f^{\natural}\mathscr{G}^{\bullet})$$

for $\mathscr{F}^{\bullet} \in \mathbf{D}_c^-(Y)$, $\mathscr{G}^{\bullet} \in \mathbf{D}_{qc}^+(Y)$, defined by replacing $\mathscr{G}^{\bullet}$ with a bounded below complex of injectives.

It is easy to check that (3.3.19) and (3.3.21) are independent of the choice of $K^{\bullet}$ (via (3.3.12)). Also, when f is finite étale, the isomorphisms e_f and d_f are compatible with the isomorphism $f^{\flat} \simeq f^{\natural}$ from (2.7.9) (for a proof, use Lemma 2.7.1).

Finally, we define the compatibility of $(\cdot)^!$ with respect to *residually stable* base change. Consider a cartesian diagram

$$\begin{array}{ccc} X' & \xrightarrow{u'} & X \\ {\scriptstyle f'}\downarrow & & \downarrow{\scriptstyle f} \\ Y' & \xrightarrow{u} & Y \end{array}$$

with u a residually stable map (in the sense defined near the end of §3.2, generalizing the definition in [**RD**, VI, §5]). Choose a residual complex $K^{\bullet}$ on Y, so $u^*K^{\bullet}$ is a residual complex on Y' [**RD**, VI, 5.3]. There is an obvious δ-functorial isomorphism

$$(3.3.23) \qquad u^* \circ D_{Y,K^{\bullet}} \simeq D_{Y',u^*K^{\bullet}} \circ u^*.$$

If we use (3.3.23) and the natural isomorphism $u'^*f^{\triangle} \simeq f'^{\triangle}u^*$ (cf. [**RD**, VI, 5.5], (3.1.8)), we define the δ-functorial isomorphism

$$(3.3.24) \qquad b_{u,f} : u'^* \circ f^! \simeq f'^! \circ u^*$$

to be

$$(3.3.25)$$

$$u'^* \circ D_{X,f^{\triangle}K^{\bullet}} \circ \mathbf{L}f^* \circ D_{Y,K^{\bullet}} \xrightarrow{\;\simeq\;} D_{X',u'^*f^{\triangle}K^{\bullet}} \circ u'^* \circ \mathbf{L}f^* \circ D_{Y,K^{\bullet}}$$

$$\downarrow{\simeq}$$

$$D_{X',f'^{\triangle}u^*K^{\bullet}} \circ \mathbf{L}f'^* \circ u^* \circ D_{Y,K^{\bullet}}$$

$$\uparrow{\simeq}$$

$$(D_{X',f'^{\triangle}u^*K^{\bullet}} \circ \mathbf{L}f'^* \circ D_{Y',u^*K^{\bullet}}) \circ u^*$$

It is easy to check that (3.3.25) is independent of the choice of $K^{\bullet}$. When u is an open immersion, this makes precise the compatibility of $f^!$ with respect to Zariski localization.

In [**RD**, VII, 3.4(a)], it is asserted (without proof) that $(\cdot)^!$ and the compatibility data (3.3.14), (3.3.19), (3.3.21), (3.3.24) satisfy six 'variance' properties denoted (VAR1)–(VAR6). Some of the proofs are straightfoward applications of previously proven compatibilities of $(\cdot)^\flat$, $(\cdot)^\sharp$, and $(\cdot)^\Delta$ with respect to composites, base change, etc., but there are some delicate points which should be noted. We consider each of the compatibilities in [**RD**, VII, 3.4(a)] separately, in increasing order of difficulty:

- (VAR1) *There are natural isomorphisms* $\mathrm{id}^! \simeq \mathrm{id}$, $c_{f,1} \simeq 1$, $c_{1,g} \simeq 1$ *and the isomorphism* (3.3.14) *is compatible with respect to triple composites.* The isomorphism $\mathrm{id}^! \simeq \mathrm{id}$ is defined using the double duality isomorphism (3.3.4), and the identifications $c_{f,1}$, $c_{1,g} \simeq 1$ use the fact that (3.3.5) is the identity. Finally, the compatibility of (3.3.14) with respect to triple composites makes essential use of the fact that (3.3.14) is independent of the choice of residual complex.
- (VAR2), (VAR3) *The isomorphisms* e_f *and* d_f *are compatible with composites of separated smooth maps and composites of finite maps respectively.* This is clear.
- (VAR6) *The isomorphisms* e_f *and* d_f *are compatible with residually stable base change (using* (3.3.24)*), and* $b_{u,f}$ *is compatible with composites in* u *and* f. The main point is that the isomorphisms in Lemma 3.2.1, which are implicit in the definition of (3.3.16), are compatible with residually stable base change. This follows from unwinding the definitions and using (3.1.5).

The remaining compatibilities (VAR4), (VAR5) are harder and so are best given in the form of a theorem.

THEOREM 3.3.1. [**RD**, VII, 3.4(a)]

1. (VAR5) *Consider a commutative diagram of noetherian schemes admitting a dualizing complex*

$$
\begin{array}{ccc}
X & \xrightarrow{\ f\ } & Y \\
& {\scriptstyle gf} \searrow & \downarrow {\scriptstyle g} \\
& & Z
\end{array}
$$

with Z *separated,* f *finite,* g *separated smooth. Then we have the following commutative diagrams of isomorphisms:*
   - *If* gf *is finite, then*

(3.3.26)
$$
\begin{array}{ccc}
(gf)^\flat & \xrightarrow{\ \sim\ } & f^\flat g^\sharp \\
{\scriptstyle d_{gf}} \downarrow & & \downarrow {\scriptstyle d_f}\Big| {\scriptstyle e_g} \\
(gf)^! & \xrightarrow{\ \sim\ } & f^! g^!
\end{array}
$$

commutes.

- *If gf is separated smooth, then*

(3.3.27)
$$
\begin{array}{ccc}
(gf)^{\sharp} & \xrightarrow{\;\simeq\;} & f^{b}g^{\sharp} \\
{\scriptstyle e_{gf}}\big\downarrow & & {\scriptstyle d_{f}}\big\downarrow{\scriptstyle e_{g}} \\
(gf)^{!} & \xrightarrow[\;\simeq\;]{} & f^{!}g^{!}
\end{array}
$$

commutes.

2. (VAR4) *Consider a cartesian commutative diagram of noetherian schemes admitting a dualizing complex*

$$
\begin{array}{ccc}
X' & \xrightarrow{\;u'\;} & Y' \\
{\scriptstyle f'}\big\downarrow & {\scriptstyle h}\searrow & \big\downarrow{\scriptstyle f} \\
X & \xrightarrow[\;u\;]{} & Y
\end{array}
$$

with f finite and u separated smooth. The diagram of isomorphisms

(3.3.28)
$$
\begin{array}{ccc}
u'^{\sharp}f^{b} & \xrightarrow{\;\;\;\;\;\simeq\;\;\;\;\;} & f'^{b}u^{\sharp} \\
{\scriptstyle e_{u'}}\big\downarrow{\scriptstyle d_{f}} & & {\scriptstyle d_{f'}}\big\downarrow{\scriptstyle e_{u}} \\
u'^{!}f^{!} & \xleftarrow[\;\simeq\;]{} h^{!} \xrightarrow[\;\simeq\;]{} & f'^{!}u^{!}
\end{array}
$$

commutes.

The separatedness hypothesis on Z in (VAR5) is only needed because, as was noted in §2.8, the proof of the well-definedness of (2.8.4) involves unwinding definitions and using the difficult second part of Theorem 2.5.2, which we were only able to prove with a separatedness condition.

PROOF. (VAR5) follows immediately from Theorem 3.2.2 and the well-definedness of (2.8.4), which amounts to a compatibility of (3.3.22) and (3.3.20) via (2.7.4) and (2.7.5). The proof of (VAR4) is a bit involved, as we now explain. The basic idea is to make the definitions sufficiently explicit so that we can eliminate all appearances of $(\cdot)^{\Delta}$ which are implicit in the definition of $(\cdot)^{!}$ and reduce ourselves to a general commutativity claim that makes sense when Y is replaced by any locally noetherian scheme and a residual complex $K^{\bullet}$ on Y is replaced by any object in $\mathbf{D}_{\mathrm{qc}}^{+}(Y)$.

Going back to the definitions of the maps in (3.3.28), (VAR4) amounts to the commutativity of the following diagram of functors on $\mathbf{D}_{\mathrm{c}}^{+}(Y)$, where we write $K_{Y'}^{\bullet} = f^{\Delta}K^{\bullet}$, $K_{X}^{\bullet} = u^{\Delta}K^{\bullet}$, $K_{X'}^{\bullet} = h^{\Delta}K^{\bullet} = u'^{\Delta}K_{Y'}^{\bullet} = f'^{\Delta}K_{X}^{\bullet}$, and we use (3.3.4), (3.3.20), (3.3.22), and (3.3.16) in each column:

(3.3.29)

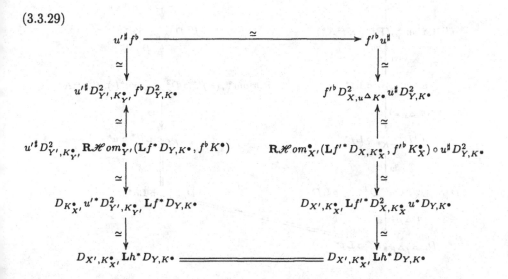

We note that in the middle of the columns on each side of (3.3.29), the
$\mathbf{R}\mathscr{H}om$ terms are functors through evaluation of the duality functors D_K in
their left arguments. All terms in (3.3.29) aside from $u'^{\sharp}f^{\flat}$ and $f'^{\flat}u^{\sharp}$ in the top
row are a composite $F \circ D_{Y,K^\bullet}$ for various functors F. Since $D_{Y,K^\bullet}$ induces an
anti-equivalence of categories between $\mathbf{D}_c^+(Y)$ and $\mathbf{D}_c^-(Y)$, with quasi-inverse

$D_{Y,K^\bullet}$ via (3.3.4), the commutativity of (3.3.29) is equivalent to the commutativity of the following diagram of functors on $\mathbf{D}_c^-(Y)$:

(3.3.30)

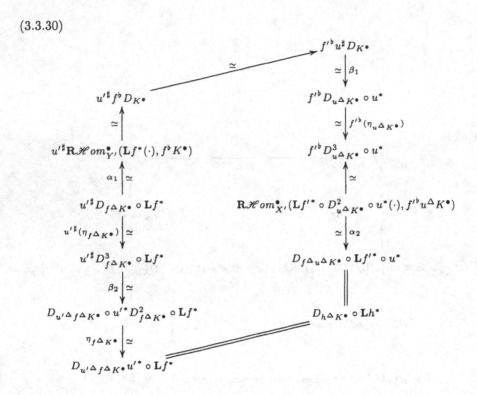

Here, the maps α_i, β_i use (3.3.16) and α_2 also uses the isomorphism $\eta_{u^\Delta K^\bullet}$ from (3.3.4).

In order to simplify the columns in (3.3.30), note that since $u' : X' \to Y'$ is a smooth separated map, for any residual complex $K'^\bullet$ on Y' the diagram of isomorphisms

(3.3.31)
$$
\begin{array}{ccc}
u'^{\sharp} D_{K'^\bullet} & \xrightarrow{\;\simeq\;} & D_{u'^\Delta K'^\bullet} \circ u'^* \\
{\scriptstyle u'^{\sharp}(\eta_{K'^\bullet})}\Big\downarrow & & \Big\uparrow{\scriptstyle \eta_{K'^\bullet}} \\
u'^{\sharp} D_{K'^\bullet}^3 & \xrightarrow[\;\simeq\;]{} & D_{u'^\Delta K'^\bullet} \circ u'^* \circ D_{K'^\bullet}^2
\end{array}
$$

commutes in $\mathbf{D}(X')$, and since $f' : X' \to X$ is a finite map, for any residual complex $K_X^\bullet$ on X the diagram of isomorphisms

$(3.3.32)$

$$
\begin{array}{ccc}
f'^b D_{K_X^\bullet} & \xrightarrow{\;\;f'^b(\eta_{K_X^\bullet})\;\;} & f'^b D_{K_X^\bullet}^3 \\[2pt]
\simeq \uparrow & & \uparrow \simeq \\[2pt]
\mathbf{R}\mathscr{H}om_{X'}^\bullet(\mathbf{L}f'^*(\cdot), f'^b K_X^\bullet) & & \mathbf{R}\mathscr{H}om_{X'}^\bullet(\mathbf{L}f'^* \circ D_{K_X^\bullet}^2(\cdot), f'^b K_X^\bullet) \\[2pt]
\simeq \downarrow & & \downarrow \simeq \\[2pt]
D_{f'^\Delta K_X^\bullet} \circ \mathbf{L}f'^* & \xleftarrow{\;\;\eta_{K_X^\bullet}\;\;} & D_{f'^\Delta K_X^\bullet} \circ \mathbf{L}f'^* \circ D_{K_X^\bullet}^2
\end{array}
$$

commutes in $\mathbf{D}(X')$. The commutativity of (3.3.31) and (3.3.32) follow easily from functoriality and the fact that (3.3.5) is the identity map.

The diagrams (3.3.31) and (3.3.32) allow us to replace (3.3.30) by the following diagram of isomorphisms between functors on $\mathbf{D}_c^-(Y)$:

$(3.3.33)$

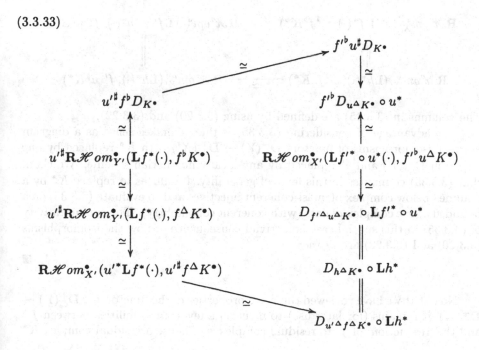

Before we verify that (3.3.33) commutes, we want to eliminate the $(\cdot)^{\Delta}$'s. By [**RD**, VI, 3.1(VAR4)] and [**RD**, VI, p.314(III)], the diagram of isomorphisms

$$(3.3.34)$$

$$
\begin{array}{ccc}
u'^{\sharp} f^{\flat} K^{\bullet} & \xrightarrow{\;\simeq\;} & f'^{\flat} u^{\sharp} K^{\bullet} \\
{\scriptstyle\simeq}\Big\downarrow & & \Big\downarrow{\scriptstyle\simeq} \\
u'^{\sharp} f^{\Delta} K^{\bullet} & & f'^{\flat} u^{\Delta} K^{\bullet} \\
{\scriptstyle\simeq}\Big\downarrow & & \Big\downarrow{\scriptstyle\simeq} \\
u'^{\Delta} f^{\Delta} K^{\bullet} & \xleftarrow[\;\simeq\;]{} h^{\Delta} K^{\bullet} \xrightarrow[\;\simeq\;]{} & f'^{\Delta} u^{\Delta} K^{\bullet}
\end{array}
$$

commutes in $\mathbf{D}(X')$. We can therefore 'identify' the top and bottom rows of (3.3.34), so the commutativity of (3.3.33) is equivalent to the commutativity of the following diagram of isomorphisms between functors from $\mathbf{D}_c^-(Y)$ to $\mathbf{D}_c^+(X')$:

$$(3.3.35)$$

$$
\begin{array}{ccc}
u'^{\sharp} f^{\flat} \mathbf{R}\mathscr{H}om_Y^{\bullet}(\cdot, K^{\bullet}) & \xrightarrow{\;\simeq\;} & f'^{\flat} u^{\sharp} \mathbf{R}\mathscr{H}om_Y^{\bullet}(\cdot, K^{\bullet}) \\
{\scriptstyle\simeq}\Big\uparrow & & \Big\downarrow{\scriptstyle\simeq} \\
u'^{\sharp} \mathbf{R}\mathscr{H}om_{Y'}^{\bullet}(\mathbf{L}f^*(\cdot), f^{\flat} K^{\bullet}) & & f'^{\flat} \mathbf{R}\mathscr{H}om_X^{\bullet}(u^*(\cdot), u^{\sharp} K^{\bullet}) \\
{\scriptstyle\simeq}\Big\downarrow & & \Big\uparrow{\scriptstyle\simeq} \\
\mathbf{R}\mathscr{H}om_{X'}^{\bullet}(u'^* \mathbf{L}f^*(\cdot), u'^{\sharp} f^{\flat} K^{\bullet}) & & \mathbf{R}\mathscr{H}om_{X'}^{\bullet}(\mathbf{L}f'^* \circ u^*(\cdot), f'^{\flat} u^{\sharp} K^{\bullet}) \\
\| & & \| \\
\mathbf{R}\mathscr{H}om_{X'}^{\bullet}(\mathbf{L}h^*(\cdot), u'^{\sharp} f^{\flat} K^{\bullet}) & \xrightarrow{\;\simeq\;} & \mathbf{R}\mathscr{H}om_{X'}^{\bullet}(\mathbf{L}h^*(\cdot), f'^{\flat} u^{\sharp} K^{\bullet})
\end{array}
$$

The columns in (3.3.35) are defined by using (3.3.20) and (3.3.22).

The advantage of considering (3.3.35) is that it makes sense as a diagram between isomorphisms of functors $\mathbf{D}_c^-(Y) \to \mathbf{D}_{qc}^+(X')$ with $K^{\bullet}$ replaced by *any* object in $\mathbf{D}_{qc}^+(Y)$ and Y replaced by any locally noetherian scheme. We claim that (3.3.35) commutes in this level of generality. It suffices to replace $K^{\bullet}$ by a bounded below complex of quasi-coherent injectives and to evaluate (3.3.35) on a bounded above complex of flats (with coherent cohomology). The commutativity of (3.3.35) in this special case is a trivial consequence of how the isomorphisms (3.3.20) and (3.3.22) are *defined*. ∎

Now that we have reviewed the basic properties of the functor $f^! : \mathbf{D}_c^+(Y) \to \mathbf{D}_c^+(X)$, it remains (for later use) to mention a few compatibilities between f^{Δ} and the 'restriction' of $f^!$ to residual complexes. Choose a residual complex $K^{\bullet}$

on Y. We define the derived category isomorphism $f^! K^\bullet \simeq f^\Delta K^\bullet$ to be

$$(3.3.36) \qquad f^! K^\bullet == f^!_{K^\bullet}(K^\bullet) == D_{X, f^\Delta K^\bullet} \circ \mathbf{L}f^* \circ \mathbf{R}\mathscr{H}om^\bullet_Y(K^\bullet, K^\bullet)$$

$$\Big\downarrow \simeq$$

$$D_{X, f^\Delta K^\bullet} \circ \mathbf{L}f^*(\mathscr{O}_Y[0])$$

$$\Big\downarrow \simeq$$

$$f^\Delta K^\bullet.$$

Using (3.3.16) and (3.3.36), it makes sense to consider the diagrams

$$(3.3.37) \qquad \begin{array}{ccc} f^! K^\bullet & \xrightarrow{\ \simeq\ } & f^\Delta K^\bullet \\ {\scriptstyle e_f}\Big\downarrow & \swarrow {\scriptstyle \simeq} & \\ f^\sharp K^\bullet & & \end{array} \qquad \begin{array}{ccc} f^! K^\bullet & \xrightarrow{\ \simeq\ } & f^\Delta K^\bullet \\ {\scriptstyle d_f}\Big\downarrow & \swarrow {\scriptstyle \simeq} & \\ f^b K^\bullet & & \end{array}$$

for smooth separated f and finite f respectively. Also, for finite type maps $X \xrightarrow{f} Y \xrightarrow{g} Z$ and a residual complex $K^\bullet$ on Z, we can consider the diagram

$$(3.3.38) \qquad \begin{array}{ccccc} (gf)^! K^\bullet & \xrightarrow{\ \simeq\ } & f^! g^! K^\bullet & \xrightarrow{\ \simeq\ } & f^! g^\Delta K^\bullet \\ {\scriptstyle \simeq}\Big\downarrow & & & \swarrow {\scriptstyle \simeq} & \\ (gf)^\Delta K^\bullet & \xrightarrow[\ \simeq\ ]{} & f^\Delta g^\Delta K^\bullet & & \end{array}$$

LEMMA 3.3.2. *The derived category diagrams* (3.3.37) *and* (3.3.38) *commute.*

PROOF. The commutativity of (3.3.37) follows immediately from the definitions of (3.3.19) and (3.3.21), as well as the fact that (3.3.5) is the identity map. The commutativity of (3.3.38) follows from the fact that in degree 0, the double duality map of complexes

$$\mathscr{O}_Y[0] \to D_{g^\Delta K^\bullet} D_{g^\Delta K^\bullet}(\mathscr{O}_Y[0])$$

from (3.3.4) is just the canonical map

$$\mathscr{O}_Y \to \prod_{q \in \mathbf{Z}} \mathscr{H}om_Y((g^\Delta K^\bullet)^q, (g^\Delta K^\bullet)^q),$$

defined without the intervention of signs, as in (1.3.17). ∎

This completes our discussion of the basic theory of the functor $f^!$ for a finite type map $f : X \to Y$ between noetherian schemes admitting a dualizing complex.

3.4. The Trace Map Tr$_f$ and Grothendieck-Serre Duality

Now we turn to the second ingredient in the duality theorem, the derived category trace map (3.3.2). Let $f : X \to Y$ be a finite type map between noetherian schemes admitting a dualizing complex. The trace map $\mathrm{Tr}_f : \mathbf{R}f_* f^! \to 1$ for *proper* f will be defined in terms of a choice of residual complex $K^\bullet$ on Y. For now, we do *not* require f to be proper. Since Y is noetherian with finite Krull dimension, there is a translation-compatible isomorphism of functors on $\mathbf{D}_c^+(Y)$

$$(3.4.1) \qquad \mathbf{R}f_* f^! \xmapsto{\quad\quad} \mathbf{R}f_* \mathbf{R}\mathscr{H}om_X^\bullet(\mathbf{L}f^* \circ D_{K^\bullet}(\cdot), f^\triangle K^\bullet)$$

$$\downarrow \simeq$$

$$\mathbf{R}\mathscr{H}om_Y^\bullet(D_{K^\bullet}(\cdot), \mathbf{R}f_*(f^\triangle K^\bullet))$$

$$\downarrow \simeq$$

$$\mathbf{R}\mathscr{H}om_Y^\bullet(D_{K^\bullet}(\cdot), f_* f^\triangle K^\bullet),$$

which is a special case of the more general δ-bifunctorial composite map [**RD**, II, 5.10]

$$(3.4.2) \qquad \mathbf{R}f_* \mathbf{R}\mathscr{H}om_X^\bullet(\mathbf{L}f^* \mathscr{F}^\bullet, \mathscr{G}^\bullet) \longrightarrow \mathbf{R}\mathscr{H}om_Y^\bullet(\mathbf{R}f_* \mathbf{L}f^* \mathscr{F}^\bullet, \mathbf{R}f_* \mathscr{G}^\bullet)$$

$$\downarrow$$

$$\mathbf{R}\mathscr{H}om_Y^\bullet(\mathscr{F}^\bullet, \mathbf{R}f_* \mathscr{G}^\bullet)$$

for $\mathscr{F}^\bullet \in \mathbf{D}^-(Y)$, $\mathscr{G}^\bullet \in \mathbf{D}^+(X)$, defined by replacing $\mathscr{F}^\bullet$ (resp. $\mathscr{G}^\bullet$) with a bounded above (resp. bounded below) complex of flats (resp. injectives). The map (3.4.2) is an isomorphism if $\mathscr{F}^\bullet$ has coherent cohomology.

By (3.3.4), there is also a translation-compatible isomorphism on $\mathbf{D}_c(Y)$

$$(3.4.3) \qquad 1 \simeq \mathbf{R}\mathscr{H}om_Y^\bullet(D_{K^\bullet}(\cdot), K^\bullet).$$

Thus, for proper f we can define $\mathrm{Tr}_f : \mathbf{R}f_* f^! \to 1$ on $\mathbf{D}_c^+(Y)$ by composing (3.4.1) and the inverse of (3.4.3) *if* we can define a morphism of complexes

$$(3.4.4) \qquad \mathrm{Tr}_{f,K^\bullet} : f_* f^\triangle K^\bullet \to K^\bullet$$

for proper f. Such a definition of Tr_f will be *independent* of the choice of $K^\bullet$ if (3.4.4) is Zariski local on Y, functorial with respect to isomorphisms between residual complexes (such as (3.3.7)), and compatible with respect to translation in $K^\bullet$ and tensoring $K^\bullet$ with an invertible sheaf.

In order to define (3.4.4) for proper f, we begin with the finite case. When f is finite, (3.4.4) is constructed in terms of the derived category 'finite' trace morphism Trf_f in [**RD**, VI, pp.335–339], as follows. Consider a finite map $f : X \to Y$ between noetherian schemes, where Y admits a dualizing complex. For any quasi-coherent $f_* \mathscr{O}_X$-module $\mathscr{F}$ on Y, let $\mathscr{F}^\sim$ denote the corresponding quasi-coherent $\mathscr{O}_X$-module (so $f_*(\mathscr{F}^\sim) \simeq \mathscr{F}$). By [**RD**, VI, 4.1], the complex of quasi-coherent sheaves $\mathscr{H}om_Y(f_*\mathcal{O}_X, K^\bullet)^\sim$ on X, which is canonically isomorphic to $f^\flat K^\bullet$ in $\mathbf{D}(X)$, is a residual complex and $f_*(\mathscr{H}om_Y(f_*\mathcal{O}_X, K^\bullet)^\sim) \simeq \mathscr{H}om_Y(f_*\mathcal{O}_X, K^\bullet)$ is a Cousin complex on Y with respect to the filtration $Z^\bullet(K^\bullet)$ on Y which is associated to $K^\bullet$.

By the theory of Cousin complexes (really Lemma 3.2.1), for any two complexes $K_1^\bullet$, $K_2^\bullet$ on X (or on Y) which are Cousin complexes with respect to the *same* filtration on the underlying topological space, any map $K_1^\bullet \to K_2^\bullet$ in the derived category is induced by a unique map $K_1^\bullet \to K_2^\bullet$ of complexes. Thus, we can represent the derived category isomorphism $f^\triangle K^\bullet \simeq f^\flat K^\bullet$ from (3.3.16) by a unique isomorphism of complexes of $\mathcal{O}_X$-modules

$$(3.4.5) \qquad f^\triangle K^\bullet \simeq \mathscr{H}om_Y(f_*\mathcal{O}_X, K^\bullet)^\sim$$

(recovering (3.3.17) when f is a closed immersion) and we can represent the derived category map

$$(3.4.6) \qquad \mathscr{H}om_Y(f_*\mathcal{O}_X, K^\bullet) \simeq \mathbf{R}f_*f^\flat K^\bullet \xrightarrow{\mathrm{Trf}_f} K^\bullet.$$

by a unique map of complexes

$$(3.4.7) \qquad \mathscr{H}om_Y(f_*\mathcal{O}_X, K^\bullet) \to K^\bullet$$

(namely, the 'evaluate at 1' map). Composing $f_*((3.4.5))$ and (3.4.7), we get a map of complexes

$$(3.4.8) \qquad \rho_{f,K^\bullet} : f_*f^\triangle K^\bullet \to K^\bullet$$

for finite f. This is the definition of (3.4.4) for *finite* f. By uniqueness in the above construction, $\rho_{f,K^\bullet}$ is compatible with translation, Zariski localization, and tensoring with an invertible sheaf.

The general definition of (3.4.4) as a map of graded sheaves (i.e., sheaf maps between terms of the same degree, not necessarily compatible with the differentials on each side) is given by the following rather non-trivial result.

THEOREM 3.4.1. [**RD**, VI, 4.2; VII, 2.1] *Let $f : X \to Y$ be a finite type map between noetherian schemes and suppose that Y (and hence X) admits a dualizing complex. There is a unique way to define the trace map (3.4.4) as a map of graded sheaves (not necessarily as complexes) for all such f so that*

1. *there is functoriality with respect to maps $K^\bullet \to K'^\bullet$ between residual complexes which have the same associated codimension function on Y,*

2. *if $g : Y \to Z$ is another such scheme morphism and $K^\bullet$ is a residual complex on Z, then*

$$\mathrm{Tr}_{gf,K^\bullet} = \mathrm{Tr}_{g,K^\bullet} \circ g_*(\mathrm{Tr}_{f,g^\triangle K^\bullet}) \circ (gf)_*(c_{f,g}).$$

3. *if f is finite, then $\mathrm{Tr}_{f,K^\bullet}$ is compatible with the derived category trace morphism Trf_f in the sense that $\mathrm{Tr}_{f,K^\bullet} = \rho_{f,K^\bullet}$ (in other words, the derived category isomorphism $f^\flat K^\bullet \simeq f^\triangle K^\bullet$ from (3.3.16) carries $\mathrm{Trf}_f(K^\bullet)$ over to $\mathrm{Tr}_{f,K^\bullet}$).*

Moreover, $\mathrm{Tr}_{f,K^\bullet}$ is compatible with residually stable base change and when f is proper, $\mathrm{Tr}_{f,K^\bullet}$ is a map of complexes.

By the uniqueness in Theorem 3.4.1, it follows that $\mathrm{Tr}_{f,K^\bullet}$ is compatible with translation and tensoring with an invertible sheaf, since $\rho_{f,K^\bullet}$ has these properties. Also, since open immersions are residually stable we conclude that $\mathrm{Tr}_{f,K^\bullet}$ respects Zariski localization. Thus, the definition of $\mathrm{Tr}_{f,K^\bullet}$ in (3.3.2) for proper f by means of (3.4.1)–(3.4.4) is *independent* of the choice of $K^\bullet$.

An important point to observe is that the general theory of the 'residual complex' trace in Theorem 3.4.1 is completely determined by functoriality requirements and the case of finite morphisms. In particular, this theory is *independent* of any arbitrary sign conventions. For the proof of the duality theorem later, it is important to also have a 'projective space' analogue of the third part of Theorem 3.4.1. The following result also ensures that our choice of sign in the definition of Trp was 'correct' (also see (2.8.5)):

LEMMA 3.4.2. [**RD**, VII, 3.2] *Let Y be a noetherian scheme and let the map $f : \mathbf{P}_Y^n \to Y$ be the projection. Assume that Y admits a dualizing complex (so Y has finite Krull dimension) and choose a residual complex $K^\bullet$ on Y. The derived category isomorphism $f^\sharp K^\bullet \simeq f^\triangle K^\bullet$ from (3.3.16) carries $\mathrm{Trp}_f(K^\bullet)$ over to $\mathrm{Tr}_{f,K^\bullet}$.*

PROOF. We refer the reader to [**RD**], which uses Theorem 3.4.1(2),(3), but we emphasize that the essential 'computational' input in the proof is that (2.8.5) is the identity, which amounts to Lemma 2.8.2. ∎

We now make some remarks on Theorem 3.4.1. The precise construction process in [**RD**] for the general trace in Theorem 3.4.1 is given in terms of direct limits of 'finite trace' maps $\rho_{f_i,K^\bullet}$ for various auxiliary finite maps f_i (and this inspired the method of proof in Chapter 4). The *explicit* details from [**RD**] are very lengthy and are not needed in what follows, so we omit a general discussion of them here. However, in §5.2 we will want to recover the classical description of Grothendieck duality on a proper reduced curve X over an algebraically closed field k in terms of Rosenlicht's sheaf of 'regular differentials' and an explicit 'trace' map defined in terms of residues of meromorphic differentials on the normalization of X. The argument will be almost entirely pure thought, except that we will need to know that the abstract trace map in the *smooth* case is related to residues of differentials. This relation is given by [**RD**, VII, 1.3], even for proper smooth curves over an *artin ring* with algebraically closed residue field, and in such generality this relationship between Grothendieck's trace map and residues happens to be the first step in the general *proof* that $\mathrm{Tr}_{f,K^\bullet}$ is a map of complexes for proper f as in Theorem 3.4.1.

Alas, the essential computation in the proof of this 'curve result' [**RD**, VII, 1.3] is [**RD**, VII, 1.2] and the proof of [**RD**, VII, 1.2] is omitted in [**RD**]. The result [**RD**, VII, 1.2] plays a prominent foundational role in the proof that for proper f, $\mathrm{Tr}_{f,K^\bullet}$ is a map of complexes. This result also implies that Grothendieck's trace map for proper smooth connected curves over an algebraically closed field k is the *negative* of the classical map defined using residues, conditional on an explication as in (2.3.8). We give the proof of [**RD**, VII, 1.2] in Appendix B in order to clear up the matter.

There are four basic properties of the trace morphism (3.3.2). We now recall these properties and make some remarks on the proofs (which are omitted in [**RD**]).

LEMMA 3.4.3. [**RD**, VII, 3.4(*b*)] *Consider noetherian schemes admitting a dualizing complex.*

- (TRA1) *If* $X \xrightarrow{f} Y \xrightarrow{g} Z$ *is a composite of proper maps, then*

$$\text{Tr}_{gf} = \text{Tr}_g \circ \mathbf{R}g_*(\text{Tr}_f) \circ \mathbf{R}(gf)_*(c_{f,g}).$$

- (TRA2) *For finite* $f : X \to Y$*, the isomorphism* $d_f : f^{\flat} \simeq f^!$ *carries* Trf_f *over to* Tr_f*.*
- (TRA3) *If* $f : \mathbf{P}_Y^n \to Y$ *is the projection and* Y *is separated, the isomorphism* $e_f : f^{\sharp} \simeq f^!$ *carries* Trp_f *over to* Tr_f*.*
- (TRA4) *If*

$$
\begin{array}{ccc}
X' & \xrightarrow{u'} & X \\
{\scriptstyle f'}\downarrow & & \downarrow{\scriptstyle f} \\
Y' & \xrightarrow{u} & Y
\end{array}
$$

is a cartesian diagram with f *proper and* u *residually stable, then the diagram*

$$
\begin{array}{ccc}
u^* \mathbf{R} f_* f^! & \xrightarrow{u^*(\text{Tr}_f)} & u^* \\
{\scriptstyle \simeq}\downarrow & & \uparrow{\scriptstyle \text{Tr}_{f'}} \\
\mathbf{R} f'_* u'^* f^! & \xrightarrow[\mathbf{R} f'_*(b_{u,f})]{} & \mathbf{R} f'_* f'^! u^*
\end{array}
$$

commutes.

Separatedness of the base should not be needed in (TRA3); the only reason we require it is that we have only proven Theorem 2.3.3 subject to this condition.

PROOF. The proof of (TRA1) amounts to unwinding definitions and using functoriality with respect to the isomorphism $1 \simeq D_Y \circ D_Y$ in order to reduce to the compatibility of (3.4.4) and (3.4.2) with respect to composites in f, as well as the fact that (3.3.5) is the identity (this is applied to the duality functor $D_{Y, g^\triangle K^\bullet}$ for a residual complex $K^\bullet$ on Z). Similarly, (TRA4) follows from the compatibility of (3.4.2) with respect to flat base change and the compatibility of (3.4.4) with respect to residually stable base change (see the end of Theorem 3.4.1).

For (TRA2), we use (3.3.16) and the third part of Theorem 3.4.1 to replace Tr_f with Trf_f and to replace $f^\triangle K^\bullet$ (which is implicit in the definition of Tr_f) with $f^\flat K^\bullet$. It is then easy to reduce to the general compatibility property [**RD**,

III, 6.9(d)] of the derived category 'finite trace' Trf$_f$, which asserts that for $\mathscr{F}^\bullet \in \mathbf{D}_c^-(Y)$, $\mathscr{G}^\bullet \in \mathbf{D}_{qc}^+(Y)$, the diagram

$$(3.4.9) \qquad \mathbf{R}f_* \mathbf{R}\mathscr{H}om_X^\bullet(\mathbf{L}f^*\mathscr{F}^\bullet, f^\flat \mathscr{G}^\bullet) \xrightarrow{\ \sim\ } \mathbf{R}f_* f^\flat \mathbf{R}\mathscr{H}om_Y^\bullet(\mathscr{F}^\bullet, \mathscr{G}^\bullet)$$

$$\downarrow \qquad\qquad\qquad\qquad\qquad \downarrow {\scriptstyle \mathrm{Trf}_f}$$

$$\mathbf{R}\mathscr{H}om_Y^\bullet(\mathscr{F}^\bullet, \mathbf{R}f_* f^\flat \mathscr{G}^\bullet) \xrightarrow[\ \mathrm{Trf}_f\ ]{} \mathbf{R}\mathscr{H}om_Y^\bullet(\mathscr{F}^\bullet, \mathscr{G}^\bullet)$$

commutes, where the left column is (3.4.2) and the top row is (3.3.20). The proof of this commutativity is easy.

The proof of (TRA3) is similar to the proof of (TRA2), except that we use Lemma 3.4.2 instead of the third part of Theorem 3.4.1 and we need the 'projective space' analogue for Trp of the (relatively easy) compatibility property (3.4.9) for Trf (using (3.3.22) in place of (3.3.20) in the top row). Such an analogue is not mentioned in [**RD**], but is given by the difficult Theorem 2.3.3 (which we only proved for a separated base).

∎

Let $f : X \to Y$ be a proper map between noetherian schemes and assume that Y admits a dualizing complex. The *duality morphism* is defined to be the composite

$$(3.4.10) \quad \theta_f : \mathbf{R}f_* \mathbf{R}\mathscr{H}om_X^\bullet(\mathscr{F}^\bullet, f^! \mathscr{G}^\bullet) \longrightarrow \mathbf{R}\mathscr{H}om_Y^\bullet(\mathbf{R}f_*\mathscr{F}^\bullet, \mathbf{R}f_* f^! \mathscr{G}^\bullet)$$

$$\downarrow {\scriptstyle \mathrm{Tr}_f}$$

$$\mathbf{R}\mathscr{H}om_Y^\bullet(\mathbf{R}f_*\mathscr{F}^\bullet, \mathscr{G}^\bullet)$$

for $\mathscr{F}^\bullet \in \mathbf{D}_{qc}^-(X)$, $\mathscr{G}^\bullet \in \mathbf{D}_c^+(Y)$. By construction, this is translation-compatible in both variables and respects *residually stable* base change (e.g., it is compatible with Zariski localization on the base).

The *Grothendieck-Serre duality theorem* is

THEOREM 3.4.4. [**RD**, VII, 3.4(c)] *The map (3.4.10) is an isomorphism.*

Briefly, here is the idea of the proof (cf. proof of [**RD**, VII, 3.3]). Since the property of a derived category map being an isomorphism is local, we may work locally on the base. By means of Chow's Lemma and noetherian induction, the problem is reduced to the case of the projection $\mathbf{P}_Y^n \to Y$ with Y separated and the case of finite maps. Then Theorem 3.4.1(3), Lemma 3.4.2, and (TRA2), (TRA3) in Lemma 3.4.3 permit reduction to previously established duality theorems [**RD**, III, 5.1, 6.7] for projective space and finite maps respectively, completing the proof. This will be extended to more general bases Y in Theorem 4.3.1 when f is CM with pure relative dimension.

In the case of a proper *smooth* map $f : X \to Y$ with pure relative dimension n, the abstract duality morphism (3.4.10) can be defined purely in terms of the sheaf $\omega_{X/Y} = \Omega_{X/Y}^n$ and the trace map

$$(3.4.11) \qquad\qquad \gamma_f : \mathbf{R}^n f_*(\omega_{X/Y}) \to \mathscr{O}_Y$$

induced on degree 0 cohomology by the derived category map

$$(3.4.12) \qquad \mathbf{R}f_* f^\sharp(\mathscr{O}_Y[0]) \xrightarrow[\simeq]{\mathbf{R}f_*(e_f)} \mathbf{R}f_* f^!(\mathscr{O}_Y[0]) \xrightarrow{\mathrm{Tr}_f} \mathscr{O}_Y[0].$$

As in (2.3.8), we identify

$$(3.4.13) \qquad \mathrm{H}^0(\mathbf{R}f_*(\omega_{X/Y}[n])) \simeq \mathbf{R}^n f_*(\omega_{X/Y})$$

as follows: if $\omega_{X/Y} \to \mathscr{I}^\bullet$ is an injective resolution chosen to compute derived functors of $\omega_{X/Y}$, then we compute the left side of (3.4.13) using the resolution $\omega_{X/Y}[n] \to \mathscr{I}^{\bullet+n}$; in other words, (3.4.13) is represented by the isomorphism

$$\mathrm{H}^0(f_* \mathscr{I}^{\bullet+n}) = \mathrm{H}^n(f_* \mathscr{I}^\bullet)$$

with the intervention of signs. Note that if we had chosen the alternative possibility of using $\mathscr{I}^\bullet[n]$ as the 'preferred' injective resolution of $\omega_{X/Y}[n]$, then the definition of (3.4.13) would change by a sign of $(-1)^{n^2} = (-1)^n$. In other words, whereas the *derived category* map (3.4.12) does not involve any sign ambiguity, the explication of it as the map γ_f is sensitive to conventions up to a sign of $(-1)^n$ (which we eliminate by the above specification of the definition of (3.4.13)). Our choice of definition of γ_f ensures that we recover (2.3.1) when $X = \mathbf{P}_Y^n$ (see Lemma 3.4.3(TRA3)).

There is a 'derived category' duality theorem for proper smooth maps f (with pure relative dimension) which is described in terms of the functor $f^\sharp$ and the trace map γ_f as in (3.4.11), and not in terms of the functor $f^!$ (whose definition uses residual complexes). Such an alternative duality theorem will suggest how to state a result over bases which do not admit a dualizing complex. The model for such a reformulation is the duality isomorphism for the projective space map $f : \mathbf{P}_Y^n \to Y$ with Y locally noetherian [**RD**, III, 5.1], in which case we use $f^\sharp$ in place of $f^!$ and

$$\mathrm{Trp}_f : \mathbf{R}f_* f^\sharp(\cdot) \simeq \mathbf{R}f_*(\omega_{\mathbf{P}_Y^n/Y}[n]) \overset{\mathbf{L}}{\otimes} (\cdot) \simeq \mathbf{R}^n f_*(\omega_{\mathbf{P}_Y^n/Y})[0] \overset{\mathbf{L}}{\otimes} (\cdot) \simeq (\cdot)$$

in place of Tr$_f$, with the middle map defined as in (2.3.8) and the right side defined using (2.3.1).

In general, for a proper smooth map $f : X \to Y$ with pure relative dimension n (and Y a noetherian scheme admitting a dualizing complex, hence having finite Krull dimension), $\mathbf{R}^i f_*$ vanishes on quasi-coherent sheaves for $i > n$, so there is a canonical map

$$(3.4.14) \qquad \mathbf{R}f_*(\omega_{X/Y}[n]) \to \mathbf{R}^n f_*(\omega_{X/Y})[0] \xrightarrow{\gamma_f} \mathscr{O}_Y[0],$$

where the map on the left is defined without the intervention of signs, using (3.4.13); note that the definitions of (3.4.13) and γ_f are both sensitive to conventions up to the *same* sign of $(-1)^n$, so if one chooses the alternative definition (3.4.13) then the definition of γ_f would also change and the composite (3.4.14) would be unaffected (but 'computing' (3.4.14) in degree 0 is of course highly sensitive to the definition of (3.4.13)).

Using the projection formula (2.1.10), as well as (3.4.14), we obtain a 'modified' δ-functorial trace map

$$(3.4.15) \qquad \mathrm{Tr}'_f : \mathbf{R}f_* f^\sharp(\mathscr{G}^\bullet) \simeq \mathbf{R}f_*(\omega_{X/Y}[n]) \overset{\mathbf{L}}{\otimes} \mathscr{G}^\bullet \to \mathscr{G}^\bullet$$

for $\mathscr{G}^\bullet \in \mathbf{D}^b_{\mathrm{qc}}(Y)$. We will show in Theorem 4.3.2 that the restriction to $\mathbf{D}^b_{\mathrm{c}}(Y)$ of the isomorphism $e_f : f^\sharp \simeq f^!$ carries Tr'_f over to Tr_f. This is not trivial. Since Y is noetherian with finite Krull dimension, we can use Tr'_f to define a δ-bifunctorial duality morphism

$$\theta'_f : \mathbf{R}f_* \mathbf{R}\mathscr{H}om^\bullet_X(\mathscr{F}^\bullet, f^\sharp \mathscr{G}^\bullet) \to \mathbf{R}\mathscr{H}om^\bullet_Y(\mathbf{R}f_*\mathscr{F}^\bullet, \mathscr{G}^\bullet)$$

for $\mathscr{F}^\bullet \in \mathbf{D}^-_{\mathrm{qc}}(X)$, $\mathscr{G}^\bullet \in \mathbf{D}^b_{\mathrm{c}}(Y)$. This map is an isomorphism, by Theorem 3.4.4, the Lemma on Way-Out Functors [**RD**, I, 7.9], and the identification of Tr_f and Tr'_f on $\mathbf{D}^b_{\mathrm{c}}(Y)$ via e_f. When $Y = \mathrm{Spec}(k)$ for a field k, $\mathscr{F}^\bullet = \mathscr{F}[0]$ for a locally free coherent sheaf $\mathscr{F}$ on X, and $\mathscr{G}^\bullet = \mathscr{O}_Y[0]$, the *isomorphism* $\mathrm{H}^{-i}(\theta'_f)$ agrees with the map

$$(3.4.16) \qquad \mathrm{H}^{n-i}(X, \mathscr{F}^\vee \otimes \Omega^n_{X/k}) \to \mathrm{Hom}_k(\mathrm{H}^i(X, \mathscr{F}), k)$$

as in the Preface, up to a universal sign depending on n and i, where we use (3.4.11) to define $t_X : \mathrm{H}^n(X, \Omega^n_{X/k}) \to k$. This kind of explicit consequence of the 'derived category' duality theory will be justified in §5.1 (see Theorem 5.1.2ff).

For a proper smooth map $f : X \to Y$ with pure relative dimension n, where Y is a noetherian scheme admitting a dualizing complex, we see that the data of (3.4.11) is enough to formulate a duality theory. We want to extend this theory to proper smooth maps with pure relative dimension over any locally noetherian base Y, and more importantly, we want to analyze the behavior of the duality theory with respect to base change. First, we need to define γ_f without assuming Y admits a dualizing complex. The idea is to work over open affines in a locally noetherian Y and use standard direct limit arguments [**EGA**, IV$_3$, §8–§11] to realize f as a base change of a proper smooth map $f_0 : X_0 \to Y_0$ with pure relative dimension n, where Y_0 is a finite type **Z**-scheme (and since the fibers of f are geometrically reduced, f_0 can also be assumed to have geometrically connected fibers if f does [**EGA**, IV$_3$, 12.2.4(vi)]). Since Y_0 is noetherian and admits a dualizing complex, so γ_{f_0} makes sense, we can try to define γ_f to be the base change of γ_{f_0}. The details of this definition of γ_f, particularly its well-definedness, are non-trivial because they make essential use of the fact that (3.4.11) respects *any* base change $Y' \to Y$ where Y' is noetherian and admits a dualizing complex.

Unfortunately, the theory of residual complexes only behaves well with respect to *residually stable* base change, and residual complexes are used in the definition of (3.4.11), so it is not even obvious that (3.4.11) is compatible with *flat* base change. Once we prove in §4.1–§4.2 that (3.4.11) is compatible with *any* base change $Y' \to Y$ between noetherian schemes admitting a dualizing complex, we will be able to extend the definitions of γ_f, Tr'_f, and θ'_f (with some mild extra constraints) to the case of a proper smooth map with pure relative dimension over an arbitrary locally noetherian base scheme. In order to prove the duality theorem in this generality, which asserts that θ'_f is an isomorphism, we can use flat base change to $\mathrm{Spec}(\widehat{\mathscr{O}_{Y,y}})$'s for $y \in Y$ to reduce to the case of a base which is a complete local noetherian ring. Such bases are noetherian and admit a dualizing complex (!), so we can try to relate θ'_f and θ_f, and then use Theorem 3.4.4. This plan is carried out more generally in §4.3 for proper CM maps with pure relative dimension.

It should be noted that the method of construction of γ_f for smooth f and general locally noetherian Y in [RD, VII, §4] is somewhat 'opposite' to the method outlined above. Instead of defining γ_f by base changing up from a theory over finite type Z-schemes (on which dualizing complexes exist), [RD] descends down from complete local noetherian rings (on which dualizing complexes exist). One must carry out a 'descent from the completion' argument to define (3.4.11) on Y by 'descent' from the $\mathrm{Spec}(\widehat{\mathscr{O}_{Y,y}})$'s, though the proof of [RD, VII, 4.2] has a minor error which can be corrected. In any case, this technique requires the base change compatibility for the trace map which is not proven in [RD].

3.5. Dualizing Sheaves and CM maps

Let $f : X \to Y$ be a proper smooth map with pure relative dimension n, where Y is noetherian and admits a dualizing complex. We want the trace map $\gamma_f : \mathrm{R}^n f_*(\Omega^n_{X/Y}) \to \mathscr{O}_Y$ in (3.4.11) to be compatible with any base change $Y' \to Y$ where Y' is noetherian and admits a dualizing complex. In §1.1, we saw that it is natural to try to generalize this statement so that it includes the case of proper CM maps with pure relative dimension (and in particular includes the case of finite flat maps). In order to state such a generalization, we need a theory of a 'dualizing sheaf' for CM maps which replaces the theory of $\Omega^n_{X/Y}$ for smooth maps with pure relative dimension n. The purpose of this section is to give enough of this theory for the proof in Chapter 4 of the Main Theorem 3.6.5, which implies the above base change compatibility of γ_f (for bases which are noetherian and admit a dualizing complex). Such a result enables us to define γ_f over any locally noetherian base (see Corollary 3.6.6). Further properties of dualizing sheaves and applications to duality theory over locally noetherian bases will be given in §4.3ff.

We begin by explaining the special role of CM maps in duality theory. Such maps turn out to be exactly the ones for which there is a good theory of a 'dualizing sheaf.' Let $f : X \to Y$ be a locally finite type map of schemes, with Y locally noetherian. Assume that f factorizes as $f = \pi \circ i$, where $i : X \to P$ is a closed immersion and $\pi : P \to Y$ is separated smooth with bounded fiber dimension. Such a factorization always exists if we shrink X around any desired point (e.g., replace X by an open affine U which maps into an open affine V in Y and take $P = \mathbf{A}^N_V$ for suitably large N). By [RD, III, 8.7] and §2.7, there is a well-defined δ-functor $f^! : \mathbf{D}^+_{\mathrm{qc}}(Y) \to \mathbf{D}^+_{\mathrm{qc}}(X)$ which is compatible with localization on X and Y, and $f^! \simeq i^\flat \pi^\sharp$. By Lemma 3.3.2, this coincides with the notion of $f^!$ defined in §3.3 if Y is noetherian and admits a dualizing complex. When f is separated smooth with pure relative dimension n, we have $f^! \mathscr{O}_Y \simeq \omega_{X/Y}[n]$, so $\mathrm{H}^j(f^! \mathscr{O}_Y)$ vanishes when $j \neq -n$ and $\mathrm{H}^{-n}(f^! \mathscr{O}_Y) \simeq \omega_{X/Y}$ is a Y-flat sheaf on X (even invertible on X). This fits into a more general picture, for which we give a proof in Theorem 3.5.1 due to lack of a suitable reference.

For the proof of the following theorem and for the rest of this book, we refer to [AK2, §1] for the details of the theory of base change maps for $\mathscr{E}xt$'s. Since this theory will be frequently used below, we remind the reader of the basic setup. Let $X \to S$ be a locally finitely presented map of schemes, $\mathscr{F}$ a finitely presented $\mathscr{O}_X$-module, $\mathscr{G}$ a quasi-coherent $\mathscr{O}_X$-module, S' any S-scheme, and $p: X' = X \times_S S' \to X$ the projection. In [AK2, §1.8], a 'base change map'

$$(3.5.1) \qquad p^* \mathscr{E}xt_X^q(\mathscr{F}, \mathscr{G}) \to \mathscr{E}xt_{X'}^q(p^* \mathscr{F}, p^* \mathscr{G})$$

is defined under the hypothesis that $\mathscr{F}$ is S-flat if $q \geq 1$ and X is S-flat if $q \geq 2$. The definition is given over open affines U in X, in terms of certain projective resolutions of $\Gamma(U, \mathscr{F})$ over $\Gamma(U, \mathscr{O}_X)$. In the affine case, with $\mathrm{Spec}(B) \to \mathrm{Spec}(A)$ the map $X \to S$ and

$$\begin{array}{ccc} \mathrm{Spec}(B') & \xrightarrow{p} & \mathrm{Spec}(B) \\ \downarrow & & \downarrow \\ \mathrm{Spec}(A') & \longrightarrow & \mathrm{Spec}(A) \end{array}$$

the cartesian diagram of interest, with $\mathscr{F}, \mathscr{G}$ associated to B-modules M and N respectively, here is the construction of (3.5.1).

We assume for simplicity that the finitely presented A-algebra B and the finitely presented B-module M are A-flat (as this holds in all applications below). To start off, $\mathscr{E}xt_X^q(\mathscr{F}, \mathscr{G})$ is isomorphic to the quasi-coherent sheaf associated to $\mathrm{Ext}_B^q(M, N)$, in a manner which is functorial and δ-functorial in both variables M and N; this is essentially [AK2, 1.6], though one needs to be careful with the proof to verify the desired functoriality (recall from [Tohoku, II, 2.3, p.144] that the δ-functoriality in the first variable of Ext can be described in terms injective resolutions in the second variable). Thus, we will describe (3.5.1) in terms of module Ext's. In this commutative algebra setting, we may even allow B to be an arbitrary flat A-algebra and M to be arbitrary A-flat B-module. Let $P^\bullet$ be a projective resolution of M as a B-module, so the A-flatness of M and B ensures that the base change $P'^\bullet \to M'$ by the functor $\otimes_B B' = \otimes_A A'$ is a projective resolution of M' as a B'-module. Then there is a canonical map of $B' = A' \otimes_A B$-modules

$$A' \otimes_A \mathrm{H}^q(\mathrm{Hom}_B^\bullet(P^\bullet, N)) \to \mathrm{H}^q(\mathrm{Hom}_{B'}^\bullet(P'^\bullet, N')).$$

This is the concrete description of (3.5.1) in the affine setting.

The map (3.5.1) is an isomorphism when S' is S-flat. Criteria for (3.5.1) to be an isomorphism for more general base changes are given in [AK2, 1.9] (which requires $\mathscr{G}$ to be S-flat). We will often leave it to the reader to check that all relevant finite presentation and flatness hypotheses are satisfied whenever we use base change maps for $\mathcal{E}xt$'s below. For example, the flatness conclusion for $H^{-n}(f_U^!\mathcal{O}_Y)$ in the following theorem is sometimes useful for this purpose.

THEOREM 3.5.1. [**RD**, III, Exercise 9.7] *Let* $f : X \to Y$ *be a flat map between locally noetherian schemes and assume that* $f = \pi \circ i$, *where* i *is a closed immersion and* π *is separated smooth with bounded fiber dimension. Then* f *is CM with pure relative dimension* n *if and only if* $H^j(f^!\mathcal{O}_Y) = 0$ *for* $j \neq -n$ *and the coherent sheaf* $H^{-n}(f^!\mathcal{O}_Y)$ *is* Y-*flat. When these conditions hold, the coherent* $\mathcal{O}_X$-*module* $H^{-n}(f^!\mathcal{O}_Y)$ *is invertible if and only if the coherent* $\mathcal{O}_{X_y}$-*modules* $H^{-n}(f_y^!\mathcal{O}_{X_y})$ *are invertible for all* $y \in Y$ *(which is equivalent to the fibers* X_y *being Gorenstein schemes for all* $y \in Y$).

PROOF. For the final remark concerning Gorenstein fibers, we refer the reader to the details in [**RD**, V, Prop 9.3, Theorem 9.1]. For the rest, we can work locally on X and Y, so without loss of generality $X = \mathrm{Spec}(B)$, $Y = \mathrm{Spec}(A)$, with B a finite type flat A-algebra. Choose a presentation $B = A[t]/I$, where $A[t]$ denotes $A[t_1, \ldots, t_N]$ and I is an ideal in $A[t]$ (N will be fixed).

First consider the case when $A = k$ is a field. For a closed point $x \in X \hookrightarrow \mathbb{A}_k^N$, $H^j(f^!\mathcal{O}_Y)_x \simeq \mathrm{Ext}_A^{N+j}(A/J, A)$, where $A = k[t]_x$, $J = I_x$. Since X is CM with pure dimension n if and only if $\mathcal{O}_{X,x}$ is CM with dimension n for every closed point $x \in X$, it suffices to show that if $(R, \mathfrak{m})$ is a regular local ring with dimension N and J is a proper ideal in R, then R/J is CM with dimension n if and only if $\mathrm{Ext}_R^{N-j}(R/J, R) = 0$ for all $j \neq n$. The dualizing complex $\mathcal{O}_Z[N]$ on the regular scheme $Z = \mathrm{Spec}(R)$ is normalized in the sense of [**RD**, p.276], so by a corollary [**RD**, V, 6.5] of Grothendieck's Local Duality Theorem it follows that $\mathrm{Ext}_R^{N-j}(R/J, R) = 0$ if and only if the local cohomology group $H_{\mathfrak{m}}^j(R/J)$ vanishes. By [**SGA2**, V, Thm 3.1$(i),(iii)$], $H_{\mathfrak{m}}^j(R/J) = 0$ for all $j \neq n$ if and only if $n = \dim R/J$ and $H_{\mathfrak{m}}^j(R/J) = 0$ for all $j < \dim R/J$. However, by [**SGA2**, V, Example 3.4] this latter condition is equivalent to R/J having R-depth at least $\dim R/J$, which in turn is equivalent to the local noetherian ring R/J being CM. This settles the case when A is a field.

For the general case, with the flat ring map $A \to B = A[t]/I$ as above, it remains to show

- for $j_0 \in \mathbb{Z}$, $\mathrm{Ext}_{k(y)[t]}^j(k(y) \otimes_A B, k(y)[t])$ vanishes for all $j \neq j_0$ and all $y \in Y = \mathrm{Spec}(A)$ if and only if $\mathrm{Ext}_{A[t]}^j(B, A[t]) = 0$ for all $j \neq j_0$ and $\mathrm{Ext}_{A[t]}^{j_0}(B, A[t])$ is A-flat,

- when these conditions hold, the A-flat $\mathrm{Ext}_{A[t]}^{j_0}(B, A[t])$ is an invertible B-module if and only if the $k(y) \otimes_A B$-module $\mathrm{Ext}_{k(y)[t]}^{j_0}(k(y) \otimes_A B, k(y)[t])$ is invertible for all $y \in Y$.

Since B is flat over A, [**EGA**, IV$_3$, 12.3.4] and 'base change for Ext' [**AK2**, Thm 1.9(ii)] implies that if

$$\mathrm{Ext}_{k(y)[t]}^j(k(y) \otimes_A B, k(y)[t]) = 0$$

for all $j \neq j_0$ and $y \in Y$, then $\mathrm{Ext}^j_{A[t]}(B, A[t]) = 0$ for all $j \neq j_0$, $\mathrm{Ext}^{j_0}_{A[t]}(B, A[t])$ is A-flat, and the natural base change map

$$(3.5.2) \qquad k(y) \otimes_A \mathrm{Ext}^{j_0}_{A[t]}(B, A[t]) \to \mathrm{Ext}^{j_0}_{k(y)[t]}(k(y) \otimes_A B, k(y)[t])$$

is an isomorphism. It is easy to see that if M is a finite B-module which is A-flat, then M is invertible as a B-module if and only if $k(y) \otimes_A M$ is invertible as a $k(y) \otimes_A B$-module for all $y \in Y$. Thus, when $\mathrm{Ext}^j_{k(y)[t]}(k(y) \otimes_A B, k(y)[t])$ vanishes for all $j \neq j_0$, the A-flat finite B-module $M = \mathrm{Ext}^{j_0}_{A[t]}(B, A[t])$ is an invertible B-module if and only if $\mathrm{Ext}^{j_0}_{k(y)[t]}(k(y) \otimes_A B, k(y)[t])$ is an invertible $k(y) \otimes_A B$-module for all $y \in Y$.

It remains to show that if $\mathrm{Ext}^j_{A[t]}(B, A[t])$ vanishes for $j \neq j_0$ and is A-flat for $j = j_0$, then the module $\mathrm{Ext}^j_{k(y)[t]}(k(y) \otimes_A B, k(y)[t])$ vanishes for all $j \neq j_0$, $y \in Y$. Let $P^\bullet \to B \to 0$ be a projective resolution by finite free $A[t]$-modules, so since B and $A[t]$ are A-flat, the complex $k(y) \otimes_A P^\bullet$ is a *resolution* of $k(y) \otimes_A B$ by finite free $k(y)[t]$-modules for all $y \in Y$. Thus, the cohomology of the dual complex

$$P^{\bullet\vee} = \mathrm{Hom}^\bullet_{A[t]}(P^\bullet, A[t])$$

computes $\mathrm{Ext}^\bullet_{A[t]}(B, A[t])$ and the cohomology of the complex $k(y) \otimes_A P^{\bullet\vee}$ computes

$$\mathrm{Ext}^\bullet_{k(y)[t]}(k(y) \otimes_A B, k(y)[t])$$

for all $y \in Y$. We are assuming that $\mathrm{H}^j(P^{\bullet\vee})$ vanishes for $j \neq j_0$ and is A-flat for $j = j_0$, and we want to conclude that $\mathrm{H}^j(k(y) \otimes_A P^{\bullet\vee}) = 0$ for $j \neq j_0$, $y \in Y$. This would be easy if the bounded below complex $P^{\bullet\vee}$ were a *bounded* complex of A-flats, since for any bounded above complex $Q^\bullet$ of flat A-modules with $\mathrm{H}^j(Q^\bullet) = 0$ for $j \neq j_0$ and $\mathrm{H}^{j_0}(Q^\bullet)$ flat over A, the kernel $Z^{j_0}(Q^\bullet)$ and image $B^{j_0}(Q^\bullet)$ are A-flat and of formation compatible with any base change over A, so the natural map

$$A' \otimes_A \mathrm{H}^{j_0}(Q^\bullet) \to \mathrm{H}^{j_0}(A' \otimes_A Q^\bullet)$$

is an isomorphism for any A-algebra A' and $\mathrm{H}^j(A' \otimes_A Q^\bullet) = 0$ for all $j \neq j_0$ and A-algebras A'. Thus, we need to find a way to 'replace' $P^{\bullet\vee}$ by a bounded complex of A-flats.

The complex $P^{\bullet\vee}$ has finite Tor-dimension over A, since it is isomorphic in $\mathbf{D}(A)$ to a translate of the *flat* A-module $\mathrm{Ext}^{j_0}_{A[t]}(B, A[t])$. Thus, we can choose a bounded complex $Q^\bullet$ of flat A-modules and a quasi-isomorphism of complexes $\varphi : Q^\bullet \to P^{\bullet\vee}$ (cf. proof of [**RD**, II, 4.2]). If $k(y) \otimes_A \varphi$ is a quasi-isomorphism for all $y \in Y$, then

$$\mathrm{H}^j(k(y) \otimes_A P^{\bullet\vee}) \simeq \mathrm{H}^j(k(y) \otimes_A Q^\bullet) = 0$$

for all $j \neq j_0$, so we'd be done. By studying the mapping cone of φ, it suffices to prove that if $F^\bullet$ is an exact bounded below complex of flat A-modules and M is an A-module, then $M \otimes_A F^\bullet$ is an exact complex. Let $K^\bullet \to M \to 0$ be a projective resolution of M, so

$$0 \simeq M \overset{L}{\otimes} F^{\bullet} \simeq K^{\bullet} \otimes F^{\bullet}$$

in $\mathbf{D}(A)$. Thus, it suffices to show that the natural augmentation map $K^{\bullet} \otimes F^{\bullet} \to M \otimes F^{\bullet}$ induces a surjection on the kernels in each degree. This is easy to check directly. ∎

COROLLARY 3.5.2. *Let* $f : X \to Y$ *be a CM map with pure relative dimension* n, *where* Y *is an arbitrary scheme, and assume that* f *factorizes as* $f = \pi \circ i$ *where* $i : X \to P$ *is a closed immersion and* $\pi : P \to Y$ *is smooth with pure relative dimension* N. *Then* $\mathscr{E}xt_P^j(i_*\mathcal{O}_X, \omega_{P/Y}) = 0$ *for* $j \neq N - n$ *and* $\mathscr{E}xt_P^{N-n}(i_*\mathcal{O}_X, \omega_{P/Y})$ *is a finitely presented* Y-*flat* $\mathcal{O}_X$-*module which commutes with arbitrary base change over* Y. *The same conclusions hold with* $\omega_{P/Y}$ *replaced by* $\mathcal{O}_P$.

PROOF. We can work locally on all schemes, so by standard direct limit arguments [**EGA**, IV_3, §8–§11] we can assume that Y is noetherian and π is separated. Then we get the vanishing and flatness assertions over Y from Theorem 3.5.1 (since $i^b \pi^\sharp \simeq f^!$), and the base change compatibility follows from [**AK2**, Lemma 1.5] and the proof of Theorem 3.5.1. Since $\omega_{P/Y}$ is locally isomorphic to $\mathcal{O}_P$, the last part is clear. ∎

Let $f : X \to Y$ be a CM map between locally noetherian schemes and assume that f has pure relative dimension n. We define the coherent sheaf ω_f on X by gluing the $\mathcal{O}_U$-modules $\mathrm{H}^{-n}(f_U^! \mathcal{O}_Y)$ for opens U covering X such that $f_U : U \hookrightarrow X \xrightarrow{f} Y$ factors as a closed immersion $i : U \to P$ followed by a separated smooth map $\pi : P \to Y$ with pure relative dimension. We could try to avoid appealing to the theory of $(\cdot)^!$ and directly glue sheaves of type $\mathscr{E}xt_P^{N-n}(i_*\mathcal{O}_U, \omega_{P/Y})$, where $P \to Y$ has pure relative dimension N, but then we would then have to redo many difficult well-definedness arguments that arise in the construction of the theory of $(\cdot)^!$ in §2.7 and [**RD**, III, §8]. By Theorem 3.5.1, ω_f is always Y-flat and ω_f is invertible if and only if the fibers of f are Gorenstein schemes (which can be checked on geometric fibers). When f is smooth, the sheaf ω_f can be canonically identified with $\Omega_{X/Y}^n$, so we sometimes write $\omega_{X/Y}$ instead of ω_f for CM maps f with pure relative dimension. In particular, ω_f is *canonically* trivial when f is étale. We will use this in §4.4 to show that for any CM map $f : X \to Y$ as above, the formation of ω_f is 'insensitive' to étale localization on X (generalizing the well-known analogue for relative differentials).

Following the terminology in [**K**], we call ω_f the *dualizing sheaf* for f (though [**DR**, I, §2.1] calls ω_f the *sheaf of regular differentials* for f, presumably out of analogy with the example of proper reduced curves over an algebraically closed field, which we will explain in §5.2). The dualizing sheaf ω_f plays an essential role in the duality theorem for proper CM maps with pure relative dimension over any locally noetherian base, as we will see in §4.3.

In order to study the base change compatibility of the trace map (3.4.11) in the proper smooth case of Grothendieck-Serre duality, we will need to work in the more general CM setting with dualizing sheaves, rather than in the more restrictive smooth setting with sheaves of relative differentials. The rest of this section is devoted to making the construction of dualizing sheaves more explicit in terms of $\mathscr{E}xt$'s so that we can use the base change theory of $\mathscr{E}xt$'s to define an intrinsic base change theory for dualizing sheaves in §3.6, in terms of which we will be able to contemplate the base change property of the trace map in the more general CM setting.

Let $f : X \to Y$ be a CM map with pure relative dimension n, and assume that Y is locally noetherian. Choose an open subscheme $U \subseteq X$ such that there is a factorization

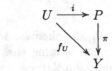

with $f_U = f|_U$, π separated smooth with pure relative dimension N, and i a closed immersion. We regard $N - n$ as the 'codimension' of i. Motivated by (2.7.3), we define the isomorphism of sheaves

$$(3.5.3) \qquad \omega_f|_U \xlongequal{\hspace{1cm}} \mathrm{H}^{-n}(f_U^! \mathscr{O}_Y) \xrightarrow{\;\simeq\;} \mathrm{H}^{-n}(i^b \pi^\sharp \mathscr{O}_Y)$$

$$\mathscr{E}xt_P^{N-n}(i_* \mathscr{O}_U, \omega_{P/Y}) = \mathrm{H}^{N-n}(i^b \omega_{P/Y}) \xleftarrow[\;\simeq\;]{} \mathrm{H}^{-n}(i^b \omega_{P/Y}[N])$$

analogously to (2.7.3). That is, if $\omega_{P/Y} \to \mathscr{I}^\bullet$ is an injective resolution used to compute $\mathscr{E}xt_P^\bullet(i_* \mathscr{O}_U, \omega_{P/Y})$, then we calculate with $\omega_{P/Y}[N] \simeq \mathscr{I}^{\bullet+N}$ in $\mathbf{D}(P)$, so the identity $\mathrm{H}^{-n} = \mathrm{H}^{N-n}$ in (3.5.3) is

$$\mathrm{H}^{-n}(\mathscr{H}om_P(i_* \mathscr{O}_U, \mathscr{I}^{\bullet+N})) = \mathrm{H}^{N-n}(\mathscr{H}om_P(i_* \mathscr{O}_U, \mathscr{I}^\bullet))$$

without the intervention of signs. This explication would change by a universal sign of $(-1)^{N(N-n)}$ if we instead had used $\omega_{P/Y}[N] \to \mathscr{I}^\bullet[N]$. The above convention for defining (3.5.3) is consistent with (2.7.3) and our use of Cartan-Eilenberg resolutions in various definitions. In practice, this issue will not matter very much because all linear maps commute with -1.

We want to make explicit the sense in which (3.5.3) is 'independent' of the choice of factorization of f_U. Any two such factorizations of f_U can be 'dominated' by a third, so the essential case to consider is that in which one factorization of f_U 'dominates' the other. That is, consider a commutative scheme

diagram

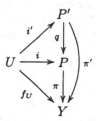

with i' a closed immersion and q a separated smooth map with pure relative dimension d, so π' is separated smooth with pure relative dimension $N' = N + d$. The abstract composite isomorphism

$$(3.5.4) \qquad \mathscr{E}xt_P^{N-n}(i_*\mathscr{O}_U, \omega_{P/Y}) \simeq \omega_f|_U \simeq \mathscr{E}xt_{P'}^{N'-n}(i'_*\mathscr{O}_U, \omega_{P'/Y})$$

encodes how the explication of $\omega_f|_U$ is 'independent' of the choice of factorization of f_U, and is the map induced by evaluating

$$(3.5.5) \qquad i^b \pi^\sharp \simeq i'^b q^\sharp \pi^\sharp \simeq i'^b \pi'^\sharp$$

on $\mathscr{O}_Y[0]$ and passing to cohomology in degree $-n$ (using the convention in (3.5.3)).

We can make (3.5.4) explicit by unwinding the definition of the isomorphism $\psi_{i',q} : i^b \simeq i'^b q^\sharp$ from (2.7.4) which is used in (3.5.5). More precisely, suppose $U = X$ and consider the scheme diagram

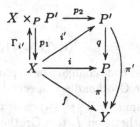

(cf. (2.7.6)) in which the graph map $\Gamma_{i'}$ is a section to the separated smooth projection map p_1 which has pure relative dimension d, so $\Gamma_{i'}$ is an lci map with pure codimension d (and is even transversally regular relative to P' with 'relative codimension d'). Thus, it makes sense to consider the 'fundamental local isomorphism'

$$\eta_{\Gamma_{i'}} : \mathscr{E}xt_{X \times_P P'}^d((\Gamma_{i'})_*\mathscr{O}_X, \cdot) \simeq \omega_{X/X \times_P P'} \otimes \Gamma_{i'}^*(\cdot)$$

from (2.5.1). Recall that if we choose an injective resolution $\omega_{P/Y} \to \mathscr{I}^\bullet$ to compute derived functors of $\omega_{P/Y}$, then the explication (2.7.3) involves the computation

$$i^b \pi^\sharp \mathscr{O}_Y \simeq \mathscr{E}xt_P^{N-n}(i_*\mathscr{O}_X, \omega_{P/Y})[-n]$$

which requires using the quasi-isomorphism $\omega_{P/Y}[N] \to \mathscr{I}^{\bullet+N}$ in $\mathbf{D}(P)$ and

$$\mathrm{H}^{-n}(\mathscr{H}om_P(i_*\mathscr{O}_X, \mathscr{I}^{\bullet+N})) = \mathrm{H}^{N-n}(\mathscr{H}om_P(i_*\mathscr{O}_X, \mathscr{I}^\bullet)).$$

Combining this with Lemma 2.6.1, which makes somewhat explicit the degeneration that occurs in a Grothendieck spectral sequence, paying attention to isomorphisms of type $C^\bullet[r] \simeq C^{\bullet+r}$ leads us to the following description of the composite (3.5.4) (or rather, (3.5.5)), up to a sign of $(-1)^{d(N-n)}$:

(3.5.6)
$$\mathscr{E}xt_P^{N-n}(i_*\mathscr{O}_X, \omega_{P/Y})$$

$$\simeq \Bigg\downarrow \zeta'_{\Gamma_{i'}, p_1}$$

$$\omega_{X/X \times_P P'} \otimes \Gamma_{i'}^*(\omega_{X \times_P P'/X} \otimes p_1^* \mathscr{E}xt_P^{N-n}(i_*\mathscr{O}_X, \omega_{P/Y}))$$

$$\simeq \Bigg\downarrow \alpha$$

$$\omega_{X/X \times_P P'} \otimes \Gamma_{i'}^*(\omega_{X \times_P P'/X} \otimes \mathscr{E}xt_{P'}^{N-n}(p_{2*}\mathscr{O}_{X \times_P P'}, q^*\omega_{P/Y}))$$

$$\Bigg\|$$

$$\omega_{X/X \times_P P'} \otimes \Gamma_{i'}^* \mathscr{E}xt_{P'}^{N-n}(p_{2*}\mathscr{O}_{X \times_P P'}, \omega_{P'/P} \otimes q^*\omega_{P/Y}))$$

$$\simeq \Bigg\uparrow \zeta'_{q,\pi}$$

$$\omega_{X/X \times_P P'} \otimes \Gamma_{i'}^* \mathscr{E}xt_{P'}^{N-n}(p_{2*}\mathscr{O}_{X \times_P P'}, \omega_{P'/Y})$$

$$\simeq \Bigg\uparrow \eta_{\Gamma_{i'}}$$

$$\mathscr{E}xt_{X \times_P P'}^d((\Gamma_{i'})_*\mathscr{O}_X, \mathscr{E}xt_{P'}^{N-n}(p_{2*}\mathscr{O}_{X \times_P P'}, \omega_{P'/Y}))$$

$$\simeq \Bigg\uparrow$$

$$\mathscr{E}xt_{P'}^{N'-n}(i'_*\mathscr{O}_X, \omega_{P'/Y})$$

The map α in (3.5.6) is the flat base change isomorphism for $\mathscr{E}xt$ and the bottom map in (3.5.6) arises from the Grothendieck spectral sequence associated to the composite functor isomorphism $i'^\flat \simeq \Gamma_{i'}^\flat p_2^\flat$ (when evaluated on $\omega_{P'/Y}$). By Theorem 3.5.1 and the CM hypothesis on f, this Grothendieck spectral sequence is very degenerate; this is why the map at the bottom of (3.5.6) is an isomorphism. Briefly, here is where the sign of $(-1)^{d(N-n)}$ comes from. If $\omega_{P'/Y} \to \mathscr{I}^\bullet$ and $\mathscr{E}xt_P^{N-n}(\mathscr{O}_X, \omega_{P/Y}) \to \mathscr{J}^\bullet$ are injective resolutions, then the isomorphisms

$$\mathscr{I}^{\bullet+N}[d] \simeq \mathscr{I}^{\bullet+N'}, \quad \mathscr{J}^{\bullet+n}[d] \simeq \mathscr{J}^{\bullet+n+d}$$

compatible with respective augmentations from

$$\omega_{P'/Y}[N'], \quad \mathscr{E}xt_P^{N-n}(\mathscr{O}_X, \omega_{P/Y})[n+d]$$

are given by multiplication by the respective signs $(-1)^{d(r+N')}$, $(-1)^{d(r+n+d)}$ in degree r. Multiplying these together gives a sign of $(-1)^{d(N-n)}$.

Assume for a moment that one of i or i' is an lci map; this need not be true in general (consider general finite f with artinian fiber rings which are not complete intersection rings). Note that by [**EGA**, IV$_4$, 19.2.4], this lci condition for i (resp. i') is equivalent to the stronger condition that i (resp. i') is transversally regular over Y and so this property is preserved by arbitrary base change on Y. Moreover, by [**EGA**, IV$_4$, 19.3.7], this condition (for either i or i') is equivalent to the *intrinsic* property that f is a relative complete intersection over Y (i.e., flat and locally finitely presented, with all local rings on the geometric fibers equal to complete intersection rings). Thus, under this hypothesis *both* maps i and i' are transversally regular over Y. We may then use (2.5.7), Lemma 2.6.2, and the flat base change compatibility of the fundamental local isomorphism to identify (3.5.6) with a diagram consisting *entirely* of various ζ' isomorphisms. This is a significant simplication, since the ζ' maps are easy to calculate in local coordinates. For example, in this situation it is *obvious* that (3.5.6) is compatible with locally noetherian base change (recall that the fundamental local isomorphism, when evaluated on a quasi-coherent sheaf flat over the base, is compatible with arbitrary base change preserving the lci condition).

In §3.6, we will show in general (i.e., without lci hypotheses on i or i') that the explicit description (3.5.6) of (3.5.4) respects locally noetherian base change. This will enable us to construct a good global theory of base change for dualizing sheaves. When the above map f is smooth, the isomorphism $f^{\sharp}\mathcal{O}_Y \simeq i^{\flat}\pi^{\sharp}\mathcal{O}_Y$ as in (2.7.5) yields an isomorphism on degree $-n$ cohomology

$$(3.5.7) \qquad \omega_{X/Y} \simeq \mathscr{E}xt_P^{N-n}(i_*\mathcal{O}_X, \omega_{P/Y})$$

which is compatible with (3.5.3) and the canonical isomorphism $\omega_f \simeq \Omega_{X/Y}^n = \omega_{X/Y}$ (and uses the injective resolution $\pi^{\sharp}\mathcal{O}_Y \to \mathscr{I}^{\bullet+N}$ if we compute the $\mathscr{E}xt$ term with the injective resolution $\mathscr{I}^{\bullet}$ of $\omega_{P/Y}$). We now show that, up to an explicit universal sign depending only on n and N, the 'abstract' (3.5.7) is the classical 'Koszul isomorphism'

$$(3.5.8) \qquad \omega_{X/Y} \xrightarrow[\simeq]{\zeta'_{i,\pi}} \omega_{X/P} \otimes i^*\omega_{P/Y} \xleftarrow[\simeq]{\eta_i} \mathscr{E}xt_P^{N-n}(i_*\mathcal{O}_X, \omega_{P/Y}),$$

whose base change compatibility is quite easy to verify:

LEMMA 3.5.3. *Let $f : X \to Y$ be a separated smooth scheme map with pure relative dimension n, factorizing as $f = \pi \circ i$ where $i : X \hookrightarrow P$ is a closed immersion and $\pi : P \to Y$ is a separated smooth map with pure relative dimension N (so i is an lci map with pure codimension $N - n$, and is even transversally regular relative to Y with 'relative codimension $N - n$'). The 'Koszul isomorphism' (3.5.8) is compatible with arbitrary base change over Y and if Y is locally noetherian, then (3.5.8) is equal to $(-1)^{n(N-n)}$ times (3.5.7).*

All that really matters in what follows is that (3.5.8) and (3.5.7) coincide up to a universal sign depending only on n and N, as this ensures that (3.5.7) is compatible with locally noetherian base change. In (3.6.11)ff, we will generalize the isomorphism (3.5.7) to CM maps f which are relative local complex intersections over Y.

PROOF. Since $\zeta'_{i,\pi}$ and η_i are locally defined in terms of Koszul resolutions and the base change map for $\mathscr{E}xt_P^{N-n}(i_*\mathcal{O}_X, \omega_{P/Y})$ can be computed locally over small open affines in P by means of a Koszul resolution of $i_*\mathcal{O}_X$ over $\mathcal{O}_P$, it is trivial to check that (3.5.8) is compatible with arbitrary base change over Y.

Now assume that Y is locally noetherian. In order to identify (3.5.8) and (3.5.7), we will reduce ourselves to an explicit calculation with exterior products and the ζ' isomorphisms. The relevant scheme diagrams are

$$
\begin{array}{ccc}
X \times_Y P & \xrightarrow{\;p_2\;} & P \\
{\scriptstyle j}\big\uparrow\big\downarrow\;\;{\scriptstyle i\;}\;\;{}^{p_1}\nearrow & {\scriptstyle \pi}\big\downarrow & \\
X & \xrightarrow{\;\;f\;\;} & Y
\end{array}
\qquad
\begin{array}{ccc}
X \times_Y X & \xrightarrow{\;i'\;} & X \times_Y P \\
{\scriptstyle \Delta}\big\uparrow\big\downarrow\;{\scriptstyle q_2}\;\;{}^{j}\nearrow & {}^{p_2}\big\downarrow & \\
X & \xrightarrow{\;\;i\;\;} & P
\end{array}
$$

where $i'(x_1, x_2) = (x_1, i(x_2))$ (cf. (2.7.6), (2.7.11)). Keeping in mind that X is Y-smooth, we find via two applications of (2.7.3) that, up to a sign of $(-1)^{n(N-n)}$, the isomorphism (3.5.7) is the composite

(3.5.9)

$$
\omega_{X/Y}
$$
$$
\simeq\Big\downarrow \zeta'_{j,\pi\circ p_2}
$$
$$
\omega_{X/X\times_Y P} \otimes j^*\omega_{X\times_Y P/Y}
$$
$$
\simeq\Big\uparrow \eta_j
$$
$$
\mathscr{E}xt^N_{X\times_Y P}(j_*\mathcal{O}_X, \omega_{X\times_Y P/Y})
$$
$$
\simeq\Big\downarrow \varphi
$$
$$
\mathscr{E}xt^n_{X\times_Y X}(\Delta_*\mathcal{O}_X, \mathscr{E}xt^{N-n}_{X\times_Y P}(i'_*\mathcal{O}_{X\times_Y X}, \omega_{X\times_Y P/Y}))
$$
$$
\simeq\Big\downarrow \zeta'_{p_2,\pi}
$$
$$
\mathscr{E}xt^n_{X\times_Y X}(\Delta_*\mathcal{O}_X, \mathscr{E}xt^{N-n}_{X\times_Y P}(i'_*\mathcal{O}_{X\times_Y X}, \omega_{X\times_Y P/P} \otimes p_2^*\omega_{P/Y}))
$$
$$
\simeq\Big\uparrow \beta
$$
$$
\mathscr{E}xt^n_{X\times_Y X}(\Delta_*\mathcal{O}_X, \omega_{X\times_Y X/X} \otimes q_2^*\mathscr{E}xt^{N-n}_P(i_*\mathcal{O}_X, \omega_{P/Y}))
$$
$$
\simeq\Big\downarrow \eta_\Delta
$$
$$
\omega_{X/X\times_Y X} \otimes \Delta^*\omega_{X\times_Y X/X} \otimes \mathscr{E}xt^{N-n}_P(i_*\mathcal{O}_X, \omega_{P/Y})
$$
$$
\simeq\Big\uparrow \zeta'_{\Delta,q_2}
$$
$$
\mathscr{E}xt^{N-n}_P(i_*\mathcal{O}_X, \omega_{P/Y})
$$

where β involves a flat base change map for $\mathscr{E}xt$, φ arises from a degenerate Grothendieck spectral sequence, and the extra sign is obtained by reasoning as in the study of (3.5.6).

All maps in (3.5.9), aside from φ, are (by definition) explicitly locally described in terms of Koszul resolutions. Fortunately, (2.6.14) gives an explicit description of φ in these terms as well (cf.(2.6.18)). Thus, the comparison of (3.5.8) and (3.5.9) is reduced to an explicit calculation, which we now work out.

By using (2.5.7), flat base change for η, compatibilities for ζ' with respect to composites, and the description of φ in terms of fundamental local isomorphisms (via Lemma 2.6.2), the above composite (3.5.9) is identified with $(-1)^{n(N-n)}$ times the composite

$$(3.5.10)$$

$$\omega_{X/Y}$$

$$\downarrow \zeta'_{j,\pi \circ p_2}$$

$$\omega_{X/X\times_Y P} \otimes j^* \omega_{X\times_Y P/Y}$$

$$\downarrow \zeta'_{\Delta,i'}$$

$$\omega_{X/X\times_Y X} \otimes \Delta^* \omega_{X\times_Y X/X\times_Y P} \otimes j^* \omega_{X\times_Y P/Y}$$

$$\downarrow \zeta'_{p_2,\pi}$$

$$\omega_{X/X\times_Y X} \otimes \Delta^* \omega_{X\times_Y X/X\times_Y P} \otimes j^* (\omega_{X\times_Y P/P} \otimes p_2^* \omega_{P/Y})$$

$$\|$$

$$\omega_{X/X\times_Y X} \otimes \Delta^* (\omega_{X\times_Y X/X\times_Y P} \otimes i'^* (\omega_{X\times_Y P/P} \otimes p_2^* \omega_{P/Y}))$$

$$\|$$

$$\omega_{X/X\times_Y X} \otimes \Delta^* (q_2^* \omega_{X/P} \otimes \omega_{X\times_Y X/X} \otimes q_2^* i^* \omega_{P/Y})$$

$$\| \text{ no sign}$$

$$\omega_{X/X\times_Y X} \otimes \Delta^* (\omega_{X\times_Y X/X} \otimes q_2^* \omega_{X/P} \otimes q_2^* i^* \omega_{P/Y})$$

$$\downarrow \zeta'_{\Delta,q_2}$$

$$\omega_{X/P} \otimes i^* \omega_{X/P}$$

where the final term at the end of (3.5.10) is identified with $\mathscr{E}xt_P^{N-n}(i_* \mathscr{O}_X, \omega_{P/Y})$ via η_i. In view of the definition of (3.5.8), we are faced with comparing (3.5.10) and $\zeta'_{i,\pi}$.

Working locally, we can suppose $t_1, \ldots, t_m$ are global functions on P cutting out $i : X \hookrightarrow P$ and $x_1, \ldots, x_n$ are global functions on P which induce étale coordinates $\bar{x}_j$ on X. Thus, we may view $t_1, \ldots, t_m, x_1, \ldots, x_n$ are étale coordinates on P (upon shrinking P around X). In particular, $N = n+m$ and $X \hookrightarrow X \times_Y P$ is cut out by the functions $1 \otimes t_j$ and $\bar{x}_j \otimes 1 - 1 \otimes x_j$. Define

$$t^\vee = t_1^\vee \wedge \cdots \wedge t_m^\vee, \quad dt = dt_m \wedge \cdots \wedge dt_1$$

(note the orderings) and define $(\overline{x} \otimes 1 - 1 \otimes x)^{\vee}$, dx, $d\overline{x}$, etc. similarly. In these terms, the first step of (3.5.10) is determined by

$$d\overline{x} \mapsto (1 \otimes t)^{\vee} \wedge (\overline{x} \otimes 1 - 1 \otimes x)^{\vee} \otimes j^{*}(d(\overline{x} \otimes 1 - 1 \otimes x) \wedge d(1 \otimes t) \wedge d(\overline{x} \otimes 1)).$$

Since the $d(\overline{x}_j \otimes 1)$ at the end cancels against the first term in $d(\overline{x}_j \otimes 1 - 1 \otimes x_j) = d(\overline{x}_j \otimes 1) - d(1 \otimes x_j)$, we can rewrite the above expression as

$$(-1)^{mn}(-1)^{n}(-1)^{n(n+m)}(\overline{x} \otimes 1 - 1 \otimes x)^{\vee} \wedge (1 \otimes t)^{\vee} \otimes j^{*}(d(\overline{x} \otimes 1) \wedge p_2^{*}(dx \wedge dt)).$$

Note the signs completely cancel out. The second, third, and fourth steps in (3.5.10) take this to

$$(\overline{x} \otimes 1 - 1 \otimes x)^{\vee} \otimes \Delta^{*}((1 \otimes t)^{\vee} \otimes i'^{*}(d(\overline{x} \otimes 1)) \otimes p_2^{*}(dx \wedge dt)).$$

Relative to $q_2 : X \times_Y X \to X$, in $\Omega^1_{X \times_Y X/X}$ we have $d(\overline{x}_j \otimes 1) = d(\overline{x}_j \otimes 1 - 1 \otimes x_j)$. Thus, following through the rest of (3.5.10) brings us to

$$t^{\vee} \otimes (dt \wedge dx) = \zeta'_{i,\pi}(d\overline{x}),$$

as desired.   ∎

Note that if we had not introduced the extra sign in the definition of (1.3.28), then η_i in (3.5.8) would change by a sign of $(-1)^{(N-n)(N-n+1)/2}$ and η_j (resp. η_Δ) in (3.5.9) would change by a sign of $(-1)^{N(N+1)/2}$ (resp. $(-1)^{n(n+1)/2}$). But using (2.6.14) to describe φ would introduce a sign of $(-1)^{n(N-n)}$ (see the remark following Theorem 2.5.1). The product of all of four of these signs is 1, so the identification of (3.5.7) and (3.5.9) is 'independent' of the sign in the definition of (1.3.28).

3.6. Base Change for Dualizing Sheaves

Since dualizing sheaves are locally given by $\mathscr{E}xt$-sheaves, we can use the theory of base change for $\mathscr{E}xt$'s to formulate a reasonable base change theory for dualizing sheaves which generalizes the well-known base change theory for relative differentials:

THEOREM 3.6.1. *Let $f : X \to Y$ be a CM map between locally noetherian schemes, with pure relative dimension n. For any cartesian square*

$$\begin{array}{ccc} X' & \xrightarrow{p'} & X \\ {\scriptstyle f'}\downarrow & & \downarrow{\scriptstyle f} \\ Y' & \xrightarrow{p} & Y \end{array}$$

with Y' a locally noetherian scheme, there is a unique way to define an isomorphism $\beta_{f,p} : p'^{*}\omega_f \simeq \omega_{f'}$ which is compatible with Zariski localization on X and, *in case f factorizes as $X \xrightarrow{i} P \xrightarrow{\pi} Y$ for a closed immersion i and a separated smooth π with pure relative dimension N, makes the diagram*

(3.6.1)

$$
\begin{array}{ccc}
p'^{*}\omega_f & \xrightarrow{\quad\beta_{f,p}\quad} & \omega_{f'} \\
\underset{\simeq}{} & & \\
\| & & \| \\
p'^{*}\mathscr{E}xt_P^{N-n}(i_{*}\mathcal{O}_X,\omega_{P/Y}) & \xrightarrow[\simeq]{} & \mathscr{E}xt_{P'}^{N-n}(i'_{*}\mathcal{O}_{X'},\omega_{P'/Y'})
\end{array}
$$

commute (i' denotes the base change of i). In particular, if f is smooth (so ω_f and $\omega_{f'}$ are canonically identified with $\Omega^n_{X/Y}$ and $\Omega^n_{X'/Y'}$ respectively and we can take $\pi = f$, $i = \mathrm{id}_X$ if f is separated), then $p^{*}\omega_f \simeq \omega_{f'}$ is the usual base change isomorphism for top degree relative differentials.

Moreover, if $q : Y'' \to Y'$ is a further locally noetherian base change, then the composite isomorphism $\beta_{f',q} \circ q'^{*}(\beta_{f,p}) : q'^{*}p'^{*}\omega_f \simeq q'^{*}\omega_{f'} \simeq \omega_{f''}$ is equal to $\beta_{f,pq} : (pq)'^{*}\omega_f \simeq \omega_{f''}$.

Since all linear maps commute with -1, if we changed (3.5.3) by a universal sign depending only on n and N, then *both* columns in (3.6.1) would change by the *same* sign. Thus, $\beta_{f,p}$ is independent of the convention used to define the explication (3.5.3). It suffices to show that the composite isomorphism (3.5.4) respects 'base change for $\mathscr{E}xt$'. In the explicit description (3.5.6), the only step whose base change compatibility is not clear is the spectral sequence isomorphism at the bottom. If f were a relative complete intersection over Y, so i is transversally regular over Y, then this base change problem is trivial, since we can use the commutative diagram in Lemma 2.6.2 and the easy base change compatibility of all other sides in that diagram. Thus, the difficulty is due entirely to the fact that we want to treat the general CM case.

For conceptual clarity and usefulness later, we prove a more general statement. Consider a commutative diagram of schemes

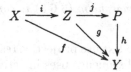

in which h is separated smooth with pure relative dimension, g is CM with pure relative dimension, and i and j are closed immersions with 'pure relative codimensions' δ and d respectively (i.e., for all $y \in Y$, $X_y \hookrightarrow Z_y$ has pure codimension δ and $Z_y \hookrightarrow P_y$ has pure codimension d); these properties are clearly preserved by arbitrary base change on Y. Finally, assume that i is transversally regular with respect to Y, so f is *automatically* CM with pure relative dimension and i is an lci map with pure codimension δ and remains so after arbitrary base change. In the context of (3.5.6), the closed immersions $\Gamma_{i'}$ and p_2 play the roles of i and j respectively, while π' plays the role of h.

Now assume Y is locally noetherian. From a degenerate Grothendieck spectral sequence (cf. Theorem 3.5.1), we have an isomorphism

$$(3.6.2) \qquad \mathscr{E}xt_P^{d+\delta}((ji)_*\mathscr{O}_X, \omega_{P/Y}) \simeq \mathscr{E}xt_Z^\delta(i_*\mathscr{O}_X, \mathscr{E}xt_P^d(j_*\mathscr{O}_Z, \omega_{P/Y})).$$

We just need to prove that (3.6.2) is of formation compatible with locally noetherian base change over Y. This claim is Zariski local on P, so we can replace $\omega_{P/Y}$ by $\mathscr{O}_P$. Also, as we mentioned earlier (by [**AK2**, Lemma 1.6]), the module of sections of any of the above $\mathscr{E}xt$'s over an open affine is canonically identified with the corresponding module Ext. Thus, it follows (via the mapping properties of Cartan-Eilenberg resolutions) that the Grothendieck spectral sequence for $\mathscr{E}xt$'s yields the Grothendieck spectral sequence for Ext's over open affines $\mathrm{Spec}(B)$ in P (so in particular, the spectral sequence for module Ext's is compatible with localization at an element of B). By working locally on P, we are therefore reduced to the following commutative algebra problem.

Let $A \to B$ be a smooth map with pure relative dimension, where A is an arbitrary ring, and let $I \subseteq J$ be two finitely generated ideals in B such that B/I is CM over A with pure relative dimension. Assume that $\mathrm{Spec}(B/J) \hookrightarrow \mathrm{Spec}(B/I)$ is transversally regular over $\mathrm{Spec}(A)$, with pure codimension δ, so B/J is CM over A with pure relative dimension. Let d be the pure codimension of $\mathrm{Spec}(B/I) \hookrightarrow \mathrm{Spec}(B)$ relative to $\mathrm{Spec}(A)$. Consider the isomorphism

$$(3.6.3) \qquad \mathrm{Ext}_B^{d+\delta}(B/J, B) \simeq \mathrm{Ext}_{B/I}^\delta(B/J, \mathrm{Ext}_B^d(B/I, B))$$

arising from the degenerate Grothendieck spectral sequence.

From the hypotheses and the end of Corollary 3.5.2, it follows that

$$\mathrm{Ext}_B^d(B/I, B)$$

is A-flat and finitely presented as a B-module, $\mathrm{Ext}_B^m(B/I, B) = 0$ for $m \neq d$, and

$$\mathrm{Ext}_{B/I}^n(B/J, \mathrm{Ext}_B^d(B/I, B)) \simeq \mathrm{Ext}_B^{n+d}(B/I, B) = 0$$

for $n \neq \delta$. Thus, by [**AK2**, Lemma 1.9], there are natural base change isomorphisms (over A) for all three Ext's in (3.6.3).

THEOREM 3.6.2. *The isomorphism* (3.6.3) *is compatible with arbitrary base change over* A.

Since base change maps for Ext use projective resolutions in the first variable and the Grothendieck spectral sequence uses injective resolutions in the second variable, Theorem 3.6.2 is not obvious.

PROOF. (of Theorem 3.6.2) The spectral sequence for Ext's is compatible with localization with respect to an element of B, so we may work Zariski locally on B and thus may assume there are elements $f_1, \ldots, f_\delta \in B/I$ which generate J/I and have the property that for $1 \leq i \leq \delta - 1$, the ring $B_i \overset{\text{def}}{=} (B/I)/(f_1, \ldots, f_i)$ is A-flat and $f_{i+1} \in B_i$ is a regular element. After any base change over A, the f_i's still have these properties (see [**EGA**, IV$_4$, §19.2]).

We want to argue by induction on $\delta \geq 1$ (the case $\delta = 0$ is trivial). Note that the transversally regular condition forces all intermediate $\mathrm{Spec}(B_i)$'s between

$\text{Spec}(B/I)$ and $\text{Spec}(B/J)$ to be CM over A with pure relative dimension, but with a smaller pure codimension δ_i inside of $\text{Spec}(B/I)$ (if $\delta > 1$). This will make it possible to carry out induction (or rather, reduction to the case $\delta = 1$). More precisely, by using the behavior of total derived functors with respect to composites (i.e., the derived category version of the Leray spectral sequence) and an enormous amount of degeneration in our setting, (3.6.3) fits into the top row of the following *commutative* diagram of isomorphisms:

(3.6.4)

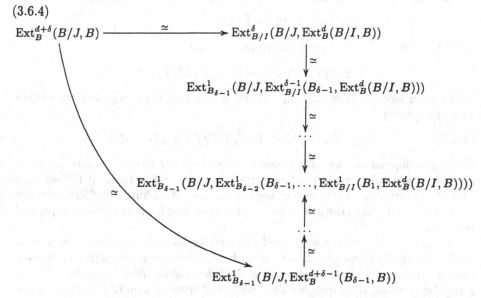

By functoriality (e.g., with respect to base change maps for Ext's), it suffices to prove that all maps in (3.6.4) aside from the top row are compatible with base change over A. These maps are all instances of a generalization of the case $\delta = 1$ which is treated by Lemma 3.6.3 below, thereby completing the proof. ∎

LEMMA 3.6.3. *Let A be a ring, B a flat A-algebra, I an ideal in B such that $B' = B/I$ is A-flat. Let $f \in B'$ be a regular element such that B'/f is A-flat and let M be an A-flat B-module such that $\text{Ext}_B^m(B', M) = 0$ for all $m \neq n$. Consider the isomorphism*

(3.6.5) $$\text{Ext}_B^{n+1}(B'/f, M) \simeq \text{Ext}_{B'}^1(B'/f, \text{Ext}_B^n(B', M))$$

arising from the degenerate Grothendieck spectral sequence. This isomorphism is compatible with base change maps over A (which need not be isomorphisms).

PROOF. In order to remove any possibility of confusion about or apparent dependence upon signs in the computation of Ext via projective resolutions in the first variable, we will use an arbitrary sign $\epsilon = \pm 1$ in our calculations and show that it completely cancels out. We will say that a map *depends on ϵ* to mean that changing ϵ changes the map by a factor of -1 (and is *independent of*

ϵ otherwise). Using the B'-projective resolution

(3.6.6) $0 \to B' \xrightarrow{\cdot f} B' \to B'/f \to 0$

of B'/f, an isomorphism depending on ϵ

$$\operatorname{Ext}_B^n(B', M)/f \simeq \operatorname{Ext}_{B'}^1(B'/f, \operatorname{Ext}_B^n(B', M))$$

arises from the δ-functoriality of $\operatorname{Ext}_{B'}^{\bullet}(\cdot, \operatorname{Ext}_B^n(B', M))$. Also, since

$$\operatorname{Ext}_B^{n+1}(B', M) = 0,$$

the δ-functor $\operatorname{Ext}_B^{\bullet}(\cdot, M)$ gives rise to an injection

$$\operatorname{Ext}_B^n(B', M)/f \hookrightarrow \operatorname{Ext}_B^{n+1}(B'/f, M)$$

which is an isomorphism and depends on ϵ. Putting these together, we obtain an isomorphism

(3.6.7) $\operatorname{Ext}_B^{n+1}(B'/f, M) \simeq \operatorname{Ext}_{B'}^1(B'/f, \operatorname{Ext}_B^n(B', M))$

which is independent of ϵ. Since base change maps for Ext are defined in terms of projective resolutions in the first variable and are δ-*functorial*, it follows that (3.6.7) is compatible with base change maps over A. Thus, it suffices to prove that (3.6.7) is equal to the isomorphism (3.6.5) arising from a degenerate spectral sequence.

In particular, we no longer need to consider base change issues, so we can work in a slightly more general setting without flatness conditions or an auxiliary ring A. More precisely, let B be a ring, M a B-module, B' a B-algebra, $f \in B'$ a regular element, n an integer. The method of construction of (3.6.7) gives rise to an injection (independent of ϵ)

(3.6.8) $\operatorname{Ext}_{B'}^1(B'/f, \operatorname{Ext}_B^n(B', M)) \hookrightarrow \operatorname{Ext}_B^{n+1}(B'/f, M).$

Meanwhile, since $\operatorname{projdim}_{B'}(B'/f) \le 1$, the spectral sequence arising from

$$\operatorname{Hom}_{B'}(B'/f, \operatorname{Hom}_B(B', \cdot)) \simeq \operatorname{Hom}_B(B'/f, \cdot)$$

has $E_2^{pq} = \operatorname{Ext}_{B'}^p(B'/f, \operatorname{Ext}_B^q(B', \cdot)) = 0$ for all $p > 1$, so we have an injection

$$d_2^{1,n} : \operatorname{Ext}_{B'}^1(B'/f, \operatorname{Ext}_B^n(B', M)) \hookrightarrow \operatorname{Ext}_B^{n+1}(B'/f, M).$$

The map $d_2^{1,n}$ is an isomorphism if $\operatorname{Ext}_B^{n+1}(B', M) = 0$, in which case its inverse is exactly (3.6.5). Thus, it suffices to show in general that (3.6.8) and $d_2^{1,n}$ are equal.

The construction of the Grothendieck spectral sequence and the method in [**Tohoku**, II, 2.3, p.144] for realizing the δ-functoriality of Ext in the first variable by means of injective resolutions in the second variable allows us to 'compute' the two maps of interest as follows. Let $M \to I^{\bullet}$ be an injective resolution in the category of B-modules, $J^{\bullet\bullet}$ a Cartan-Eilenberg resolution of $\operatorname{Hom}_B(B', I^{\bullet})$ in the category of B'-modules, and $J'^{\bullet\bullet}$ the canonical truncation of $J^{\bullet\bullet}$ in rows ≤ 1. The complex $J'^{\bullet\bullet}$ consists of $\operatorname{Hom}_{B'}(B'/f, \cdot)$-acyclics and the complex $H_h^n(J^{\bullet\bullet})$ of horizontal cohomology in degree n is an injective B'-module resolution of $\mathrm{Ext}_B^n(B', M)$. Thus, $d_2^{1,n}$ is equal to the composite

$$H^1(\mathrm{Hom}_{B'}(B'/f, H_h^n(J^{\bullet\bullet})))$$

$$\Big\uparrow \simeq$$

$$\ker(H^{n+1}(\mathrm{Hom}_{B'}(B'/f, \mathrm{Tot}^{\oplus}(J^{\bullet\bullet}))) \to H^0(\mathrm{Hom}_{B'}(B'/f, H_h^{n+1}(J^{\bullet\bullet}))))$$

$$\Big\downarrow$$

$$H^{n+1}(\mathrm{Hom}_{B'}(B'/f, \mathrm{Tot}^{\oplus}(J^{\bullet\bullet})))$$

$$\Big\uparrow \simeq$$

$$H^{n+1}(\mathrm{Hom}_{B'}(B'/f, \mathrm{Hom}_B(B', I^{\bullet})))$$

$$\Big\downarrow \simeq$$

$$H^{n+1}(\mathrm{Hom}_B(B'/f, I^{\bullet}))$$

and (3.6.8) is the composite

$$(3.6.9) \qquad \begin{array}{ccc} H^1(\mathrm{Hom}_{B'}(B'/f, H_h^n(J^{\bullet\bullet}))) & \xleftarrow{\;\simeq\;} & H^0(H_h^n(J^{\bullet\bullet}))/f \\[2mm] \Big\| & & \Big\| \\[2mm] H^{n+1}(\mathrm{Hom}_B(B'/f, I^{\bullet})) & \longleftarrow & H^n(\mathrm{Hom}_B(B', I^{\bullet}))/f \end{array}$$

where the top map (depending on ϵ) arises from the snake lemma after applying $\mathrm{Hom}_{B'}(\cdot, H_h^n(J^{\bullet\bullet}))$ to (3.6.6), the equality in the right column arises from the definition of Cartan-Eilenberg resolutions, and the bottom map (depending on ϵ) arises from the snake lemma after applying $\mathrm{Hom}_B(\cdot, I^{\bullet})$ to (3.6.6). In particular, this composite map in (3.6.9) is independent of ϵ.

Thus, it suffices to prove the commutativity of the following diagram (where the two maps δ_1, δ_2 are coboundary maps from the snake lemma and depend on

ϵ):

(3.6.10)

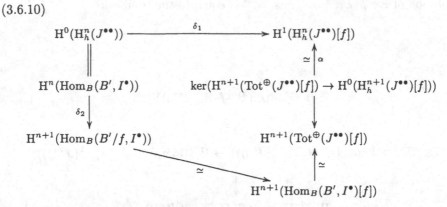

where $N'[f]$ denotes the f-torsion submodule in a B'-module N'. The proof of the commutativity of (3.6.10) is a slightly non-trivial diagram chase, as we now explain. Choose any $\varphi \in \operatorname{Hom}_B(B', I^n)$ which dies in $\operatorname{Hom}_B(B', I^{n+1})$, so we can view φ as an element in $\operatorname{H}^n(\operatorname{Hom}_B(B', I^\bullet)) = \operatorname{H}^0(\operatorname{H}^n_h(J^{\bullet\bullet}))$. Since I^n is an injective B-module and $f \in B$ is a regular element, we can pick a $\psi \in \operatorname{Hom}_B(B', I^n)$ such that $\epsilon f \psi = \varphi$. In particular, the composite of ψ with $\mathrm{d}^n : I^n \to I^{n+1}$ gives rise to an element $\overline{\psi}$ in $\operatorname{Hom}_B(B'/f, I^{n+1})$ which represents $\delta_2(\varphi)$. We define $\widetilde{\psi}$ to be the image of $\overline{\psi}$ under

$$\operatorname{Hom}_B(B'/f, I^{n+1}) = \operatorname{Hom}_B(B', I^{n+1})[f] \hookrightarrow J^{n+1,0}[f] \subseteq J^{n+1,0},$$

so $\widetilde{\psi}$ is killed by f and coincides with the image of $\mathrm{d}^n \circ \psi \in \operatorname{Hom}_B(B', I^{n+1})$ in $J^{n+1,0}$. In particular, the cohomology class $[\widetilde{\psi}] \in \operatorname{H}^{n+1}(\operatorname{Tot}^\oplus(J^{\bullet\bullet})[f])$ vanishes in

$$\operatorname{H}^0(\operatorname{H}^{n+1}_h(J^{\bullet\bullet})[f]) = \operatorname{Ext}^{n+1}_B(B', M)[f] \subseteq \operatorname{H}^{n+1}(\operatorname{Hom}_B(B', I^\bullet)).$$

We just have to show that $\alpha([\widetilde{\psi}]) = \delta_1(\varphi)$.

Let $\psi' \in J^{n,0}$ be the image of $\psi \in \operatorname{Hom}_B(B', I^n)$ and let

$$\mathrm{d}^n_h : J^{n,0} \to J^{n+1,0}, \quad \mathrm{d}^n_v : J^{n,0} \to J^{n,1}$$

be the horizontal and vertical differentials respectively. Although ψ' is probably not f-torsion, $\mathrm{d}^n_h(\psi') = \widetilde{\psi}$ is killed by f, so $\mathrm{d}^n_h(f\psi') = 0$. By the Cartan-Eilenberg construction, the B'-module $\ker \mathrm{d}^n_h$ is injective, so the element $\epsilon f \psi'$ in this kernel can be written as $\epsilon f \psi' = \epsilon f \psi''$ for some $\psi'' \in \ker \mathrm{d}^n_h$ (independent of ϵ). Thus, $\psi' - \psi'' \in \operatorname{Hom}_{B'}(B'/f, J^{n,0})$ satisfies $\mathrm{d}^n_h(\psi' - \psi'') = \widetilde{\psi}$, so

$$[\widetilde{\psi}] = [-\mathrm{d}^n_v(\psi' - \psi'')] = [\mathrm{d}^n_v(\psi'')].$$

It follows that $\alpha([\widetilde{\psi}])$ is represented by $\mathrm{d}^n_v(\psi'')$, so ψ'' represents a class in $\operatorname{H}^n(J^{\bullet,0})$ whose image $\mathrm{d}^n_v(\psi'')$ in $\operatorname{H}^n(J^{\bullet,1})$ is killed by f and lies in the kernel of the map

$$\operatorname{Hom}_{B'}(B'/f, \operatorname{H}^n(J^{\bullet,1})) \to \operatorname{Hom}_{B'}(B'/f, \operatorname{H}^n(J^{\bullet,2})).$$

Since $\epsilon f \psi = \varphi \in \mathrm{H}^n(\mathrm{Hom}_B(B', I^{\bullet})) \simeq \mathrm{H}^0(\mathrm{H}^n_h(J^{\bullet \bullet}))$ is represented by $\epsilon f \psi' = \epsilon f \psi''$, the coboundary map δ_1 takes φ to the cohomology class represented by $d^n_v(\psi'')$, which we have seen is $\alpha([\widetilde{\psi}])$. This completes the proof. ∎

Although we have now proven Theorem 3.6.1, our discussion of base change for dualizing sheaves requires one further compatibility observation. Note that in case $f : X \to Y$ is a CM map with pure relative dimension n and Y admits a dualizing complex, we have a derived category isomorphism $\omega_f[n] \simeq f^! \mathscr{O}_Y$. For any *residually stable* base change $u : Y' \to Y$ where Y' is noetherian with a dualizing complex, there is a base change isomorphism $b_{u,f} : u^*(f^! \mathscr{O}_Y) \simeq f'^! \mathscr{O}_{Y'}$ from (3.3.24). From the definition of $b_{u,f}$ and the construction of base change isomorphism for ω_f in the proof of Theorem 3.6.1, we easily deduce:

COROLLARY 3.6.4. *The isomorphism $b_{u,f}$ is compatible with the isomorphism $u^* \omega_f[n] \simeq \omega_{f'}[n]$ from Theorem 3.6.1.*

Now that we have a good notion of base change for dualizing sheaves via base change for $\mathscr{E}xt$-sheaves, at least in the locally noetherian case, we want to generalize the relation between dualizing sheaves and $\mathscr{E}xt$-sheaves and formulate the basic properties of a trace map γ_f for proper CM maps f generalizing the trace map (3.4.11) in the smooth case. This will be important in the study of base change and duality for proper CM maps in §4.3.

Let $f : X \to Y$ be a CM map with pure relative dimension n, with Y a locally noetherian scheme, and suppose that $f = \pi \circ i$ where $i : X \to P$ is a closed immersion which is *transversally regular over* Y and $\pi : P \to Y$ is CM with pure relative dimension N (so i is an lci map with pure codimension $N - n$ and remains so after any base change over Y). The point is that π is no longer assumed to be smooth (in particular, the property that i is transversally regular over Y is no longer intrinsic to f, unless of course P is a relative complete intersection over Y). We want to define an isomorphism of $\mathscr{O}_X$-modules

$$(3.6.11) \qquad \omega_f \simeq \mathscr{E}xt_P^{N-n}(i_* \mathscr{O}_X, \omega_{P/Y})$$

for all such f so that the following properties hold:

- in case π is separated smooth, this is the isomorphism (3.5.7) (coinciding up to universal sign with (3.5.8)),
- (3.6.11) is Zariski local on P,
- (3.6.11) is compatible with any locally noetherian base change on Y.

It is clearly enough to consider the case in which all scheme maps are affine, or more generally when π factorizes as $\pi = \pi' \circ i'$, with $i' : P \hookrightarrow P'$ a closed immersion and $\pi' : P' \to Y$ separated smooth with pure relative dimension N'.

We define (3.6.11) to be the composite

$$(3.6.12) \qquad \omega_{X/Y} =\!=\!=\!=\!=\!=\!= \mathscr{E}xt_{P'}^{N'-n}((i'i)_*\mathscr{O}_X, \omega_{P'/Y})$$

$$\downarrow \simeq$$

$$\mathscr{E}xt_P^{N-n}(i_*\mathscr{O}_X, \mathscr{E}xt_{P'}^{N'-N}(i'_*\mathscr{O}_{X'}, \omega_{P'/Y}))$$

$$\|$$

$$\mathscr{E}xt_P^{N-n}(i_*\mathscr{O}_X, \omega_{P/Y})$$

where the first and last identifications are special cases of (3.5.3). In order to show that this composite is independent of the factorization of π (and therefore globalizes), it suffices to consider the case in which $i' = \pi'' \circ i''$ for a closed immersion i'' and a separated smooth π'' with pure relative dimension. In this case, we need to compare the definitions of (3.6.12) based on the two factorizations $\pi' \circ i'$ and $(\pi'\pi'') \circ i''$ of π. It is straightfoward to reduce to the commutativity of the diagram

$$\begin{array}{ccc} (i'i)^b & \xrightarrow[\simeq]{\psi_{i''i,\pi''}} & (i''i)^b\pi''^\sharp \\ \simeq \downarrow & & \downarrow \simeq \\ i^b i'^b & \xrightarrow[i^b(\psi_{i'',\pi''})]{\simeq} & i^b i''^b \pi''^\sharp \end{array}$$

(this is applied to $\pi'^\sharp(\cdot)$, and universal sign issues in the explication of $\mathscr{E}xt$'s are irrelevant). This commutativity is a special case of Theorem 2.7.2(1). Thus, (3.6.11) is well-defined. The base change compatibility of (3.6.11) follows from our study of (3.6.2) (cf. Theorem 3.6.2). This is where we use the hypothesis that i is transversally regular. Although the above definition of (3.6.11) clearly makes sense when i is merely a closed immersion, and in this generality is Zariski local on P and recovers (3.5.7) when π is separated and smooth, we only see how to verify base change compatibility over Y in the case of transversally regular i (and this is the only case we will need later, in the proof of Lemma 4.2.2).

We end this section by discussing the trace map over a general locally noetherian base. As usual, let $f : X \to Y$ be a CM map with pure relative dimension n, and now assume that f is also *proper*. If Y admits a dualizing complex (e.g., Y is of finite type over $\mathbf{Z}$), then there is a trace map

$$(3.6.13) \qquad \gamma_f : \mathrm{R}^n f_*(\omega_f) \simeq \mathrm{H}^0(\mathbf{R}f_*(f^!\mathscr{O}_Y)) \xrightarrow{\mathrm{H}^0(\mathrm{Tr}_f)} \mathscr{O}_Y$$

(where the first isomorphism is defined as in (3.4.13)). Using Theorem 3.6.1, it makes sense to ask if (3.6.13) is compatible with base change to another base admitting a dualizing complex (e.g., another finite type $\mathbf{Z}$-scheme, or a complete local noetherian ring). The main result of this book is:

THEOREM 3.6.5. *If $f : X \to Y$ is a proper CM map with pure relative dimension and Y is noetherian and admits a dualizing complex, then the map*

γ_f *is compatible with any base change* $Y' \to Y$, *where* Y' *is noetherian and admits a dualizing complex.*

This will be proven in Chapter 4; note that multiplying γ_f by $(-1)^n$ is harmless here. The one case which is obvious is when $Y' \to Y$ is residually stable, since the isomorphism $\omega_f \simeq f^! \mathscr{O}_Y$ is compatible with residually stable base change (see Corollary 3.6.4) and Tr_f is compatible with residually stable base change (see Lemma 3.4.3(TRA4)). The following corollary generalizes [**RD**, III, 11.2].

COROLLARY 3.6.6. *There is a unique way to define an* $\mathscr{O}_Y$-*linear*

$$\gamma_f : \mathrm{R}^n f_*(\omega_f) \to \mathscr{O}_Y$$

for proper CM maps $f : X \to Y$ *with pure relative dimension* n *over locally noetherian bases* Y *so that* γ_f *is compatible with arbitrary locally noetherian base change and recovers* (3.6.13) *in case* Y *is noetherian and admits a dualizing complex.*

Moreover, if f *is smooth and* $g : Y \to Z$ *is a proper smooth map with pure relative dimension* m *and* Z *is locally noetherian, then the diagram*

(3.6.14)

$$
\begin{array}{ccc}
\mathrm{R}^{n+m}(gf)_*(\omega_{gf}) & \xrightarrow{\;\zeta'_{f,g}\;} & \mathrm{R}^m g_* \, \mathrm{R}^n f_*(\omega_f \otimes f^* \omega_g) \\
\Big\downarrow{\scriptstyle \gamma_{gf}} & & \Big\uparrow{\scriptstyle \simeq} \\
& & \mathrm{R}^m g_*(\mathrm{R}^n f_*(\omega_f) \otimes \omega_g) \\
& & \Big\downarrow{\scriptstyle \gamma_f} \\
\mathscr{O}_Z & \xleftarrow[\;\gamma_g\;]{} & \mathrm{R}^m g_*(\omega_g)
\end{array}
$$

commutes. When f *is smooth with geometrically connected fibers,* γ_f *is an isomorphism.*

Since $(-1)^n(-1)^m = (-1)^{n+m}$, the commutativity of (3.6.14) is unaffected by using the alternative convention for defining (3.4.13) in the general definition of γ_f.

PROOF. By standard direct limit arguments, over any open affine in Y we can realize f as the base change of a proper CM map $f_0 : X_0 \to Y_0$ with pure relative dimension n, where Y_0 is a finite type **Z**-scheme, and f_0 is smooth with geometrically connected fibers if f is [**EGA**, IV$_3$, 9.9.1, 9.9.2(*ix*), 12.2.4(*vi*)]. We define γ_f by gluing the base changes of the maps γ_{f_0}. By Theorem 3.6.5, such gluing makes sense and is independent of all choices. By Lemma 3.3.2 and Lemma 3.4.3(TRA1), the commutativity of (3.6.14) is clear, by reduction to the case of finite type **Z**-schemes.

To show that γ_f is an isomorphism when f is smooth with geometrically connected fibers, we may again assume that Y is a finite type **Z**-scheme. By base change to $\mathrm{Spec}(k(y))$ for $y \in Y$ and the duality theorem for proper smooth geometrically connected schemes over fields as in (3.4.16), we see that $\gamma_f \otimes_{\mathscr{O}_Y} k(y)$

is an isomorphism for all $y \in Y$. By Nakayama's Lemma, it follows that γ_f is surjective and $R^n f_*(\omega_f)$ is locally a quotient of $\mathcal{O}_Y$, so γ_f is an isomorphism. ∎

In §4.3 we will generalize the definition of $\zeta'_{f,g} : \omega_{gf} \simeq \omega_f \otimes f^* \omega_g$ to the case where f and g are CM rather than smooth, and in §4.4 we will see that (3.6.14) commutes in the CM case.

By passage to limits, the definitions of ω_f for CM morphisms f with pure relative dimension and γ_f for such *proper* f can be uniquely extended to the case of an arbitrary base scheme in a manner which is compatible with base change. Since the base change map for ω_f coincides with the one for top differentials when f is smooth (see Theorem 3.6.1), in the proper smooth case we thereby obtain the desired base change compatibility in §1.1. Observe that in order to prove γ_f is an isomorphism for smooth f with geometrically connected fibers, we need the base change compatibility of γ_f! Also, the final part of Theorem 3.5.1 carries over to the case of an arbitrary base: the Y-flat ω_f is invertible if and only if the (geometric) fibers of f are Gorenstein schemes. Indeed, Theorem 3.5.1 ensures that f has Gorenstein (geometric) fibers if and only if $\omega_f|_{X_y}$ is invertible for all $y \in Y$, and since ω_f is Y-flat of finite presentation this is equivalent to ω_f being invertible on X.

CHAPTER 4

Proof of Main Theorem

Fix a cartesian diagram of schemes

$$
\begin{array}{ccc}
X' & \xrightarrow{\ u'\ } & X \\
{\scriptstyle f'}\downarrow & & \downarrow{\scriptstyle f} \\
Y' & \xrightarrow{\ u\ } & Y
\end{array}
$$

in which Y and Y' are noetherian schemes admitting a dualizing complex and f is a proper CM map with pure relative dimension n. We want to prove Theorem 3.6.5, which asserts that the diagram

$$
\begin{array}{ccc}
u^* \operatorname{R}^n f_*(\omega_f) & \longrightarrow & \operatorname{R}^n f'_*(\omega_{f'}) \\
{\scriptstyle u^*(\gamma_f)}\downarrow & & \downarrow{\scriptstyle \gamma_{f'}} \\
u^* \mathcal{O}_Y & =\!=\!=\!= & \mathcal{O}_{Y'}
\end{array}
$$

commutes; of course, this is unaffected by multiplying γ_f and $\gamma_{f'}$ by $(-1)^n$. By the last part of Theorem 3.6.1, if u factorizes into a composite $u = u_1 \circ u_2$ of maps between noetherian schemes admitting a dualizing complex, then it suffices to treat the base changes u_1 and u_2 separately. We frequently use this below without comment.

The only case we know so far is when the base change map u is residually stable (for reasons given after the statement of Theorem 3.6.5). This includes the case of the canonical map $\operatorname{Spec}(\mathcal{O}_{Y,y}) \to Y$ for $y \in Y$, so we may assume that $Y = \operatorname{Spec}(A)$, $Y' = \operatorname{Spec}(A')$ for local noetherian rings A, A' which admit dualizing complexes and the map $\varphi : A \to A'$ corresponding to u is a *local* map. By the Krull Intersection Theorem, we may assume that A' is a local artin ring, so φ factorizes as $A \twoheadrightarrow A/\ker(\varphi) \hookrightarrow A'$, where the local noetherian ring $A/\ker(\varphi)$ has nilpotent maximal ideal and so is artinian. Thus, it suffices to treat the two cases where A' is an artinian quotient of A and where $A \to A'$ is a local map between artin local rings.

4.1. Case of an Artinian Quotient

We first consider the case where $A' = A/\mathfrak{a}$ for an $\mathfrak{m}$-primary ideal $\mathfrak{a}$ in the local noetherian ring A (with maximal ideal $\mathfrak{m}$). Our arguments work for any proper ideal $\mathfrak{a}$ of A, but the case of $\mathfrak{m}$-primary $\mathfrak{a}$ is the only one we will need. In order to remember that we are considering a quotient of A which is artinian in the application, we change the notation and write X_0, Y_0, A_0 instead of X', Y', $A' = A/\mathfrak{a}$ and i, j, f_0 instead of u, u', f'. Thus, the relevant scheme diagram is

$$
\begin{array}{ccc}
X_0 & \xrightarrow{\;j\;} & X \\
{\scriptstyle f_0}\downarrow & & \downarrow{\scriptstyle f} \\
Y_0 & \xrightarrow{\;i\;} & Y
\end{array}
$$

For conceptual clarity (and technical necessity later), we assume for now that the CM map f is separated and finite type, but *not necessarily proper*.

Fix a choice of residual complex $K_Y^\bullet$ on the local noetherian scheme $Y = \mathrm{Spec}(A)$. We define the residual complexes $K_{Y_0}^\bullet = i^\Delta K_Y^\bullet$, $K_X^\bullet = f^\Delta K_Y^\bullet$, $K_{X_0}^\bullet = j^\Delta K_X^\bullet$. For example, when A_0 is artinian (the case we care about) then $K_{Y_0}^\bullet$ is an injective hull of $k = A/\mathfrak{m}$ over A_0, supported in some degree (which can be computed by (3.2.4)). Back in the general setting, by (3.3.17) we have equalities $K_{Y_0}^\bullet = \mathscr{H}om_Y^\bullet(\mathcal{O}_Y/\mathfrak{a}, K_Y^\bullet)$ and $K_{X_0}^\bullet = \mathscr{H}om_X^\bullet(\mathcal{O}_X/\mathfrak{a}, K_X^\bullet)$. Also, since $if_0 = jf$, we can define an isomorphism of complexes

(4.1.1)
$$
f_0^\Delta K_{Y_0}^\bullet = f_0^\Delta i^\Delta K_Y^\bullet \simeq j^\Delta f^\Delta K_Y^\bullet = j^\Delta K_X^\bullet = K_{X_0}^\bullet = \mathscr{H}om_X^\bullet(\mathcal{O}_X/\mathfrak{a}, K_X^\bullet).
$$

The flatness of f gives rise to an isomorphism of complexes

$$
f_0^* K_{Y_0}^\bullet \simeq \mathscr{H}om_X^\bullet(\mathcal{O}_X/\mathfrak{a}, f^* K_Y^\bullet),
$$

so we obtain a natural map of complexes of *flasque* sheaves (which are acyclic for total direct images)

(4.1.2)
$$
\mathscr{H}om_X^\bullet(f^* K_Y^\bullet, f^\Delta K_Y^\bullet)
$$
$$
\|
$$
$$
\mathscr{H}om_X^\bullet(f^* K_Y^\bullet, K_X^\bullet)
$$
$$
\downarrow
$$
$$
j_* \mathscr{H}om_{X_0}^\bullet(\mathscr{H}om_X(\mathcal{O}_X/\mathfrak{a}, f^* K_Y^\bullet), \mathscr{H}om_X(\mathcal{O}_X/\mathfrak{a}, K_X^\bullet))
$$
$$
\downarrow{\scriptstyle\simeq}
$$
$$
j_* \mathscr{H}om_{X_0}^\bullet(f_0^* K_{Y_0}^\bullet, f_0^\Delta K_{Y_0}^\bullet)
$$

where the vertical right arrow is the map "pass to the induced map on $\mathfrak{a}$-torsion" and the bottom map uses (4.1.1). The flasqueness of the terms in the complexes in (4.1.2) rests on Lemma 2.1.3 and the general fact that for any ringed space Z and any injective $\mathcal{O}_Z$-module $\mathscr{I}$, the functor $\mathscr{H}om_Z(\cdot, \mathscr{I})$ takes flasque sheaves to flasque sheaves (observe the proof involves "extension by zero" sheaves, and so really requires $\mathscr{I}$ to be an injective object in the category of $\mathcal{O}_Z$-modules and not just in some full subcategory such as quasi-coherent sheaves in the case where Z is a scheme). The complex $\mathscr{H}om_X^\bullet(f^*K_Y^\bullet, f^\triangle K_Y^\bullet)$ (resp. $\mathscr{H}om_{X_0}^\bullet(f_0^*K_{Y_0}^\bullet, f_0^\triangle K_{Y_0}^\bullet)$) in (4.1.2) represents $f^!\mathcal{O}_Y$ (resp. $f_0^!\mathcal{O}_{Y_0}$) and so is isomorphic to $\omega_{X/Y}[n]$ (resp. $\omega_{X_0/Y_0}[n]$) in the derived category. The flasqueness of the terms in $\mathscr{H}om_X^\bullet(f^*K_Y^\bullet, f^\triangle K_Y^\bullet)$, $\mathscr{H}om_{X_0}^\bullet(f_0^*K_{Y_0}^\bullet, f_0^\triangle K_{Y_0}^\bullet)$ implies that these complexes can (and will) be used to compute the hypercohomology of $\omega_{X/Y}[n]$ and $\omega_{X_0/Y_0}[n]$ respectively. Having made this choice, in view of our definitions of the explications such as (3.4.11) we see that the computations of γ_f, γ_{f_0} will require computing ordinary higher direct images of $\omega_{X/Y}$ with the flasque resolution $\mathscr{H}om_X^{\bullet-n}(f^*K_Y^\bullet, f^\triangle K_Y^\bullet)$, and likewise for ω_{X_0/Y_0} on X_0. Beware that in other contexts which are not going to be considered here (such as (3.1.7)), we are required to use $\mathscr{H}om_X^\bullet(f^*K_Y^\bullet, f^\triangle K_Y^\bullet)[-n]$ as the 'preferred' injective resolution of $\omega_{X/Y}$ (and likewise for ω_{X_0/Y_0}); this will become a relevant issue in Appendix A.

The map induced by (4.1.2) on cohomology in degree $-n$ is a map of sheaves

(4.1.3) $$\omega_{X/Y} \to j_*\omega_{X_0/Y_0}$$

Since base change maps in sheaf cohomology are defined via the adjointness of pushfoward and pullback on the level of acyclic (e.g., flasque) resolutions, the importance of the construction of (4.1.3) by means of the map (4.1.2) between flasque resolutions is:

LEMMA 4.1.1. *The map* (4.1.3) *is adjoint to the base change isomorphism* $j^*\omega_{X/Y} \simeq \omega_{X_0/Y_0}$.

The proof will require working locally on X, so it is essential that we have not yet required f to be proper. Let us grant that Lemma 4.1.1 is true and use it to deduce Theorem 3.6.5 for base change to a closed subscheme. *Now assume that f is proper*, so $\mathrm{Tr}_f : f_*K_X^\bullet = f_*f^\triangle K_Y^\bullet \to K_Y^\bullet$ is a map of complexes, as is $\mathrm{Tr}_{f_0} : f_{0*}K_{X_0}^\bullet = f_{0*}f_0^\triangle K_{Y_0}^\bullet \to K_{Y_0}^\bullet$. Consider the diagram of *complexes*

(4.1.4)

$$f_*\mathscr{H}om_X^\bullet(f^*K_Y^\bullet, K_X^\bullet) \longrightarrow f_*j_*\mathscr{H}om_{X_0}^\bullet(f_0^*K_{Y_0}^\bullet, K_{X_0}^\bullet)$$

$$f_*\mathscr{H}om_X^\bullet(f^*K_Y^\bullet, K_X^\bullet) \qquad\qquad i_*f_{0*}\mathscr{H}om_{X_0}^\bullet(f_0^*K_{Y_0}^\bullet, K_{X_0}^\bullet)$$

$$\mathscr{H}om_Y^\bullet(K_Y^\bullet, f_*K_X^\bullet) \longrightarrow i_*\mathscr{H}om_{Y_0}^\bullet(K_{Y_0}^\bullet, f_{0*}K_{X_0}^\bullet)$$

$$\mathscr{H}om_Y^\bullet(K_Y^\bullet, K_Y^\bullet) \longrightarrow i_*\mathscr{H}om_{Y_0}^\bullet(K_{Y_0}^\bullet, K_{Y_0}^\bullet)$$

$$\mathcal{O}_Y[0] \longrightarrow i_*\mathcal{O}_{Y_0}[0]$$

The top row uses (4.1.2) and the next two rows are the maps "pass to the induced map on a-torsion." The bottom vertical maps in (4.1.4) are special cases of the quasi-isomorphism (3.3.4) and the top and bottom squares in (4.1.4) are clearly commutative. If we verify the commutativity of the middle square in (4.1.4), then Lemma 4.1.1 and the flasqueness of the terms in $\mathcal{H}om_X^\bullet(f^*K_Y^\bullet, f^\triangle K_Y^\bullet)$ and $\mathcal{H}om_{X_0}^\bullet(f_0^*K_{Y_0}^\bullet, f_0^\triangle K_{Y_0}^\bullet)$ ensure that passing to the 0th *cohomology* sheaves in (4.1.4) yields the commutative diagram

$$
\begin{array}{ccc}
R^n f_*(\omega_{X/Y}) & \longrightarrow & i_* R^n f_{0*}(\omega_{X_0/Y_0}) \\
\Big\downarrow{\scriptstyle\gamma_f} & & \Big\downarrow{\scriptstyle i_*(\gamma_{f_0})} \\
\mathcal{O}_Y & \longrightarrow & i_* \mathcal{O}_{Y_0}
\end{array}
$$

where the rows are adjoints of the canonical base change maps. By the adjointness of i_* and i^*, we obtain Theorem 3.6.5 in the case of base change to a closed subscheme.

The commutativity of the middle square in (4.1.4) follows from the assertion that

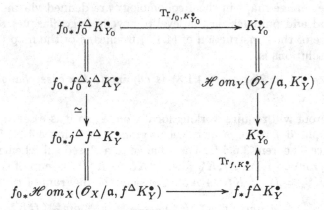

commutes, where the middle of the left column uses that $if_0 = fj$. By Theorem 3.4.1(3), this commutativity is equivalent to the commutativity of the diagram

$$(4.1.5) \qquad
\begin{array}{ccc}
i_* f_{0*} f_0^\triangle i^\triangle K_Y^\bullet & \xrightarrow{\ \simeq\ } & f_* j_* j^\triangle f^\triangle K_Y^\bullet \\
{\scriptstyle \mathrm{Tr}_{f_0, i^\triangle K_Y^\bullet}}\Big\downarrow & & \searrow{\scriptstyle \mathrm{Tr}_{j, f^\triangle K_Y^\bullet}} \\
i_* i^\triangle K_Y^\bullet \xrightarrow[{\scriptstyle \mathrm{Tr}_{i,K_{Y^\bullet}}}]{} & K_Y^\bullet \xleftarrow[{\scriptstyle \mathrm{Tr}_{f,K_Y^\bullet}}]{} & f_* f^\triangle K_Y^\bullet
\end{array}
$$

Since $if_0 = fj$, the commutativity of (4.1.5) follows from Theorem 3.4.1(2) (which does not require f to be proper). It remains to prove Lemma 4.1.1.

PROOF. (of Lemma 4.1.1) The statement of the lemma is local on X. Since we have *not* required f to be proper, we may work locally on X and can even assume that that X and Y are affine, so there is a closed immersion $h : X \hookrightarrow P$ over Y, where P is a smooth affine Y-scheme with pure relative dimension N. Let $\pi : P \to Y$ be the structure map, $\pi_0 : P_0 \to Y_0$ the base change to A_0, and

$J : P_0 \hookrightarrow P$ the canonical closed immersion. In what follows, keep in mind the basic scheme diagram

$$\begin{array}{ccc} X_0 & \xrightarrow{\;j\;} & X \\ & & \\ f_0 \left(\begin{array}{ccc} \downarrow h_0 & & \downarrow h \\ & & \\ P_0 & \xrightarrow{\;J\;} & P \\ & & \\ \downarrow \pi_0 & & \downarrow \pi \\ & & \\ Y_0 & \xrightarrow[\;i\;]{} & Y \end{array} \right) f \end{array}$$

Choose a resolution $E^\bullet \to h_* \mathcal{O}_X$ by locally free coherent $\mathcal{O}_P$-modules. By the flatness of f and π, the naturally induced map $E_0^\bullet = E^\bullet \otimes_A A_0 \to h_{0*} \mathcal{O}_{X_0}$ is an analogous such resolution on P_0. The following trivial auxiliary lemma applies to the smooth map π and the residual complex $K_Y^\bullet$, as well as to the smooth map π_0 and the residual complex $K_{Y_0}^\bullet$.

Let $q : P \to Y$ be a separated smooth map between noetherian schemes which admit dualizing complexes, and assume that q has pure relative dimension N. Choose a residual complex $K^\bullet$ on Y. There is a derived category isomorphism $q^\sharp K^\bullet \simeq q^\triangle K^\bullet$ from (3.3.16), so since $q^\triangle K^\bullet$ is a bounded complex of injectives we can represent this isomorphism by a quasi-isomorphism of complexes $q^\sharp K^\bullet \to q^\triangle K^\bullet$, unique up to homotopy.

LEMMA 4.1.2. *The composite map of complexes*

$$(4.1.6) \qquad \omega_{P/Y}[N] \to \mathcal{H}om_P^\bullet(q^* K^\bullet, q^\sharp K^\bullet) \to \mathcal{H}om_P^\bullet(q^* K^\bullet, q^\triangle K^\bullet)$$

is a quasi-isomorphism and represents the canonical derived category map

$$\omega_{P/Y}[N] \xrightarrow{\;\simeq\;} \omega_{P/Y}[N] \overset{\mathbf{L}}{\otimes} q^* \mathbf{R}\mathcal{H}om_Y^\bullet(K^\bullet, K^\bullet)$$

$$\downarrow$$

$$\mathbf{R}\mathcal{H}om_P^\bullet(q^* K^\bullet, q^\sharp K^\bullet) \xleftarrow[\;\simeq\;]{} \omega_{P/Y}[N] \overset{\mathbf{L}}{\otimes} \mathbf{R}\mathcal{H}om_P^\bullet(q^* K^\bullet, q^* K^\bullet)$$

PROOF. It is easy to check that the composite map of complexes is the canonical one we expect in the derived category. Since q is flat, Y is locally noetherian, and $K^\bullet \in \mathbf{D}_c^b(Y)$ has bounded above coherent cohomology sheaves, [**RD**, II, 5.8] asserts that this canonical derived category map is an isomorphism. ∎

Using Lemma 4.1.2, we can consider the following diagram of complexes of $\mathcal{O}_P$-modules, where the top terms are $f_{K_Y^\bullet}^! \mathcal{O}_Y$ and $j_* f_{0 K_{Y_0}^\bullet}^! \mathcal{O}_{Y_0}$ respectively, and where $K_X^\bullet \overset{\text{def}}{=} f^\Delta K_Y^\bullet$, $K_{X_0}^\bullet \overset{\text{def}}{=} f_0^\Delta K_{Y_0}^\bullet$, $K_P^\bullet \overset{\text{def}}{=} \pi^\Delta K_Y^\bullet$, $K_{P_0}^\bullet \overset{\text{def}}{=} \pi_0^\Delta K_{Y_0}^\bullet$:

(4.1.7)

$$
\begin{array}{ccc}
\mathcal{H}om_X^\bullet(K_X^\bullet, f^\Delta K_Y^\bullet) & \longrightarrow & j_*\mathcal{H}om_{X_0}^\bullet(K_{X_0}^\bullet, f_0^\Delta K_{Y_0}^\bullet) \\
\simeq \downarrow & & \downarrow \simeq \\
\mathcal{H}om_X^\bullet(f^*K_Y^\bullet, K_P^\bullet) & & j_*\mathcal{H}om_{X_0}^\bullet(f_0^*K_{Y_0}^\bullet, K_{P_0}^\bullet) \\
\| & & \| \\
\mathcal{H}om_X^\bullet(f^*K_Y^\bullet, \mathcal{H}om_P^\bullet(h_*\mathcal{O}_X, K_P^\bullet)) & \overset{\alpha_1}{\longrightarrow} & j_*\mathcal{H}om_{X_0}^\bullet(f_0^*K_{Y_0}^\bullet, \mathcal{H}om_{P_0}^\bullet(h_{0*}\mathcal{O}_{X_0}, K_{P_0}^\bullet)) \\
\downarrow & & \downarrow \\
\mathcal{H}om_P^\bullet(\pi^*K_Y^\bullet, \mathcal{H}om_P^\bullet(E^\bullet, K_P^\bullet)) & \overset{\alpha_2}{\longrightarrow} & J_*\mathcal{H}om_{P_0}^\bullet(\pi_0^*K_{Y_0}^\bullet, \mathcal{H}om_{P_0}^\bullet(E_0^\bullet, K_{P_0}^\bullet)) \\
\text{signs!}\| & & \|\text{signs!} \\
\mathcal{H}om_P^\bullet(E^\bullet, \mathcal{H}om_P^\bullet(\pi^*K_Y^\bullet, K_P^\bullet)) & \overset{\alpha_3}{\longrightarrow} & J_*\mathcal{H}om_{P_0}^\bullet(E_0^\bullet, \mathcal{H}om_{P_0}^\bullet(\pi_0^*K_{Y_0}^\bullet, K_{P_0}^\bullet)) \\
\varphi' \uparrow & & \uparrow \varphi_0' \\
\mathcal{H}om_P^\bullet(E^\bullet, \mathcal{H}om_P^\bullet(\pi^*K_Y^\bullet, \pi^\sharp K_Y^\bullet)) & \overset{\alpha_4}{\longrightarrow} & J_*\mathcal{H}om_{P_0}^\bullet(E_0^\bullet, \mathcal{H}om_{P_0}^\bullet(\pi_0^*K_{Y_0}^\bullet, \pi_0^\sharp K_{Y_0}^\bullet)) \\
\varphi \uparrow & & \uparrow \varphi_0 \\
\mathcal{H}om_P^\bullet(E^\bullet, \omega_{P/Y}[N]) & \overset{\beta}{\longrightarrow} & J_*\mathcal{H}om_{P_0}^\bullet(E_0^\bullet, \omega_{P_0/Y_0}[N])
\end{array}
$$

Let's describe the maps in this diagram. The top horizontal map is (4.1.2) and the map α_1 is the "pass to $\mathfrak{a}$-torsion" map, which makes sense since the $\mathfrak{a}$-torsion on $f^*K_Y^\bullet$ is canonically isomorphic to $f_0^*K_{Y_0}^\bullet$ (by the flatness of f) and the $\mathfrak{a}$-torsion in $K_P^\bullet = \pi^\Delta K_Y^\bullet$ is

$$J^\Delta\pi^\Delta K_Y^\bullet \simeq \pi_0^\Delta i^\Delta K_Y^\bullet = \pi_0^\Delta K_{Y_0}^\bullet = K_{P_0}^\bullet.$$

The maps α_2, α_3, α_4 are defined similarly. The map β is the canonical base change map. The vertical composites $\varphi' \circ \varphi$ and $\varphi_0' \circ \varphi_0$ are quasi-isomorphisms because of Lemma 4.1.2 and the fact that $E^\bullet$ and $E_0^\bullet$ are bounded above complexes of locally free coherent sheaves. Finally, the 'equalities' in the middle of each column involve an intervention of signs (more precisely, a sign of $(-1)^{pq}$ on $\mathcal{H}om_P(\pi^*K_Y^p, \mathcal{H}om_P(E^q, (K_P^\bullet)^r))$, and likewise over P_0; note that the analogous 'equalities' of complexes when $E^\bullet$ is replaced by $\mathcal{F}[0]$ involve no signs).

The derived category composites along the columns in (4.1.7) identify the $-n$th cohomology sheaves $\omega_{X/Y}$ and ω_{X_0/Y_0} on the top of each column with $\mathcal{E}xt_P^{N-n}(h_*\mathcal{O}_X, \omega_{P/Y})$ and $\mathcal{E}xt_{P_0}^{N-n}(h_{0*}\mathcal{O}_{X_0}, \omega_{P_0/Y_0})$ respectively. These identifications are the ones in (3.5.3), up to a universal sign depending only on N and n; indeed, it is easy to check that the columns in (4.1.7) are an explicit version of (3.5.3) up to such a sign (see (4.3.17) for a variation on this). By the commutativity of (3.6.1), it is enough to show that (4.1.7) commutes in the derived category (so the induced diagram on $-n$th cohomology sheaves is commutative). We will prove that (4.1.7) is commutative up to homotopy. This is the most we can expect, since the maps $\pi^\sharp K_Y^\bullet \to \pi^\Delta K_Y^\bullet$ and $\pi_0^\sharp K_{Y_0}^\bullet \to \pi_0^\Delta K_{Y_0}^\bullet$ used in the columns of (4.1.7) are chosen (as in Lemma 4.1.2) to represent certain derived category isomorphisms and so are only well-defined up to homotopy. Among the five rectangles in (4.1.7), all aside from the top and the second from the bottom are clearly commutative.

In order to prove the commutativity (not just up to homotopy) of the top rectangle in (4.1.7), we use the diagram of complexes of $\mathscr{O}_{X_0}$-modules

(4.1.8)

$$f_0^\Delta i^\Delta K_Y^\bullet \qquad \mathscr{H}om_{P_0}^\bullet(h_{0*}\mathscr{O}_{X_0}, \pi_0^\Delta i^\Delta K_Y^\bullet)$$

$$\mathscr{H}om_X^\bullet(\mathscr{O}_X/\mathfrak{a}, f^\Delta K_Y^\bullet) =\!=\!=\!=\!= j^\Delta f^\Delta K_Y^\bullet \qquad \overset{\simeq}{} \qquad h_0^\Delta \pi_0^\Delta i^\Delta K_Y^\bullet$$

$$\simeq \Big\downarrow \qquad\qquad \simeq \Big\downarrow$$

$$\mathscr{H}om_X^\bullet(\mathscr{O}_X/\mathfrak{a}, h^\Delta \pi^\Delta K_Y^\bullet) =\!=\!=\!=\!= j^\Delta h^\Delta \pi^\Delta K_Y^\bullet =\!=\!=\!=\!= h_0^\Delta J^\Delta \pi^\Delta K_Y^\bullet \quad =$$

$$\Big\| \qquad\qquad \uparrow{\simeq} \qquad\qquad \simeq \Big\uparrow$$

$$\mathscr{H}om_X^\bullet(\mathscr{O}_X/\mathfrak{a}, \mathscr{H}om_P^\bullet(h_*\mathscr{O}_X, \pi^\Delta K_Y^\bullet)) \overset{\simeq}{\longleftarrow} (hj)^\Delta \pi^\Delta K_Y^\bullet =\!=\!=\!=\!= (Jh_0)^\Delta \pi^\Delta K_Y^\bullet \quad =$$

$$\Big\| \qquad\qquad \simeq \Big\downarrow$$

$$\mathscr{H}om_P^\bullet(h_*\mathscr{O}_X, \mathscr{H}om_P^\bullet(\mathscr{O}_P/\mathfrak{a}, \pi^\Delta K_Y^\bullet)) =\!=\!=\!=\!= \mathscr{H}om_{P_0}^\bullet(h_{0*}\mathscr{O}_{X_0}, J^\Delta \pi^\Delta K_Y^\bullet)$$

The commutativity of (4.1.8) is easy to check, by using the description (3.3.17) of $(\cdot)^\Delta$ for closed immersions. By applying

$$\mathscr{H}om_X^\bullet(f^* K_Y^\bullet, \cdot), \quad \mathscr{H}om_{X_0}^\bullet(f_0^* K_{Y_0}^\bullet, \cdot)$$

in the appropriate places in (4.1.8), the commutativity of the outside edge of (4.1.8) implies the commutativity of the top square in (4.1.7).

It remains to study the second square from the bottom in (4.1.7), whose commutativity up to homotopy is implied by that of the diagram

(4.1.9) $$\mathscr{H}om_P(\mathscr{O}_P/\mathfrak{a}, \pi^\Delta K_Y^\bullet) =\!=\!= J^\Delta \pi^\Delta K_Y^\bullet =\!=\!= \pi_0^\Delta i^\Delta K_Y^\bullet$$

$$\uparrow \qquad\qquad\qquad \Big\|$$

$$\mathscr{H}om_P(\mathscr{O}_P/\mathfrak{a}, \pi^\sharp K_Y^\bullet) \xrightarrow{\ \simeq\ } \pi_0^\sharp K_{Y_0}^\bullet \longrightarrow \pi_0^\Delta K_{Y_0}^\bullet$$

The homotopy-commutativity of (4.1.9) amounts to the equality, up to homotopy, of the two maps from $\mathscr{H}om_P(\mathscr{O}_P/\mathfrak{a}, \pi^\sharp K_Y^\bullet)$ to $\mathscr{H}om_P(\mathscr{O}_P/\mathfrak{a}, \pi^\Delta K_Y^\bullet)$ in (4.1.9). This can be checked by working in the derived category, since the complex

$$\mathscr{H}om_P(\mathscr{O}_P/\mathfrak{a}, \pi^\Delta K_Y^\bullet)$$

is a bounded complex of injectives on P_0. In the derived category $\mathbf{D}(P_0)$, (4.1.9) is equivalent to the diagram

$$
\begin{array}{ccc}
J^b \pi^\Delta K_Y^\bullet & \overset{\simeq}{\longrightarrow} & J^\Delta \pi^\Delta K_Y^\bullet \mathrel{=\!=\!=} \pi_0^\Delta i^\Delta K_Y^\bullet \\
{\scriptstyle\simeq}\big\uparrow & & \nearrow {\scriptstyle\simeq} \\
J^b \pi^\sharp K_Y^\bullet & \mathrel{=\!=\!=} & \pi_0^\sharp i^b K_Y^\bullet
\end{array}
$$

The commutativity of this last diagram follows immediately from Lemma 3.3.2 and the non-trivial Theorem 3.3.1(VAR4). This completes the proof of Lemma 4.1.1.

∎

4.2. Case of Artin Local Base Schemes

It remains to prove Theorem 3.6.5 in the case of a base change $Y' \to Y$ in which $Y = \operatorname{Spec}(A)$, $Y' = \operatorname{Spec}(A')$ with A and A' artin local rings. Let k and k' be the respective residue fields. We emphasize that in the application to the proof of Theorem 3.6.5, k' is often not algebraic over k. The first step is to reduce to the case in which $A \to A'$ is *flat* and the artin local fiber ring $A' \otimes_A k$ is Gorenstein.

LEMMA 4.2.1. *Let $R \to S$ be a local map between local artin rings, and let $\mathfrak{m}_R$, $\mathfrak{m}_S$ be the respective maximal ideals of R and S. There is a factorization $f = g \circ h$ such that $h : R \to R'$ is a flat local map between artin local rings with Gorenstein fiber ring $R'/\mathfrak{m}_R R'$ and $g : R' \to S$ is a surjection.*

PROOF. A similar result was proven by Avramov, Foxby, and Herzog [**AFH**, Theorem 1.1] in the context of complete local noetherian rings rather than artin local rings, where they even get $R'/\mathfrak{m}_R R'$ to be a regular local ring. We are interested in the case in which all local rings are required to be artinian, and the proof of [**AFH**, Theorem 1.1] easily adapts to this case as long as the 'regular' condition is relaxed to the 'Gorenstein' condition. For the convenience of the reader, we give the modified argument.

Recall that for a field F, a *Cohen ring for F* is defined to be F if F has characteristic 0 and is defined to be a complete characteristic 0 discrete valuation ring with residue field F and uniformizer p in case F has positive characteristic p; Cohen rings are unique up to isomorphism, and the isomorphism is unique if F is perfect. Let C and D be respective Cohen rings for $R/\mathfrak{m}_R$ and $S/\mathfrak{m}_S$. By the Cohen Structure Theorem, we can choose maps $C \to R$, $D \to S$ which induce the identity on the residue fields. Choosing a finite set of generators $y_1, \ldots, y_t$ of $\mathfrak{m}_S$ and an integer m such that $\mathfrak{m}_S^m = 0$, we can define a *surjection* of complete local noetherian D-algebras

$$ D[\![Y]\!]/(Y^m) \overset{\text{def}}{=} D[\![Y_1, \ldots, Y_t]\!]/(Y_1^m, \ldots, Y_t^m) \twoheadrightarrow S $$

via $Y_j \mapsto y_j$. By [**EGA**, 0_{IV}, 19.8.6(i)], we can choose a lift $\varphi : C \to D[\![Y]\!]/(Y^m)$ of $C \to R \to S$, and φ is clearly flat. Define $R' = R \otimes_C (D[\![Y]\!]/(Y^m))$. Since R

is an artinian quotient of a formal power series ring over C, the local ring R' is 0-dimensional noetherian, hence artinian. The fiber ring $R'/\mathfrak{m}_R R'$ is the artin ring $(S/\mathfrak{m}_S)[Y]/(Y^m)$, which is Gorenstein.

■

Applying Lemma 4.2.1 to $A \to A'$, we get a factorization $A \to B \twoheadrightarrow A'$ where $A \to B$ is flat with Gorenstein fiber ring $B \otimes_A k$ and $B \twoheadrightarrow A'$ is surjective. The result of §4.1 applies to $B \twoheadrightarrow A'$, so we may assume that the base change $A \to A'$ is flat and the fiber ring $A' \otimes_A k$ is Gorenstein. Under this assumption on the artin local base change $A \to A'$, we will reduce the compatibility of γ_f with respect to the base change $A \to A'$ to the case of γ_g's for certain auxiliary *finite* CM maps $g : Z \to Y = \mathrm{Spec}(A)$ (and the base change $A \to A'$). These special cases will be treated by a direct calculation which makes essential use of the fact that the local artin ring $A' \otimes_A k$ is Gorenstein.

Let

$$Z \xrightarrow{\ i\ } X$$
$$g \searrow \quad \downarrow f$$
$$Y$$

be a commutative diagram, with g finite CM and i *transversally regular* over Y (we will soon construct many such Z's). By Lemma 3.4.3(TRA1), we have a commutative diagram

(4.2.1)
$$\mathbf{R}g_*(\omega_{Z/Y}[0]) \longrightarrow \mathbf{R}f_*(\omega_{X/Y}[n])$$
$$\mathrm{Tr}_g \searrow \quad \downarrow \mathrm{Tr}_f$$
$$\mathcal{O}_Y[0]$$

where the top map is $\mathbf{R}f_*$ applied to the derived category map

(4.2.2)
$$\mathbf{R}i_*\omega_{Z/Y}[0] \xrightarrow{\ \simeq\ } i_*\omega_{Z/Y}[0] \xrightarrow{\ \simeq\ } i_*g^!\mathcal{O}_Y \xrightarrow{\ \simeq\ } i_*i^!f^!\mathcal{O}_Y \xrightarrow{\ \simeq\ } i_*i^!\omega_{X/Y}[n]$$
$$\downarrow \mathrm{Tr}_i$$
$$\omega_{X/Y}[n]$$

In more concrete terms, the derived category composite map

$$i_*\omega_{Z/Y}[0] \to \omega_{X/Y}[n]$$

in (4.2.2) is represented by the map of complexes of flasques

(4.2.3) $i_* \mathcal{H}om_Z(g^* \mathcal{I}, g^\Delta \mathcal{I}) = i_* \mathcal{H}om_Z(i^* f^* \mathcal{I}, i^\Delta f^\Delta \mathcal{I})$

$$\| $$

$$\mathcal{H}om_X(f^* \mathcal{I}, i_* i^\Delta f^\Delta \mathcal{I})$$

$$\downarrow {\scriptstyle \mathrm{Tr}_i}$$

$$\mathcal{H}om_X(f^* \mathcal{I}, f^\Delta \mathcal{I}).$$

where the residual complex $\mathcal{I} = \mathcal{I}[0]$ on the local *artin* scheme $Y = \mathrm{Spec}(A)$ is the (quasi-)coherent sheaf associated to an injective hull of k over A. By (3.2.2) and (3.2.4), the residual complex $f^\Delta \mathcal{I}$ on X is a complex of injectives supported in degrees from $-n$ to 0 and has degree 0 term

(4.2.4) $(f^\Delta \mathcal{I})^0 = \bigoplus_{x \in X^0} \mathcal{I}(x),$

where X^0 is the set of closed points of X (i.e., codimension n irreducible closed subsets of X) and $\mathcal{I}(x)$ is a quasi-coherent sheaf supported at x whose stalk $\mathcal{I}(x)_x$ an injective hull of $k(x)$ over $\mathcal{O}_{X,x}$.

By passing to degree 0 cohomology in (4.2.1), we get a commutative diagram

(4.2.5)
$$g_* \omega_{Z/Y} \xrightarrow{\alpha_Z} \mathrm{R}^n f_*(\omega_{X/Y})$$
$$\searrow{\scriptstyle \gamma_g} \qquad \downarrow{\scriptstyle \gamma_f}$$
$$\mathcal{O}_Y$$

LEMMA 4.2.2. *The map α_Z is of formation compatible with the flat base change $u : Y' \to Y$.*

The choice of convention in the definition of (3.6.13) affects α_Z up to a universal sign of $(-1)^n$, which is harmless for Lemma 4.2.2.

PROOF. It suffices to prove more generally that the derived category *composite* map (4.2.2) commutes with the *flat* base change to Y'. By using the compatibility between $i^!$, Tr_i and i^b, Trf_i in Lemma 3.4.3(TRA3), as well as the general flat base change compatibility of the derived category 'finite trace' Trf [**RD**, III, 6.6(2)], it suffices to show that the composite derived category isomorphism

(4.2.6) $\omega_{Z/Y}[0] \simeq g^! \mathcal{O}_Y \simeq i^! f^! \mathcal{O}_Y \simeq i^! \omega_{X/Y}[n] \simeq i^b \omega_{X/Y}[n]$

is compatible with the flat base change $Y' \to Y$. The maps in (4.2.6) are isomorphisms, so the complexes in (4.2.6) have all cohomology vanishing outside of degree 0 (since this is obvious for the left term $\omega_{Z/Y}[0]$). Thus, it is enough to prove that the map of degree 0 cohomology sheaves

(4.2.7) $\omega_{Z/Y} \simeq \mathcal{E}xt_X^n(i_* \mathcal{O}_Z, \omega_{X/Y})$

induced by (4.2.6) is compatible with the flat base change $Y' \to Y$ (recall that in the case of flat base change, the general base change maps for $\mathscr{E}xt^\bullet$ as defined in [**AK2**, §1] are equal to the maps on cohomology induced by flat base change for $\mathbf{R}\mathscr{H}om^\bullet$). For concreteness, we have computed this $\mathscr{E}xt$ according to the convention in (2.7.3), as this is the convention used in the definition of (3.6.11).

We claim that (4.2.7) is equal to (3.6.11), whose base change compatibility was verified in Theorem 3.6.2 (since i is *transversally regular* over Y). The definition of (4.2.6), and hence of (4.2.7), does not involve any trace maps and clearly makes sense under the weaker assumption that the CM map $X \to Y$ is separated and finite type, rather than proper. In this generality, it is easy to check that the formation of (4.2.6), and hence (4.2.7), is compatible with Zariski localization on X. Thus, to identify (4.2.7) and (3.6.11), we may assume that X is affine, so $X \to Y$ can be factorized as a closed immersion followed by a separated smooth map with pure relative dimension. In this case, by recalling how (3.6.11) is *defined*, the equality of (4.2.7) and (3.6.11) follows from Theorem 3.3.1 and the properties (VAR1)–(VAR3) preceding Theorem 3.3.1. ∎

By [**EGA**, IV$_4$, 19.2.9], for each closed point $x \in X$ there is an 'increasing' set of closed subschemes $i_{m,x} : Z_{m,x} \hookrightarrow X$ for $m \geq 1$, with $Z_{m,x}$ supported at x, $g_{m,x} : Z_{m,x} \to Y$ a finite CM map, $i_{m,x}$ transversally regular over Y, and the $Z_{m,x}$'s inducing a cofinal set of infinitesimal neighborhoods of x in $\overline{X} = X \times_A k$. More explicitly, $\mathscr{O}_{\overline{X},x}$ is a CM local ring with dimension n, so if $t_1, \ldots, t_n \in \Gamma(U, \mathscr{O}_X)$ lift a system of parameters in $\mathscr{O}_{\overline{X},x}$, with $U \subseteq X$ a sufficiently small neighborhood of $x \in X$, then we take $Z_{m,x}$ to be defined by $t_1^m, \ldots, t_n^m$ (to see that this cuts out the *closed* point $\{x\}$ on X, we may work on $\overline{X}$, which has the same underlying topological space as X). Thus, if $\mathscr{I}_{Z_{m,x}}$ denotes the coherent ideal sheaf of $Z_{m,x}$ on X, then each power of the maximal ideal $\mathfrak{m}_x$ of $\mathscr{O}_{X,x}$ contains the stalk ideal $(\mathscr{I}_{Z_{m,x}})_x$ for sufficiently large m.

Consider the map of sheaves

$$\alpha = \bigoplus \alpha_{Z_{m,x}} : \bigoplus_{m,x} g_{m,x*} \omega_{Z_{m,x}/Y} \to \mathbf{R}^n f_*(\omega_{X/Y}).$$

We now prove that α is *surjective* (the idea to consider proving this surjectivity was inspired by the analogue [**Verd**, Prop 2, p.399] in the smooth case). Note that the question of whether or not α is surjective has nothing to do with any base change questions. Since the sheaves involved are quasi-coherent and Y is affine (or since Y is a 1-point scheme), the surjectivity of α can be checked by working with global sections and sheaf cohomology instead of pushforwards to Y and higher direct image sheaves.

Recall that the complexes in (4.2.3) consist of flasque sheaves, which are suitable for computing hypercohomology. Moreover, these complexes are supported in degrees between $-n$ and 0, and we want to study the induced map on cohomology in degree 0 after taking global sections. Thus, it suffices to show that the map

$$\bigoplus_{m,x} \mathrm{Tr}_{i_{m,x}}$$

defined by

(4.2.8)

$$\bigoplus_{m,x} \mathrm{Hom}_X(f^*\mathscr{I}, i_{m,x*}(i_{m,x}^{\Delta} f^{\Delta}\mathscr{I})^0) \longrightarrow \mathrm{Hom}_X(f^*\mathscr{I}, (f^{\Delta}\mathscr{I})^0)$$

$$\downarrow \simeq$$

$$\bigoplus_{x\in X^0} \mathrm{Hom}_X(f^*\mathscr{I}, \mathscr{I}(x))$$

is surjective (for the isomorphism at the end of (4.2.8), we use (4.2.4) and the fact that $f^*\mathscr{I}$ is a coherent sheaf on the noetherian scheme X). By applying (3.3.17) to the closed immersions $i_{m,x}$ and using Theorem 3.4.1(3), we need to prove the surjectivity of the canonical map

$$\bigoplus_{m\geq 1} \mathrm{Hom}_X(f^*\mathscr{I}, \mathscr{H}om_X(\mathscr{O}_{Z_{m,x}}, \mathscr{I}(x))) \to \mathrm{Hom}_X(f^*\mathscr{I}, \mathscr{I}(x))$$

for all $x \in X^0$. The sheaf $f^*\mathscr{I}$ is coherent on X and $\mathscr{I}(x)$ is a quasi-coherent sheaf supported at the closed point x, so it suffices to prove that every element of the stalk $\mathscr{I}(x)_x$ is annihilated by $(\mathscr{I}_{Z_{m,x}})_x$ for some $m \geq 1$. But $\mathscr{I}(x)_x$ is an injective hull of $k(x)$ over $\mathscr{O}_{X,x}$, so every element of $\mathscr{I}(x)_x$ is killed by a power of the maximal ideal $\mathfrak{m}_x$ of $\mathscr{O}_{X,x}$. Since we have seen that every power of $\mathfrak{m}_x$ contains $(\mathscr{I}_{Z_{m,x}})_x$ for sufficiently large m, we deduce the surjectivity of α. This implies that $u^*(\alpha)$ is also surjective.

By Lemma 4.2.2 and the surjectivity of $u^*(\alpha)$, the assertion of Theorem 3.6.5 for the proper CM map f and the base change u is reduced to the case of the finite flat maps $g_{m,x} : Z_{m,x} \to Y$ and the base change u. This completes the reduction of Theorem 3.6.5 to the case of the base change $u : Y' \to Y$ and a *finite flat* map $f : X \to Y$. Note that so far, we have only used the fact that the local artin base change $A \to A'$ is flat. We have not used the fact that the the fiber ring $A' \otimes_A k$ is Gorenstein.

Now assume that f is finite flat, so $X = \mathrm{Spec}(B)$ for a *finite flat* A-algebra B. Let $I = \Gamma(Y, \mathscr{I})$, an injective hull of k over A. The isomorphism (3.3.16) yields a canonical isomorphism of sheaves $(f^{\Delta}\mathscr{I})[0] \simeq \mathrm{H}^0(f^{\flat}\mathscr{I}) \simeq \mathscr{H}om_Y(f_*\mathscr{O}_X, \mathscr{I})$. The 'finite trace' Trf for finite flat maps is essentially the 'evaluate at 1' map, so we have a B-linear identification

(4.2.9) $\Gamma(X, \omega_{X/Y}) =\!=\!=\!= \mathrm{Hom}_X(f^*\mathscr{I}, f^{\Delta}\mathscr{I})[0]$

$$\|$$

$$\mathrm{Hom}_B(B \otimes_A I, \mathrm{Hom}_A(B, I))$$

$$\|$$

$$\mathrm{Hom}_A(I, \mathrm{Hom}_A(B, I))$$

and the map $\gamma_f : \Gamma(X, \omega_{X/Y}) = \Gamma(Y, f_*\omega_{X/Y}) \to \Gamma(Y, \mathcal{O}_Y)$ is the 'evaluate at 1' map

$$(4.2.10) \qquad \operatorname{Hom}_A(I, \operatorname{Hom}_A(B, I)) \longrightarrow \operatorname{Hom}_A(I, I) \xleftarrow{\cong} A.$$

There are analogous descriptions of $\Gamma(X', \omega_{X'/Y'})$ and $\gamma_{f'}$ in terms of A', $B' = A' \otimes_A B$, and $I' = A' \otimes_A I$, since I' is an injective hull of k' over A', thanks to the fact that the local artin ring $A' \otimes_A k$ is *Gorenstein*:

LEMMA 4.2.3. *Let $(R, \mathfrak{m})$ be a local noetherian ring, I an injective hull of the residue field $R/\mathfrak{m}$ as an R-module. Let $R \to R'$ be a flat map to another local noetherian ring $(R', \mathfrak{m}')$ such that the fiber ring $R'/\mathfrak{m}R'$ is a Gorenstein artin ring. Then $R' \otimes_R I$ is an injective hull for the residue field $R'/\mathfrak{m}'$ as an R'-module.*

PROOF. Recall from [**RD**, VI, 5.2] that if $(S, \mathfrak{n})$ is a local noetherian ring and J is an S-module, then J is an injective hull of the residue field if and only if

- J is $\mathfrak{n}$-power torsion (i.e., supported at the maximal ideal),
- $\operatorname{Hom}_S(S/\mathfrak{n}, J)$ is 1-dimensional over $S/\mathfrak{n}$,
- there is a base of open ideals $\mathfrak{a}_1 \supseteq \mathfrak{a}_2 \supseteq \ldots$ for the local topology of S such that for all n, the S-module $\operatorname{Hom}_S(S/\mathfrak{a}_n, J)$ has the same finite S-length as $S/\mathfrak{a}_n$.

The proof that $R' \otimes_R I$ satisfies the above three criteria to be an injective hull of $R'/\mathfrak{m}'$ over R' is easy and is exactly the argument in [**RD**, p.353], which even indicates the essential role of the Gorenstein condition: the $\mathfrak{m}'$-torsion in $R'/\mathfrak{m}R'$ is 1-dimensional over $R'/\mathfrak{m}'$.   ∎

The usefulness of the identification (4.2.9) is indicated by the following critical lemma.

LEMMA 4.2.4. *The canonical base change map $u'^* \omega_{X/Y} \simeq \omega_{X'/Y'}$ is identified via (4.2.9) with the natural base change map*

$$(4.2.11) \qquad A' \otimes_A \operatorname{Hom}_A(I, \operatorname{Hom}_A(B, I)) \simeq \operatorname{Hom}_{A'}(I', \operatorname{Hom}_{A'}(B', I'))$$

(this base change map is an isomorphism because I and B are finite A-modules).

PROOF. Instead of (4.2.11), we can study

$$B' \otimes_B \operatorname{Hom}_B(B \otimes_A I, \operatorname{Hom}_A(B, I)) \to \operatorname{Hom}_{B'}(B' \otimes_{A'} I', \operatorname{Hom}_{A'}(B', I')),$$

which is canonically identified with the global sections of $\mathbf{H}^0$ of the flat base change map

$$u'^* \mathbf{R}\mathcal{H}om_X^\bullet(f^*K^\bullet, f^\flat K^\bullet) \to \mathbf{R}\mathcal{H}om_{X'}^\bullet(f'^*K'^\bullet, f'^\flat K'^\bullet),$$

where $K^\bullet = \mathscr{I}[0]$ is a residual complex on Y and $K'^\bullet = u^*K^\bullet$ is a residual complex on Y' (see Lemma 4.2.3).

Choose a closed immersion $h : X \hookrightarrow P = \mathrm{Spec}(R)$ into a smooth affine Y-scheme with pure relative dimension N (e.g., we can take $R = A[t_1, \ldots, t_N]$ for suitably large N). Let $\pi : P \to Y$ denote the structure map and define $P' = \mathrm{Spec}(R')$, h', π' by base change via $u : Y' \to Y$. Using (3.3.16) and the left column of (4.1.7), we obtain an identification

$$(4.2.12) \qquad h_*\mathbf{R}\mathscr{H}om_X^{\bullet}(f^*K^{\bullet}, f^{\flat}K^{\bullet}) \simeq \mathbf{R}\mathscr{H}om_P^{\bullet}(h_*\mathscr{O}_X, \omega_{P/Y}[N])$$

Passing to degree 0 cohomology and global sections yields an R-module identification

$$(4.2.13) \qquad \mathrm{Hom}_B(B \otimes_A I, \mathrm{Hom}_A(B, I)) \simeq \mathrm{Ext}_R^N(B, \Omega_{R/A}^N),$$

where R acts on the left through the quotient map $R \twoheadrightarrow B$ corresponding to h. The isomorphism (4.2.13) is a concrete realization of the inverse of the abstract isomorphism (4.2.9). There are obvious analogues of (4.2.12) in $\mathbf{D}(P')$ and of (4.2.13) over R'. By using the commutativity of (3.6.1), it suffices to show that (4.2.13) and its analogue over R' are compatible via the flat base change $R \to R' = A' \otimes_A R$.

We will prove more generally that (4.2.12) and its analogue in $\mathbf{D}(P')$ are compatible via the flat base change $P' \to P$. The essential point is to reformulate the *definition* of (4.2.12) so that it makes no implicit use of $(\cdot)^{\Delta}$'s, since there is no meaningful theory of base change for $(\cdot)^{\Delta}$ with respect to a base change which might not be residually stable (e.g., the above map $R \to R'$ is residually stable if and only if k' is algebraic over k).

Since P is affine, we can choose a resolution $E^{\bullet} \to h_*\mathscr{O}_X$ by locally free coherent $\mathscr{O}_P$-modules, so $E'^{\bullet} \stackrel{\mathrm{def}}{=} u^*E^{\bullet}$ is an analogous such resolution of $h'_*\mathscr{O}_{X'}$. By the commutativity of (3.3.26) (note the separatedness condition there is satisfied in the present setting) and the commutativity of (3.3.38), it is easy to use the left column of (4.1.7) to show that (4.2.12) is equal to the derived category composite

$$h_*\mathbf{R}\mathscr{H}om_X^{\bullet}(f^*K^{\bullet}, f^{\flat}K^{\bullet}) \xrightarrow[\simeq]{\psi_{h,\pi}} h_*\mathbf{R}\mathscr{H}om_X^{\bullet}(\mathbf{L}h^*\pi^*K^{\bullet}, h^{\flat}\pi^{\sharp}K^{\bullet})$$

$$\downarrow \simeq$$

$$\mathbf{R}\mathscr{H}om_P^{\bullet}(\pi^*K^{\bullet}, h_*h^{\flat}\pi^{\sharp}K^{\bullet})$$

$$\downarrow \simeq$$

$$\mathbf{R}\mathscr{H}om_P^{\bullet}(\pi^*K^{\bullet}, \mathbf{R}\mathscr{H}om_P^{\bullet}(E^{\bullet}, \pi^{\sharp}K^{\bullet}))$$

$$\Big\| \text{ signs!}$$

$$\mathbf{R}\mathscr{H}om_P^{\bullet}(E^{\bullet}, \omega_{P/Y}[N]) \xrightarrow[\simeq]{} \mathbf{R}\mathscr{H}om_P^{\bullet}(E^{\bullet}, \mathbf{R}\mathscr{H}om_P^{\bullet}(\pi^*K^{\bullet}, \pi^{\sharp}K^{\bullet}))$$

where the top map in the right column is the canonical map from [**RD**, II, 5.10] (which is an isomorphism due to the boundedness of the complexes and the finiteness of the Krull dimensions of the noetherian schemes involved). It is trivial to check that every map in this composite is naturally compatible with the *flat* base change $P' = Y' \times_Y P \to P$.

∎

From Lemma 4.2.4 and the explicit description of γ_f in (4.2.10), we deduce the compatibility of γ_f with respect to the artin local base change $A \to A'$. This completes the proof of Theorem 3.6.5.

4.3. Duality for Proper CM Maps in the Locally Noetherian Case

Let $f : X \to Y$ be a proper CM map with pure relative dimension n, and assume that Y is locally noetherian. In this section, we use γ_f from Corollary 3.6.6 to define a duality morphism θ'_f and we state and partially prove a duality theorem which asserts that θ'_f is an isomorphism (the proof will be completed in §4.4). When Y is noetherian and admits a dualizing complex, this will essentially recover Theorem 3.4.4.

Define

$$(4.3.1) \qquad f^{\sharp}(\mathscr{F}^{\bullet}) = \omega_f[n] \overset{\mathbf{L}}{\otimes} f^* \mathscr{F}^{\bullet},$$

for $\mathscr{F}^{\bullet} \in \mathbf{D}^+_{\mathrm{qc}}(Y)$. This coincides with the old notion of $f^{\sharp}$ when f is smooth. Since the functors $\mathbf{R}^q f_*$ preserve quasi-coherence, from the spectral sequence

$$\mathbf{R}^p f_*(\mathbf{H}^q(\mathscr{F}^{\bullet})) \Longrightarrow \mathbf{H}^{p+q}(\mathbf{R}f_*(\mathscr{F}^{\bullet}))$$

for $\mathscr{F}^{\bullet} \in \mathbf{D}^+(X)$ it follows that $\mathbf{R}f_* : \mathbf{D}^+(X) \to \mathbf{D}^+(Y)$ takes $\mathbf{D}^+_{\mathrm{qc}}(X)$ into $\mathbf{D}^+_{\mathrm{qc}}(Y)$. Moreover, f_* has finite cohomological dimension on $\mathrm{Qco}(X)$, so we can define a special case of the projection formula (2.1.10) on $\mathbf{D}^b(\mathrm{Qco}(X)) \simeq \mathbf{D}^b_{\mathrm{qc}}(X)$ without any finite Krull dimension conditions.

To be precise, for $\mathscr{G}^{\bullet} \in \mathbf{D}^b_{\mathrm{qc}}(Y)$, choose a quasi-isomorphism $\mathscr{G}'^{\bullet} \to \mathscr{G}^{\bullet}$ with $\mathscr{G}'^{\bullet}$ a bounded above complex of flats. Let $\omega_f[n] \to \mathscr{F}^{\bullet}$ be a quasi-isomorphism to a bounded complex of f_*-acyclic quasi-coherents. Since

$$\mathscr{F}^{\bullet} \otimes f^* \mathscr{G}'^{\bullet} \simeq \omega_f[n] \overset{\mathbf{L}}{\otimes} f^* \mathscr{G}^{\bullet} \in \mathbf{D}^b_{\mathrm{qc}}(X),$$

there exists a quasi-isomorphism

$$\mathscr{F}^{\bullet} \otimes f^* \mathscr{G}'^{\bullet} \to \mathscr{I}^{\bullet}$$

to a bounded below complex of *quasi-coherent* injectives (by Lemma 2.1.6). Then

$$f_*(\mathscr{F}^{\bullet}) \otimes \mathscr{G}'^{\bullet} \to f_*(\mathscr{F}^{\bullet} \otimes f^* \mathscr{G}'^{\bullet}) \to f_*(\mathscr{I}^{\bullet})$$

defines a δ-functorial map

$$(4.3.2) \qquad \mathbf{R}f_*(\omega_f[n]) \overset{\mathbf{L}}{\otimes} \mathscr{G}^{\bullet} \to \mathbf{R}f_* f^{\sharp}(\mathscr{G}^{\bullet})$$

in $\mathbf{D}^b_{\mathrm{qc}}(Y)$ which is of formation compatible with locally noetherian flat base change. When Y is noetherian with finite Krull dimension, (4.3.2) is a special case of the usual projection formula (2.1.10). Thus, by base change to local rings, we conclude that (4.3.2) is an isomorphism.

Using (4.3.2) and the map

$$(4.3.3) \qquad \mathbf{R}f_*(\omega_f[n]) \to \mathbf{R}^n f_*(\omega_f)[0] \xrightarrow{\gamma_f} \mathscr{O}_Y[0]$$

(in which the first step is represented by $\mathrm{H}^0(f_*\mathscr{I}^{\bullet+n}) = \mathrm{H}^n(f_*\mathscr{I}^\bullet)$ as is implicit in (3.6.13), where $\mathscr{I}^\bullet$ is an f_*-acyclic resolution of ω_f), we can define a δ-functorial trace map

$$(4.3.4) \qquad \mathrm{Tr}'_f : \mathbf{R}f_* f^\sharp(\mathscr{G}^\bullet) \to \mathscr{G}^\bullet$$

for $\mathscr{G}^\bullet \in \mathbf{D}^b_{\mathrm{qc}}(Y)$, as in (3.4.15), using our modified projection formula (4.3.2) instead of (2.1.10). For $\mathscr{G}^\bullet = \mathscr{O}_Y[0]$, this recovers (4.3.3). We will have to check later that when Y is noetherian and admits a dualizing complex, this is closely related to the Grothendieck-Serre trace map Tr_f. For now, we make the trivial observation that when Y is noetherian and admits a dualizing complex, then (4.3.3) coincides with $\mathrm{Tr}_f(\mathscr{O}_Y[0])$ because of how γ_f is *defined* in (3.6.13).

Since γ_f commutes with any locally noetherian base change, it is clear that Tr'_f is of formation compatible with locally noetherian *flat* base change over Y. The functor f_* has finite cohomological dimension on the category of quasi-coherent $\mathscr{O}_X$-modules, so the construction of a translation-compatible canonical map

$$\mathbf{R}f_* \mathbf{R}\mathscr{H}om^\bullet_X(\mathscr{F}^\bullet, \mathscr{G}^\bullet) \to \mathbf{R}\mathscr{H}om^\bullet_Y(\mathbf{R}f_*\mathscr{F}^\bullet, \mathbf{R}f_*\mathscr{G}^\bullet)$$

can be given for $\mathscr{F}^\bullet \in \mathbf{D}^b_c(X)$ and $\mathscr{G}^\bullet \in \mathbf{D}^b_{\mathrm{qc}}(X)$, via the method in [**RD**, II, 5.5] (the coherence and boundedness conditions on $\mathscr{F}^\bullet$ ensure the quasi-coherence of the cohomology sheaves of $\mathbf{R}\mathscr{H}om^\bullet_X(\mathscr{F}^\bullet, \mathscr{G}^\bullet)$ and $\mathbf{R}\mathscr{H}om^\bullet_Y(\mathbf{R}f_*\mathscr{F}^\bullet, \mathbf{R}f_*\mathscr{G}^\bullet)$).

As in (3.4.10), we can use Tr'_f to define the translation-compatible *duality morphism*

$$(4.3.5) \qquad \theta'_f : \mathbf{R}f_* \mathbf{R}\mathscr{H}om^\bullet_X(\mathscr{F}^\bullet, f^\sharp \mathscr{G}^\bullet) \to \mathbf{R}\mathscr{H}om^\bullet_Y(\mathbf{R}f_*\mathscr{F}^\bullet, \mathscr{G}^\bullet),$$

but only for $\mathscr{F}^\bullet \in \mathbf{D}^b_c(X)$ and $\mathscr{G}^\bullet \in \mathbf{D}^b_{\mathrm{qc}}(Y)$. The duality morphism is obviously compatible with any locally noetherian flat base change over Y. The duality theorem in the locally noetherian case is:

THEOREM 4.3.1. *Let $f : X \to Y$ be a proper CM morphism with pure relative dimension, and assume that Y is locally noetherian. The duality morphism θ'_f is an isomorphism.*

The proof begins with some preliminary reductions. It suffices to prove that θ'_f is an isomorphism after the flat base change to $\mathrm{Spec}(\widehat{\mathscr{O}_{Y,y}})$ for all $y \in Y$. Thus, we may assume that Y is noetherian and admits a dualizing complex, in which case γ_f coincides with the map (3.6.13). By the Lemma on Way-Out Functors [**RD**, I, 7.1], it suffices to consider the case where $\mathscr{F}^\bullet$ in (4.3.5) is a coherent sheaf concentrated in degree 0 and $\mathscr{G}^\bullet$ in (4.3.5) is a quasi-coherent sheaf concentrated in degree 0.

We claim that it is enough to consider cases in which $\mathscr{G}$ is a *coherent* sheaf in degree 0. Since any quasi-coherent sheaf on a noetherian scheme is the direct limit of its coherent subsheaves by Lemma 2.1.8, the reduction to coherent $\mathscr{G}$ follows from the assertion that for a fixed coherent sheaf $\mathscr{F}$ on X and a variable quasi-coherent sheaf $\mathscr{G}$ on Y, the cohomology sheaves of $\mathbf{R}f_* \mathbf{R}\mathscr{H}om^\bullet_X(\mathscr{F}, f^\sharp \mathscr{G})$

and $\mathbf{R}\mathscr{H}om_Y^{\bullet}(\mathbf{R}f_*\mathscr{F}, \mathscr{G})$ are of formation compatible with direct limits in $\mathscr{G}$. To prove this, we can assume that Y is affine, so since $\mathbf{R}f_*\mathscr{F} \in \mathbf{D}_c^b(Y)$, the case of $\mathbf{R}\mathscr{H}om_Y^{\bullet}(\mathbf{R}f_*\mathscr{F}, \cdot)$ is clear, by a distinguished triangle argument (since $\mathscr{E}xt_Y^{\bullet}(\mathscr{H}, \cdot)$ for coherent $\mathscr{H}$ is compatible with formation of direct limits in a quasi-coherent variable). The other case follows from a Leray spectral sequence argument (since $\mathbf{R}^{\bullet}f_*(\cdot)$ and $\mathscr{E}xt_X^{\bullet}(\mathscr{F}, \cdot)$ commute with direct limits in a quasi-coherent variable).

Thus, in order to prove Theorem 4.3.1, it is enough to consider the case in which Y is a noetherian scheme which admits a dualizing complex and $\mathscr{F}^{\bullet} \in \mathbf{D}_c^b(X)$, $\mathscr{G}^{\bullet} \in \mathbf{D}_c^b(Y)$. By Theorem 3.4.4, it suffices to prove:

THEOREM 4.3.2. *Let $f : X \to Y$ be a proper CM map with pure relative dimension and assume that Y is a noetherian scheme which admits a dualizing complex. There is an isomorphism $f^{\sharp} \simeq f^!$ of δ-functors on $\mathbf{D}_c^+(Y)$ which coincides with e_f when f is smooth and which takes Tr'_f over to Tr_f when we restrict to $\mathbf{D}_c^b(Y)$.*

The rest of this chapter is concerned with the proof of Theorem 4.3.2. *For the rest of this section, all schemes are assumed to be noetherian and admit a dualizing complex.*

Let $f : X \to Y$ be a flat map of finite type. For conceptual clarity, we do not yet assume that f is CM. It is also important that we make no properness assumptions yet, because in some proofs we will need to work locally on X. Let $g : Y \to Z$ be another flat map of finite type. When f and g are smooth, with relative dimensions n and m (which are locally constant functions on X and Y respectively), there is an isomorphism

$$\zeta'_{f,g} : \omega_{X/Z}[n+m] \simeq \omega_{X/Y}[n] \otimes f^*\omega_{Y/Z}[m]$$

and this is compatible with any base change and with triple composites (in the sense of (2.2.4)). Moreover, this isomorphism underlies the definition of $(gf)^{\sharp} \simeq f^{\sharp}g^{\sharp}$ in the smooth case. For finite type flat morphisms, we want to define generalizations of $f^{\sharp}g^{\sharp} \to (gf)^{\sharp}$ and the inverse of $\zeta'_{f,g}$. These generalizations will be shown to be isomorphisms when f and g are CM.

Let $K^{\bullet}$ be a residual complex on Z, so $g^{\triangle}K^{\bullet}$ and $f^{\triangle}g^{\triangle}K^{\bullet} \simeq (gf)^{\triangle}K^{\bullet}$ are residual complexes on Y and X respectively. Thus, $f^! = f^!_{g^{\triangle}K^{\bullet}}$, $g^! = g^!_{K^{\bullet}}$, and $(gf)^! = (gf)^!_{K^{\bullet}}$. Define the derived category map

(4.3.6)
$$\xi_{f,g} : f^!\mathcal{O}_Y \overset{\mathbf{L}}{\otimes} f^*g^!\mathcal{O}_Y \to (gf)^!\mathcal{O}_Z$$

to be the composite

$$f^!_{g^\Delta K^\bullet} \cdot \mathcal{O}_Y \overset{\mathbf{L}}{\otimes} f^* g^!_{K^\bullet} \cdot \mathcal{O}_Z$$

$$\|$$

$$\mathcal{H}om^\bullet_X(f^* g^\Delta K^\bullet, f^\Delta g^\Delta K^\bullet) \overset{\mathbf{L}}{\otimes} f^* \mathcal{H}om^\bullet_Y(g^* K^\bullet, g^\Delta K^\bullet)$$

$$\downarrow$$

$$\mathcal{H}om^\bullet_X(f^* g^\Delta K^\bullet, f^\Delta g^\Delta K^\bullet) \otimes \mathcal{H}om^\bullet_X(f^* g^* K^\bullet, f^* g^\Delta K^\bullet)$$

$$\downarrow$$

$$\mathcal{H}om^\bullet_X((gf)^* K^\bullet, (gf)^\Delta K^\bullet)$$

where the middle map is the canonical map and the bottom map is defined in the obvious way (by 'composition') without the intervention of signs. It is easy to check that $\xi_{f,g}$ is independent of the choice of $K^\bullet$ and is compatible with residually stable base change over Z, as well as with triple composites in the sense that for a third flat map of finite type $h : Z \to W$ to a noetherian scheme admitting a dualizing complex, we have the equality of derived category maps

$$(4.3.7) \qquad \xi_{f,hg} \circ (1 \overset{\mathbf{L}}{\otimes} f^*(\xi_{g,h})) = \xi_{gf,h} \circ (\xi_{f,g} \overset{\mathbf{L}}{\otimes} 1),$$

analogous to (2.2.4).

For $\mathcal{G}^\bullet \in \mathbf{D}^+_c(Y)$, we define (as in the smooth case) the functor

$$f^\sharp : \mathbf{D}^+_c(Y) \to \mathbf{D}^+_c(X)$$

to be

$$f^\sharp(\mathcal{G}^\bullet) = f^! \mathcal{O}_Y \overset{\mathbf{L}}{\otimes} f^* \mathcal{G}^\bullet.$$

This agrees with (4.3.1) when f is a proper CM map with pure relative dimension n. The map $\xi_{f,g}$ allows us to define a natural map

$$(4.3.8) \qquad\qquad\qquad f^\sharp g^\sharp \to (gf)^\sharp$$

in the obvious manner, compatible with triple composites and residually stable base change. We will check in Theorem 4.3.3 below that this is the inverse of the usual isomorphism when f and g are smooth. More generally, we will show that $\xi_{f,g}$ is an isomorphism when f and g are CM and that $\xi^{-1}_{f,g} = \zeta'_{f,g}$ when f and g are smooth. In order to do this, it is convenient to study the following more general problem.

For a finite type flat $f : X \to Y$ and a residual complex $K^\bullet$ on Y, we define the derived category map

$$(4.3.9) \qquad\qquad t_{f,K^\bullet} : f^!_{K^\bullet} \cdot \mathcal{O}_Y \overset{\mathbf{L}}{\otimes} f^* K^\bullet \to f^\Delta K^\bullet$$

to be the composite

(4.3.10)

$$\mathcal{H}om_X^\bullet(f^*K^\bullet, f^\triangle K^\bullet) \overset{\mathbf{L}}{\otimes} f^*K^\bullet \longrightarrow \mathcal{H}om_X^\bullet(f^*K^\bullet, f^\triangle K^\bullet) \otimes f^*K^\bullet$$

$$\downarrow \text{signs!}$$

$$f^\triangle K^\bullet \underset{\text{no signs!}}{\longleftarrow} \mathcal{H}om_X^\bullet(\mathcal{H}om_X^\bullet(f^*K^\bullet, f^*K^\bullet), f^\triangle K^\bullet)$$

It is not difficult to check that (4.3.9) is compatible with residually stable base change, isomorphisms between different choices of residual complexes, tensoring $K^\bullet$ with an invertible sheaf, and composites in f (using (4.3.6)). Moreover, it is easy to check that the signs intervening in the middle map in (4.3.10) do *not* affect the composite (since $(-1)^{p(p+1)} = 1$ for all $p \in \mathbf{Z}$) and that (4.3.9) is compatible with translation in $K^\bullet$, where we recall that the definition of (3.3.12) uses (3.3.11), which rests on the isomorphism

$$\mathcal{H}om_X^\bullet(\mathcal{F}^\bullet[n], \mathcal{G}^\bullet[n]) \simeq \mathcal{H}om_X^\bullet(\mathcal{F}^\bullet, \mathcal{G}^\bullet)$$

as in (1.3.11), involving a sign of $(-1)^{nm}$ in degree m. Thus, by (3.3.7), we see that $t_{f,K^\bullet}$ is 'independent' of the choice of $K^\bullet$. For $\mathcal{G}^\bullet \in \mathbf{D}_c^-(Y)$, we define the map

(4.3.11) $$f_{K^\bullet}^! \mathcal{O}_Y \overset{\mathbf{L}}{\otimes} f^* D_{Y,K^\bullet}(\mathcal{G}^\bullet) \to D_{X,f^\triangle K^\bullet} f^* \mathcal{G}^\bullet$$

to be the composite

$$f_{K^\bullet}^! \mathcal{O}_Y \overset{\mathbf{L}}{\otimes} f^* D_{K^\bullet}(\mathcal{G}^\bullet) \longrightarrow f_{K^\bullet}^! \mathcal{O}_Y \otimes f^* D_{K^\bullet}(\mathcal{G}^\bullet)$$

$$\downarrow$$

$$f_{K^\bullet}^! \mathcal{O}_Y \otimes \mathcal{H}om_X^\bullet(f^*\mathcal{G}^\bullet, f^*K^\bullet)$$

$$\downarrow \text{no signs!}$$

$$\mathcal{H}om_X^\bullet(f^*\mathcal{G}^\bullet, f_{K^\bullet}^! \mathcal{O}_Y \otimes f^*K^\bullet)$$

$$\downarrow t_{f,K^\bullet}$$

$$\mathcal{H}om_X^\bullet(f^*\mathcal{G}^\bullet, f^\triangle K^\bullet)$$

$$\| $$

$$D_{f^\triangle K^\bullet} f^*\mathcal{G}^\bullet$$

Note that (4.3.11) recovers (4.3.9) when $\mathcal{G}^\bullet = \mathcal{O}_Y[0]$.

For $\mathcal{G}^\bullet \in \mathbf{D}_c^+(Y)$, using (4.3.11) with the complex $D_{K^\bullet}(\mathcal{G}^\bullet) \in \mathbf{D}_c^-(Y)$ enables us to define the δ-functorial

(4.3.12) $$e_f : f^\sharp \mathcal{G}^\bullet \to f^! \mathcal{G}^\bullet$$

to be the composite

$$f_K^! \bullet \mathcal{O}_Y \overset{\mathbf{L}}{\otimes} f^* \mathcal{G}^\bullet \simeq f_K^! \bullet \mathcal{O}_Y \overset{\mathbf{L}}{\otimes} f^* D_{K\bullet}(D_{K\bullet}(\mathcal{G}^\bullet)) \to D_{f^\triangle K} \bullet f^*(D_{K\bullet}(\mathcal{G}^\bullet)).$$

The map e_f is independent of the choice of $K^\bullet$, is translation-compatible and functorial in $\mathcal{G}^\bullet$, and respects residually stable base change. It is easy to check that e_f coincides with (3.3.21) in the smooth case and is compatible with composites in f (using (4.3.6)). The first part of Theorem 4.3.2 follows from the next result.

THEOREM 4.3.3. When $f : X \to Y$ is a finite type CM map with pure relative dimension and Y is a noetherian scheme which admits a dualizing complex, the map $e_f : f^\sharp \to f^!$ is an isomorphism of δ-functors on $\mathbf{D}_c^+(Y)$. Moreover, when f and g are CM maps of finite type with pure relative dimension, the map $\xi_{f,g}$ is an isomorphism (so $f^\sharp g^\sharp \to (gf)^\sharp$ is also an isomorphism). If f and g are also assumed to be smooth, the inverse of $\xi_{f,g}$ is $\zeta'_{f,g}$.

In [**Verd**, Cor 2, p.396] a related result (ignoring the final part about $\zeta'_{f,g}$ in the smooth case) is proven in greater generality, but with less 'explicit' control on the maps of functors. Unfortunately, it does not seem that the method of proof of [**Verd**, Cor 2, p.396] is helpful if one wants to prove that in the CM case $\xi_{f,g}$ is of formation compatible with base change in the category of noetherian schemes admitting a dualizing complex. We certainly want to know this fact in order to contemplate the compatibility of Tr'_f and θ'_f with respect to composites in the case of locally noetherian schemes (see Corollary 4.4.5). The base change compatibility of $\xi_{f,g}$ in the CM case will be proven in Theorem 4.4.4 below.

PROOF. (of Theorem 4.3.3). We begin by proving the last part. Let $f : X \to Y$ and $g : Y \to Z$ be finite type smooth maps with pure relative dimensions n and m respectively. In order to compare the isomorphism $\zeta'_{f,g}$ with the map

$$\omega_{X/Y}[n] \otimes f^* \omega_{Y/Z}[m] \to \omega_{X/Z}[n+m]$$

obtained from $\xi_{f,g}$, we can work locally and so it is enough to choose global sections ω_1 and ω_2 of $\omega_{X/Y}$ and $\omega_{Y/Z}$ respectively, and to chase $\omega_1 \otimes f^* \omega_2$ under $\xi_{f,g}$ and $\zeta'_{f,g}$. In order to carry out such a diagram chase, we slightly reformulate the definition of $\xi_{f,g}$, as follows.

Choose a residual complex $K^\bullet$ on Y, and choose a map of complexes $f^\sharp K^\bullet \to f^\triangle K^\bullet$ which represents the derived category isomorphism (3.3.16). We use this choice to define the composite map

$$(4.3.13) \quad \omega_{X/Y}[n] \to \mathcal{H}om_X^\bullet(f^* K^\bullet, f^\sharp K^\bullet) \to \mathcal{H}om_X^\bullet(f^* K^\bullet, f^\triangle K^\bullet) = f_K^! \bullet \mathcal{O}_Y$$

as in Lemma 4.1.2. By Lemma 4.1.2 this composite is a quasi-isomorphism, so it gives a flat resolution of $f_K^! \bullet \mathcal{O}_Y$. Since f is smooth, the map (4.3.13) fits into the commutative diagram of complexes

(4.3.14)

$$\omega_{X/Y}[n] \otimes f^*K^\bullet =\!=\!=\!= f^\sharp K^\bullet$$

$$f^!_{K^\bullet} \mathscr{O}_Y \otimes f^*K^\bullet \longrightarrow f^\triangle K^\bullet$$

where the right column is the same choice of representative for (3.3.16) which we use to define (4.3.13) and the bottom row is as in (4.3.10). There are analogues of (4.3.13) and (4.3.14) for g in place of f. Using these, it is not hard to check that $\xi_{f,g}$ is represented by the diagram of maps of complexes

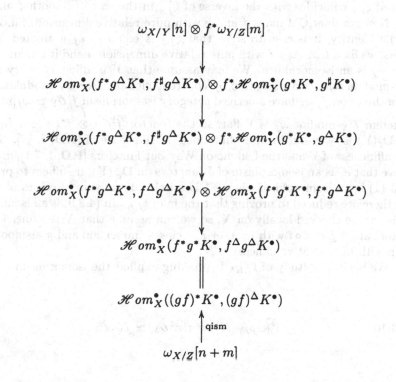

$$\omega_{X/Y}[n] \otimes f^*\omega_{Y/Z}[m]$$

$$\mathscr{H}om^\bullet_X(f^*g^\triangle K^\bullet, f^\sharp g^\triangle K^\bullet) \otimes f^*\mathscr{H}om^\bullet_Y(g^*K^\bullet, g^\sharp K^\bullet)$$

$$\mathscr{H}om^\bullet_X(f^*g^\triangle K^\bullet, f^\sharp g^\triangle K^\bullet) \otimes f^*\mathscr{H}om^\bullet_Y(g^*K^\bullet, g^\triangle K^\bullet)$$

$$\mathscr{H}om^\bullet_X(f^*g^\triangle K^\bullet, f^\triangle g^\triangle K^\bullet) \otimes \mathscr{H}om^\bullet_X(f^*g^*K^\bullet, f^*g^\triangle K^\bullet)$$

$$\mathscr{H}om^\bullet_X(f^*g^*K^\bullet, f^\triangle g^\triangle K^\bullet)$$

$$\mathscr{H}om^\bullet_X((gf)^*K^\bullet, (gf)^\triangle K^\bullet)$$

$$\uparrow \text{qism}$$

$$\omega_{X/Z}[n+m]$$

If we chase $\omega_1 \otimes f^*\omega_2$ through the definition of $\zeta'_{f,g}$ and through the above reformulation of the definition of $\xi_{f,g}$, the identification of $\xi_{f,g}$ and the inverse of $\zeta'_{f,g}$ is reduced to the commutativity of the outside edge of the diagram of complexes

(4.3.15)

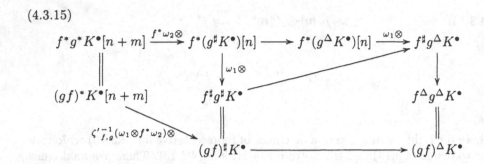

By using (VAR3) above Theorem 3.3.1, we see that (4.3.15) is commutative, so indeed $\xi_{f,g}$ coincides with the inverse of $\zeta'_{f,g}$ in the case of smooth f and g.

Now consider CM maps f and g with pure relative dimension. Since (3.3.5) is the identity, it is easy to show that $\xi_{f,g}$ is equal to e_f evaluated on $g^!\mathscr{O}_Y$. Thus, we fix a CM map f with pure relative dimension n and it remains to show that e_f is an isomorphism. We may assume that Y is affine, so every coherent $\mathscr{O}_Y$-module admits a resolution by coherent globally free $\mathscr{O}_Y$-modules. By the hypotheses on f, we have a derived category isomorphism $f^!\mathscr{O}_Y \simeq \omega_f[n]$ and the coherent $\mathscr{O}_X$-module ω_f is Y-flat, so the functor $f^!\mathscr{O}_Y \overset{\mathbf{L}}{\otimes} f^*(\cdot) \simeq \omega_f[n] \overset{\mathbf{L}}{\otimes}_Y (\cdot)$ on $\mathbf{D}_c(Y)$ is 'way-out in both directions' in the sense of [**RD**, I, §7]. Thus, by the affineness of Y and the Lemma on Way-out Functors [**RD**, I, 7.1], in order to prove that e_f is an isomorphism of δ-functors on $\mathbf{D}_c^+(Y)$, it suffices to prove that (4.3.11) is an isomorphism when $\mathscr{G}^\bullet$ is a finite direct sum of copies of $\mathscr{O}_Y$. We are therefore reduced to proving that the map $t_{f,K^\bullet}$ in (4.3.9) is an isomorphism. This can be checked locally on X, so we can assume that X is affine. Choose a factorization $f = \pi \circ i$ with $i : X \to P$ a closed immersion and π a smooth affine map with pure relative dimension N.

We begin the study of $t_{f,K^\bullet}$ by making explicit the isomorphism

(4.3.16) $i^\flat(\omega_{P/Y}[N]) = i^\flat\pi^\sharp\mathscr{O}_Y \simeq f^!_{K^\bullet}\mathscr{O}_Y.$

Since $i^\flat\pi^\triangle K^\bullet$ is a bounded complex of injectives on X, we have

$$\mathscr{H}om_X^\bullet(\cdot, i^\flat\pi^\triangle K^\bullet) = \mathbf{R}\mathscr{H}om_X^\bullet(\cdot, i^\flat\pi^\triangle K^\bullet)$$

as functors $\mathbf{D}(X) \to \mathbf{D}(X)$. The isomorphism (4.3.16) in $\mathbf{D}(X)$ is equal to the composite

(4.3.17)

$$i^b(\omega_{P/Y}[N]) \xrightarrow[\text{signs!}]{\cong} i^b\,\mathcal{H}om_P^\bullet(\mathcal{H}om_P^\bullet(\omega_{P/Y}[N], \pi^\Delta K^\bullet), \pi^\Delta K^\bullet)$$

$$\uparrow {\scriptstyle\text{qism}}$$

$$\mathcal{H}om_X^\bullet(\mathbf{L}i^*\,\mathcal{H}om_P^\bullet(\omega_{P/Y}[N], \pi^\Delta K^\bullet), i^b\pi^\Delta K^\bullet)$$

$$\downarrow {\scriptstyle\cong}$$

$$\mathcal{H}om_X^\bullet(\mathbf{L}i^*\,\mathcal{H}om_P^\bullet(\omega_{P/Y}[N], \pi^\Delta K^\bullet), i^\Delta\pi^\Delta K^\bullet)$$

$$\downarrow {\scriptstyle\beta_1}$$

$$\mathcal{H}om_X^\bullet(\mathbf{L}i^*\,\mathcal{H}om_P^\bullet(\mathcal{H}om_P^\bullet(\pi^*K^\bullet, \pi^\sharp K^\bullet), \pi^\Delta K^\bullet), i^\Delta\pi^\Delta K^\bullet)$$

$$\downarrow {\scriptstyle\beta_2}$$

$$\mathcal{H}om_X^\bullet(\mathbf{L}i^*\,\mathcal{H}om_P^\bullet(\mathcal{H}om_P^\bullet(\pi^*K^\bullet, \pi^\Delta K^\bullet), \pi^\Delta K^\bullet), i^\Delta\pi^\Delta K^\bullet)$$

$$\downarrow {\scriptstyle\text{signs!}}\ {\scriptstyle\cong}$$

$$\mathcal{H}om_X^\bullet(\mathbf{L}i^*\pi^*K^\bullet, i^\Delta\pi^\Delta K^\bullet)$$

$$\|$$

$$f_K^!\cdot\mathcal{O}_Y =\!=\!=\!=\!=\!=\!=\!= \mathcal{H}om_X^\bullet(f^*K^\bullet, f^\Delta K^\bullet)$$

The composite $\beta_2 \circ \beta_1$ is a quasi-isomorphism since π is smooth (see Lemma 4.1.2), the map labelled 'qism' is a special case of the derived category isomorphism (3.3.20) for the finite map i, and the two maps labelled 'signs!' are special cases of the map (3.3.4) and so involve an intervention of signs.

In order to simplify (4.3.17), choose a quasi-isomorphism of complexes ψ : $\pi^\sharp K^\bullet \to \pi^\Delta K^\bullet$ which represents the derived category isomorphism (3.3.16). Use this choice to define the composite map of complexes

(4.3.18) $\quad j : \omega_{P/Y}[N] \to \mathcal{H}om_P^\bullet(\pi^*K^\bullet, \pi^\sharp K^\bullet) \xrightarrow{\psi} \mathcal{H}om_P^\bullet(\pi^*K^\bullet, \pi^\Delta K^\bullet)$

as in (4.3.13), which we know is a quasi-isomorphism. By means of (4.3.18), it is easy to check that the derived category composite (4.3.17) is equal to the composite

(4.3.19)

$$i^b(\omega_{P/Y}[N]) \xrightarrow{\ \cong\ } i^b\,\mathcal{H}om_P^\bullet(\pi^*K^\bullet, \pi^\Delta K^\bullet)$$

$$\uparrow {\scriptstyle\cong}$$

$$\mathcal{H}om_X^\bullet(f^*K^\bullet, f^\Delta K^\bullet = i^\Delta\pi^\Delta K^\bullet) =\!=\!=\!= \mathcal{H}om_X^\bullet(\mathbf{L}i^*\pi^*K^\bullet, i^b\pi^\Delta K^\bullet)$$

where the right column is a special case of (3.3.20) for the finite map i. Since Y is affine noetherian and $K^\bullet$ is a bounded complex of quasi-coherent sheaves with coherent cohomology, we may choose a quasi-isomorphism $\rho : F^\bullet \to K^\bullet$ for a bounded above complex $F^\bullet$ of locally free coherent sheaves. We will now use the quasi-isomorphism ρ and the composite (4.3.19) to show that $t_{f,K^\bullet}$ is an isomorphism.

There is an obvious composite map of complexes

$$\alpha : \omega_{P/Y}[N] \to \mathcal{H}om_P^\bullet(\pi^* K^\bullet, \pi^\sharp K^\bullet) \to \mathcal{H}om_P^\bullet(\pi^* F^\bullet, \pi^\sharp K^\bullet)$$

which we claim is a quasi-isomorphism. To prove this, consider the commutative diagram of complexes

$$(4.3.20) \quad
\begin{array}{ccccc}
\omega_{P/Y}[N] & \longrightarrow & \mathcal{H}om_P^\bullet(\pi^* K^\bullet, \pi^\sharp K^\bullet) & \xrightarrow{\psi} & \mathcal{H}om_P^\bullet(\pi^* K^\bullet, \pi^\triangle K^\bullet) \\
 & \searrow_{\alpha} & & & \downarrow_{\pi^*(\rho)} \\
 & & \mathcal{H}om_P^\bullet(\pi^* F^\bullet, \pi^\sharp K^\bullet) & \xrightarrow{\psi} & \mathcal{H}om_P^\bullet(\pi^* F^\bullet, \pi^\triangle K^\bullet)
\end{array}$$

In the diagram (4.3.20), the composite across the top row is a quasi-isomorphism by Lemma 4.1.2, the right column is a quasi-isomorphism because $\pi^\triangle K^\bullet$ is a bounded complex of injectives, and the bottom row is a quasi-isomorphism because $\pi^* F^\bullet$ is a bounded above complex of locally free coherent sheaves and ψ is a quasi-isomorphism between bounded complexes. Thus, we deduce that α is a quasi-isomorphism.

The canonical map $\beta : \mathcal{H}om_P^\bullet(\pi^* F^\bullet, \pi^\sharp K^\bullet) \to \mathbf{R}\mathcal{H}om_P^\bullet(\pi^* F^\bullet, \pi^\sharp K^\bullet)$ is a quasi-isomorphism because $\pi^\sharp K^\bullet$ is a bounded complex of quasi-coherent sheaves and $\pi^* F^\bullet$ is a bounded above complex of locally free coherent sheaves. The quasi-isomorphism

$$i^\flat \pi^\sharp K^\bullet \xrightarrow{i^\flat(\psi)} i^\flat \pi^\triangle K^\bullet = i^\triangle \pi^\triangle K^\bullet \simeq f^\triangle K^\bullet$$

gives an injective resolution of $i^\flat \pi^\sharp K^\bullet$. It is therefore easy to show that the composite (4.3.19) is equal to the derived category composite in the following diagram:

$$(4.3.21) \quad
\begin{array}{ccc}
i^\flat(\omega_{P/Y}[N]) & \xrightarrow[\simeq]{i^\flat(\alpha)} & i^\flat \mathcal{H}om_P^\bullet(\pi^* F^\bullet, \pi^\sharp K^\bullet) \\
 & & \downarrow{\simeq}{\scriptstyle i^\flat(\beta)} \\
 & & i^\flat \mathbf{R}\mathcal{H}om_P^\bullet(\pi^* F^\bullet, \pi^\sharp K^\bullet) \\
 & & \uparrow{\simeq} \\
 & & \mathbf{R}\mathcal{H}om_X^\bullet(\mathbf{L}i^* \pi^* F^\bullet, i^\flat \pi^\sharp K^\bullet) \\
 & & \| \\
\mathcal{H}om_X^\bullet(f^* K^\bullet, f^\triangle K^\bullet) & =\!=\!=\!= & \mathcal{H}om_X^\bullet(i^* \pi^* F^\bullet, f^\triangle K^\bullet)
\end{array}$$

The middle of the right column is a special case of (3.3.20).

There is an obvious analogue

(4.3.22)

$$i^\flat(\omega_{P/Y}[N]) \xrightarrow[\simeq]{i^\flat(\alpha)} i^\flat \mathscr{H}om_P^\bullet(\pi^*F^\bullet, \pi^\Delta K^\bullet)$$

$$\simeq \Big\downarrow i^\flat(\beta)$$

$$i^\flat \mathbf{R}\mathscr{H}om_P^\bullet(\pi^*F^\bullet, \pi^\Delta K^\bullet)$$

$$\Big\uparrow \simeq$$

$$\mathbf{R}\mathscr{H}om_X^\bullet(Li^*\pi^*F^\bullet, i^\flat \pi^\Delta K^\bullet)$$

$$\mathscr{H}om_X^\bullet(f^*K^\bullet, f^\Delta K^\bullet) \text{=====} \mathscr{H}om_X^\bullet(i^*\pi^*F^\bullet, f^\Delta K^\bullet)$$

of (4.3.21) with $\pi^\Delta K^\bullet$ in place of $\pi^\sharp K^\bullet$, and the choice of quasi-isomorphism $\psi : \pi^\sharp K^\bullet \to \pi^\Delta K^\bullet$ used in the definition of (4.3.18) gives rise to a commutative diagram of the form (4.3.21) $\to$ (4.3.22) such that *all* maps in this 'composite' derived category diagram are isomorphisms. Thus, we deduce the commutativity of the derived category diagram

(4.3.23)

$$i^\flat(\omega_{P/Y}[N]) \overset{\mathbf{L}}{\otimes} f^*F^\bullet \xrightarrow{\ \simeq\ } f_{K^\bullet}^! \mathcal{O}_Y \overset{\mathbf{L}}{\otimes} f^*K^\bullet$$

$$\simeq \Big\downarrow \qquad\qquad\qquad\qquad \Big\downarrow t_{f,K^\bullet}$$

$$i^\flat(\mathscr{H}om_P^\bullet(\pi^*F^\bullet, \pi^\Delta K^\bullet)) \overset{\mathbf{L}}{\otimes} f^*F^\bullet \qquad\qquad f^\Delta K^\bullet$$

$$\varphi \Big\uparrow \simeq \qquad\qquad\qquad\qquad \Big\uparrow \simeq$$

$$\mathscr{H}om_X^\bullet(f^*F^\bullet, i^\flat \pi^\Delta K^\bullet) \otimes_X f^*F^\bullet \xrightarrow[\text{no signs!}]{} i^\flat \pi^\Delta K^\bullet$$

where the map labelled φ is a special case of the isomorphism (3.3.20).

We want to show that $t_{f,K^\bullet}$ in (4.3.23) is an isomorphism. Since (4.3.23) commutes, it suffices to show that the bottom row is a quasi-isomorphism. This bottom row is the 'evaluation pairing,' which is identified with

(4.3.24)

$$\mathscr{H}om_P^\bullet(\pi^*F^\bullet, \mathscr{H}om_P^\bullet(i_*\mathcal{O}_X, \pi^\Delta K^\bullet)) \otimes_P \pi^*F^\bullet \to \mathscr{H}om_P^\bullet(i_*\mathcal{O}_X, \pi^\Delta K^\bullet),$$

and does not involve an intervention of signs.

Choose a resolution $E^\bullet \to i_*\mathcal{O}_X$ by locally free coherent $\mathcal{O}_P$-modules. Since $\pi^\Delta K^\bullet$ is a bounded complex of injectives and $F^\bullet$ is a bounded above complex of coherent locally free sheaves, we can identify (4.3.24) up to *quasi-isomorphism* with the analogous map of complexes

(4.3.25) $\quad \mathscr{H}om_P^\bullet(\pi^*F^\bullet, \mathscr{H}om_P^\bullet(E^\bullet, \pi^\Delta K^\bullet)) \otimes \pi^*F^\bullet \to \mathscr{H}om_P^\bullet(E^\bullet, \pi^\Delta K^\bullet)$

in which $i_*\mathcal{O}_X$ is replaced by $E^\bullet$ and there is no intervention of signs. We must show that (4.3.25) is a quasi-isomorphism, so it suffices to check that the composite of (4.3.25) with a well-chosen quasi-isomorphism is a quasi-isomorphism.

By means of an intervention of signs, we have an isomorphism of complexes
(4.3.26)
$$\mathcal{H}om_P^\bullet(E^\bullet, \mathcal{H}om_P^\bullet(\pi^*F^\bullet, \pi^\Delta K^\bullet)) \simeq \mathcal{H}om_P^\bullet(\pi^*F^\bullet, \mathcal{H}om_P^\bullet(E^\bullet, \pi^\Delta K^\bullet)).$$

Applying $\mathcal{H}om_P^\bullet(E^\bullet, \cdot) \otimes \pi^*F^\bullet$ to the composite quasi-isomorphism

$$\omega_{P/Y}[N] \xrightarrow{j} \mathcal{H}om_P^\bullet(\pi^*K^\bullet, \pi^\Delta K^\bullet) \xrightarrow{\pi^*(\rho)} \mathcal{H}om_P^\bullet(\pi^*F^\bullet, \pi^\Delta K^\bullet)$$

we get a quasi-isomorphism

$$\mathcal{H}om_P^\bullet(E^\bullet, \omega_{P/Y}[N]) \otimes \pi^*F^\bullet \to \mathcal{H}om_P^\bullet(E^\bullet, \mathcal{H}om_P^\bullet(\pi^*F^\bullet, \pi^\Delta K^\bullet)) \otimes \pi^*F^\bullet$$

whose composite with the isomorphism $(4.3.26) \otimes \pi^*F^\bullet$ and the map $(4.3.25)$ is a map of complexes

(4.3.27) $$\mathcal{H}om_P^\bullet(E^\bullet, \omega_{P/Y}[N]) \otimes \pi^*F^\bullet \to \mathcal{H}om_P^\bullet(E^\bullet, \pi^\Delta K^\bullet)$$

which involves a sign of $(-1)^{nm}$ on $\mathcal{H}om_P(E^n, \omega_{P/Y}) \otimes_P \pi^*F^m$. Obviously $(4.3.25)$ is a quasi-isomorphism if and only if $(4.3.27)$ is a quasi-isomorphism. Our proof that $(4.3.27)$ is a quasi-isomorphism will ultimately depend upon the fact that $f = \pi \circ i$ is a CM map with pure relative dimension and so satisfies the characterization in Theorem 3.5.1.

The composite map

$$\xi : \omega_{P/Y}[N] \otimes \pi^*F^\bullet = \pi^\sharp F^\bullet \xrightarrow{\pi^*(\rho)} \pi^\sharp K^\bullet \xrightarrow{\psi} \pi^\Delta K^\bullet$$

is a composite of quasi-isomorphisms and so is an injective resolution of the complex $\omega_{P/Y}[N] \otimes \pi^*F^\bullet$. We can use this resolution to identify $(4.3.27)$ with a map

(4.3.28) $$\mathcal{H}om_P^\bullet(E^\bullet, \omega_{P/Y}[N]) \otimes \pi^*F^\bullet \longrightarrow \mathcal{H}om_P^\bullet(E^\bullet, \omega_{P/Y}[N] \otimes \pi^*F^\bullet)$$
$$\downarrow{\xi}$$
$$\mathcal{H}om_P^\bullet(E^\bullet, \pi^\Delta K^\bullet)$$

in which the first part involves an intervention of signs. It is easy to identify $(4.3.28)$ in $\mathbf{D}(P)$ with a special case of the more general map (which involves an intervention of signs, unlike $(2.1.9)$)

(4.3.29)

$$\mathbf{R}\mathcal{H}om_P^\bullet(i_*\mathcal{O}_X, \omega_{P/Y}[N]) \overset{\mathbf{L}}{\otimes} \pi^*\mathcal{F}^\bullet \to \mathbf{R}\mathcal{H}om_P^\bullet(i_*\mathcal{O}_X, \omega_{P/Y}[N] \overset{\mathbf{L}}{\otimes} \pi^*\mathcal{F}^\bullet)$$

for $\mathcal{F}^\bullet \in \mathbf{D}_{qc}^b(Y) \simeq \mathbf{D}^b(\mathrm{Qco}(Y))$, defined by means of a resolution of $i_*\mathcal{O}_X$ by locally free coherent sheaves. We will use the CM condition on $f = \pi \circ i$ to prove that $(4.3.29)$ is an isomorphism in $\mathbf{D}(P)$. In particular, $(4.3.27)$ is a quasi-isomorphism, thereby completing the proof of Theorem 4.3.3.

The map $(4.3.29)$ is compatible with translation in $\mathcal{F}^\bullet$, so by the Lemma on Way-Out Functors [$\mathbf{RD}$, I, 7.1(i)] we only need to consider $(4.3.29)$ in the special case $\mathcal{F}^\bullet = \mathcal{M}[0]$ for a single quasi-coherent sheaf $\mathcal{M}$ on Y. The cohomology of $\mathbf{R}\mathcal{H}om_P^\bullet(i_*\mathcal{O}_X, \omega_{P/Y}[N])$ in degree d is $\mathcal{E}xt_P^{d+N}(i_*\mathcal{O}_X, \omega_{P/Y})$, which is Y-flat for all d: it vanishes for $d \neq -n$ and coincides with the Y-flat dualizing sheaf ω_f

if $d = -n$. Here we have used the hypothesis that f is CM with pure relative dimension n. Thus, the induced cohomology map in degree d in (4.3.29) with $\mathscr{F}^\bullet = \mathscr{M}[0]$ is a map

$$(4.3.30) \qquad \mathscr{E}xt_P^{d+N}(i_*\mathscr{O}_X, \omega_{P/Y}) \otimes_Y \mathscr{M} \to \mathscr{E}xt_P^{d+N}(i_*\mathscr{O}_X, \omega_{P/Y} \otimes_Y \mathscr{M}).$$

When $\mathscr{M} = \mathscr{O}_Y$, this is the identity map. Thus, by functoriality in $\mathscr{M}$ and chasing sections, this map must coincide with the 'base change' map denoted $b^{d+N}(\mathscr{M})$ in [**AK2**, §1.8]. Thus, the Y-flatness of all sheaves $\mathscr{E}xt^\bullet(i_*\mathscr{O}_X, \omega_{P/Y})$ implies that (4.3.30) is an isomorphism, by [**AK2**, Theorem 1.9(*ii*),(3)$\Rightarrow$(2)]. ∎

4.4. Conclusion of Proof of Duality Theorem

The remaining step in the proof of Theorem 4.3.1 is the last part of Theorem 4.3.2, concerning the identification of Tr'_f and Tr_f on $\mathbf{D}_c^b(Y)$, for a proper CM map $f : X \to Y$ with pure relative dimension, where Y is a noetherian scheme admitting a dualizing complex. This is a special case of the following more general result for proper flat maps, *not* required to be CM.

THEOREM 4.4.1. *Let Y be a noetherian scheme admitting a dualizing complex and let $f : X \to Y$ be a proper flat map with pure relative dimension. For $\mathscr{G}^\bullet \in \mathbf{D}_c^b(Y)$, the diagram*

$$(4.4.1)$$

$$
\begin{array}{ccc}
\mathbf{R}f_* f^\sharp \mathscr{G}^\bullet & \xrightarrow{\ e_f\ } & \mathbf{R}f_* f^! \mathscr{G}^\bullet \\
{\scriptstyle \simeq}\uparrow & & \downarrow{\scriptstyle \mathrm{Tr}_f(\mathscr{G}^\bullet)} \\
\mathbf{R}f_*(f^! \mathscr{O}_Y) \overset{\mathbf{L}}{\otimes} \mathscr{G}^\bullet & \xrightarrow[\mathrm{Tr}_f(\mathscr{O}_Y)\overset{\mathbf{L}}{\otimes}1]{} & \mathscr{G}^\bullet
\end{array}
$$

commutes, where the left column is the projection formula.

The last part of Theorem 4.3.2 is a special case of (4.4.1) because we have already observed below (4.3.4) that $\mathrm{Tr}_f(\mathscr{O}_Y) = \mathrm{Tr}'_f(\mathscr{O}_Y)$, due to how γ_f is defined.

PROOF. Fix a residual complex $K^\bullet$ on Y, so $K^\bullet \in \mathbf{D}_c^b(Y)$ is a *bounded* complex of quasi-coherent injectives on Y and likewise for $f^\triangle K^\bullet$ on X. We write D_Y and η_Y (resp. D_X and η_X) instead of $D_{Y,K^\bullet}$ and $\eta_{K^\bullet}$ (resp. $D_{X,f^\triangle K^\bullet}$ and $\eta_{f^\triangle K^\bullet}$) respectively. The functor $f^!$ is computed as $f_{K^\bullet}^!(\cdot) = D_X \circ f^* \circ D_Y$. Note that for any $\mathscr{F}^\bullet \in \mathbf{D}_c(X)$, there is a natural quasi-isomorphism

$$\eta_X(\mathscr{F}^\bullet) : \mathscr{F}^\bullet \to D_X^2(\mathscr{F}^\bullet) = \mathscr{H}om_X^\bullet(\mathscr{H}om_X^\bullet(\mathscr{F}^\bullet, f^\triangle K^\bullet), f^\triangle K^\bullet)$$

with the right side a complex of *flasque* sheaves, and hence suitable for computing $\mathbf{R}f_*(\mathscr{F}^\bullet)$. For example, for any complex $\mathscr{G}^\bullet \in \mathbf{D}_c^b(Y)$, we have the identity

$$\mathbf{R}f_* f^! \mathscr{G}^\bullet = f_* D_X f^* D_Y(\mathscr{G}^\bullet).$$

Fix a choice of complex $\mathscr{G}^{\bullet} \in \mathbf{D}_c^b(Y)$, and without loss of generality assume that $\mathscr{G}^{\bullet}$ is a bounded above complex of (perhaps non-quasi-coherent) *flat* $\mathcal{O}_Y$-modules, so $f^{\sharp}\mathscr{G}^{\bullet} = f^!\mathcal{O}_Y \otimes f^*\mathscr{G}^{\bullet}$, without any $\overset{\mathbf{L}}{\otimes}$. Note that this step requires $\mathscr{G}^{\bullet} \in \mathbf{D}^-(Y)$, whereas the definition of e_f requires $\mathscr{G}^{\bullet} \in \mathbf{D}_c^+(Y)$; this is the most important reason why we require $\mathscr{G}^{\bullet} \in \mathbf{D}_c^b(Y)$. The diagram (4.4.1) is identified with the outside edge of the diagram of complexes

(4.4.2)

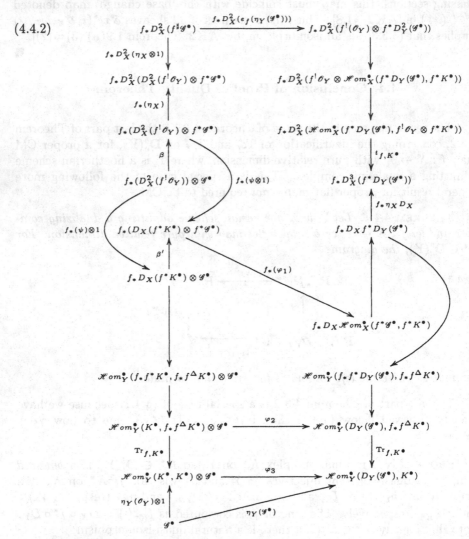

In this diagram, the terms $f_*D_X^2(f^{\sharp}\mathscr{G}^{\bullet})$ and $f_*D_X(f^*K^{\bullet}) \otimes \mathscr{G}^{\bullet}$ in the left column are equal to $\mathbf{R}f_*f^{\sharp}\mathscr{G}^{\bullet}$ and $\mathbf{R}f_*(f^!\mathcal{O}_Y) \overset{\mathbf{L}}{\otimes} \mathscr{G}^{\bullet}$ respectively, while the term $f_*D_Xf^*D_Y(\mathscr{G}^{\bullet})$ in the right column is equal to $\mathbf{R}f_*f^!\mathscr{G}^{\bullet}$. We need to describe some of the labelled maps in (4.4.2), and we must check that $f_*(\eta_X)$ and β in the left column are quasi-isomorphisms, so then it *does* make sense to consider the commutativity of the outside edge of (4.4.2) as a (large) diagram in $\mathbf{D}(Y)$. To show that $f_*(\eta_X)$ and β are quasi-isomorphisms, we may assume that Y is affine, in which case we can replace $\mathscr{G}^\bullet$ with a bounded above complex $\mathscr{G}'^\bullet$ of locally free coherent sheaves. This makes it obvious that β is a quasi-isomorphism. The local freeness of the coherent $\mathscr{G}'^m$'s ensures that

$$\eta_X : D_X^2(f^!\mathscr{O}_Y) \otimes f^*\mathscr{G}'^\bullet \to D_X^2(D_X^2(f^!\mathscr{O}_Y) \otimes f^*\mathscr{G}'^\bullet)$$

is a quasi-isomorphism between complexes of *flasque* sheaves, so from the f_*-acyclicity of flasques we deduce that $f_*(\eta_X)$ is a quasi-isomorphism, as desired.

Since (3.3.5) is the identity in the derived category, the *derived category* inverse of the quasi-isomorphism $f_*\eta_X D_X$ in the right column of (4.4.2) is $f_* D_X(\eta_X)$. The maps φ_1, φ_2, φ_3 in (4.4.2) are special cases of the map

$$\mathscr{H}om_Y^\bullet(K_1^\bullet, K_2^\bullet) \otimes \mathscr{F}^\bullet \to \mathscr{H}om_Y^\bullet(\mathscr{H}om_Y^\bullet(\mathscr{F}^\bullet, K_1^\bullet), K_2^\bullet)$$

defined for bounded complexes $K_1^\bullet$, $K_2^\bullet$ and any complex $\mathscr{F}^\bullet$, with an extra sign of $(-1)^{(p+1)r} = (-1)^{r(p-r)}$ on $\mathscr{H}om_Y(K_1^p, K_2^q) \otimes_Y \mathscr{G}^r$. Finally, the map ψ which appears twice in (4.4.2) is defined to be the composite

(4.4.3)

$$D_X^2(f^!\mathscr{O}_Y) \longrightarrow D_X^2(f^!\mathscr{O}_Y \otimes \mathscr{H}om_X^\bullet(f^*K^\bullet, f^*K^\bullet))$$

$$\downarrow \text{no signs!}$$

$$D_X^2(\mathscr{H}om_X^\bullet(f^*K^\bullet, f^!\mathscr{O}_Y \otimes f^*K^\bullet))$$

$$\downarrow t_{f,K^\bullet}$$

$$D_X^2\mathscr{H}om_X^\bullet(f^*K^\bullet, f^\triangle K^\bullet)$$

$$\|$$

$$D_X(f^*K^\bullet) \xleftarrow{\quad D_X(\eta_X) \quad} D_X \circ D_X^2(f^*K^\bullet,)$$

It is easy to check that all subdiagrams in (4.4.2) aside from the 'upper rectangular' part (with bottom edge given by the diagonal map $f_*(\varphi_1)$) are commutative on the level of complexes. Since flasques are f_*-acyclic and $D_X(\cdot)$ is always a complex of flasques, it suffices to check the commutativity in $\mathbf{D}(X)$ of the diagram obtained from the 'top' part of (4.4.2) by removing the f_*'s in front. In other words, by using the definition of ψ via (4.4.3), it suffices to prove

the commutativity of the outside edge of the diagram
(4.4.4)

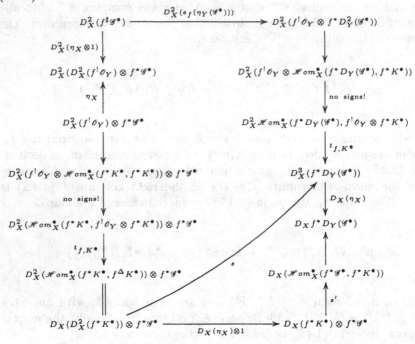

In this diagram, the lower right map s' is defined by using the canonical map of complexes

$$(4.4.5) \qquad \mathscr{H}om^\bullet(\mathscr{F}^\bullet, \mathscr{E}^\bullet) \otimes \mathscr{H}^\bullet \to \mathscr{H}om^\bullet(\mathscr{H}om^\bullet(\mathscr{H}^\bullet, \mathscr{F}^\bullet), \mathscr{E}^\bullet)$$

for bounded below $\mathscr{E}^\bullet$, bounded $\mathscr{F}^\bullet$, and bounded above $\mathscr{H}^\bullet$, with an intervention of the sign $(-1)^{r(p+1)}$ on the term $\mathscr{H}om(\mathscr{F}^p, \mathscr{E}^q) \otimes \mathscr{H}^r$. Meanwhile, the 'diagonal' map s in (4.4.4) is the composite

$$D_X^3(f^*K^\bullet) \otimes f^*\mathscr{G}^\bullet \xrightarrow{t} D_X^3(\mathscr{H}om_X^\bullet(f^*\mathscr{G}^\bullet, f^*K^\bullet)) \to D_X^3(f^*D_Y(\mathscr{G}^\bullet)),$$

with t defined to be the composite
(4.4.6)

$$D_X^3(f^*K^\bullet) \otimes f^*\mathscr{G}^\bullet \longrightarrow D_X(\mathscr{H}om_X^\bullet(f^*\mathscr{G}^\bullet, D_X^2(f^*K^\bullet)))$$

$$\downarrow$$

$$D_X(\mathscr{H}om_X^\bullet(D_X(f^*K^\bullet) \otimes f^*\mathscr{G}^\bullet, f^\triangle K^\bullet))$$

$$\|$$

$$D_X^3(\mathscr{H}om_X^\bullet(f^*\mathscr{G}^\bullet, f^*K^\bullet)) \longleftarrow D_X^2(D_X(f^*K^\bullet) \otimes f^*\mathscr{G}^\bullet)$$

defined by using (4.4.5) and two applications of the canonical map of complexes

$$\mathscr{H}om^\bullet(\mathscr{H}^\bullet \otimes \mathscr{F}^\bullet, \mathscr{E}^\bullet) \to \mathscr{H}om^\bullet(\mathscr{F}^\bullet, \mathscr{H}om^\bullet(\mathscr{H}^\bullet, \mathscr{E}^\bullet))$$

for bounded below $\mathcal{E}^{\bullet}$, bounded above $\mathcal{H}^{\bullet}$, and arbitrary $\mathcal{F}^{\bullet}$, with a sign of $(-1)^{pq}$ on $\mathcal{H}om(\mathcal{H}^p \otimes \mathcal{F}^q, \mathcal{E}^r)$.

The key fact which will enable us to prove the commutativity of both subdiagrams in (4.4.4) is the following lemma. We emphasize that this lemma takes place in the derived category (it is *false* on the level of complexes).

LEMMA 4.4.2. *The diagram of complexes*

(4.4.7)
$$
\begin{array}{ccc}
D_X^3(f^*K^{\bullet}) \otimes f^*\mathcal{G}^{\bullet} & \xrightarrow{\;t\;} & D_X^3(\mathcal{H}om_X^{\bullet}(f^*\mathcal{G}^{\bullet}, f^*K^{\bullet})) \\
{\scriptstyle D_X(\eta_X)\otimes 1}\Big\downarrow & & \Big\downarrow{\scriptstyle D_X(\eta_X)} \\
D_X(f^*K^{\bullet}) \otimes f^*\mathcal{G}^{\bullet} & \xrightarrow{\;s'\;} & D_X(\mathcal{H}om_X^{\bullet}(f^*\mathcal{G}^{\bullet}, f^*K^{\bullet}))
\end{array}
$$

commutes in $\mathbf{D}(X)$.

PROOF. Let $D = D_X$, $\eta = \eta_X$. By recalling the definitions of s' and t in (4.4.5) and (4.4.6), we can identify (4.4.7) with the outside edge of

(4.4.8)
$$
\begin{array}{ccc}
D^3(f^*K^{\bullet}) \otimes f^*\mathcal{G}^{\bullet} & \xrightarrow{\hspace{2cm}} & D^2(D(f^*K^{\bullet}) \otimes f^*\mathcal{G}^{\bullet}) \\
{\scriptstyle D(\eta)\otimes 1}\Big\downarrow & {\scriptstyle \eta}\nearrow & \Big\downarrow \\
D(f^*K^{\bullet}) \otimes f^*\mathcal{G}^{\bullet} & D^2(D(\mathcal{H}om_X^{\bullet}(f^*\mathcal{G}^{\bullet}, f^*K^{\bullet}))) \\
\Big\downarrow & {\scriptstyle \eta}\nearrow & \Big\| \\
D(\mathcal{H}om_X^{\bullet}(f^*\mathcal{G}^{\bullet}, f^*K^{\bullet})) & \xleftarrow{\;D(\eta)\;} & D(D^2\mathcal{H}om_X^{\bullet}(f^*\mathcal{G}^{\bullet}, f^*K^{\bullet}))
\end{array}
$$

The lower triangle in (4.4.8) commutes in $\mathbf{D}(X)$ because (3.3.5) is the identity map and η is an isomorphism in $\mathbf{D}(X)$. The middle parallelogram in (4.4.8) commutes by functoriality. It remains to check the top triangle in (4.4.8), which does *not* commute on the level of complexes. However, since (3.3.5) is the identity, the derived category inverse of $D(\eta) \otimes 1$ in the left column of (4.4.8) is the map $(\eta D) \otimes 1$. Since $\mathcal{G}^{\bullet}$ is a bounded above complex of flats, so $(\cdot) \otimes f^*\mathcal{G}^{\bullet} = (\cdot) \overset{\mathbf{L}}{\otimes} f^*\mathcal{G}^{\bullet}$ on $\mathbf{D}(X)$, we may instead consider the triangular diagram

$$
\begin{array}{ccc}
D^3(f^*K^{\bullet}) \otimes f^*\mathcal{G}^{\bullet} & \xrightarrow{\hspace{1.5cm}} & D^2(D(f^*K^{\bullet}) \otimes f^*\mathcal{G}^{\bullet}) \\
{\scriptstyle (\eta D)\otimes 1}\Big\uparrow & {\scriptstyle \eta}\nearrow & \\
D(f^*K^{\bullet}) \otimes f^*\mathcal{G}^{\bullet} & &
\end{array}
$$

This diagram even commutes on the level of complexes, thanks to the sign calculation

$$
(-1)^{p(r-p)+q(r-p+1)+pq} = (-1)^{(p+q)(r-(p+q))}.
$$
∎

Continuing with the proof of Theorem 4.4.1, if we write out the definition of s in (4.4.4) and use Lemma 4.4.2, it is easy to see that the bottom triangle of (4.4.4) is commutative in $\mathbf{D}(X)$. In order to analyze the top part of (4.4.4), we use functoriality with respect to the isomorphism $\eta_X : 1 \simeq D_X^2$ on $\mathbf{D}_c(X)$ in order to identify the top part of (4.4.4) (i.e., the part above the bottom triangle) with the outside edge of

(4.4.9)

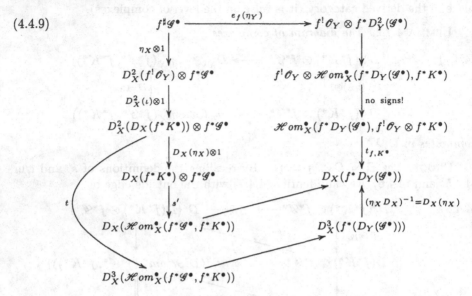

where the map $\iota : f^! \mathcal{O}_Y \to D_X(f^* K^\bullet)$ appearing near the top of the left column is the composite

$$f^! \mathcal{O}_Y \longrightarrow f^! \mathcal{O}_Y \otimes \mathcal{H}om_X^\bullet(f^* K^\bullet, f^* K^\bullet) \underset{\text{no signs!}}{\xrightarrow{\simeq}} \mathcal{H}om_X^\bullet(f^* K^\bullet, f^! \mathcal{O}_Y \otimes f^* K^\bullet)$$

$$\downarrow t_{f, K^\bullet}$$

$$D_X(f^* K^\bullet) =\!=\!=\!=\!=\!=\!=\!= \mathcal{H}om_X^\bullet(f^* K^\bullet, f^\triangle K^\bullet)$$

which we easily compute is the *identity* map (via $f^! = f_{K^\bullet}^!$ and the *definition* of $t_{f, K^\bullet}$ in (4.3.10)).

The commutativity of the 'top' part of (4.4.9) on the level of complexes is an easy exercise in chasing signs, and by using functoriality we see that the commutativity of the 'lower left' part of (4.4.9) follows from Lemma 4.4.2. This completes the proof of Theorem 4.4.1.   ∎

We conclude with some remarks on compatibilities of duality with respect to *composites* of CM maps. Consider two CM maps $f : X \to Y$ and $g : Y \to Z$ with pure relative dimensions n and m respectively, where X, Y, and Z are noetherian schemes admitting a dualizing complex. By Theorem 4.3.3, there is an isomorphism $\xi_{f,g} : \omega_f \otimes f^* \omega_g \simeq \omega_{gf}$ which enables us to define an isomorphism $c_{f,g} : (gf)^\sharp \simeq f^\sharp g^\sharp$ as in (4.3.8). Using $c_{f,g}$, it makes sense to assert:

THEOREM 4.4.3. *For proper CM maps f and g as above (where X, Y, Z are noetherian schemes admitting a dualizing complex), the trace maps Tr'_f, Tr'_g, and Tr'_{gf} are compatible (by means of the isomorphisms $\xi_{f,g}$ and $c_{f,g}$), and likewise for the duality morphisms θ'_f, θ'_g, and θ'_{gf}.*

PROOF. By the definition of the duality morphism, it is enough to consider the trace map. In order to handle the case of trace maps, it is straightfoward to use Theorem 4.3.2, Theorem 4.4.1, the compatibility of $e_f : f^{\sharp} \simeq f^{!}$ with respect to composites in f, and the compatibility of the projection formula (2.1.10) with respect to composite scheme maps and iterated $\overset{L}{\otimes}$'s to reduce to verifying the commutativity of

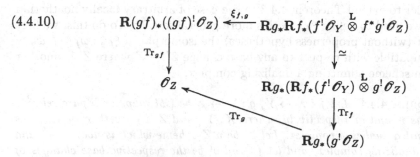

$$(4.4.10)$$

Recall that the *definition* of $\xi_{f,g}$ in (4.3.6) and the definition of the trace maps are given in terms of a choice of residual complex $K^{\bullet}$ on Z. Using these explicit definitions, the identity in Lemma 3.4.3(TRA1), and the canonical map $\mathscr{F}_1^{\bullet} \overset{L}{\otimes} \mathscr{F}_2^{\bullet} \to \mathscr{F}_1^{\bullet} \otimes \mathscr{F}_2^{\bullet}$ for two complexes of sheaves $\mathscr{F}_1^{\bullet}$, $\mathscr{F}_2^{\bullet}$ (at least one of which is bounded above), a slightly long (but not difficult) diagram chase reduces us to verifying the commutativity of the derived category diagram

$$(4.4.11)$$

$$
\begin{array}{ccc}
\mathscr{H}om_Y^{\bullet}(g^{\Delta}K^{\bullet}, g^{\Delta}K^{\bullet}) \overset{L}{\otimes} \mathscr{H}om_Y^{\bullet}(g^{*}K^{\bullet}, g^{\Delta}K^{\bullet}) & \overset{\simeq}{\longleftarrow} & \mathscr{O}_Y[0] \overset{L}{\otimes} \mathscr{H}om_Y^{\bullet}(g^{*}K^{\bullet}, g^{\Delta}K^{\bullet}) \\
\downarrow & & \downarrow \simeq \\
\mathscr{H}om_Y^{\bullet}(g^{\Delta}K^{\bullet}, g^{\Delta}K^{\bullet}) \otimes \mathscr{H}om_Y^{\bullet}(g^{*}K^{\bullet}, g^{\Delta}K^{\bullet}) & \longrightarrow & \mathscr{H}om_Y^{\bullet}(g^{*}K^{\bullet}, g^{\Delta}K^{\bullet})
\end{array}
$$

in $\mathbf{D}(Y)$. We can even make the more general commutativity claim in which $g^{*}K^{\bullet}$ is replaced by any bounded below complex $\mathscr{F}^{\bullet}$ on Y and $g^{\Delta}K^{\bullet}$ is replaced by any bounded complex $\mathscr{G}^{\bullet}$ on Y. This commutativity is trivial to check, since upon choosing a commutative diagram

$$
\begin{array}{ccc}
\mathscr{H}_1^{\bullet} & \longrightarrow & \mathscr{H}_2^{\bullet} \\
\downarrow & & \downarrow \\
\mathscr{O}_Y[0] & \longrightarrow & \mathscr{H}om_Y^{\bullet}(\mathscr{G}^{\bullet}, \mathscr{G}^{\bullet})
\end{array}
$$

with the vertical maps quasi-isomorphisms from bounded above complexes of flats (and the bottom row the canonical map), our problem reduces to proving

the commutativity of

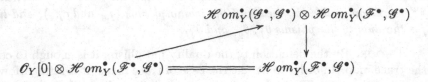

$$\mathcal{H}om_Y^\bullet(\mathcal{G}^\bullet, \mathcal{G}^\bullet) \otimes \mathcal{H}om_Y^\bullet(\mathcal{F}^\bullet, \mathcal{G}^\bullet)$$

$$\mathcal{O}_Y[0] \otimes \mathcal{H}om_Y^\bullet(\mathcal{F}^\bullet, \mathcal{G}^\bullet) = \mathcal{H}om_Y^\bullet(\mathcal{F}^\bullet, \mathcal{G}^\bullet)$$

This is obviously commutative, even on the level of complexes. ∎

In order to extend Theorem 4.4.3 to the case of arbitrary locally noetherian schemes, we need to define $\xi_{f,g}$ in this more general setting. To do this, we first prove that (without properness hypotheses) the isomorphism $\xi_{f,g} : \omega_f \otimes f^*\omega_g \simeq \omega_{gf}$ is compatible with respect to any base change $Z' \to Z$ where Z' is another noetherian scheme admitting a dualizing complex.

THEOREM 4.4.4. *Let $f : X \to Y$, $g : Y \to Z$ be CM maps with pure relative dimensions n and m respectively, where X, Y, and Z are noetherian schemes which admit a dualizing complex. Let Z' be a Z-scheme which is noetherian and admits a dualizing complex, and let f' and g' be the respective base changes of f and g with respect to $Z' \to Z$. The base change of the isomorphism $\xi_{f,g}$ is the isomorphism $\xi_{f',g'}$.*

PROOF. We may certainly assume that X, Y, Z, Z' are affine. It is easy to construct a commutative diagram

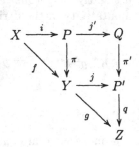

in which the upper right square is cartesian, i and j are closed immersions, and π', q are smooth affine maps with pure relative dimensions M, N respectively. We can express $\xi_{f,g}$ as a map

$$(4.4.12) \qquad \mathcal{E}xt_P^{N-n}(i_*\mathcal{O}_X, \omega_{P/Y}) \otimes_P \pi^*\mathcal{E}xt_{P'}^{M-m}(j_*\mathcal{O}_Y, \omega_{P'/Z})$$

$$\downarrow$$

$$\mathcal{E}xt_Q^{(N+M)-(n+m)}((j'i)_*\mathcal{O}_X, \omega_{Q/Z})$$

Since our schemes are affine, we can 'identify' this pairing of $\mathscr{E}xt$'s with a map of groups

$$(4.4.13) \qquad \operatorname{Ext}_P^{N-n}(i_*\mathscr{O}_X, \omega_{P/Y}) \otimes \operatorname{Ext}_{P'}^{M-m}(j_*\mathscr{O}_Y, \omega_{P'/Z})$$

$$\downarrow$$

$$\operatorname{Ext}_Q^{(N+M)-(n+m)}((j'i)_*\mathscr{O}_X, \omega_{Q/Z})$$

Without the intervention of signs, we can view (4.4.13) as a map of groups

$$(4.4.14) \qquad \operatorname{Hom}_{\mathbf{D}(P)}(i_*\mathscr{O}_X[n], \omega_{P/Y}[N]) \otimes \operatorname{Hom}_{\mathbf{D}(P')}(j_*\mathscr{O}_Y[m], \omega_{P'/Z}[M])$$

$$\downarrow$$

$$\operatorname{Hom}_{\mathbf{D}(Q)}((j'i)_*\mathscr{O}_X[n+m], \omega_{Q/Z}[N+M])$$

(if $\omega_{P/Y}[N] \to \mathscr{I}^\bullet$ is a chosen injective resolution, we identify a morphism $i_*\mathscr{O}_X[n] \to \omega_{P/Y}[N]$ with a morphism $i_*\mathscr{O}_X \to \mathscr{I}^{\bullet-n}$).

There is another definition of a map such as (4.4.14), as follows. Let $\varphi : i_*\mathscr{O}_X[n] \to \omega_{P/Y}[N]$ and $\psi : j_*\mathscr{O}_Y[m] \to \omega_{P'/Z}[M]$ be morphisms in $\mathbf{D}(P)$ and $\mathbf{D}(P')$ respectively. Without the intervention of signs, we then get maps in $\mathbf{D}(Q)$ given by

$$\varphi' : (j'i)_*\mathscr{O}_X[n+m] \longrightarrow j'_*\omega_{P/Y}[N+m] =\!=\!= j'_*{j'}^*\omega_{Q/P'}[N+m]$$

$$\|$$

$$\omega_{Q/P'}[N] \overset{L}{\otimes} j'_*\mathscr{O}_P[m]$$

(since $\omega_{Q/P'}$ is $\mathscr{O}_Q$-flat) and

$$\psi' : \omega_{Q/P'}[N] \overset{L}{\otimes} j'_*\mathscr{O}_P[m] \xrightarrow{1 \overset{L}{\otimes} {\pi'}^*(\psi)} \omega_{Q/P'}[N] \overset{L}{\otimes} {\pi'}^*\omega_{P'/Z}[M] \xrightarrow{\zeta'_{\pi',q}} \omega_{Q/Z}[N+M]$$

By composition, we arrive at a map $\psi' \circ \varphi' : (j'i)_*\mathscr{O}_X[n+m] \to \omega_{Q/Z}[N+M]$ in $\mathbf{D}(Q)$. This law of composition gives another way to define a map as in (4.4.14). We claim that these two definitions coincide (though we only need this up to a universal sign depending only on n, m, N, M).

Before giving the details of the comparison between the two constructions of (4.4.14), we make some general remarks. For a residual complex $K^\bullet$ on Y (such as g^Δ applied to a residual complex on Z), the quasi-isomorphism $\omega_{P/Y}[N] \to \mathscr{H}om_P^\bullet(\pi^*K^\bullet, \pi^\Delta K^\bullet)$ from Lemma 4.1.2 can be used to compute $i^b(\omega_{P/Y}[N])$ (and likewise for a residual complex on Z, with j and q in place of i and π). In other words, we want the canonical map

$$(4.4.15) \qquad \mathscr{H}om_P^\bullet(i_*\mathscr{O}_X, \mathscr{H}om_P^\bullet(\pi^*K^\bullet, \pi^\Delta K^\bullet)) \to i^b(\mathscr{H}om_P^\bullet(\pi^*K^\bullet, \pi^\Delta K^\bullet))$$

to be a quasi-isomorphism. Although $\mathscr{H}om_P^\bullet(\pi^*K^\bullet, \pi^\Delta K^\bullet)$ is a bounded complex with coherent cohomology sheaves, the individual terms of the complex are usually not even quasi-coherent, so it is not at all clear if these terms are $\mathscr{H}om_P(i_*\mathscr{O}_X, \cdot)$-acyclic. Thus, some argument is needed in order to show that (4.4.15) is a quasi-isomorphism. By (3.3.20) and the flatness of $f = \pi \circ i$ and π (so $\pi^*(\cdot)$ is i^*-acyclic), there is an isomorphism in $\mathbf{D}(X)$ given by

(4.4.16)

$$i^\flat(\mathscr{H}om_P^\bullet(\pi^*K^\bullet, \pi^\Delta K^\bullet)) \xleftarrow{\;\simeq\;} \mathbf{R}\mathscr{H}om_X^\bullet(Li^*\pi^*K^\bullet, i^\flat\pi^\Delta K^\bullet)$$

$$\rotatebox{90}{=}$$

$$\mathscr{H}om_X^\bullet(f^*K^\bullet, \mathscr{H}om_P^\bullet(i_*\mathscr{O}_X, \pi^\Delta K^\bullet))$$

$$\rotatebox{90}{=}$$

$$\mathscr{H}om_P^\bullet(\pi^*K^\bullet, \mathscr{H}om_P^\bullet(i_*\mathscr{O}_X, \pi^\Delta K^\bullet))$$

$$\rotatebox{90}{=}$$

$$\mathscr{H}om_P^\bullet(i_*\mathscr{O}_X, \mathscr{H}om_P^\bullet(\pi^*K^\bullet, \pi^\Delta K^\bullet))$$

and this is readily checked to be an inverse to the map (4.4.15).

Another fact we need in the comparison of the two constructions of (4.4.14) is that the diagram of complexes

(4.4.17)

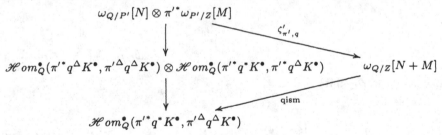

commutes up to homotopy (where $K^\bullet$ is a choice of residual complex on Z). To prove this, we use Lemma 3.3.2 (for separated smooth maps) and the definition of the top and bottom maps in the left column via (4.3.18) to deduce that

$$
\begin{CD}
\pi'^\sharp q^\sharp K^\bullet @>>> (q\pi')^\sharp K^\bullet \\
@VVV \\
\pi'^\sharp q^\Delta K^\bullet @>>> \pi'^\Delta q^\Delta K^\bullet @>>> (q\pi')^\Delta K^\bullet
\end{CD}
$$

commutes in the derived category, and thus commutes up to homotopy (by Lemma 2.1.4). The homotopy-commutativity of (4.4.17) now follows, since

$$\mathrm{Hom}^\bullet(C^\bullet, \cdot)$$

takes homotopic maps to homotopic maps.

A related homotopy-commutative fact we need is that

(4.4.18)

$$\omega_{Q/P'}[N] \xrightarrow{\text{qism}} \mathscr{H}om_Q^\bullet(\pi'^*q^\Delta K^\bullet, \pi'^\Delta q^\Delta K^\bullet)$$

$$j'_*\omega_{P/Y}[N] \xrightarrow[\text{qism}]{} \mathscr{H}om_P^\bullet(\pi^*g^\Delta K^\bullet, \pi^\Delta g^\Delta K^\bullet)$$

is homotopy-commutative, where the left column is defined by adjointness to the base change morphism for relative differentials and the right column is induced by "pass to map on $\mathscr{I}_P$-torsion" (via $g^\Delta \simeq j^\Delta q^\Delta$ and (3.3.17)). To verify this homotopy-commutativity, we are essentially in the situation of Lemma 4.1.1, except that we want to make sure that (4.4.18) is homotopy-commutative and not just commutative in $\mathbf{D}(Q)$. This follows if we verify that the outside edge of the diagram of complexes

(4.4.19)

$$j'_*\pi^\sharp j^\Delta K'^\bullet \longrightarrow j'_*\pi^\Delta j^\Delta K'^\bullet$$

$$\mathscr{H}om_Q(j'_*\mathscr{O}_P, \pi'^\sharp K'^\bullet) \longrightarrow j'_*j'^\Delta\pi'^\Delta K'^\bullet$$

$$\pi'^\sharp K'^\bullet \longrightarrow \pi'^\Delta K'^\bullet$$

is homotopy-commutative, where $K'^\bullet$ is a residue complex on P' (such as $q^\Delta K^\bullet$ above). But $\pi'^\Delta K'^\bullet$ is a bounded complex of injectives, so we just need to check that (4.4.19) commutes in $\mathbf{D}(Q)$. This follows from Lemma 3.3.2 and Theorem 3.3.1(2), which ensure the commutativity of

$$\pi^\sharp j^\flat \longrightarrow \pi^\Delta j^\Delta$$

$$j'^\flat\pi'^\sharp \longrightarrow j'^\Delta\pi'^\Delta$$

in $\mathbf{D}(P)$, when evaluated on any residual complex on P'.

We now come to the comparison of the two proposed definitions of (4.4.14). Fix a residual complex $K^\bullet$ on Z. Note that we may work Zariski locally, since the conclusion concerns a comparison of maps of sheaves. The idea is to realize descriptions of both constructions as the sides of a diagram which commutes in the derived category. Since $\pi^*(\cdot)$ is always i^*-acyclic (as π and $\pi \circ i = f$ are flat), we may use (4.3.6), (4.3.19), and (4.4.15) to see that, upon replacing Q and P' by suitable Zariski opens (without loss of generality), the natural composite map

$$i_*\mathscr{O}_X[n] \xrightarrow{\varphi} \omega_{P/Y}[N] \xrightarrow{\text{qism}} \mathscr{H}om_P^\bullet(\pi^*g^\Delta K^\bullet, \pi^\Delta g^\Delta K^\bullet)$$

in $\mathbf{D}(P)$ is represented by a map of sheaves

$$\varphi' : i_* \mathcal{O}_X \to \mathcal{H}om_P^{-n}(\pi^* g^\Delta K^\bullet, \pi^\Delta g^\Delta K^\bullet)$$

in degree $-n$. This may be identified with a global section of

$$\mathcal{H}om_Q^{-n}(j'_* \pi^* g^\Delta K^\bullet, (j'i)_* i^\Delta \pi^\Delta g^\Delta K^\Delta).$$

In a similar manner, we may suppose that the derived category map

$$\psi : j_* \mathcal{O}_Y[m] \to \omega_{P'/Z}[M]$$

is represented by a map of sheaves

$$\psi' : j_* \mathcal{O}_Y \to \mathcal{H}om_{P'}^{-m}(q^* K^\bullet, q^\Delta K^\bullet)$$

in degree $-m$, so $\pi'^* \psi'$ gives rise to a section of

$$\mathcal{H}om_Q^{-m}(\pi'^* q^* K^\bullet, j'_* \pi^* g^\Delta K^\bullet),$$

using (3.3.17) and the isomorphism $j^\Delta q^\Delta \simeq g^\Delta$. If we appeal to the triple composite compatibility of (3.2.3), then we obtain the commutativity of

$$
\begin{array}{ccccc}
f^\Delta g^\Delta & \!\!=\!\!\!=\!\! & i^\Delta \pi^\Delta j^\Delta q^\Delta & & \\
\| & & & \searrow & \\
(gf)^\Delta & \!\!=\!\!\!=\!\! & (j'i)^\Delta (q\pi')^\Delta & \!\!=\!\!\!=\!\! & i^\Delta j'^\Delta \pi'^\Delta q^\Delta
\end{array}
$$

It then follows that the natural pairing

$$\mathcal{H}om_Q^{-n}(j'_* \pi^* g^\Delta K^\bullet, (j'i)_* i^\Delta \pi^\Delta g^\Delta K^\bullet) \otimes \mathcal{H}om_Q^{-m}(\pi'^* q^* K^\bullet, j'_* \pi^* g^\Delta K^\bullet)$$

$$\downarrow$$

$$\mathcal{H}om_Q^{-n-m}(\pi'^* q^* K^\bullet, (j'i)_* i^\Delta \pi^\Delta g^\Delta K^\bullet)$$

$$\downarrow \simeq$$

$$\mathcal{H}om_Q((j'i)_* \mathcal{O}_X, \mathcal{H}om_Q^{-n-m}((q\pi')^* K^\bullet, (q\pi')^\Delta K^\bullet))$$

takes $j'_*(\varphi) \otimes \pi'^*(\psi')$ to a representative of the image of $\varphi \otimes \psi$ under (4.4.13). We now wish to compare this somewhat more explicit description of (4.4.13) with the 'derived category Hom' construction of (4.4.14); this comparison will require the homotopy-commutativity of (4.4.17).

Let $\mathcal{F}^\bullet \to j_* \mathcal{O}_Y[m]$ be a flat resolution, so the $\mathcal{O}_Q$-flatness of $\omega_{Q/P'}$ and the isomorphism of sheaves

$$\omega_{Q/P'} \otimes j'_* \mathcal{O}_P \simeq j'_* j'^* \omega_{Q/P'} \simeq j'_* \omega_{P/Y}$$

leads to a quasi-isomorphism

$$\omega_{Q/P'}[N] \otimes \pi'^* \mathcal{F}^\bullet \to j'_* \omega_{P/Y}[N+m]$$

(defined without the intervention of signs). Also, the derived category composite

$$\lambda : \mathscr{F}^{\bullet} \xrightarrow{\;\text{qism}\;} j_{*}\mathscr{O}_{Y}[m] \xrightarrow{\;\psi\;} \omega_{P'/Z}[M] \xrightarrow{\;\text{qism}\;} \mathscr{H}om_{P'}^{\bullet}(q^{*}K^{\bullet}, q^{\Delta}K^{\bullet})$$

gives rise to a derived category morphism

$$\pi'^{*}(\lambda) : \pi'^{*}\mathscr{F}^{\bullet} \to \pi'^{*}\mathscr{H}om_{P'}^{\bullet}(q^{*}K^{\bullet}, q^{\Delta}K^{\bullet}) \to \mathscr{H}om_{Q}^{\bullet}(\pi'^{*}q^{*}K^{\bullet}, \pi'^{*}q^{\Delta}K^{\bullet}).$$

Now consider the derived category diagram in $\mathbf{D}(Q)$

(4.4.20)

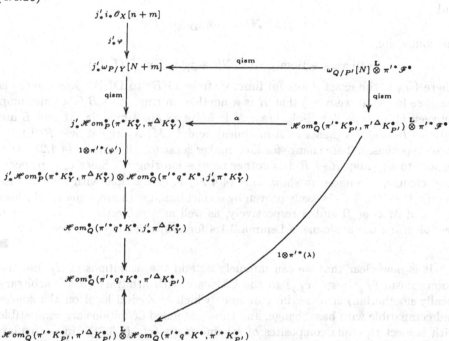

where $K_{Y}^{\bullet} = g^{\Delta}K^{\bullet}$, $K_{P'}^{\bullet} = q^{\Delta}K^{\bullet}$, the map α is induced by "pass to map on $\mathscr{I}_{P}$-torsion," the bottom vertical maps are defined as in (4.4.17), and we identify $\pi'^{*}(\psi')$ as an $\mathscr{I}_{P}$-torsion section of a $\mathscr{H}om_{Q}^{-m}$-sheaf. The top square commutes because (4.4.18) is homotopy commutative, and the bottom triangle easily commutes. Using the homotopy-commutative (4.4.17) we see that going clockwise around (4.4.20) from $j_{*}'\omega_{P/Y}[N+m]$ to

$$\mathscr{H}om_{Q}^{\bullet}(\pi'^{*}q^{*}K^{\bullet}, \pi'^{\Delta}K_{P'}^{\bullet})$$

represents (4.4.14). Meanwhile, going down the left column in (4.4.20) between the same two endpoints represents (4.4.13).

We now have two (equivalent) definitions of (4.4.13): an abstract definition using $\xi_{f,g}$ and residual complexes, and a direct definition in terms of $\mathbf{R}\operatorname{Hom}$'s and composition of maps in the derived category. For the proof of the theorem, we may work in the affine case, but with Q and P' so small that $\omega_{Q/P'}$ and $\omega_{P'/Z}$ are trivial. Choosing trivializations to convert $\zeta_{\pi',q}'$ into the identity on $\mathscr{O}_{Q}$

and using the compatibility of $\mathscr{E}xt$ with respect to flat base change (such as π'), the $\mathbf{R}\operatorname{Hom}$ construction of $\xi_{f,g}$ is, up to a universal sign, a special case of the general pairing

$$(4.4.21) \qquad \operatorname{Ext}_B^i(L, M) \otimes_A \operatorname{Ext}_A^j(M, N) \to \operatorname{Ext}_A^{i+j}(L, N)$$

for a ring map $A \to B$, an A-module N, and B-modules M, L (viewed as A-modules via $A \to B$) which associates to any

$$\varphi \in \operatorname{Ext}_B^i(L, M) = \operatorname{Hom}_{\mathbf{D}(B)}(L, M[i])$$

and

$$\psi \in \operatorname{Ext}_A^j(M, N) = \operatorname{Hom}_{\mathbf{D}(A)}(M, N[j])$$

the composite

$$\psi[i] \circ \varphi_A \in \operatorname{Hom}_{\mathbf{D}(A)}(L, N[i+j]) = \operatorname{Ext}_A^{i+j}(L, N),$$

where $(\cdot)_A$ is the exact 'forgetful functor' from $\mathbf{D}(B)$ to $\mathbf{D}(A)$. Suppose (as is the case for what we need) that R is a noetherian ring, $A \to B$ is a finite map between finite type flat R-algebras, N is finite as an A-module, M and L are finite as B-modules (hence as A-modules), and L, M, N are flat over R. Under these hypotheses, 'base change for Ext' makes sense for all terms in (4.4.21) with respect to any map $R \to R'$ to another noetherian ring R'. Since $\zeta'_{\pi',q}$ respects base change, it suffices to show that (4.4.21) is compatible with such a base change $R \to R'$. This is easily proven by a calculation with *projective* resolutions of L and M over B and A respectively, as well as a projective resolution of L over A, using the analogue of Lemma 2.1.4 for projective modules over a ring. ∎

It is now clear that we can uniquely extend the definitions of ω_f and the isomorphism $\xi_{f,g}$ (resp. $c_{f,g}$) to the CM case with arbitrary (resp. arbitrary locally noetherian) schemes in a manner which is Zariski local on the source and compatible with base change, and these extended definitions are compatible with respect to triple composites of scheme maps (cf. (4.3.7)). In particular, for CM maps $f : X \to Y$ with pure relative dimension, the formation of the dualizing sheaf ω_f is *canonically* 'insensitive' to étale localization on X, generalizing the well-known analogue in case f is smooth. By means of base change and direct limit arguments, we can uniquely define $\gamma_f : \mathbf{R}^n f_*(\omega_f) \to \mathscr{O}_Y$ for arbitrary proper CM maps $f : X \to Y$ with pure relative dimension n over an arbitrary base, compatibly with base change and the earlier definition in the locally noetherian case. This leads us to the following generalization of Corollary 3.6.6.

COROLLARY 4.4.5. *If $f : X \to Y$ and $g : Y \to Z$ are proper CM maps with pure relative dimensions m and n, then the diagram (3.6.14) commutes, where we use $\xi_{f,g}$ is the top row. The map γ_f is always surjective and when f has geometrically reduced, geometrically connected fibers then γ_f is an isomorphism.*

PROOF. The commutativity of (3.6.14) with $\xi_{f,g}$ in place of $\zeta'_{f,g}$ over an arbitrary base follows by direct limits and base change from the case of finite type $\mathbf{Z}$-schemes (or more generally, noetherian schemes admitting a dualizing complex), which is exactly the result of applying H^0 to the diagram (4.4.10) studied in the proof of Theorem 4.4.3. The surjectivity of γ_f follows by base change to the case where Y is the spectrum of a field. In order to show that γ_f is an isomorphism when f has geometrically connected, geometrically reduced fibers, we just need to show that $R^n f_*(\omega_f)_y$ is a quotient of $\mathscr{O}_{Y,y}$ for all $y \in Y$. By base change from the noetherian case and a right exactness argument, $R^n f_*(\omega_f)$ is at least finitely presented on Y. Thus, by Nakayama's Lemma it suffices to establish the result on fibers. That is, when $Y = \mathrm{Spec}(k)$ for a field k, we want $\dim_k H^n(X, \omega_f) = 1$. By Theorem 5.1.2 (whose proof does not use Corollary 4.4.5), $H^n(X, \omega_f)$ is canonically isomorphic to $H^0(X, \mathscr{O}_X)^\vee$, which is 1-dimensional by our geometric connectedness and geometric reducedness hypotheses. ∎

We end this chapter by briefly discussing the relation between fiber products and trace maps. Since the examples in the next chapter and in the appendices do not make use of this relationship, and this book is already long enough, we leave it to the interested reader to fill in the details along the lines indicated below (also see [D, Appendix $(a),(d)$]).

Let S be a scheme and let $f_i : X_i \to S$ be a finite set of CM S-schemes with pure relative dimensions n_i. Let $\pi : X \to S$ be an S-scheme equipped with S-maps $\pi_i : X \to X_i$ identifying X with the fiber product of the X_i's over S. In particular, X is proper CM over S with pure relative dimension $n = \sum n_i$. Since $\xi_{f,g} : \omega_f \otimes f^*\omega_g \simeq \omega_{gf}$ is compatible with triple composites as in (4.3.7), by the method of proof of the 'general associativity law' for abstract groups one checks that upon *ordering* the indices as $i_1, \ldots, i_m$, there is an unambigously defined isomorphism

$$(4.4.22) \qquad \pi_{i_1}^*(\omega_{f_{i_1}}) \otimes \cdots \otimes \pi_{i_m}^*(\omega_{f_{i_m}}) \simeq \omega_\pi.$$

In the smooth case, this is exactly the canonical map obtained from wedge products of relative differential forms (by the final part of Theorem 4.3.3). Using (4.4.22), pullback and cup product give rise to a canonical map of $\mathscr{O}_S$-modules

$$(4.4.23) \qquad q : R^{n_{i_1}} f_{i_1*}(\omega_{f_{i_1}}) \otimes \cdots \otimes R^{n_{i_m}} f_{i_m*}(\omega_{f_{i_m}}) \to R^n \pi_*(\omega_\pi).$$

This is all compatible with base change on S. When the f_i's are all proper, it makes sense to ask about the relationship between $\gamma_f \circ q$ and $\gamma_{f_{i_1}} \otimes \cdots \otimes \gamma_{f_{i_m}}$. Even without properness, it makes sense to ask about the dependence of (4.4.22) on the ordering of the factors on the left side (think of the smooth case). For both questions the essential case is $m = 2$, so we now suppose this is the case (with the 'factor' schemes labelled as X_1 and X_2 relative to our chosen ordering).

Based on a calculation with differential forms in the smooth case, we expect that permutation of the factors in (4.4.22) with $m = 2$ should introduce a sign of $(-1)^{n_1 n_2}$. The explicit description of $\xi_{f,g}$ (in the proof of Theorem 4.4.4) as a pairing of derived category Hom's is well-suited to analyzing this point. Working locally, we may choose closed immersions $X_j \hookrightarrow P_j$ over S, where P_j is smooth and separated over S with pure relative dimension N_j. Since the diagram

$$\omega_{P_1 \times_S P_2/P_1}[N_2] \overset{\mathbf{L}}{\otimes} \omega_{P_1 \times_S P_2/P_2}[N_1] \longrightarrow \omega_{P_1 \times_S P_2/P_2}[N_1] \overset{\mathbf{L}}{\otimes} \omega_{P_1 \times_S P_2/P_1}[N_2]$$

$$\omega_{P_1 \times_S P_2/S}[N_1 + N_2] \xrightarrow{\quad (-1)^{N_1 N_2} \quad} \omega_{P_1 \times_S P_2/S}[N_1 + N_2]$$

commutes, we can apply the 'derived category Hom' description of $\xi_{f,g}$ to the special case of the composites

$$X \to X_1 \to S, \quad X \to X_2 \to S$$

so as to reduce ourselves to chasing signs when 'composing' two morphisms $\varphi_j : \mathscr{O}_{X_j}[n_j] \to \mathscr{O}_{P_j}[N_j]$ in the sense of the second definition of (4.4.14). The two maps

$$\mathscr{O}_{X_1 \times_S X_2}[n_1 + n_2] \to \mathscr{O}_{P_1 \times_S P_2}[N_1 + N_2]$$

obtained in this way depend on the order of composition, and the problem is to show these two maps agree up to an explicitly determined universal sign. Meanwhile, the crucial step required in the analysis of (4.4.23) is to prove that, under suitable flatness and noetherian hypotheses, cup products on higher direct images can be recovered from projection formulas in the derived category (so we can think about things entirely in the derived category context and Corollary 4.4.5 becomes useful). The idea is to use derived category projection formulas to define a general 'pairing' of total direct images which recovers a general pairing of higher direct image sheaves satisfying the axioms uniquely characterizing the classical cup product. The details are lengthy (as far as the author is aware).

CHAPTER 5

Examples

In this final chapter, we make the abstract derived category duality theorem (i.e., Theorem 4.3.1) somewhat concrete. More precisely, we recover from the general theory some of the most widely used consequences for duality of higher direct images; a little care even gives us results without any noetherian hypotheses on the base scheme. A special case of this duality result for higher direct image sheaves is the fact that for a proper CM map $f : X \to \mathrm{Spec}(k)$ which has pure relative dimension n over a field k, the cup product pairing

$$\mathrm{H}^i(X, \mathscr{F}^\vee) \times \mathrm{H}^{n-i}(X, \mathscr{F} \otimes \omega_f) \to \mathrm{H}^n(X, \omega_f) \xrightarrow{\gamma_f} k$$

is a perfect duality of finite dimensional vector spaces for any locally free coherent $\mathscr{F}$ on X. With $\mathscr{F} = \mathscr{O}_X$ and $i = 0$, this justifies the final step in the proof of Corollary 4.4.5. In general, we want to recover that (1.1.2) is an isomorphism under local freeness hypotheses on the higher direct images, as well as a suitable analogue with smoothness relaxed to the CM condition. This issue is handled in §5.1.

For reduced proper curves over an algebraically closed field, we also want to deduce the classical results of Rosenlicht that describe the dualizing sheaf and trace map in terms of 'regular differentials' and residues. Although this special case of duality theory can be established by direct arguments of a more elementary nature [AK1, VIII], there are some subtle technical points which must be verified in order to justify the use of this concrete formulation "on the non-smooth geometric fibers" when one begins life in a relative situation, such as a semistable curve over a discrete valuation ring, with smooth and geometrically connected generic fiber (e.g., as in [M], [R]). This matter will be explained in greater detail in §5.2.

5.1. Higher Direct Images

Before we state the duality theorem in terms of higher direct images, we digress to record a lemma concerning cohomology and base change without noetherian assumptions. Let $f : X \to Y$ be a proper, finitely presented map of schemes, let $\mathscr{F}$ be a Y-flat finitely presented $\mathscr{O}_X$-module, and let $m \in \mathbf{Z}$ be an integer. We are interested in the property that the $R^j f_*(\mathscr{F})$'s are locally free of finite rank as $\mathscr{O}_Y$-modules for all $j > m$. By Grothendieck's theory of cohomology and base change [EGA, III$_1$, §7], for locally noetherian Y these higher direct images are locally free of finite rank on Y for all $j > m$ if and only if the natural base change maps

(5.1.1) $$R^j f_*(\mathscr{F})_y \otimes_{\mathscr{O}_{Y,y}} k(y) \to H^j(X_y, \mathscr{F}_y)$$

are surjective for all $j \geq m$ and all $y \in Y$, in which case formation of the $R^j f_*(\mathscr{F})$'s commute with locally noetherian base change for all $j \geq m$ (so in particular the maps (5.1.1) are isomorphisms for all $j \geq m$). Actually, we have compatibility with *arbitrary* base change for $j \geq m$, since Čech cohomology can be used to compute the cohomology of quasi-coherent sheaves on quasi-compact separated schemes, as well as 'pullback maps' in cohomology between such schemes, so direct limit arguments with Čech complexes can be used to deduce compatibility with arbitrary base change for $j \geq m$. In (5.1.1) we indulge in a slight abuse of notation, with $R^j f_*(\mathscr{F})_y$ denoting the stalk of the direct image sheaf at y and $\mathscr{F}_y$ denoting the pullback of $\mathscr{F}$ to the fiber $X_y = f^{-1}(y)$; this should cause no confusion. Grothendieck's theory also ensures (for locally noetherian Y and fixed j) the *openness* of the locus of $y \in Y$ where (5.1.1) is surjective. One can also extract these results from the arguments in [**H**, III, §12], once one inputs the coherence of higher direct images without projectivity assumptions [**EGA**, III$_1$, 3.2.1].

With a locally noetherian base there are more precise results known; e.g., relations between the surjectivity of (5.1.1) for fixed j, y and the local freeness of $R^{j+1} f_*(\mathscr{F})$ near y. However, the above situation that simultaneously considers all $j \geq m$ (or all $j > m$) is the one in which we are interested. We first want to extend, to arbitrary base schemes, the surjectivity criterion (5.1.1) for local freeness of higher direct images. This is essentially an exercise in using direct limits to reduce to the noetherian case, but we need to be a little careful in order to 'descend' the local freeness condition on higher direct images through direct limits:

LEMMA 5.1.1. *Let $m \in \mathbf{Z}$, $f : X \to Y$, and $\mathscr{F}$ be as above, with Y an arbitrary scheme. The higher direct images $R^j f_*(\mathscr{F})$ are Y-flat for all $j > m$ if and only if the natural maps (5.1.1) are surjective for all $y \in Y$ and all $j \geq m$. When these conditions hold, then $R^j f_*(\mathscr{F})$ is locally free of finite rank on Y for all $j > m$ and is of formation compatible with arbitrary base change on Y for $j \geq m$. In such a situation, there exists a Zariski covering $\{U_\alpha\}$ of Y, proper maps $f_\alpha : X_\alpha \to Y_\alpha$ of noetherian schemes, and Y_α-flat coherent $\mathscr{O}_{X_\alpha}$-modules $\mathscr{F}_\alpha$ such that $R^j f_{\alpha *}(\mathscr{F}_\alpha)$ is locally free on Y_α for all $j > m$ and the setup over U_α is a base change of the data over Y_α (so $R^m f_*(\mathscr{F})$ is of finite presentation).*

PROOF. We may assume $Y = \mathrm{Spec}(A)$ is affine and then standard direct limit arguments from [**EGA**, IV$_3$, §8–§11] ensure that for a sufficiently large noetherian subring $A_0 \subseteq A$ there exists a proper

$$f_0 : X_0 \to \mathrm{Spec}(A_0)$$

and an A_0-flat coherent $\mathscr{O}_{X_0}$-module $\mathscr{F}_0$ inducing f and $\mathscr{F}$ after base change. Let $\{A_i\}$ denote the set of finitely generated A_0-subalgebras of A, partially ordered by inclusion. Let $f_i : X_i \to Y_i$ and $\mathscr{F}_i$ denote the base change by $\mathrm{Spec}(A_i) \to \mathrm{Spec}(A_0)$. For any $y \in Y$, let $y_i \in \mathrm{Spec}(A_i)$ be the image of y. Since

Čech cohomology can be used to compute both the cohomology of quasi-coherent sheaves on quasi-compact, separated schemes, as well as 'pullback maps', it is clear that the direct limit of the maps

$$(5.1.2) \quad \mathrm{H}^j(X_i, \mathscr{F}_i) \otimes_{A_i} k(y_i) = \mathrm{R}^j f_{i*}(\mathscr{F}_i)_y \otimes_{\mathcal{O}_{Y_i,y_i}} k(y_i) \to \mathrm{H}^j((X_i)_{y_i}, (\mathscr{F}_i)_{y_i})$$

(which are compatible with change in i) is exactly the map (5.1.1). Combining this with the compatible isomorphisms

$$\mathrm{H}^j((X_i)_{y_i}, (\mathscr{F}_i)_{y_i}) \otimes_{k(y_i)} k(y_{i'}) \simeq \mathrm{H}^j((X_{i'})_{y_{i'}}, (\mathscr{F}_{i'})_{y_{i'}})$$

for $i' \geq i$, if we let V_i (resp. V) denote the locus of points on Y_i (resp. Y) where the jth higher direct image surjects onto the jth cohomology of the fiber for all $j \geq m$, then we conclude that $y \in V$ if and only if $y_i \in V_i$ for some (hence all) large i, possibly depending on y; here we have used that the jth fiber cohomologies vanish for large j uniform in all fibers over the quasi-compact Y and Y_i's. Similarly, we see that the preimage of V_i under $Y_{i'} \to Y_i$ (for $i' \geq i$) is contained in $V_{i'}$.

Since the V_i's are open, hence constructible, it follows from [**EGA**, IV$_3$, 8.3.4] that $V_i = Y_i$ for large i if and only if $V = Y$. Thus, when $V = Y$ we can use Grothendieck's theory over the noetherian $V_i = Y_i$ for large i to deduce the desired local freeness (of finite rank) and base change compatibility for higher direct images over Y from the analogous result over Y_i. Conversely, assuming $\mathrm{R}^j f_*(\mathscr{F})$ is Y-flat for all $j > m$, we want to prove that (5.1.1) is surjective for all $j \geq m$ and all $y \in Y$. More generally, we will prove directly that the $\mathrm{R}^j f_*(\mathscr{F})$'s commute with arbitrary base change on Y for $j \geq m$. This kind of problem is perhaps most efficiently proven using Spaltenstein's formulation of the theory of total derived functors in derived categories without boundedness conditions [**Sp**], but we give an ad hoc argument that works in our situation.

With $Y = \mathrm{Spec}(A)$, we want to prove that for any A-algebra A', the natural map

$$A' \otimes_A \mathrm{H}^j(X, \mathscr{F}) \to \mathrm{H}^j(X', \mathscr{F}')$$

is an isomorphism for all $j \geq m$ when $\mathrm{H}^j(X, \mathscr{F})$ is A-flat for all $j > m$; here X' and $\mathscr{F}'$ denote the base change by $\mathrm{Spec}(A') \to \mathrm{Spec}(A)$. Since $\pi : X' \to X$ is an affine map, there is a natural isomorphism

$$\mathrm{H}^j(X', \mathscr{F}') \simeq \mathrm{H}^j(X, \pi_* \mathcal{O}_{X'} \otimes_{\mathcal{O}_X} \mathscr{F}).$$

Thus, it suffices (following the method of Grothendieck) to prove the natural A-linear map

$$(5.1.3) \qquad M \otimes_A \mathrm{H}^j(X, \mathscr{F}) \to \mathrm{H}^j(X, M \otimes_A \mathscr{F})$$

is an isomorphism for all A-modules M, where $M \otimes_A \mathscr{F}$ denotes the obvious quasi-coherent sheaf on X. By using Čech theory, we see that both sides of (5.1.3) vanish for large j indendepent of M and are of formation compatible with direct limits in M. We can therefore try to prove by descending induction that (5.1.3) is an isomorphism for all $j \geq m$ and all M, and for any fixed j it is sufficient to consider just finitely generated A-modules M.

Fix $j_0 > m$ such that (5.1.3) is an isomorphism for all $j \geq j_0$ and all M. We want to prove that

(5.1.4) $$M \otimes_A \mathrm{H}^{j_0-1}(X, \mathscr{F}) \to \mathrm{H}^{j_0-1}(X, M \otimes \mathscr{F})$$

is an isomorphism for all M. It suffices to consider finitely generated M, so there is an exact sequence of A-modules

$$0 \to K \to F \to M \to 0$$

where F is a finite free A-module (and K might not be finitely generated). Since $\mathscr{F}$ is Y-flat, there is a short exact sequence of quasi-coherent $\mathcal{O}_X$-modules

$$0 \to K \otimes_A \mathscr{F} \to F \otimes_A \mathscr{F} \to M \otimes_A \mathscr{F} \to 0.$$

For $j \geq m$, consider the commutative diagram with exact columns

(5.1.5)
$$
\begin{array}{ccc}
K \otimes_A \mathrm{H}^j(X, \mathscr{F}) & \longrightarrow & \mathrm{H}^j(X, K \otimes_A \mathscr{F}) \\
\downarrow & & \downarrow \\
F \otimes_A \mathrm{H}^j(X, \mathscr{F}) & \longrightarrow & \mathrm{H}^j(X, F \otimes_A \mathscr{F}) \\
\downarrow & & \downarrow \\
M \otimes_A \mathrm{H}^j(X, \mathscr{F}) & \longrightarrow & \mathrm{H}^j(X, M \otimes_A \mathscr{F}) \\
\downarrow & & \downarrow{\scriptstyle \delta} \\
0 & & \mathrm{H}^{j+1}(X, K \otimes_A \mathscr{F})
\end{array}
$$

where the left column is injective on top when $j > m$, due to the assumed A-flatness of $\mathrm{H}^j(X, \mathscr{F})$ for $j > m$. The middle horizontal arrow in (5.1.5) is always an isomorphism. When $j = j_0 > m$ in (5.1.5), the horizontal maps are all isomorphisms by hypothesis, so by injectivity at the top of the left column of (5.1.5) when $j = j_0 > m$, we deduce the injectivity at the top of the right column in this case. Thus, when $j = j_0 - 1$ in (5.1.5) the map δ is 0. We thereby obtain the surjectivity of the bottom horizontal map in (5.1.5) when $j = j_0 - 1$, so by passage to the limit, the map (5.1.4) is surjective for all A-modules M. Applying this to the A-module K, we get the surjectivity of the top horizontal map in (5.1.5) with $j = j_0 - 1$. Now an easy diagram chase yields the injectivity of the bottom horizontal map in (5.1.5) with $j = j_0 - 1$, so we are done. ∎

With Lemma 5.1.1 established, we can prove the higher direct image version of Grothendieck's duality theorem, valid over an arbitrary base scheme.

THEOREM 5.1.2. *Let $f : X \to Y$ be a proper CM map with pure relative dimension n and let $\mathscr{F}$ be a locally free $\mathcal{O}_X$-module with finite rank (so $\mathscr{F}$ is Y-flat). Assume that $\mathrm{R}^j f_*(\mathscr{F})$ is locally free of finite rank on Y for all $j > m$ (so $\mathrm{R}^m f_*(\mathscr{F})$ is of finite presentation). Then for all quasi-coherent $\mathscr{G}$ on Y, the natural map*

(5.1.6) $$\mathrm{R}^{n-j} f_*(\mathscr{F}^\vee \otimes \omega_f \otimes f^* \mathscr{G}) \to \mathscr{H}om_Y(\mathrm{R}^j f_*(\mathscr{F}), \mathscr{G})$$

induced by

$$(5.1.7) \qquad \mathrm{R}^j f_*(\mathscr{F}) \otimes \mathrm{R}^{n-j} f_*(\mathscr{F}^\vee \otimes \omega_f \otimes f^*\mathscr{G}) \longrightarrow \mathrm{R}^n f_*(\omega_f \otimes f^*\mathscr{G})$$

$$\Big\downarrow \simeq$$

$$\mathscr{G} \xleftarrow{\ \gamma_f \otimes 1\ } \mathrm{R}^n f_*(\omega_f) \otimes \mathscr{G}$$

is an isomorphism for all $j \geq m$. In particular, the sheaves $\mathrm{R}^{n-j} f_*(\mathscr{F}^\vee \otimes \omega_f)$ are of formation compatible with arbitrary base change on Y for $j \geq m$ and are locally free for $j > m$.

PROOF. Working over open affines in Y, we can consider the analogous assertion for cohomology modules rather than higher direct image sheaves. Cup products are compatible with pullback, so by Corollary 4.4.5, Lemma 5.1.1, and direct limit arguments we may assume Y is noetherian and admitting a dualizing complex, such as a finite type $\mathbf{Z}$-scheme.

Consider the duality isomorphism

$$\theta_f : \mathrm{R}f_* \mathbf{R}\mathscr{H}om_X^\bullet(\mathscr{F}[0], f^\sharp(\mathscr{G}[0])) \to \mathbf{R}\mathscr{H}om_Y^\bullet(\mathrm{R}f_*(\mathscr{F}[0]), \mathscr{G}[0])$$

from Theorem 3.4.4 (with $\mathscr{F}^\bullet = \mathscr{F}[0]$, $\mathscr{G}^\bullet = \mathscr{G}[0]$). This is defined using γ_f, and by our flatness (or equivalently, local freeness of finite rank) hypotheses for the coherent $\mathscr{F}$ on X and the coherent $\mathrm{R}^j f_*(\mathscr{F})$'s on Y for $j > m$, we conclude that for $j \geq m$ the map $\mathrm{H}^{-j}(\theta_f)$ is an isomorphism

$$\mathrm{H}^{-j}(\theta_f) : \mathrm{R}^{n-j} f_*(\mathscr{F}^\vee \otimes \omega_f \otimes f^*\mathscr{G}) \simeq \mathscr{H}om_Y(\mathrm{R}^j f_*(\mathscr{F}), \mathscr{G}).$$

It suffices to show that, up to a universal sign depending only on j and n, this is exactly the map (5.1.6).

Without loss of generality, we may work over an affine base $\mathrm{Spec}(A)$ and use Hom groups and cohomology modules rather than $\mathscr{H}om$ sheaves and higher direct image sheaves. In these terms, we may allow $\mathscr{F}$ to be an arbitrary quasi-coherent $\mathscr{O}_X$-module and by the above reasoning we arrive at an isomorphism

$$\mathrm{H}^{-j}(\theta_f) : \mathrm{Ext}_X^{n-j}(\mathscr{F}, \omega_f \otimes f^*\mathscr{G}) \simeq \mathrm{Hom}(\mathrm{H}^j(X, \mathscr{F}), \mathscr{G}).$$

This is the map that, up to a universal sign depending only on j and n, associates to

$$\varphi \in \mathrm{Ext}_X^{n-j}(\mathscr{F}, \omega_f \otimes f^*\mathscr{G}) = \mathrm{Hom}_{\mathbf{D}(X)}(\mathscr{F}[0], (\omega_f \otimes f^*\mathscr{G})[n-j])$$

and

$$\psi \in \mathrm{H}^j(X, \mathscr{F}) = \mathrm{Hom}_{\mathbf{D}(X)}(\mathscr{O}_X[0], \mathscr{F}[j])$$

the element $\gamma_f(\varphi[j] \circ \psi) \in \Gamma(Y, \mathscr{G})$, where $\varphi[j] \circ \psi$ is an element in

$$\mathrm{Hom}_{\mathbf{D}(X)}(\mathscr{O}_X[0], (\omega_f \otimes f^*\mathscr{G})[n]) = \mathrm{H}^n(X, \omega_f \otimes f^*\mathscr{G}) \simeq \mathrm{H}^n(X, \omega_f) \otimes_A \Gamma(Y, \mathscr{G}).$$

One checks this by a routine analysis on the level of double complexes, and the undetermined universal sign arises at this step (i.e., one can see by pure thought with double complexes that such a universal sign exists).

Ignoring the composition with the trace map γ_f and the universal sign, this is a special case of the analogously defined canonical pairing

$$(5.1.8) \qquad \mathrm{Ext}^i_Z(\mathscr{H}_1, \mathscr{H}_2) \times \mathrm{H}^j(Z, \mathscr{H}_1) \to \mathrm{H}^{i+j}(Z, \mathscr{H}_2)$$

for $\mathscr{O}_Z$-modules $\mathscr{H}_1$, $\mathscr{H}_2$ on any ringed space Z. By δ-functoriality in $\mathscr{H}_2$ and a universal δ-functor argument, it is clear (by reduction to the case $i = 0$) that this general pairing recovers the cup product when $\mathscr{H}_1$ is locally free of finite rank (via the δ-functorial $\mathrm{Ext}^\bullet_Z(\mathscr{H}_1, \cdot) \simeq \mathrm{H}^\bullet(Z, \mathscr{H}_1^\vee \otimes (\cdot))$). Here we have used the compatibility of (1.3.3) and the snake lemma. Thus, the map $\mathrm{H}^{-j}(\theta_f)$ is indeed the cup product in (5.1.7), up to a universal sign depending only on j and n.

∎

It is a meaningful (but not very interesting) question to ask whether the ordering of the cup product in (5.1.7) is 'correct' in the sense that the corresponding map (5.1.6) is equal to the map defined by the duality morphism $\mathrm{H}^{-j}(\theta_f)$ (over a noetherian base admitting a dualizing complex). This amounts to determining the universal sign implicit in the proof of Theorem 5.1.2. This appears to be a rather overwhelming exercise in keeping track of translations and explications of derived category maps, so we content ourselves here with describing the method.

In order to determine the universal sign, we claim it suffices to consider a single example where $Y = \mathrm{Spec}(k)$ with k a field not of characteristic 2 and where X is a proper CM k-scheme with pure dimension n (e.g., $X = \mathbf{P}^n_k$). Choose one such example. Since all vector spaces are free, we have two maps

$$(5.1.9) \qquad \mathrm{Ext}^{n-j}_X(\mathscr{F}, \omega_f) \to \mathrm{H}^j(X, \mathscr{F})^\vee$$

for quasi-coherent $\mathscr{F}$ on X, one defined by the *isomorphism* $\mathrm{H}^{-j}(\theta_f)$ and one defined by (5.1.8). For $0 \leq j \leq n$, let $\epsilon_{j,n}$ denote the universal sign relating these two maps. By Lemmas 2.1.3 and 2.1.5, sheaf cohomology on quasi-coherent $\mathscr{O}_X$-modules can be computed in terms of derived functors on the category of quasi-coherent $\mathscr{O}_X$-modules. Since $\mathrm{H}^n(X, \omega_f) \neq 0$, we conclude that the functors H^j on quasi-coherent sheaves are *non-zero* for all $0 \leq j \leq n$. Moreover, if $j > 0$ and $\mathscr{F}$ is a quasi-coherent sheaf for which $\mathrm{H}^j(X, \mathscr{F}) \neq 0$, then upon choosing a short exact sequence

$$(5.1.10) \qquad 0 \to \mathscr{F} \to \mathscr{I} \to \mathscr{F}' \to 0,$$

via δ-functoriality we get a surjection $\mathrm{H}^{j-1}(X, \mathscr{F}') \twoheadrightarrow \mathrm{H}^j(X, \mathscr{F})$. Thus,

$$\mathrm{H}^{j-1}(X, \mathscr{F}') \neq 0$$

and the natural map

$$\mathrm{H}^j(X, \mathscr{F})^\vee \to \mathrm{H}^{j-1}(X, \mathscr{F}')^\vee$$

is injective. Notice that both sides of (5.1.9) are δ-functors in $\mathscr{F}$ as we let j vary. Also, it is clear that the two definitions of (5.1.9) determined by θ_f and (5.1.8) are each δ-functorial up to some universal signs depending only on j and n. Paying close attention to translations, (1.3.7) and (1.3.11) are all that is required for the determination of the universal signs measuring the failure of such δ-functoriality (i.e., this has nothing to do with algebraic geometry).

However, there are too many translations floating around for me to keep track of them with confidence. Granting that one has carried out this calculation correctly, we may then apply these signed δ-functoriality results to (5.1.10) to conclude that $\epsilon_{j-1,n}$ is explicitly determined by $\epsilon_{j,n}$. Recalling that k does not have characteristic 2, the calculation of the $\epsilon_{j,n}$'s in general is reduced to the calculation of $\epsilon_{n,n}$. Taking $\mathscr{F} = \omega_f$, we see that $\epsilon_{n,n}\gamma_f$ is the image of 1 under the map

$$\operatorname{Hom}_X(\omega_f, \omega_f) \to \operatorname{H}^n(X, \omega_f)^\vee$$

arising as the H^{-n}-map of

$$\theta_f : \mathbf{R}\operatorname{Hom}^{\bullet}_X(\omega_f, \omega_f[n]) \to \mathbf{R}\operatorname{Hom}^{\bullet}_k(\mathbf{R}\Gamma(\omega_f), k).$$

Thus, computing $\epsilon_{n,n}$ is now just a matter of keeping straight how we explicate total derived functors in terms of ordinary derived functors and how we defined the isomorphism $\mathbf{R}\Gamma(\omega_f[n]) \simeq \operatorname{H}^n(X, \omega_f)[0]$ in the definition of γ_f. We leave it to the energetic reader to determine what happens.

If we weaken the hypothesis on $\mathscr{F}$ in Theorem 5.1.2 to just Y-flatness and finite presentation on X (rather than local freeness of finite rank on X), then we instead get abstract isomorphisms

$$(5.1.11) \qquad \mathscr{E}xt^{n-j}_f(\mathscr{F}, \omega_f \otimes f^*\mathscr{G}) \simeq \mathscr{H}om_Y(\mathbf{R}^j f_*(\mathscr{F}), \mathscr{G})$$

for $j \geq m$, where $\mathscr{E}xt^{\bullet}_f$ denotes the derived functors of $f_*\mathscr{H}om_X(\mathscr{F}, \cdot)$. When $Y = \operatorname{Spec}(k)$ for a field k, we may take $m = 0$ and then (5.1.11) is Serre duality in the CM case (without projectivity assumptions). In the projective case (with arbitrary relative dimension) these isomorphisms are studied in greater detail in [K], where more precise results are obtained. Taking $m = n$ in Theorem 5.1.2, the same method of proof leads us to the following result, which will be useful in our study of curves below.

COROLLARY 5.1.3. *Let* $f : X \to Y$ *be a proper CM map with pure relative dimension* n. *Then for any quasi-coherent* $\mathscr{F}$ *on* X *and any quasi-coherent* $\mathscr{G}$ *on* Y, *the natural composite map*

$$(5.1.12) \quad f_*\mathscr{H}om_X(\mathscr{F}, \omega_f \otimes f^*\mathscr{G}) \longrightarrow \mathscr{H}om_Y(\mathbf{R}^n f_*(\mathscr{F}), \mathbf{R}^n f_*(\omega_f \otimes f^*\mathscr{G}))$$

$$\downarrow^{\gamma_f}$$

$$\mathscr{H}om_Y(\mathbf{R}^n f_*(\mathscr{F}), \mathscr{G})$$

is an isomorphism.

PROOF. Working locally on Y, we can express $\mathscr{F}$ as a direct limit of finitely presented quasi-coherent sheaves. Since $\mathscr{H}om$ takes direct limits in the first variable to inverse limits and $\mathbf{R}^{\bullet}f_*$ (resp. f_*) commutes with direct limits of quasi-coherent sheaves (resp. inverse limits of sheaves), we may assume $\mathscr{F}$ is finitely presented. But the evaluation of $\mathbf{R}^n f_*$ on quasi-coherent sheaves commutes with arbitrary base change (by right exactness), so $\mathbf{R}^n f_*(\mathscr{F})$ is also finitely presented. Now expressing $\mathscr{G}$ as a direct limit of finitely presented quasi-coherent sheaves, we can reduce to the case where $\mathscr{G}$ is also finitely presented, since $\mathscr{H}om_X(\mathscr{H},\cdot)$ takes direct limits in the second variable to direct limits when $\mathscr{H}$ is finitely presented. Now with $\mathscr{F}$ and $\mathscr{G}$ both finitely presented, the compatibility of $\mathrm{R}^n f_*$ with respect to arbitrary base change on quasi-coherent sheaves permits us to use a direct limit argument to reduce to the noetherian case.

In the locally noetherian case, it suffices to show that, up to a universal sign depending only on n, the duality isomorphism

$$\theta'_f : \mathrm{R}f_* \mathrm{R}\mathscr{H}om^{\bullet}_X(\mathscr{F}[0], f^{\sharp}(\mathscr{G}[0])) \simeq \mathrm{R}\mathscr{H}om^{\bullet}_Y(\mathrm{R}f_*(\mathscr{F}[0]), \mathscr{G}[0])$$

induces (5.1.12) on degree $-n$ cohomology. As both maps in question are functorial in $\mathscr{G}$, by left exactness we may assume $\mathscr{G}$ is a quasi-coherent injective. Also, we may assume the base is $Y = \mathrm{Spec}(A)$ and can then work with Hom groups (rather than $\mathscr{H}om$ sheaves) and cohomology modules (rather than higher direct image sheaves). If $\mathscr{F} \to \mathscr{I}^{\bullet}$ and $f^{\sharp}\mathscr{G} \to \mathscr{J}^{\bullet}$ are injective resolutions, then any choice of element

$$\varphi \in \mathrm{Hom}_X(\mathscr{F}, \omega_f \otimes f^*\mathscr{G})$$

'lifts' to a morphism of injective resolutions. This gives rise to an element in $\mathrm{Hom}^{-n}_X(\mathscr{I}^{\bullet}, \mathscr{J}^{\bullet})$ which induces φ on the $(-n)$th homology of

$$\mathrm{R}\,\mathrm{Hom}^{\bullet}_X(\mathscr{F}[0], f^{\sharp}(\mathscr{G}[0])).$$

Since $\mathscr{G}$ is an injective resolution of itself, $\mathrm{H}^{-n}(\theta'_f)$ gives us a map

$$\mathrm{H}^n(X,\mathscr{F}) \to \mathrm{H}^n(X, \omega_f \otimes f^*\mathscr{G}) \simeq \mathrm{H}^n(X, \omega_f) \otimes_A \Gamma(Y, \mathscr{G})$$

in which the first step of this is $\mathrm{H}^n(\varphi)$, up to a universal sign depending only on n. Thus, $\mathrm{H}^{-n}(\theta'_f)$ is as expected. $\blacksquare$

Taking $\mathscr{G} = \mathcal{O}_Y$ in Corollary 5.1.3, we arrive at an isomorphism

$$f_*\mathscr{H}om_X(\mathscr{F}, \omega_f) \simeq \mathscr{H}om_Y(\mathrm{R}^n f_*(\mathscr{F}), \mathcal{O}_Y)$$

which is natural in the quasi-coherent $\mathscr{F}$ and compatible with base change on Y. Passing to global sections, we obtain an isomorphism

$$\mathrm{Hom}_X(\mathscr{F}, \omega_f) \simeq \mathrm{Hom}_Y(\mathrm{R}^n f_*(\mathscr{F}), \mathcal{O}_Y)$$

which is natural in $\mathscr{F}$, compatible with base change on Y, and with $\mathscr{F} = \omega_f$ sends 1 to γ_f. Thus, the sheaf ω_f and the trace map

$$\gamma_f : \mathrm{R}^n f_*(\omega_f) \to \mathcal{O}_Y$$

constitute a universal object for $\mathrm{Hom}_Y(\mathrm{R}^n f_*(\cdot), \mathcal{O}_Y)$ on quasi-coherent $\mathcal{O}_X$-modules, compatible with base change on Y in the obvious manner. This point of view suggests a method for giving 'explicit' constructions of duality theory for proper CM maps: all one needs to do is construct an explicit representing object for the functor $\mathrm{Hom}_Y(\mathrm{R}^n f_*(\cdot), \mathcal{O}_Y)$ on quasi-coherent $\mathcal{O}_X$-modules (or even finitely presented, quasi-coherent $\mathcal{O}_X$-modules). This is carried out in §5.2 for proper reduced curves over an algebraically closed field k, where we show that Rosenlicht's explicit sheaf of 'regular differentials', equipped with a suitable 'residue map', constitutes a universal object. However, one cannot let the abstraction get out of hand, since we must make sure that on the k-smooth locus of the curve this recovers the *canonical* description of the dualizing sheaf in terms of differential forms, as this is essential in applications (see the discussion near (5.2.7) below).

5.2. Curves

Let $Y = \operatorname{Spec}(k)$ for a field k and let $f : X \to Y$ be a proper, generically smooth, reduced k-scheme with pure dimension $n = 1$. For technical reasons, it is convenient to allow that X can be disconnected. Since Serre's conditions (R_0), (S_1) are equivalent to being reduced, we conclude that the 1-dimensional X is CM over k, so the preceding theory applies. In this case, (ω_f, γ_f) is universal for the functor $\mathrm{H}^n(X, \cdot)^\vee$ on quasi-coherent $\mathcal{O}_X$-modules. When k is algebraically closed, we want to make duality on X explicit by using Grothendieck's general theory to derive Rosenlicht's classical description of a universal object in terms of meromorphic differential forms and residues on the normalization $\tilde{X}$ of X ($\tilde{X}$ is a proper, *smooth*, possibly disconnected curve over k).

Let $j : U \hookrightarrow X$ denote the dense open smooth locus, so the general theory gives a canonical isomorphism

$$\omega_f|_U \simeq \Omega^1_{U/k}.$$

The identification of ω_f with a suitable sheaf of 'meromorphic differentials' is suggested by:

LEMMA 5.2.1. *The natural map of quasi-coherent $\mathcal{O}_X$-modules*

$$\omega_f \to j_*(\omega_f|_U) = j_*(\Omega^1_{U/k})$$

is injective. In particular, the $\mathcal{O}_X$-module ω_f has no non-zero sections with non-zero annihilator.

PROOF. The quasi-coherent kernel $\mathcal{K}$ of this map is certainly coherent and is supported in the finitely many closed points of $X - U$. Thus, $\mathrm{R}^1 f_*(\mathcal{K}) = 0$, so

$$\operatorname{Hom}_X(\mathcal{K}, \omega_f) \simeq \operatorname{Hom}_Y(\mathrm{R}^1 f_*(\mathcal{K}), \mathcal{O}_Y) = 0.$$

This forces $\mathcal{K} = 0$.   ∎

Now assume k is algebraically closed. Let $\operatorname{Spec}(K)$ denote the scheme of generic points of X, so $K = \prod K_i$ is just the product of the residue fields $K_i = k(\xi_i)$ at the finitely many generic points ξ_i of X. Let $j_K : \operatorname{Spec}(K) \to X$ be the canonical map and let $\underline{\Omega}^1_{K/k} = (j_K)_* \Omega^1_{K/k}$ denote the quasi-coherent pushforward of the K-module $\Omega^1_{K/k}$; we call $\underline{\Omega}^1_{K/k}$ the sheaf of *meromorphic differentials* on X (recovering the usual notion in the smooth case). It is easy to verify the stalk calculation

$$\underline{\Omega}^1_{K/k,x} = \prod_{x \in \overline{\{\xi_i\}}} \Omega^1_{K_i/k}$$

for all $x \in X$. By Lemma 5.2.1, there is a canonical inclusion of quasi-coherent $\mathscr{O}_X$-modules

$$\omega_f \hookrightarrow \underline{\Omega}^1_{K/k}.$$

What is the image? This will be described in terms of certain meromorphic differentials on the (k-smooth) normalization of X.

Let $\pi : \widetilde{X} \to X$ denote the normalization and $\widetilde{f} = f \circ \pi$, so $\widetilde{U} = \pi^{-1}(U) \to U$ is an isomorphism and there is a *canonical* isomorphism

$$\omega_{\widetilde{f}} \simeq \Omega^1_{\widetilde{X}/k}.$$

From our point of view, the significance of this isomorphism on $\widetilde{X}$ is that its restriction to an isomorphism $\omega_{\widetilde{f}}|_{\widetilde{U}} \simeq \Omega^1_{\widetilde{U}/k}$ is compatible with $\omega_f|_U \simeq \Omega^1_{U/k}$ via the isomorphism $\pi : \widetilde{U} \simeq U$ (this compatibility is an immediate consequence of the general definition of the dualizing sheaf).

Let $\mathrm{Spec}(\widetilde{K})$ denote the scheme of generic points on $\widetilde{X}$, so we can define the quasi-coherent sheaf of meromorphic differentials $\Omega^1_{\widetilde{K}/k}$ on $\widetilde{X}$ and there is a canonical isomorphism of $\mathscr{O}_X$-modules

$$\pi_* \Omega^1_{\widetilde{K}/k} \simeq \Omega^1_{K/k}.$$

We define the $\mathscr{O}_X$-module ω_f^{reg} of *regular differentials* on X by the condition that for open $V \subseteq X$, $\omega_f^{\mathrm{reg}}(V)$ is the $\mathscr{O}_X(V)$-module of meromorphic differentials η on $\pi^{-1}(V) \subseteq \widetilde{X}$ such that that for all $x \in V(k)$ and all $s \in \mathscr{O}_{X,x}$,

$$(5.2.1) \qquad\qquad \sum_{\widetilde{x} \in \pi^{-1}(x)} \mathrm{res}_{\widetilde{x}}(s\eta) = 0,$$

where $\mathrm{res}_{\widetilde{x}}$ denotes the classical residue of a meromorphic differential on a smooth (possibly disconnected) curve over an algebraically closed field; see Appendix B for a discussion of residues on smooth curves over an algebraically closed field. For example, when X has an ordinary double point singularity at x — that is, $\widehat{\mathscr{O}}_{X,x} \simeq k[\![y,z]\!]/(yz)$, or equivalently the strict henselization of $\mathscr{O}_{X,x}$ satisfies $\mathscr{O}_{X,x}^{\mathrm{sh}} \simeq (k[y,z]/(yz))_{(y,z)}^{\mathrm{sh}}$ [**FK**, III, Prop 2.7] — a calculation over the completion or (strict) henselization of the excellent local ring $\mathscr{O}_{X,x}$ shows that (5.2.1) at x requires exactly that η have at worst simple poles at the two points $\widetilde{x}_1$, $\widetilde{x}_2$ over x, with

$$\mathrm{res}_{\widetilde{x}_1}(\eta) + \mathrm{res}_{\widetilde{x}_2}(\eta) = 0.$$

Obviously in general ω_f^{reg} is an $\mathscr{O}_X$-submodule of $\pi_* \Omega^1_{\widetilde{K}/k} = \Omega^1_{K/k}$ and as such we have

$$\pi_* \Omega^1_{\widetilde{X}/k} \subseteq \omega_f^{\mathrm{reg}}.$$

In order to prove that ω_f^{reg} is *coherent* (which is not immediate from the definition), we prove a more precise statement which will be useful later. Viewing $\mathscr{O}_X$ as a subsheaf of $\pi_*(\mathscr{O}_{\widetilde{X}})$ in the canonical manner, we define the *conductor ideal*

$$\mathscr{C} = \mathrm{ann}_{\pi_* \mathscr{O}_{\widetilde{X}}}(\pi_* \mathscr{O}_{\widetilde{X}}/\mathscr{O}_X).$$

This is a coherent ideal sheaf in $\pi_*(\mathcal{O}_{\widetilde{X}})$ which happens to lie *inside* of $\mathcal{O}_X$ and as an ideal sheaf on X it cuts out the non-smooth locus $X - U$. We write $\widetilde{\mathscr{C}}$ for the corresponding ideal sheaf on $\widetilde{X}$, so $\mathscr{C} = \pi_*\widetilde{\mathscr{C}}$ and $\pi^*\mathscr{C} \to \mathcal{O}_{\widetilde{X}}$ has image $\widetilde{\mathscr{C}}$. Since $\widetilde{\mathscr{C}}$ is a coherent ideal sheaf on the smooth curve $\widetilde{X}$ and is generically non-zero, it is an *invertible* sheaf. Thus, there is an exact sequence

$$(5.2.2) \qquad 0 \to \Omega^1_{\widetilde{X}/k} \to \Omega^1_{\widetilde{X}/k}(\widetilde{C}) \to \mathcal{O}_{\widetilde{C}} \to 0,$$

where $\mathcal{O}_{\widetilde{C}} = \mathcal{O}_{\widetilde{X}}/\widetilde{\mathscr{C}}$ is the structure sheaf of the finite closed subscheme $\widetilde{C} \subseteq \widetilde{X}$ defined by $\widetilde{\mathscr{C}}$ and $\Omega^1_{\widetilde{X}/k}(\widetilde{C}) = \Omega^1_{\widetilde{X}/k} \otimes \widetilde{\mathscr{C}}^{-1}$ is the sheaf of meromorphic differentials on $\widetilde{X}$ with 'poles no worse than $\widetilde{C}$' (where we view $\widetilde{C}$ as an effective Weil divisor on $\widetilde{X}$). Applying pushfoward by the finite map π, we get an inclusion

$$\pi_*\Omega^1_{\widetilde{X}/k} \hookrightarrow \pi_*(\Omega^1_{\widetilde{X}/k}(\widetilde{C}))$$

whose cokernel is supported in the finitely many (closed) points of $X - U = \pi(\widetilde{C})$.

Recall from [**EGA**, III$_1$, 1.4.17] that on a scheme Z, any abstract $\mathcal{O}_Z$-module extension of one quasi-coherent $\mathcal{O}_Z$-module by another is automatically quasi-coherent (and thus coherent if we work with coherent sheaves on a locally noetherian Z). In particular, since any coherent $\mathcal{O}_X$-module $\mathscr{F}$ supported on finitely closed points obviously has the property that all of its $\mathcal{O}_X$-submodules are coherent, the coherence of ω_f^{reg} follows from the following more precise result which we will need later:

LEMMA 5.2.2. *As subsheaves of $\underline{\Omega}^1_{K/k}$, we have inclusions*

$$\pi_*\Omega^1_{\widetilde{X}/k} \subseteq \omega_f^{\text{reg}} \subseteq \pi_*(\Omega^1_{\widetilde{X}/k}(\widetilde{C})).$$

PROOF. The first inclusion has already been noted. For the second inclusion, let V be open in X and $\eta \in \omega_f^{\text{reg}}(V)$, so η is a meromorphic differential on $\pi^{-1}(V)$. We want to show that η has 'poles no worse than $\widetilde{C}$' on $\pi^{-1}(V)$. Shrinking V if necessary, we may assume $\widetilde{\mathscr{C}}|_{\pi^{-1}(V)}$ is free [**EGA**, IV$_4$, 21.8.1]. Let t be a generator of $\widetilde{\mathscr{C}}|_{\pi^{-1}(V)}$, so

$$t \in H^0(V, \pi_*\widetilde{\mathscr{C}}) = \mathscr{C}(V).$$

Since $\Omega^1_{\widetilde{X}/k}(\widetilde{C}) = \Omega^1_{\widetilde{X}/k} \otimes \widetilde{\mathscr{C}}^{-1}$, we want $t\eta$ to have no poles on $\pi^{-1}(V)$. Choose a closed point $x \in V$ and let $\widetilde{x}_0 \in \pi^{-1}(x)$. Let $d = \text{ord}_{\widetilde{x}_0}(t\eta)$, and assume $d < 0$. Choose a rational function t' on $\widetilde{X}$ so that t', $t't\eta$ are regular at all $\widetilde{x} \in \pi^{-1}(x)$ distinct from $\widetilde{x}_0$ and $\text{ord}_{\widetilde{x}_0}(t') = -1 - d \geq 0$. Thus, $t' \in (\pi_*\mathcal{O}_{\widetilde{X}})_x$, so $tt' \in \mathcal{O}_{X,x}$ since $t \in \mathscr{C}(V)$. We conclude that

$$0 = \sum_{\widetilde{x} \in \pi^{-1}(\widetilde{x}_0)} \text{res}_{\widetilde{x}}(tt'\eta) = \text{res}_{\widetilde{x}_0}(tt'\eta) \neq 0,$$

a contradiction. Thus, $t\eta$ is indeed holomorphic on all of $\pi^{-1}(V)$. ∎

Clearly $\underline{\Omega}^1_{K/k}$ is a quasi-coherent *flasque* sheaf on X and the inclusion $\omega^{\mathrm{reg}}_f \hookrightarrow$ $\underline{\Omega}^1_{K/k}$ is an isomorphism at the generic points, so we have a quasi-coherent flasque resolution

$$(5.2.3) \qquad 0 \to \omega^{\mathrm{reg}}_f \to \underline{\Omega}^1_{K/k} \to \bigoplus_{x \in X^0} i_{x*}(\underline{\Omega}^1_{K/k,x}/\omega^{\mathrm{reg}}_{f,x}) \to 0,$$

where X^0 denotes the set of closed points of X and $i_x : \mathrm{Spec}(\mathscr{O}_{X,x}) \to X$ is the canonical map for all $x \in X$. The cohomology long exact sequence of (5.2.3) gives an exact sequence of k-vector spaces

$$(5.2.4) \qquad \Omega^1_{K/k} \to \bigoplus_{x \in X^0} \underline{\Omega}^1_{K/k,x}/\omega^{\mathrm{reg}}_{f,x} \to \mathrm{H}^1(X, \omega^{\mathrm{reg}}_f) \to 0.$$

For each $x \in X^0$, the k-linear map

$$\mathrm{res}_x : \underline{\Omega}^1_{K/k,x} \to k$$

defined by

$$\eta \mapsto \sum_{\widetilde{x} \in \pi^{-1}(x)} \mathrm{res}_{\widetilde{x}}(\eta)$$

kills $\omega^{\mathrm{reg}}_{f,x}$ and the composite

$$(5.2.5) \qquad \Omega^1_{K/k} \longrightarrow \bigoplus_{x \in X^0}(\underline{\Omega}^1_{K/k,x}/\omega^{\mathrm{reg}}_{f,x}) \xrightarrow{\sum \mathrm{res}_x} k$$

is 0, by the residue theorem on the connected components of $\widetilde{X}$ (which can be deduced from Grothendieck's general theory, as we explain in Appendix B).

Putting together (5.2.4) and (5.2.5), we arrive at a k-linear map

$$\mathrm{res}_X : \mathrm{H}^1(X, \omega^{\mathrm{reg}}_f) \to k,$$

and it is clear that this map is non-zero. When X is smooth, this is the classical residue map. Rosenlicht's 'explicit' description of Grothendieck duality for $X \to \mathrm{Spec}(k)$ is that $(\omega^{\mathrm{reg}}_f, \mathrm{res}_X)$ is a universal object for the functor $\mathrm{H}^1(X, \cdot)^\vee$ on quasi-coherent $\mathscr{O}_X$-modules. More precisely, by the universal property of (ω_f, γ_f), there is a unique (necessarily non-zero) map of $\mathscr{O}_X$-modules

$$(5.2.6) \qquad \omega^{\mathrm{reg}}_f \to \omega_f$$

which carries res_X over to γ_f and we claim this is an isomorphism. Actually, one can *directly* construct a duality theory on X in terms of ω^{reg}_f and res_X [**AK1**, VIII], from which it follows by universality that (5.2.6) is an isomorphism. But even if uses the results in [**AK1**, VIII] (which we will not need to do below), there is a rather subtle point lurking in the background which needs to be cleared up in order for this 'explicit' description to be truly useful.

Here is the problem. By the definition of ω^{reg}_f, there is a canonical identification

$$\omega^{\mathrm{reg}}_f|_U \simeq \Omega^1_{U/k},$$

and by Grothendieck's general theory we have a canonical isomorphism

$$\omega_f|_U \simeq \Omega^1_{U/k}.$$

Thus, (5.2.6) induces a non-zero $\mathscr{O}_U$-linear endomorphism of $\Omega^1_{U/k}$, which must be multiplication by some non-zero $s \in \Gamma(U, \mathscr{O}_X) \subseteq K$. Even if we knew that (5.2.6) were an isomorphism (e.g., by appealing to an independent construction of duality theory on X as in [**AK1**, VIII]), all we can formally conclude is that $s \in \Gamma(U, \mathscr{O}_X^\times)$. It does *not* appear to follow formally that $s \in k^\times$ holds (let alone the more precise fact one expects, namely that $s = \pm 1$ is a universal sign, independent of X).

Another way of putting this problem is that we have a priori inclusions of $\mathscr{O}_X$-modules

$$\omega_f, \omega_f^{\text{reg}} \subseteq j_*\Omega^1_{U/k}$$

and when X is connected, the assertion that $s \in k^\times$ (or equivalently, $s \in k$, since $s \neq 0$) is exactly the statement that these subsheaves of $j_*\Omega^1_{U/k}$ are *the same*; here we have used the fact that $\text{Hom}_X(\omega_f, \omega_f) \simeq \text{H}^1(X, \omega_f)^\vee \simeq \text{H}^0(X, \mathscr{O}_X)$ is 1-dimensional over k when X is connected, so $\text{Hom}_X(\omega_f, \omega_f) = k$. We will prove in Theorem 5.2.3 that $s = -1$, so (5.2.6) induces multiplication by -1 on $\Omega^1_{U/k}$ over U.

Let us give an example of why it is absolutely essential in applications to know that $s \in k^\times$ (although in practice one certainly wants $s = \pm 1$ to be a universal sign independent of $X \to \text{Spec}(k)$). Let R be a discrete valuation ring with fraction field K and residue field k (not necessarily algebraically closed). Let $f : X \to \text{Spec}(R)$ be a proper CM curve (e.g., X normal and proper flat over R with pure relative dimension 1). The most common example of interest is a semistable curve. Assume that f has a geometrically reduced closed fiber and a smooth, geometrically connected generic fiber, so the closed fiber is geometrically connected and the R-smooth locus $j : U \hookrightarrow X$ is the complement of finitely many closed points in the closed fiber of f. Consider the natural map

$$(5.2.7) \qquad \omega_f \to j_*(\omega_f|_U) = j_*\Omega^1_{U/R},$$

where we have used the isomorphism $\omega_f|_U \simeq \Omega^1_{U/R}$ from the general theory of the dualizing sheaf. Since (5.2.7) is an isomorphism on the generic fiber, by the R-flatness of ω_f this map is injective.

In fact, (5.2.7) is an isomorphism. To see this, by [**SGA2**, III, Cor 2.5] it suffices to show

$$\text{depth}_{\mathscr{O}_{X,x}}(\omega_{f,x}) \geq 2$$

for all $x \in X - U$. Since ω_f is R-flat and of formation compatible with base change, it suffices to show that for each non-smooth (closed) point $\overline{x}$ on the closed fiber $\overline{f} : \overline{X} \to \text{Spec}(k)$ we have

$$\text{depth}_{\mathscr{O}_{\overline{X},\overline{x}}}(\omega_{\overline{f},\overline{x}}) \geq 1.$$

This follows from the fact that

$$\text{Ext}^0_{\mathscr{O}_{\overline{X},\overline{x}}}(k(\overline{x}), \omega_{\overline{f},\overline{x}}) = \text{Hom}_{\overline{X}}(\mathscr{O}_{\overline{X}}/\mathfrak{m}_{\overline{x}}, \omega_{\overline{f}}) = \text{H}^1(\overline{X}, \mathscr{O}_{\overline{X}}/\mathfrak{m}_{\overline{x}})^\vee = 0$$

vanishes.

Since (5.2.7) is an isomorphism, we can view $j_*(\Omega^1_{U/R})$ as an 'explicit' description of ω_f in this case. Upon passing to the closed fiber, this description gives an identification of $\omega_{\bar{f}}$ with a certain sheaf of meromorphic differentials on the closed fiber $\overline{X}$. When k is algebraically closed, which meromorphic differentials do we get? It is quite essential in applications (e.g., [M], [R]) to know that we get *exactly* the sheaf ω_f^{reg} of regular differentials. Since the concrete description of ω_f in terms of $j_*\Omega_{U/R}^1$ in (5.2.7) recovers Grothendieck's canonical isomorphism $\omega_f|_U \simeq \Omega_{U/R}^1$ over the smooth locus U of X, and hence recovers the canonical isomorphism $\omega_{\bar{f}}|_{\overline{U}} \simeq \Omega_{\overline{U}/k}^1$ over the smooth locus $\overline{U}$ of $\overline{X}$ (by (3.6.1)), clearly our question once again comes down to the problem of showing $s \in k^\times$. With the motivation now settled, we prove:

THEOREM 5.2.3. (Rosenlicht) *Let $f : X \to Y = \mathrm{Spec}(k)$ be a proper, reduced curve over an algebraically closed field k. Let $j : U \hookrightarrow X$ be the dense smooth locus and $\mathrm{Spec}(K)$ the scheme of generic points. The coherent subsheaves*

$$\omega_f, \omega_f^{\mathrm{reg}} \subseteq j_*\Omega_{U/k}^1 \quad (\subseteq \underline{\Omega}_{K/k}^1)$$

coincide and this equality identifies

$$\gamma_f : \mathrm{H}^1(X, \omega_f) \to k, \quad -\mathrm{res}_X : \mathrm{H}^1(X, \omega_f^{\mathrm{reg}}) \to k.$$

Note the minus sign. This rests on our definition of (3.6.13), which rests on the same explication convention as (2.3.8). This definition will give Theorem B.4.1 without any sign. The origin of the sign in Theorem 5.2.3 is in the smooth case (see Theorems B.2.1, B.2.2). There is one case where we can 'directly' calculate that $\gamma_f = -\mathrm{res}_X$, namely $X = \mathbf{P}_k^1$. Let T_0, T_1 denote the standard homogenous coordinates, and let U_j be the open locus where $T_j \neq 0$. For the standard *ordered* open covering $\mathfrak{U} = \{U_0, U_1\}$ and

$$\eta = dt_1/t_1 \in \mathrm{H}^0(U_0 \cap U_1, \Omega_{X/k}^1) = \check{\mathrm{C}}^1(\mathfrak{U}, \omega_f)$$

(with $t_1 = T_1/T_0$), the cohomology class $[\eta] \in \mathrm{H}^1(X, \omega_f)$ satisfies $\gamma_f([\eta]) = (-1)^{1(1+1)/2} = -1$ by Lemma 3.4.3(TRA3) and the remarks about projective space following (3.4.13). Meanwhile, (B.4.8)–(B.4.11) imply (using $x_0 = 0$, $x_1 = \infty$) that $\mathrm{res}_X([\eta]) = \mathrm{res}_\infty(\eta) = 1$.

PROOF. The smooth *connected* case is explained in Appendix B, from which the general smooth case follows, since the trace map in the proper CM case is clearly 'additive' with respect to formation of finite disjoint unions on the source. Ultimately, the general case will be reduced to the smooth case by comparison with the normalization. The smooth case of the theorem is an essential step in the proof of the important theoretical fact that the 'residual complex trace map' for proper morphisms is a map of complexes (cf. Theorem 3.4.1).

Let $\pi : \widetilde{X} \to X$ be the normalization, $\widetilde{f} = f \circ \pi$, and let $\mathscr{C}$, $\widetilde{\mathscr{C}}$ be the coherent conductor ideal sheaves on X and $\widetilde{X}$ respectively, as in the discussion above (5.2.2). Recall that Grothendieck's general theory gives an isomorphism in $\mathbf{D}(\widetilde{X})$

$$\omega_{\widetilde{f}}[1] \simeq \pi^\flat(\omega_f[1]),$$

which is to say an isomorphism of $\pi_*(\mathcal{O}_{\widetilde{X}})$-modules

$$(5.2.8) \qquad \pi_*\omega_{\widetilde{f}} \simeq \mathcal{H}om_X(\pi_*\mathcal{O}_{\widetilde{X}}, \omega_f).$$

On the smooth locus U, (5.2.8) recovers the canonical isomorphism $\omega_f|_U \simeq \Omega^1_{U/k}$ via the canonical isomorphism $\omega_{\widetilde{f}} \simeq \Omega^1_{\widetilde{X}/k}$. Since $\widetilde{f} = f \circ \pi$ and the trace map Trf for finite morphisms is defined via 'evaluation at 1', the compatibility of Grothendieck's trace map with respect to composite morphisms implies that the diagram

$$(5.2.9) \qquad H^1(\widetilde{X}, \omega_{\widetilde{f}}) \xrightarrow{\sim} H^1(X, \pi_*\omega_{\widetilde{f}}) \xrightarrow{\sim} H^1(X, \mathcal{H}om(\pi_*\mathcal{O}_{\widetilde{X}}, \omega_f))$$

$$\gamma_{\widetilde{f}} \downarrow$$

$$k \xleftarrow[\gamma_f]{} H^1(X, \omega_f)$$

commutes.

Our first step in the proof of Theorem 5.2.3 for X is to show that ω_f^{reg} has a property similar to (5.2.8). More precisely, the 'evaluate at 1' map gives an $\mathcal{O}_X$-linear map

$$(5.2.10) \qquad \mathcal{H}om_X(\pi_*\mathcal{O}_{\widetilde{X}}, \omega_f^{\mathrm{reg}}) \to \Omega^1_{K/k}.$$

We claim that (5.2.10) is injective and its image is exactly the subsheaf

$$\pi_*\omega_{\widetilde{f}} = \pi_*\Omega^1_{\widetilde{X}/k} \subseteq \pi_*\underline{\Omega}^1_{\widetilde{K}/k} = \underline{\Omega}^1_{K/k}.$$

Over any open $V \subseteq X$, an $\mathcal{O}_V$-linear map $\varphi : \pi_*\mathcal{O}_{\widetilde{X}}|_V \to \omega_f^{\mathrm{reg}}|_V$ induces a K-linear map $K \to \Omega^1_{K/k}$ at the generic points, so $\varphi(s) = s\varphi(1)$ as meromorphic differentials on $\pi^{-1}(V)$. Thus, (5.2.10) is injective and φ is determined by the meromorphic differential $\eta = \varphi(1) \in \Omega^1_{K/k}$. Which η arise in this way? Clearly the η's we get are those for which

$$\sum_{\widetilde{x} \in \pi^{-1}(x)} \mathrm{res}_{\widetilde{x}}(ts\eta) = 0$$

for all $s \in \mathcal{O}_{X,x}$, $t \in (\pi_*\mathcal{O}_{\widetilde{X}})_x$, $x \in V(k)$. Since $\mathcal{O}_{X,x} \subseteq (\pi_*\mathcal{O}_{\widetilde{X}})_x$, this says

$$(5.2.11) \qquad \sum_{\widetilde{x} \in \pi^{-1}(x)} \mathrm{res}_{\widetilde{x}}(t\eta) = 0$$

for all $t \in (\pi_*\mathcal{O}_{\widetilde{X}})_x$, $x \in V(k)$. This condition is clearly satisfied by $\eta \in \Omega^1_{\widetilde{X}/k}(\pi^{-1}(V))$. Conversely, suppose $\eta \in \Omega^1_{K/k}$ has a pole of order $d \geq 1$ at some $\widetilde{x}_0 \in \pi^{-1}(x)$ with $x \in V(k)$. We contradict (5.2.11) by following the proof of Lemma 5.2.2. Choose a rational function t on the smooth $\widetilde{X}$ so that $t\eta$ is regular at all $\widetilde{x} \in \pi^{-1}(x)$ distinct from $\widetilde{x}_0$ and $\mathrm{ord}_{\widetilde{x}_0}(t) = d - 1 \geq 0$. Then $t \in (\pi_*\mathcal{O}_{\widetilde{X}})_x$ and computing the left side of (5.2.11) gives the non-zero $\mathrm{res}_{\widetilde{x}_0}(t\eta)$.

We now have a canonical isomorphism of $\mathcal{O}_X$-modules

$$\pi_*(\Omega^1_{\widetilde{X}/k}) \simeq \mathcal{H}om_X(\pi_*\mathcal{O}_{\widetilde{X}}, \omega_f^{\mathrm{reg}}),$$

so this is even $\pi_* \mathcal{O}_{\widetilde{X}}$-linear (as both sides are 'torsion-free'). This is the desired analogue of the abstract isomorphism (5.2.8). Using 'evaluation at 1' recovers the canonical map of $\mathcal{O}_X$-modules

$$\pi_* \Omega^1_{\widetilde{X}/k} \to \omega^{\mathrm{reg}}_f$$

which restricts to the identity on $\Omega^1_{U/k}$ over U and is the usual inclusion as subsheaves of $\underline{\Omega}^1_{K/k}$. We use this inclusion to define a k-linear map

$$\mathrm{H}^1(\widetilde{X}, \Omega^1_{\widetilde{X}/k}) = \mathrm{H}^1(X, \pi_* \Omega^1_{\widetilde{X}/k}) \to \mathrm{H}^1(X, \omega^{\mathrm{reg}}_f)$$

which is readily checked to fit into a commutative diagram

$$
\begin{array}{ccccccc}
\Omega^1_{\widetilde{K}/k} & \longrightarrow & \bigoplus_{\widetilde{x} \in \widetilde{X}^0} \underline{\Omega}^1_{\widetilde{K}/k,\widetilde{x}} / \Omega^1_{\widetilde{X}/k,\widetilde{x}} & \longrightarrow & \mathrm{H}^1(\widetilde{X}, \Omega^1_{\widetilde{X}/k}) & \longrightarrow & 0 \\
\| & & \downarrow & & \downarrow & & \\
\Omega^1_{K/k} & \longrightarrow & \bigoplus_{x \in X^0} \underline{\Omega}^1_{K/k,x} / \omega^{\mathrm{reg}}_{f,x} & \longrightarrow & \mathrm{H}^1(X, \omega^{\mathrm{reg}}_f) & \longrightarrow & 0
\end{array}
$$

By the very definition of res_X and $\mathrm{res}_{\widetilde{X}}$ in terms of the rows in this diagram, we deduce the commutativity of

(5.2.12)
$$
\begin{array}{ccc}
\mathrm{H}^1(\widetilde{X}, \Omega^1_{\widetilde{X}/k}) & \xrightarrow{\mathrm{res}_{\widetilde{X}}} & k \\
\downarrow & & \| \\
\mathrm{H}^1(X, \omega^{\mathrm{reg}}_f) & \xrightarrow[\mathrm{res}_X]{} & k
\end{array}
$$

By universality, there is a unique map of $\mathcal{O}_X$-modules

(5.2.13) $\omega^{\mathrm{reg}}_f \to \omega_f$

which makes the diagram

(5.2.14)
$$
\begin{array}{ccc}
\mathrm{H}^1(X, \omega^{\mathrm{reg}}_f) & \xrightarrow{-\mathrm{res}_X} & k \\
\downarrow & & \| \\
\mathrm{H}^1(X, \omega_f) & \xrightarrow[\gamma_f]{} & k
\end{array}
$$

commute. Note the sign in the top row of (5.2.14). Theorem 5.2.3 asserts exactly that (5.2.13) is an isomorphism and it restricts to the *identity* on $\Omega^1_{U/k}$ over U. In order to at least see what happens over U, it is harmless to apply $\mathcal{H}om_X(\pi_* \mathcal{O}_{\widetilde{X}}, \cdot)$ to (5.2.13) and study

(5.2.15) $\mathcal{H}om_X(\pi_* \mathcal{O}_{\widetilde{X}}, \omega^{\mathrm{reg}}_f) \to \mathcal{H}om_X(\pi_* \mathcal{O}_{\widetilde{X}}, \omega_f).$

We will prove that (5.2.15) is an isomorphism which restricts to the identity map on $\Omega^1_{U/k}$ over U. Once this is shown, then (5.2.13) is the identity over U, so it is at least injective with cokernel supported at the non-smooth points of

X. Our analysis of (5.2.15) comes down to proving that the diagram of quasi-coherent $\pi_* \mathcal{O}_{\widetilde{X}}$-modules

$$\mathcal{H}om_X(\pi_*\mathcal{O}_{\widetilde{X}}, \omega_f^{\mathrm{reg}}) \longrightarrow \mathcal{H}om_X(\pi_*\mathcal{O}_{\widetilde{X}}, \omega_f)$$

$$\Big\| \qquad\qquad\qquad\qquad \Big\downarrow \simeq$$

$$\pi_*\Omega^1_{\widetilde{X}/k} =\!=\!=\!=\!=\!=\!=\!=\!=\!=\!=\!=\!= \pi_*\omega_{\widetilde{f}}$$

commutes (where the right column is (5.2.8)), or equivalently that the diagram of quasi-coherent $\mathcal{O}_{\widetilde{X}}$-modules

(5.2.16)
$$\mathcal{H}om_X(\pi_*\mathcal{O}_{\widetilde{X}}, \omega_f^{\mathrm{reg}})^\sim \longrightarrow \mathcal{H}om_X(\pi_*\mathcal{O}_{\widetilde{X}}, \omega_f)^\sim$$

$$\Big\| \qquad\qquad\qquad\qquad \Big\downarrow$$

$$\Omega^1_{\widetilde{X}/k} =\!=\!=\!=\!=\!=\!=\!=\!=\!=\!=\!=\!= \omega_{\widetilde{f}}$$

commutes, where $(\cdot)^\sim$ denotes the functor giving the equivalence between quasi-coherent $\pi_*\mathcal{O}_{\widetilde{X}}$-modules on X and quasi-coherent $\mathcal{O}_{\widetilde{X}}$-modules on $\widetilde{X}$.

The universal property of $(\omega_{\widetilde{f}}, \gamma_{\widetilde{f}})$ ensures that the commutativity of (5.2.16) follows if we have commutativity after applying $H^1(\widetilde{X}, \cdot)$ and composing with

$$\gamma_{\widetilde{f}} : H^1(\widetilde{X}, \omega_{\widetilde{f}}) \to k.$$

In other words, by using the commutativity of (5.2.9), we are faced with verifying the commutativity of the outer part of the diagram

(5.2.17)
$$H^1(\widetilde{X}, \Omega^1_{\widetilde{X}/k}) \longrightarrow H^1(X, \omega_f^{\mathrm{reg}}) \longrightarrow H^1(X, \omega_f)$$

$$\Big\| \qquad {}^{-\mathrm{res}_{\widetilde{X}}}\!\!\searrow \quad \Big\downarrow {}^{-\mathrm{res}_X} \qquad\qquad \Big\downarrow {}^{\gamma_f}$$

$$H^1(\widetilde{X}, \omega_{\widetilde{f}}) \xrightarrow[\gamma_{\widetilde{f}}]{} k =\!=\!=\!=\!=\!=\!=\!= k$$

But the top triangle and right square in (5.2.17) commute by (5.2.12) and (5.2.14) respectively, while the commutativity of the lower triangle in (5.2.17) is exactly the smooth case of the theorem, which we have already established (or rather, deferred to Appendix B).

We conclude that (5.2.13) is injective and extends the identity on $\Omega^1_{U/k}$ over U, so we have an inclusion

$$\pi_*\Omega^1_{\widetilde{X}/h} \subseteq \omega_f^{\mathrm{reg}} \subset \omega_f$$

as *coherent subsheaves* of $\underline{\Omega}^1_{K/k}$ on X, and these inclusions are equalities over the smooth locus of X. In order to check that the inclusion $\omega_f^{\mathrm{reg}} \subseteq \omega_f$ is an equality, it suffices to establish the inequality between finite lengths:

$$\mathrm{length}(\omega_f/\pi_*\Omega^1_{\widetilde{X}/k}) \leq \mathrm{length}(\omega_f^{\mathrm{reg}}/\pi_*\Omega^1_{\widetilde{X}/k}),$$

or equivalently

$$\dim_k \mathrm{H}^0(X, \omega_f / \pi_* \Omega^1_{\widetilde{X}/k}) \leq \dim_k \mathrm{H}^0(X, \omega_f^{\mathrm{reg}} / \pi_* \Omega^1_{\widetilde{X}/k}).$$

However, the composite inclusion $\pi_* \Omega^1_{\widetilde{X}/k} \hookrightarrow \omega_f^{\mathrm{reg}} \hookrightarrow \omega_f$ is *exactly* the abstract map

$$\pi_* \omega_{\widetilde{f}} \simeq \mathscr{H}om_X(\pi_* \mathcal{O}_{\widetilde{X}}, \omega_f) \hookrightarrow \omega_f$$

arising from (5.2.8), since (5.2.13) restricts to the identity on $\Omega^1_{U/k}$ over U. Thus, we are really trying to compare the lengths (or equivalently, k-dimensions of H^0's) of the cokernels of the *canonical* inclusions

$$\pi_* \Omega^1_{\widetilde{X}/k} \simeq \mathscr{H}om_X(\pi_* \mathcal{O}_{\widetilde{X}}, \omega_f^{\mathrm{reg}}) \hookrightarrow \omega_f^{\mathrm{reg}}, \quad \pi_* \omega_{\widetilde{f}} \simeq \mathscr{H}om_X(\pi_* \mathcal{O}_{\widetilde{X}}, \omega_f) \hookrightarrow \omega_f.$$

It now suffices to construct a k-linear injection and k-linear isomorphism:

(5.2.18)
$$\mathrm{H}^0(X, \omega_f / \mathscr{H}om_X(\pi_* \mathcal{O}_{\widetilde{X}}, \omega_f)) \hookrightarrow \mathrm{H}^0(X, \pi_*(\mathcal{O}_{\widetilde{X}})/\mathcal{O}_X)^\vee \simeq \mathrm{H}^0(X, \omega_f^{\mathrm{reg}} / \pi_* \Omega^1_{\widetilde{X}/k}).$$

We have succeeded in reducing ourselves to a statement in Grothendieck's abstract theory and a statement about differential forms, without having to worry about the relation between Grothendieck's theory and differential forms. Let's first construct the injection in (5.2.18). Applying $\mathscr{H}om_X(\cdot, \omega_f)$ to the short exact sequence

$$0 \to \mathcal{O}_X \to \pi_* \mathcal{O}_{\widetilde{X}} \to \pi_*(\mathcal{O}_{\widetilde{X}})/\mathcal{O}_X \to 0$$

gives an exact sequence

$$0 \to \mathscr{H}om_X(\pi_* \mathcal{O}_{\widetilde{X}}, \omega_f) \to \omega_f \to \mathscr{E}xt^1_X(\pi_*(\mathcal{O}_{\widetilde{X}})/\mathcal{O}_X, \omega_f),$$

so we get an injection

(5.2.19) $$\mathrm{H}^0(X, \omega_f / \mathscr{H}om_X(\pi_* \mathcal{O}_{\widetilde{X}}, \omega_f)) \hookrightarrow \mathrm{H}^0(X, \mathscr{E}xt^1_X(\pi_*(\mathcal{O}_{\widetilde{X}})/\mathcal{O}_X, \omega_f)).$$

Since $\pi_*(\mathcal{O}_{\widetilde{X}})/\mathcal{O}_X$ is supported at finitely many closed points of X, the local-to-global Ext spectral sequence degenerates and we get a canonical k-linear isomorphism

(5.2.20) $$\mathrm{H}^0(X, \mathscr{E}xt^1_X(\pi_*(\mathcal{O}_{\widetilde{X}})/\mathcal{O}_X, \cdot)) \simeq \mathrm{Ext}^1_X(\pi_*(\mathcal{O}_{\widetilde{X}})/\mathcal{O}_X, \cdot)$$

as functors on quasi-coherent $\mathcal{O}_X$-modules. But $f_* = \mathrm{H}^0(X, \cdot)$ since $\mathrm{Spec}(k)$ has a single point, so by the freeness of vector spaces we can use (5.1.11) to get a k-linear isomorphism

(5.2.21) $$\mathrm{Ext}^1_X(\pi_*(\mathcal{O}_{\widetilde{X}})/\mathcal{O}_X, \omega_f) \simeq \mathrm{H}^0(X, \pi_*(\mathcal{O}_{\widetilde{X}})/\mathcal{O}_X)^\vee.$$

Putting together (5.2.19), (5.2.20), (5.2.21) yields the injection in (5.2.18).

To complete the proof of Theorem 5.2.3, it remains to construct the isomorphism in (5.2.18), and we will do this by considering the pairing of finite-dimensional k-vector spaces

$$\langle\,,\,\rangle : \mathrm{H}^0(X, \omega_f^{\mathrm{reg}} / \pi_* \Omega^1_{\widetilde{X}/k}) \times \mathrm{H}^0(X, \pi_*(\mathcal{O}_{\widetilde{X}})/\mathcal{O}_X) \to k$$

defined by

$$(\eta, s) \mapsto \sum_{\widetilde{x} \in \widetilde{X}^0} \mathrm{res}_{\widetilde{x}}(s\eta)$$

(where the terms for $\tilde{x} \in \pi^{-1}(U)$ are automatically 0). We will show Rosen-licht's result that this is a perfect pairing, thereby giving the desired k-linear isomorphism in (5.2.18). In order to get control over $\langle\,,\,\rangle$, it is convenient to use the invertible conductor ideal $\widetilde{\mathscr{C}}$ on $\widetilde{X}$, or equivalently the closed subscheme $\widetilde{C}$ it defines in $\widetilde{X}$. Recall from Lemma 5.2.2 that $\omega_f^{\text{reg}} \subseteq \pi_*(\Omega^1_{\widetilde{X}/k}(\widetilde{C}))$. The usefulness of $\widetilde{C}$ is that

$$\pi_*(\mathscr{O}_{\widetilde{X}}(-\widetilde{C})) = \pi_*(\widetilde{\mathscr{C}}) = \mathscr{C} \subseteq \mathscr{O}_X,$$

so it makes sense to consider the 'residue pairing' of (possibly) larger finite-dimensional k-vector spaces

$$\langle\,,\,\rangle' : \text{H}^0(X, \pi_*(\Omega^1_{\widetilde{X}/k}(\widetilde{C}))/\pi_*\Omega^1_{\widetilde{X}/k}) \times \text{H}^0(X, \pi_*\mathscr{O}_{\widetilde{X}}/\pi_*(\mathscr{O}_{\widetilde{X}}(-\widetilde{C}))) \to k.$$

By definition, the annihilator of $\mathscr{O}_X/\pi_*(\mathscr{O}_{\widetilde{X}}(-\widetilde{C}))$ under $\langle\,,\,\rangle'$ is exactly

$$\omega_f^{\text{reg}}/\pi_*\Omega^1_{\widetilde{X}/k}.$$

Thus, it suffices to prove that $\langle\,,\,\rangle'$ is a perfect pairing.

Since π_* is exact on quasi-coherent $\mathscr{O}_{\widetilde{X}}$-modules, it suffices to prove more generally that for any effective Weil divisors $D \leq D'$ on a proper smooth curve $\widetilde{X}$ over an algebraically closed field k, the residue pairing between finite-dimensional k-vector spaces

$$(5.2.22) \qquad \text{H}^0(\widetilde{X}, \Omega^1_{\widetilde{X}/k}(D')/\Omega^1_{\widetilde{X}/k}(D)) \times \text{H}^0(\widetilde{X}, \mathscr{O}_{\widetilde{X}}(-D)/\mathscr{O}_{\widetilde{X}}(-D')) \to k$$

is a perfect pairing. This only depends on $D' - D$, and if $D \leq D' \leq D''$ then the pairings (5.2.22) for (D, D'), (D, D''), (D', D'') are compatible with the obvious *exact* sequences of H^0's. Since the dimension of each H^0 in (5.2.22) is 'additive' in $D' - D$ and we can get from D to D' by adding one point at a time, by inducting on the degree of $D' - D$ we are therefore reduced to the case $D' = D + P$. In this special case we may also assume $D = 0$. This reduces us to the obvious claim that for any closed point $P \in \widetilde{X}$, there exists a meromorphic differential (resp. meromorphic function) on $\widetilde{X}$ with order -1 (resp. 1) at P. ∎

APPENDIX A

Residues and Cohomology with Supports

A.1. Statement of Results

In this appendix, we address the topic of the residue symbol. The basic definition and properties of this concept are given (without proof) in [**RD**, III, §9], and we review these below. Once one uses the isomorphism (2.2.3) instead of (2.2.1), the definition of the residue symbol in [**RD**] must be slightly changed in order to avoid sign problems. We want certain concrete numerical 'normalization' assertions such as

(A.1.1)
$$\mathrm{Res}_{\mathbf{A}_Y^n/Y} \left[\begin{array}{c} \mathrm{d}T_1 \wedge \cdots \wedge \mathrm{d}T_n \\ T_1, \ldots, T_n \end{array} \right] = 1$$

for all $n \geq 1$ and all schemes Y (with $T_1, \ldots, T_n$ the standard coordinates on $\mathbf{A}_Y^n$), which leaves no room for ambiguity of signs. Without the sign in the definition of the residue symbol in (A.1.4) below, one gets $(-1)^{n(n-1)/2}$ on the right side of (A.1.1). This is essentially because of the general relation

$$\mathrm{d}t_1 \wedge \cdots \wedge \mathrm{d}t_n = (-1)^{n(n-1)/2} \mathrm{d}t_n \wedge \cdots \wedge \mathrm{d}t_1,$$

as we will see.

Let us first make everything explicit by giving the definition of the residue symbol in 'concrete' terms. Let $f : X \to Y$ be a smooth separated morphism of schemes, with pure relative dimension n, and let $t_1, \ldots, t_n \in \Gamma(X, \mathcal{O}_X)$ generate a quasi-coherent ideal sheaf $\mathcal{I}$ whose associated closed subscheme $Z \hookrightarrow X$ is finite (hence locally free) over Y. Thus, we have a diagram

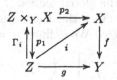

with g and p_2 finite locally free and i a closed immersion; in fact, i is an lci map with pure codimension n (just as Γ_i is also). Consider the composite isomorphism of $\mathcal{O}_Z$-modules

(A.1.2)

$$\mathcal{H}om_Y(g_*\mathcal{O}_Z, \mathcal{O}_Y) \xrightarrow{\zeta'_{\Gamma_i, p_1}} \omega_{Z/Z\times_Y X} \otimes \Gamma_i^*(\omega_{Z\times_Y X/Z} \otimes p_1^*\mathcal{H}om_Y(g_*\mathcal{O}_Z, \mathcal{O}_Y)) .$$

$$\big\|$$

$$\omega_{Z/Z\times_Y X} \otimes \Gamma_i^*(\mathcal{H}om_X(p_{2*}\mathcal{O}_{Z\times_Y X}, \omega_{X/Y}))$$

$$\simeq \uparrow \eta_{\Gamma_i}$$

$$\mathcal{E}xt^n_{Z\times_Y X}(\mathcal{O}_Z, \mathcal{H}om_X(p_{2*}\mathcal{O}_{Z\times_Y X}, \omega_{X/Y}))$$

$$\big\|$$

$$\mathcal{E}xt^n_X(\mathcal{O}_Z, \omega_{X/Y})$$

$$\simeq \downarrow \eta_i$$

$$\omega_{Z/X} \otimes i^*\omega_{X/Y}$$

Note that the definitions of η_i and η_{Γ_i} via (2.5.1) both involve the same implicit sign $(-1)^{n(n+1)/2}$ from (1.3.28), so the composite (A.1.2) is 'independent' of this sign.

   In fancier terms in the locally noetherian case, this is the induced map on H^0's of the derived category isomorphism

(A.1.3) $g^\flat(\mathcal{O}_Y) \xrightarrow{\psi_{i,f}} i^\flat f^\sharp(\mathcal{O}_Y) = i^\flat(\omega_{X/Y}[n]) \xrightarrow{\eta_i} \omega_{Z/X}[-n] \otimes i^*\omega_{X/Y}[n].$

The diligent reader may wonder why there is no sign of $(-1)^{n^2} = (-1)^n$ in the η_i part of (A.1.3), as might be expected from translation-compatibility for η_i. The reason there is no such sign is that if we calculate derived functors of $\omega_{X/Y}$ with the injective resolution $\mathcal{I}^\bullet$, then the explication (2.7.3) of (2.7.7) and the 'Cartan-Eilenberg' explication (2.5.5) of (2.5.3) are *both* given in terms of using $\mathcal{I}^{\bullet+n}$ as the injective resolution of the derived category object $\omega_{X/Y}[n]$. Alternatively, if we use $\omega_{X/Y}[n] = \mathcal{I}^\bullet[n]$ in the derived category, then the explicit description of $\psi_{i,f}$ would acquire a sign of $(-1)^n$ (since the isomorphism $\mathcal{I}^\bullet[n] \simeq \mathcal{I}^{\bullet+n}$ lifting the identity in degree $-n$ induces $(-1)^n$ in degree 0) and the translation-compatibility of η_i (which is now relevant) introduces another sign of $(-1)^n$.

   For any $\omega \in \Gamma(X, \omega_{X/Y})$, the global section $(t_1^\vee \wedge \cdots \wedge t_n^\vee) \otimes i^*\omega$ of the sheaf $\omega_{Z/X} \otimes i^*\omega_{X/Y}$ corresponds under the isomorphism (A.1.2) to an $\mathcal{O}_Y$-linear homomorphism $\varphi_\omega : g_*\mathcal{O}_Z \to \mathcal{O}_Y$. We define the *residue symbol* by

(A.1.4) $\operatorname{Res}_{X/Y}\begin{bmatrix} \omega \\ t_1, \ldots, t_n \end{bmatrix} = (-1)^{n(n-1)/2}\varphi_\omega(1) \in \Gamma(Y, \mathcal{O}_Y).$

The definition in [**RD**, III, §9] omits the sign because it uses (2.2.1) instead of (2.2.3). It is clear from the definition that (A.1.4) is $\mathscr{O}_Y$-linear in ω, local over Y, and unaffected by replacing X by any open around Z. Also, if Z is a disjoint union of $Z_1, \ldots, Z_m$ and X_i is an open in X with $X_i \cap Z = Z_i$, then

$$(\text{A.1.5}) \qquad \operatorname{Res}_{X/Y} \begin{bmatrix} \omega \\ t_1, \ldots, t_n \end{bmatrix} = \sum_{i=1}^{m} \operatorname{Res}_{X_i/Y} \begin{bmatrix} \omega \\ t_1, \ldots, t_n \end{bmatrix}.$$

There is a list of other properties (R1)–(R10) of the residue symbol in [**RD**, III, §9]. In this appendix, we use Theorem 2.8.1, a cohomological observation due to Berthelot, and explicit calculations of Tate in [**MR**, Appendix] to give proofs of all of these properties by means of the theory developed in [**RD**] and this book.

We begin by listing the properties to be proven (with some minor corrections to the statements in [**RD**]). In all properties aside from (R3) and (R4) below, the notation $X, Y, Z, f, t_1, \ldots, t_n, \omega$ is as above.

(R1) Let $s_i = \sum c_{ij} t_j$ with $c_{ij} \in \Gamma(X, \mathscr{O}_X)$. Assume that the s_i's also cut out a finite Y-scheme. Then

$$\operatorname{Res}_{X/Y} \begin{bmatrix} \omega \\ t_1, \ldots, t_n \end{bmatrix} = \operatorname{Res}_{X/Y} \begin{bmatrix} \det(c_{ij})\omega \\ s_1, \ldots, s_n \end{bmatrix}.$$

(R2) (*localization*) The residue symbol is stable under suitable étale localization on X around Z. More precisely, consider a scheme diagram

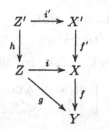

where the top square is cartesian, with f' separated étale and h finite (hence finite locally free). Let $t'_j = f'^*(t_j) \in \Gamma(X', \mathscr{O}_{X'})$, so $t'_1, \ldots, t'_n$ cut out Z' in X'. Assume that there is a locally constant function $\operatorname{rk}(Z'/Z)$ on a Zariski open neighborhood V of Z in X which restricts to the function

$$z \mapsto \operatorname{rank}_{\mathscr{O}_{Z,z}} h_*(\mathscr{O}_{Z'})_z$$

on Z (e.g., the rank of $h_* \mathscr{O}_{Z'}$ is constant). For all $\omega \in \Gamma(X, \omega_{X/Y})$,

$$\operatorname{Res}_{V/Y} \begin{bmatrix} \omega \cdot \operatorname{rk}(Z'/Z) \\ t_1, \ldots, t_n \end{bmatrix} = \operatorname{Res}_{X'/Y} \begin{bmatrix} \omega' \\ t'_1, \ldots, t'_n \end{bmatrix},$$

where ω' is the pullback of ω to X'.

(R3) (*restriction*) Let X, X' be smooth Y-schemes with pure relative dimensions $n + p$ and n respectively, and $i : X' \hookrightarrow X$ an lci map over Y, necessarily with pure codimension p. Assume that

i is cut out by $s_1, \ldots, s_p \in \Gamma(X, \mathscr{O}_X)$. Let $t_1, \ldots, t_n \in \Gamma(X, \mathscr{O}_X)$ have restrictions $t'_1, \ldots, t'_n \in \Gamma(X', \mathscr{O}_{X'})$ which cut out a finite Y-scheme on X', so $\{t_i, s_j\}$ cuts out a finite Y-scheme on X. Then for any $\omega \in \Gamma(X, \Omega^n_{X/Y})$,

$$\mathrm{Res}_{X'/Y} \begin{bmatrix} i^*\omega \\ t'_1, \ldots, t'_n \end{bmatrix} = \mathrm{Res}_{X/Y} \begin{bmatrix} \omega \wedge ds_1 \wedge \cdots \wedge ds_p \\ t_1, \ldots, t_n, s_1, \ldots, s_p \end{bmatrix}.$$

(R4) (*transitivity*) Let $X \xrightarrow{f} Y \xrightarrow{g} Z$ be two smooth maps with pure relative dimensions n and p respectively, and choose global functions $t'_1, \ldots, t'_n$ on X (resp. $s_1, \ldots s_p$ on Y) cutting out a finite Y-scheme (resp. finite Z-scheme). Let $s'_i = f^*(s_i) \in \Gamma(X, \mathscr{O}_X)$. Then for any $\omega \in \Gamma(X, \Omega^n_{X/Y})$, $\omega' \in \Gamma(Y, \Omega^p_{Y/Z})$, we have

$$\mathrm{Res}_{X/Z} \begin{bmatrix} \omega \otimes f^*\omega' \\ t'_1, \ldots, t'_n, s'_1, \ldots, s'_p \end{bmatrix}$$

$$= \mathrm{Res}_{Y/Z} \begin{bmatrix} \mathrm{Res}_{X/Y} \begin{bmatrix} \omega \\ t'_1, \ldots, t'_n \end{bmatrix} \omega' \\ s_1, \ldots, s_p \end{bmatrix}.$$

(R5) (*base change*) Formation of the residue symbol commutes with arbitrary base change on Y.

(R6) (*trace formula*) For any $\varphi \in \Gamma(X, \mathscr{O}_X)$,

$$\mathrm{Res}_{X/Y} \begin{bmatrix} \varphi \, dt_1 \wedge \cdots \wedge dt_n \\ t_1, \ldots, t_n \end{bmatrix} = \mathrm{Tr}_{Z/Y}(\varphi|_Z),$$

(R7) For any positive integers $k_1, \ldots, k_n$ not all equal to 1,

$$\mathrm{Res}_{X/Y} \begin{bmatrix} dt_1 \wedge \cdots \wedge dt_n \\ t_1^{k_1}, \ldots, t_n^{k_n} \end{bmatrix} = 0.$$

(R8) (*duality*) If $\omega|_Z = 0$, then

$$\mathrm{Res}_{X/Y} \begin{bmatrix} \omega \\ t_1, \ldots, t_n \end{bmatrix} = 0.$$

Conversely, let $\{Y_j\}$ be an étale covering of Y and such that $Z_j = Z \times_Y Y_j$ decomposes into a finite disjoint union of Z_{jk}'s with each Z_{jk} inside of an open $X_{jk} \subseteq X_j = X \times_Y Y_j$ that does not meet Z_{jm} for $m \neq k$. Also, assume that $\Gamma(X_{jk}, \mathscr{O}_{X_{jk}}) \to \Gamma(Z_{jk}, \mathscr{O}_{Z_{jk}})$ is surjective (such Y_j's, Z_{jk}'s, and X_{jk}'s always exist, by standard direct limit arguments). For any $\omega \in \Gamma(X, \Omega^n_{X/Y})$, if

$$\mathrm{Res}_{X_{jk}/Y_j} \begin{bmatrix} f\omega \\ t_1, \ldots, t_n \end{bmatrix} = 0$$

for all $f \in \Gamma(X_{jk}, \mathscr{O}_{X_{jk}})$, then $\omega|_Z = 0$.

(R9) (*exterior differentiation*) For $\eta \in \Gamma(X, \Omega_{X/Y}^{n-1})$ and positive integers $k_1, \ldots, k_n$,

$$\text{Res}_{X/Y}\left[\begin{array}{c} d\eta \\ t_1^{k_1}, \ldots, t_n^{k_n} \end{array}\right]$$

$$= \sum_{i=1}^{n} k_i \cdot \text{Res}_{X/Y}\left[\begin{array}{c} dt_i \wedge \eta \\ t_1^{k_1}, \ldots, t_i^{k_i+1}, \ldots, t_n^{k_n} \end{array}\right].$$

(R10) (*residue formula*) Let $g : X' \to X$ be a finite map, with X' smooth over Y of pure relative dimension n (so g is finite locally free). Choose $\omega' \in \Gamma(X', \Omega_{X'/Y}^n)$ and let $t_j' = g^*(t_j) \in \Gamma(X', \mathcal{O}_{X'})$. Then

$$\text{Res}_{X'/Y}\left[\begin{array}{c} \omega' \\ t_1', \ldots, t_n' \end{array}\right] = \text{Res}_{X/Y}\left[\begin{array}{c} \text{Tr}_g(\omega') \\ t_1, \ldots, t_n \end{array}\right],$$

where $\text{Tr}_g : g_* \Omega_{X'/Y}^n \to \Omega_{X/Y}^n$ is the map (2.7.36).

A.2. Proofs

The base change property (R5) follows from our observation in §2.5 that (2.5.1) is compatible with *arbitrary* base change (preserving the 'lci of pure codimension n' property) when it is evaluated on a quasi-coherent sheaf flat over the base. We will repeatedly use (R5) in many of our proofs, to justify reduction to the case where Y is local henselian, or noetherian (by direct limit arguments), or artin local (by the Krull Intersection Theorem). For example, by base change to henselizations, (R8) is an easy consequence of faithfully flat descent and the definition of the residue symbol. We now prove (R6), which will allow us to ignore certain sign issues in subsequent proofs.

PROOF. (of (R6)) First base change to henselizations on Y, so we can assume that Y is local henselian. This makes Z the finite disjoint union of local schemes. We can treat the connected components of Z separately by (A.1.5), so we may assume that Z is also local. Replacing X by an affine open around the closed point of Z, we are reduced to the affine case $Y = \text{Spec}(A)$, $X = \text{Spec}(B)$, $Z = \text{Spec}(C)$.

In this affine setting, we will use calculations of Tate. Let I be the kernel of $\pi : B \twoheadrightarrow C$ (i.e., the ideal cutting out Z on X) and let J be the kernel of the canonical projection $s : C \otimes_A B \twoheadrightarrow C$ (i.e., the ideal cutting out Z in $Z \times_Y X$). By assumption, I is generated by a regular sequence $\{t_1, \ldots, t_n\}$. Assume that J is generated by a regular sequence $\{g_1, \ldots, g_n\}$ (this is automatic when C is local, which we have seen is a case to which we can reduce ourselves for the proof of (R6)). For any $\varphi \in \text{Hom}_A(C, A)$, let $\dot{\varphi}$ denote the base change to $\text{Hom}_B(C \otimes_A B, B)$.

Choose $\omega \in \omega_{B/A}$ and pick $b_{ij} \in C \otimes_A B$ such that $t_i \otimes 1 = \sum b_{ij} g_j$. It is obvious that $s^*(dg_n \wedge \cdots \wedge dg_1)$ generates $s^* \omega_{C \otimes_A B/C}$ over C, so there is a unique $\alpha \in C$ such that the canonical isomorphism

$$\pi^* \omega_{B/A} \simeq s^* \omega_{C \otimes_A B/C}$$

takes $\pi^*\omega$ to $\alpha \cdot s^*(\mathrm{d}g_n \wedge \cdots \wedge \mathrm{d}g_1)$; note the ordering of the $\mathrm{d}g_i$'s! By using the calculations of Tate in [**MR**, Appendix] and a reduction to the case where $\omega_{B/A}$ is globally free over B, it is easy to see that there is a unique $\varphi_\omega \in \mathrm{Hom}_A(C, A)$ such that

$$\alpha = \pi(\widetilde{\varphi}_\omega(\det(b_{ij}))).$$

This is the φ_ω obtained from (A.1.2), so $(-1)^{n(n-1)/2}\varphi_\omega$ corresponds to the unique $\alpha' = (-1)^{n(n-1)/2}\alpha \in C$ for which $\pi^*\omega = \alpha' s^*(\mathrm{d}g_1 \wedge \cdots \wedge \mathrm{d}g_n)$.

Since $\pi(t_i) = s(t_i \otimes 1) = \sum s(b_{ij})s(g_j)$, for $b \in B$ we have

$$
\begin{aligned}
\pi^*(b\,\mathrm{d}t_1 \wedge \cdots \wedge \mathrm{d}t_n) &= \pi(b)s(\det(b_{ij}))s^*(\mathrm{d}g_1 \wedge \cdots \wedge \mathrm{d}g_n) \\
&= (-1)^{n(n-1)/2}\pi(b)s(\det(b_{ij}))s^*(\mathrm{d}g_n \wedge \cdots \wedge \mathrm{d}g_1).
\end{aligned}
$$

Thus, by [**MR**, Appendix, Thm A3], for $b \in B$ and $\omega_b = b\,\mathrm{d}t_1 \wedge \cdots \wedge \mathrm{d}t_n$,

$$\varphi_{\omega_b} = (-1)^{n(n-1)/2}\mathrm{Tr}_{C/A}(\pi(b)(\cdot)).$$

Evaluating this at 1 and multiplying by $(-1)^{n(n-1)/2}$, we obtain (R6). ∎

By means of (R5) and (R6), for many proofs it suffices to prove the result up to a 'universal sign' in the noetherian case (in which case (A.1.3) applies and Theorem 2.8.1 becomes useful). This is illustrated by our proofs of (R3), (R2), and (R4) below. The proofs of (R1), (R7), (R9), and (R10) will require a complete different method, based on cohomology with supports.

PROOF. (of (R3)). As in the proof of (R6), we can reduce to the case where all schemes are affine. Then by direct limit arguments and base change we can assume that all schemes are noetherian, so we may use the description (A.1.3). Also, by (R6), if we can prove (R3) up to a universal sign $\epsilon_{n,p} = \pm 1$, then the case of $X' = \mathbf{A}_Y^n \hookrightarrow \mathbf{A}_Y^{n+p} = X$ and $\omega = \mathrm{d}T_1 \wedge \cdots \wedge \mathrm{d}T_n$ shows that $\epsilon_{n,p} = 1$. Thus, we may ignore universal signs depending upon n and p.

The relevant scheme diagram is

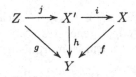

Consider the two derived category diagrams (with the right column of (A.2.1) equal to the left column of (A.2.2))

(A.2.1)

$$(ij)^\flat f^\sharp \mathcal{O}_Y =\!=\!=\!=\!=\!=\!= (ij)^\flat(\omega_{X/Y}[n+p])$$

$$\uparrow_{\psi_{ij,f}} \qquad\qquad\qquad \downarrow_{\simeq}$$

$$g^\flat \mathcal{O}_Y \qquad\qquad\qquad j^\flat i^\flat(\omega_{X/Y}[n+p])$$

$$\downarrow_{\psi_{j,h}} \qquad\qquad\qquad \downarrow_{\eta_i}$$

$$j^\flat h^\sharp \mathcal{O}_Y \qquad j^\flat(\omega_{X'/X}[-p] \otimes i^*\omega_{X/Y}[n+p])$$

$$\downarrow_{\zeta'_{i,f}}$$

$$j^\flat(\omega_{X'/Y}[n])$$

(A.2.2)

$$(ij)^\flat(\omega_{X/Y}[n+p]) \xrightarrow{\ \eta_{ij}\ } \omega_{Z/X}[-n-p] \otimes (ij)^*\omega_{X/Y}[n+p]$$

$$\downarrow_{\simeq} \qquad\qquad\qquad\qquad \downarrow_{\zeta'_{j,i}}$$

$$j^\flat i^\flat(\omega_{X/Y}[n+p]) \qquad \omega_{Z/X'}[-n] \otimes j^*(\omega_{X'/X}[-p] \otimes i^*\omega_{X/Y}[n+p])$$

$$\downarrow_{\eta_i} \qquad\qquad\qquad\qquad \uparrow_{\simeq}$$

$$j^\flat(\omega_{X'/X}[-p] \otimes i^*\omega_{X/Y}[n+p]) \qquad \omega_{Z/X'}[-n] \otimes j^*\omega_{X'/Y}[n]$$

$$\uparrow_{\zeta'_{i,f}} \qquad\qquad\nearrow_{\eta_j}$$

$$j^\flat(\omega_{X'/Y}[n])$$

By functoriality and Theorem 2.5.1, (A.2.1) commutes. Thus, by explicitly calculating the right column of (A.2.1) up to a universal sign (depending on n and p), it is easy to see that the equality of (R3) up to a universal sign $\epsilon_{n,p}$ is a consequence of the commutativity of (A.2.2) up to a universal sign $\epsilon'_{n,p}$.

By Theorem 2.8.1 with $S = Y$, we deduce the commutativity of

$$g^\flat \xrightarrow{\ \psi_{j,h}\ } j^\flat h^\sharp$$

$$\downarrow_{\psi_{ij,f}} \qquad\qquad \downarrow_{\psi_{i,f}}$$

$$(ij)^\flat f^\sharp \xrightarrow[\simeq]{} j^\flat i^\flat f^\sharp$$

so we easily reduce to proving the commutativity (up to a universal sign depending upon n and p) of the diagram of isomorphisms

$$(A.2.3) \qquad h^\sharp \mathscr{O}_Y =\!=\!=\!= \omega_{X'/Y}[n] \xrightarrow{\ \zeta'_{i,f}\ } \omega_{X'/X}[-p] \otimes i^* \omega_{X/Y}[n+p]$$

$$\psi_{i,f} \downarrow \qquad\qquad \nearrow \eta_i$$

$$i^b f^\sharp \mathscr{O}_Y =\!=\!= i^b(\omega_{X/Y}[n+p])$$

(applying j^b to (A.2.3) will finish the proof). Since all terms in (A.2.3) have cohomology supported in degree $-n$, we can check the commutativity (up to a universal sign) of (A.2.3) by looking in degree $-n$ cohomology. Lemma 3.5.3 implies the commutativity of the diagram of degree $-n$ cohomology sheaves, up to a universal sign depending only on n and p. ∎

PROOF. (of (R2)). Without loss of generality, Y is a strictly henselian local artin ring. By (A.1.5), we can reduce to the case $Z' = Z$, so $h = 1$ and $i = f' \circ i'$. In this case, by using the definition of the residue symbol via (A.1.3) and the flat base change compatibility of η_i with respect to the base change by f, we reduce ourselves to proving the commutativity of the diagram of isomorphisms

$$(A.2.4) \qquad\qquad g^b \mathscr{O}_Y \xrightarrow{\ \psi_{i,f}\ } i^b f^\sharp \mathscr{O}_Y$$

$$\psi_{i',ff'} \downarrow \qquad\qquad\qquad \downarrow \psi_{i',f'}$$

$$i'^b (ff')^\sharp \mathscr{O}_Y \xrightarrow[\ \simeq\ ] i'^b f'^\sharp f^\sharp \mathscr{O}_Y$$

This follows from (2.7.17). ∎

PROOF. (of (R4)). By using (R6) and the special case

$$\mathbf{A}_Z^{n+p} \simeq \mathbf{A}_Z^n \times_Z \mathbf{A}_Z^p \to \mathbf{A}_Z^p \to Z$$

with $\omega = dT_1 \wedge \cdots \wedge dT_n$, $\omega' = ds_1 \wedge \cdots \wedge ds_p$, it suffices to prove (R4) up to a universal sign $\epsilon_{n,p}$. Also, we may (and do) assume that Z is locally noetherian.

The relevant scheme diagram is

$$(A.2.5)$$

in which all horizontal maps are closed immersions, with i cut out by $t_1, \ldots, t_n$, k cut out by $s_1, \ldots, s_p$, and j cut out by $s'_1, \ldots, s'_p$. The upper left 'parallelogram' in (A.2.5) is cartesian and h, h', h'' are finite locally free. Let $t^\vee$ (resp. $s^\vee$) denote the global section of $\omega_{U/X}$ (resp. $\omega_{W/Y}$) defined as in §2.2.

By explicit calculation, we see that up to a universal sign which depends only on n and p, the composite isomorphism

(A.2.6)

$$j^*(\omega_{U/X} \otimes i^*\omega_{X/Y}) \otimes h'^*(\omega_{W/Y} \otimes k^*\omega_{Y/Z})$$

$$\uparrow \simeq$$

$$\omega_{V/X}[-n-p] \otimes (ij)^*(\omega_{X/Z}[n+p])$$

$$\uparrow \eta_{ij}$$

$$(ij)^\flat(\omega_{X/Z}[n+p])$$

$$\|$$

$$(ij)^\flat(gf)^\sharp \mathscr{O}_Z$$

$$\uparrow \psi_{ij,gf}$$

$$(hh')^\flat \mathscr{O}_Z$$

$$\uparrow \simeq$$

$$\mathscr{H}om_Z((hh')_*\mathscr{O}_V, \mathscr{O}_Z)$$

induces a map on H^0's which sends the global section $j^*(t^\vee \otimes i^*\omega) \otimes h'^*(s^\vee \otimes k^*\omega')$ of the top sheaf to an $\mathscr{O}_Z$-linear map $\varphi : (hh')_*\mathscr{O}_V \to \mathscr{O}_Z$ that satisfies

$$\varphi(1) = \mathrm{Res}_{X/Z} \begin{bmatrix} \omega \otimes f^*\omega' \\ t_1, \ldots, t_n, s_1, \ldots, s_p \end{bmatrix},$$

and the composite isomorphism

(A.2.7)
$$j^*(\omega_{U/X} \otimes i^*\omega_{X/Y}) \otimes h'^*(\omega_{W/Y} \otimes k^*\omega_{Y/Z})$$

$$\uparrow \eta_i$$

$$\mathbf{L}j^*(i^{\flat}\omega_{X/Y}[n]) \overset{\mathbf{L}}{\otimes} h'^*(\omega_{W/Y} \otimes k^*\omega_{Y/Z})$$

$$\uparrow \psi_{i,f}$$

$$\mathbf{L}j^*h''^{\flat}\mathcal{O}_Y \overset{\mathbf{L}}{\otimes} h'^*(\omega_{W/Y} \otimes k^*\omega_{Y/Z})$$

$$\|$$

$$j^*\mathscr{H}om_Y(h''_*\mathcal{O}_U, \mathcal{O}_Y) \otimes h'^*(\omega_{W/Y} \otimes k^*\omega_{Y/Z})$$

$$\|$$

$$\mathscr{H}om_W(h'_*\mathcal{O}_V, \mathcal{O}_W) \otimes_W (\omega_{W/Y} \otimes k^*\omega_{Y/Z})$$

$$\downarrow \simeq$$

$$\mathscr{H}om_W(h'_*\mathcal{O}_V, \omega_{W/Y} \otimes k^*\omega_{Y/Z})$$

induces a map on H^0's which sends the global section $j^*(t^\vee \otimes i^*\omega) \otimes h'^*(s^\vee \otimes k^*\omega')$ to an $\mathcal{O}_W$-linear map $\varphi': h'_*\mathcal{O}_V \to \omega_{W/Y} \otimes k^*\omega_{Y/Z}$ that satisfies

$$\varphi'(1) = s^\vee \otimes k^* \left(\mathrm{Res}_{X/Y} \begin{bmatrix} \omega \\ t_1, \ldots, t_n \end{bmatrix} \omega' \right).$$

Using Theorem 2.5.1 and functoriality with respect to η_i, the above calculations of (A.2.6) and (A.2.7) easily imply that (R4) is equivalent to the commutativity, up to a universal sign $\epsilon'_{n,p}$, of the diagram

(A.2.8)

$$
\begin{array}{ccc}
h'^b h^b \mathcal{O}_Z & \xleftarrow{\quad\simeq\quad} & (hh')^b \mathcal{O}_Z \\[2pt]
\Big\downarrow{\scriptstyle\psi_{k,g}} & & \Big\downarrow{\scriptstyle\psi_{ij,gf}} \\[2pt]
h'^b k^b g^\sharp \mathcal{O}_Z & & (ij)^b (gf)^\sharp \mathcal{O}_Z \\[2pt]
\Big\downarrow{\scriptstyle h^b(\eta_k)} & & \Big\downarrow{\scriptstyle\simeq} \\[2pt]
h'^b(\omega_{W/Y}[-p] \otimes k^* g^\sharp \mathcal{O}_Z) & & j^b i^b f^\sharp g^\sharp \mathcal{O}_Z \\[2pt]
\beta\Big\uparrow{\scriptstyle\simeq} & & \Big\downarrow{\scriptstyle\eta_j} \\[2pt]
h'^b \mathcal{O}_W \overset{\mathbf{L}}{\otimes} h'^*(\omega_{W/Y}[-p] \otimes k^* g^\sharp \mathcal{O}_Z) & & \omega_{V/U}[-p] \overset{\mathbf{L}}{\otimes} (\mathbf{L}j^*) i^b f^\sharp g^\sharp \mathcal{O}_Z \\[2pt]
\simeq\Big\downarrow & & \Big\| \\[2pt]
h'^*\omega_{W/Y}[-p] \overset{\mathbf{L}}{\otimes} h'^b \mathcal{O}_W \overset{\mathbf{L}}{\otimes} h'^* k^* g^\sharp \mathcal{O}_Z & \xleftarrow[1\otimes\beta']{} & h'^*\omega_{W/Y}[-p] \overset{\mathbf{L}}{\otimes} (\mathbf{L}j^*) i^b f^\sharp g^\sharp \mathcal{O}_Z
\end{array}
$$

where the definition of $\beta : h'^b \mathcal{O}_W \overset{\mathbf{L}}{\otimes} h'^* \simeq h'^b$ uses the fact that h' is finite locally free and the isomorphism $\beta' : (\mathbf{L}j^*) i^b f^\sharp \simeq h'^b \mathcal{O}_W \overset{\mathbf{L}}{\otimes} h'^* k^*(\cdot)$ is given by the composite isomorphism

$$
\begin{array}{ccc}
(\mathbf{L}j^*) i^b f^\sharp & \xleftarrow{\quad\psi'_{i,f}\quad} (\mathbf{L}j^*) h''^b & \xleftarrow{\quad\simeq\quad} \mathbf{L}j^*(h''^b(\mathcal{O}_Y) \overset{\mathbf{L}}{\otimes} h''^*(\cdot)) \\[6pt]
 & & \Big\| \\[6pt]
 & h'^b \mathcal{O}_W \overset{\mathbf{L}}{\otimes} h'^* \mathbf{L}k^*(\cdot) \xleftarrow{\quad\simeq\quad} & \mathcal{H}om_W(h'_* \mathcal{O}_V, \mathcal{O}_W) \overset{\mathbf{L}}{\otimes} (\mathbf{L}j^*) h''^*(\cdot)
\end{array}
$$

Since η_j is of formation compatible with flat base change [**RD**, III, 7.4(b)], the diagram of isomorphisms

$$
\begin{array}{ccc}
h'^* k^b & \xrightarrow{\quad\simeq\quad} & j^b h''^* \\[2pt]
\Big\downarrow{\scriptstyle h'^*(\eta_k)} & & \Big\downarrow{\scriptstyle\eta_j} \\[2pt]
h'^*(\omega_{W/Y}[-p] \overset{\mathbf{L}}{\otimes} \mathbf{L}k^*(\cdot)) & \xrightarrow{\quad\simeq\quad} & \omega_{V/U}[-p] \overset{\mathbf{L}}{\otimes} (\mathbf{L}j^*) h''^*(\cdot)
\end{array}
$$

commutes. Thus, we can 'replace' $h^b(\eta_k)$ in (A.2.8) by a suitable composite involving η_j. More precisely, we are reduced to checking the commutativity, up to a universal sign depending only on n and p, of the outside edge of the concatenation of following diagrams of isomorphisms, in which all functors are understood to be evaluated on the object $\mathscr{O}_Z[0]$ and we glue the right side of the top diagram to the left side of the bottom diagram:

(A.2.9)

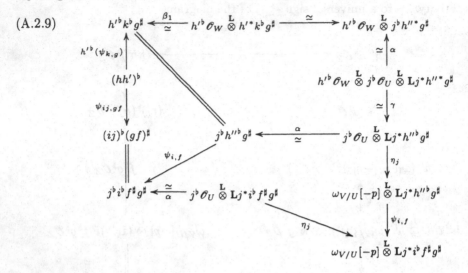

(A.2.10)

The maps denoted α use the isomorphism $j^\flat \mathscr{O}_U \overset{\mathbf{L}}{\otimes} \mathbf{L}j^*(\cdot) \simeq j^\flat$ on *bounded complexes of locally free sheaves of finite rank* and the isomorphism γ uses the isomorphism

$$h'^\flat \mathscr{O}_W \overset{\mathbf{L}}{\otimes} \mathbf{L}j^* h''^* g^\sharp \mathscr{O}_Z \overset{\simeq}{\longrightarrow} \mathscr{H}om_W(h'_*\mathscr{O}_V, \mathscr{O}_W) \otimes_W k^* g^\sharp \mathscr{O}_Z$$

$$\downarrow \simeq$$

$$j^* \mathscr{H}om_Y(h''_*\mathscr{O}_U, g^\sharp \mathscr{O}_Z)$$

$$\parallel$$

$$\mathbf{L}j^*(h''^\flat g^\sharp \mathscr{O}_Z)$$

All other maps in (A.2.9), (A.2.10) are defined on the level of natural transformations of functors (which are then evaluated at $\mathscr{O}_Z[0]$). The isomorphisms β_1, β_2, β_3 use the fact that h' and h'' are finite locally free. More precisely, the definition of β_1 uses the isomorphism $h'^b \simeq h'^b \mathscr{O}_W \overset{\mathbf{L}}{\otimes} h'^*$, the definition of β_2 uses the isomorphism of locally free coherent $\mathscr{O}_V$-modules

(A.2.11) $$\mathscr{H}om_W(h'_*\mathscr{O}_V, \mathscr{O}_W) \simeq j^*\mathscr{H}om_Y(h''_*\mathscr{O}_U, \mathscr{O}_Y)$$

(where $\mathscr{H}om_Y(h''_*\mathscr{O}_U, \mathscr{O}_Y)$ is viewed as an $\mathscr{O}_U$-module), and the definition of β_3 uses a 'flip' and the isomorphism

$$h''^b \mathscr{O}_Y \overset{\mathbf{L}}{\otimes} h''^*(\mathscr{F}^\bullet) \simeq \mathscr{H}om_Y(h''_*\mathscr{O}_U, \mathscr{O}_Y) \otimes_Y \mathscr{F}^\bullet \simeq \mathscr{H}om_Y(h''_*\mathscr{O}_U, \mathscr{F}^\bullet).$$

Beware that the $\mathscr{O}_U$-module $\mathscr{H}om_Y(h''_*\mathscr{O}_U, \mathscr{O}_Y)$ in (A.2.11) is generally not free, so the right side of (A.2.11) generally does not compute $\mathbf{L}j^*h''^b\mathscr{O}_Y$. The reason this is not a problem is that

$$\omega_{V/U}[-p] \overset{\mathbf{L}}{\otimes} \mathbf{L}j^*h''^*(\cdot) \simeq h'^*(\omega_{W/Y}[-p] \overset{\mathbf{L}}{\otimes} \mathbf{L}k^*(\cdot))$$

is a pullback from W and the left side of (A.2.11) is $\mathscr{O}_W$-free.

The diagram (A.2.10) and all but one of the subdiagrams in (A.2.9) are easily seen to commute up to a universal sign depending only on n and p. The only non-trivial subdiagram is the triangle on the left in (A.2.9), which is

$$
\begin{array}{ccccc}
(hh')^b & \xrightarrow{\;\simeq\;} & h'^b h^b & \xrightarrow{\;\psi_{k,g}\;} & h'^b k^b g^\sharp \\
{\scriptstyle \psi_{ij,gf}}\downarrow & & & & \| \\
(ij)^b (gf)^\sharp & \xrightarrow[\;\simeq\;]{} & j^b i^b f^\sharp g^\sharp & \xleftarrow[\;\psi_{i,f}\;]{} & j^b h''^b g^\sharp
\end{array}
$$

and this commutes by Theorem 2.8.1. ∎

It remains to prove (R1), (R7), (R9), (R10). Since (2.7.36) commutes with any base change (which is relevant for (R10)), it is clearly enough (by the Krull Intersection Theorem and a faithful flatness argument) to prove these four properties when the base is noetherian, and even an artin local ring with an algebraically closed residue field [**EGA**, 0_{III}, 10.3.1].

We begin by showing that (R7) is a consequence of (R10) and other properties we have already proven.

PROOF. (of (R7), conditional on (R10)) Without loss of generality, Y is a local artin scheme with an algebraically closed residue field. Thus, the closed subscheme Z on X is a finite set of points. Since it suffices to treat the connected components of Z separately (by (A.1.5)), we may assume that Z consists of a single closed point $z_0 \in X$. Let $X \to \mathbf{A}_Y^n$ be the map which pulls T_i back to $t_i \in \Gamma(X, \mathscr{O}_X)$ for all $1 \le i \le n$. By openness of the quasi-finite locus [**EGA**, IV$_3$, 13.1.4], shrinking X around z_0 allows us to assume that $X \to \mathbf{A}_Y^n$ is quasi-finite and separated. By (R2) and the fact that $k(z_0)$ is algebraically closed, we can replace X by an étale neighborhood of z_0 which is finite flat of constant rank d over an étale neighborhood U of $0 \in \mathbf{A}_Y^n$ (to see this, use direct limit arguments and the structure theorem for quasi-finite separated schemes over a henselian local base [**EGA**, IV$_4$, 18.5.11]).

Thus, by (R10), (R2), and (2.7.41) (with $b = 1$), it suffices to prove

$$\mathrm{Res}_{\mathbf{A}_Y^n/Y} \left[\begin{array}{c} dT_1 \wedge \cdots \wedge dT_n \\ T_1^{k_1}, \ldots, T_n^{k_n} \end{array} \right] = 0$$

for positive integers k_i not all equal to 1. By (R4), we are reduced to showing

$$(\mathrm{A.2.12}) \qquad\qquad \mathrm{Res}_{\mathbf{A}_Y^1/Y} \left[\begin{array}{c} dT \\ T^k \end{array} \right] = 0$$

for $k > 1$ and any scheme Y, which we may assume is affine. Define $R = \mathrm{H}^0(Y, \mathscr{O}_Y)$ and let $C = R[T]/(T^k)$. We will prove (A.2.12) by using the explicit description of the residue symbol via Tate's calculations, as in the proof of (R6).

Consider the commutative diagram of R-algebras

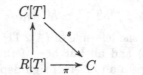

where s is C-linear and $s(T) = \pi(T) \in C$. The kernel of s is generated by $g = T - \pi(T)$. Since $\mathrm{d}g = \mathrm{d}T$, we want to find the unique R-linear $\varphi : C \to R$ such that

$$\pi(\widetilde{\varphi}(b)) = 1,$$

where $\widetilde{\varphi} : C[T] \to R[T]$ is the extension of scalars by φ and $T^k = b \cdot (T - \pi(T))$ in $C[T]$. It will then remain to show that $\varphi(1) = 0$.

Since $\pi(T)^k = 0$, obviously $b = T^{k-1} + \pi(T)T^{k-2} + \cdots + \pi(T)^{k-1}$. Thus, $\pi(\widetilde{\varphi}(b)) = 1$ is equivalent to $\varphi(\pi(T)^i) = 0$ for $0 \le i < k-1$ and $\varphi(\pi(T)^{k-1}) = 1$. Since $0 < k-1$, we deduce $\varphi(1) = 0$, as desired.   ∎

Our method of proof of the remaining properties (R1), (R9), and (R10) is based on a cohomological result from Berthelot's thesis. Before stating this result, we first want to show that in order to prove (R1), (R9), and (R10), it suffices to consider the case in which $Y = \mathrm{Spec}(R)$ for a local Gorenstein artin ring R (so $\mathscr{O}_Y[0]$ is a *residual complex* on Y). We can at least assume that $Y = \mathrm{Spec}(R)$ where R is a local artin ring with an algebraically closed residue field. There is a regular local ring surjecting onto R, so there is clearly a local Gorenstein artin ring $\widetilde{R}$ surjecting onto R. Since Z is a finite set of closed points on X, it follows from (A.1.5) that we may assume Z is a single closed point z_0 on X and then we may replace X by any desired open around Z.

Since smooth maps always locally lift across nilpotent thickenings of the base [**SGA1**, III, Thm 4.1], so we can shrink X around the point z_0 so that there is a smooth $\widetilde{R}$-scheme $\widetilde{X}$ whose base change to R is X. Moreover, if we shrink X to be affine, so $\widetilde{X}$ is affine, then we may assume that the t_j's and the differential forms under consideration all lift to $\widetilde{X}$. Thus, by (R5), it suffices for the proofs of (R1) and (R9) to consider the case where the base is a local Gorenstein artin scheme.

We want to apply a similar argument to the proof of (R10). Since (2.7.36) is compatible with base change, we can use (R2) to justify replacing X with any étale neighborhood U of z_0 and replacing X' with $X' \times_X U$. By direct limit arguments and the structure theorem for quasi-finite separated schemes over a henselian local base [**EGA**, IV$_4$, 18.5.11], we can find a suitable U so that we may assume that X' is a finite disjoint union of open subschemes, each of which contains at most one point over z_0. By (A.1.5), we are reduced to the case in which there is a unique point z_0' on X' over $z_0 \in X$. It remains to check that the *finite* map g in (R10) can also be lifted across a nilpotent thickening of Y, after shrinking X around z_0.

There is a Zariski open V (resp. V') around z_0 in X (resp. z_0' in X') so that V (resp. V') lifts to a smooth scheme over $\widetilde{Y} = \mathrm{Spec}(\widetilde{R})$. Since $g^{-1}(z_0) = \{z_0'\}$, we can shrink V and V' so that V is affine and $V' = g^{-1}(V)$. Replacing X by V, we can assume that there are smooth $\widetilde{Y}$-schemes $\widetilde{X}$ and $\widetilde{X}'$ lifting the smooth Y-schemes X and X' respectively. Since $\widetilde{X}'$ is affine and $\widetilde{X}$ is smooth over $\widetilde{Y} = \mathrm{Spec}(\widetilde{R})$, by the lifting property of smooth maps we can lift $X' \to X \hookrightarrow \widetilde{X}$ to a map $\widetilde{g} : \widetilde{X}' \to \widetilde{X}$. This is the desired finite lifting of g, and completes the reduction of the proof of (R10) to the case of a local Gorenstein artin base Y.

Thus, we now may (and do) assume for the proofs of (R1), (R9) and (R10) that $Y = \mathrm{Spec}(R)$ is a local Gorenstein artin scheme and X is smooth, separated, and quasi-compact over Y with pure relative dimension n. These hypotheses can even be strengthened to require that X is affine, Z is a single point, and Y has an algebraically closed residue field, but we avoid these extra assumptions unless they are needed.

Over a local artin Gorenstein base, Berthelot observed that residual complexes can be used to construct a nice relation between the residue symbol and cohomology with supports, as we now explain. Let $f : X \to Y$ be a quasi-compact, smooth map with pure relative dimension n and $t_1, \ldots, t_n \in \Gamma(X, \mathscr{O}_X)$ functions cutting out a finite Y-scheme Z. Let $i : Z \hookrightarrow X$ denote the canonical closed immersion. Assume that Y is a local *Gorenstein* artin scheme, so $\mathscr{O}_Y = \mathscr{O}_Y[0]$ is a *residual complex* on Y and therefore the complex $f^\Delta \mathscr{O}_Y$ is a residual complex on X. More precisely, by (3.2.4) the complex of *injectives* $f^\Delta \mathscr{O}_Y = E(f^\sharp \mathscr{O}_Y)$ is supported in degrees between $-n$ and 0, and by Lemma 3.2.1 this complex is canonically an injective resolution of $\omega_{X/Y}[n]$. From (3.1.4), (3.1.7), and the definition of the functor $E_{Z\bullet}$ in §3.1, the degree 0 term of $f^\Delta \mathscr{O}_Y$ is

(A.2.13) $\qquad (f^\Delta \mathscr{O}_Y)^0 = E(\omega_{X/Y})^n = \bigoplus_{x \in X^0} \underline{H}^n_{\{x\}}(\omega_{X/Y}),$

where X^0 is the set of closed points of X (i.e., the closed irreducible sets with codimension n) and for each $x \in X^0$, $\underline{H}^\bullet_{\{x\}}$ is the derived functor of the functor "sections supported at x". In accordance with (3.1.7), the identitification (A.2.13) involves no intervention of signs in the sense that if $\omega_{X/Y} \to \mathscr{I}^\bullet$ is an injective resolution used to compute $E(\omega_{X/Y})$, then

$$f^{\sharp}\mathscr{O}_Y = \omega_{X/Y}[n] \to \mathscr{I}^{\bullet}[n]$$

is used to compute $f^{\Delta}\mathscr{O}_Y = E(f^{\sharp}\mathscr{O}_Y)$. The relevant example below is the injective resolution

$$\omega_{X/Y} = \mathrm{H}^0(E(\omega_{X/Y})) \to E(\omega_{X/Y}),$$

whose n-fold translation is the 'preferred' resolution $\omega_{X/Y}[n] \to f^{\Delta}\mathscr{O}_Y$.

By Theorem 3.4.1, there is an $\mathscr{O}_Y$-linear 'trace' map

$$(A.2.14) \qquad \mathrm{Tr}_{f,\mathscr{O}_Y} : f_* f^{\Delta}\mathscr{O}_Y \to \mathscr{O}_Y.$$

This is usually just a map of graded sheaves, though it is a map of complexes when f is proper. In degree 0, we can explicate (A.2.14) as a map

$$(A.2.15) \qquad \bigoplus_{x \in X^0} \mathrm{H}^n_{\{x\}}(X, \omega_{X/Y}) \to R$$

where the local cohomology is computed with respect to the injective resolution $E(\omega_{X/Y}) = (f^{\Delta}\mathscr{O}_Y)[-n]$ of $\omega_{X/Y}$ (note that the spectral sequence relating $\mathrm{H}^{\bullet}_{\{x\}}$ and $\underline{\mathrm{H}}^{\bullet}_{\{x\}}$ degenerates since all $\underline{\mathrm{H}}^{\bullet}_{\{x\}}$ are supported on $\{x\}$ and so have vanishing higher cohomology, and f_* commutes with direct sums since X is noetherian). Since

$$\mathrm{H}^n_Z \simeq \bigoplus_{z \in Z} \mathrm{H}^n_{\{z\}},$$

we obtain an R-linear trace map [**Be**, VII, (1.2.5)]

$$(A.2.16) \qquad \mathrm{Tr}_{f,Z} : \mathrm{H}^n_Z(X, \omega_{X/Y}) \to R.$$

which depends on Z only through its underlying topological space and is unaffected by replacing X with an open subscheme containing Z.

Assume that there is an open affine U in X around Z (e.g., X is affine, or Z consists of a single point, which was the case considered by Berthelot). Using U and the t_j's which define the scheme structure on Z, we can also define a natural map

$$(A.2.17) \qquad \Gamma(Z, i^*\omega_{X/Y}) \to \mathrm{H}^n_Z(X, \omega_{X/Y})$$

which is independent of U (but depends on the *ordered set* $\{t_1, \ldots, t_n\}$). In fact, such a natural map can be defined with $\omega_{X/Y}$ replaced by an arbitrary quasi-coherent $\mathscr{O}_X$-module $\mathscr{F}$, so we consider this more general setting for conceptual clarity. To define (A.2.17), we need to recall how to compute the 'cohomology with supports' of quasi-coherent sheaves on an affine scheme in terms of direct limits of the cohomology of 'dual' Koszul complexes, as in [**EGA**, III$_1$, §1]. In order to remove any confusion about signs, we explain the construction in detail (since the natural reference [**SGA2**, II, Prop 5] involves the interference of signs in the coboundary maps).

Choose a positive integer k. Let Z_k denote the infinitesimal thickening of Z defined by $\{t_1^k, \ldots, t_n^k\}$, so the canonical map of functors of a *quasi-coherent* variable

$$\varinjlim \mathrm{Hom}_X(\mathscr{O}_{Z_k}, \cdot) \to \Gamma_Z(X, \cdot)$$

is an isomorphism (the limit taken over increasing k). Thus, by Lemmas 2.1.3, 2.1.6 and a universal δ-functor argument, the natural δ-functorial map

$$(A.2.18) \qquad \varinjlim \operatorname{Ext}_X^{\bullet}(\mathscr{O}_{Z_k}, \cdot) \to \operatorname{H}_Z^{\bullet}(X, \cdot)$$

is an isomorphism on the category of quasi-coherent $\mathscr{O}_X$-modules on the noetherian scheme X. The sheaf version is that the natural δ-functorial map of $\mathscr{O}_X$-modules

$$\varinjlim \mathscr{E}xt_X^{\bullet}(\mathscr{O}_{Z_k}, \cdot) \to \underline{\operatorname{H}}_Z^{\bullet}(X, \cdot)$$

is an isomorphism on the category of quasi-coherent $\mathscr{O}_X$-modules, recovering the isomorphism (A.2.18) on the level of global sections (as X is noetherian and the underlying closed subspace Z is finite and discrete). Of course, if we replace X by any open containing Z then the analogous statements remain true and are compatible with the (isomorphism) restriction maps from the setup over X.

The key point is that the Ext terms can be calculated using Koszul complexes, since Z is contained in an affine open U of X. Fix such a choice of U (the end result will be independent of this choice, essentially because of the 'restriction' compatibility mentioned at the end of the previous paragraph). Choose any quasi-coherent $\mathscr{O}_U$-module $\mathscr{F}$ (e.g., $\mathscr{F} = \omega_{X/Y}|_U$). Consider the 'dual' Koszul complex $K^{\bullet}(\mathbf{t}^k, \mathscr{F})$ as in (1.3.25), defined in terms of the module $\Gamma(U, \mathscr{F})$ and the sequence $t_1^k|_U, \ldots, t_n^k|_U$ in $\Gamma(U, \mathscr{O}_U)$. The cohomology groups of this complex are denoted $\operatorname{H}^{\bullet}(\mathbf{t}^k, \mathscr{F})$. Since we may δ-*functorially* compute $\operatorname{Ext}_U^{\bullet}(\mathscr{O}_{Z_k}, \mathscr{F})$ using the complex $\operatorname{Hom}_U^{\bullet}(K_{\bullet}(\mathbf{t}^k), \mathscr{F})$, which is identified with $K^{\bullet}(\mathbf{t}^k, \mathscr{F})$ via multiplication by $(-1)^{p(p+1)/2}$ in degree p (lifting the identity in degree 0), we get a δ-functorial isomorphism

$$(A.2.19) \qquad \operatorname{H}^{\bullet}(\mathbf{t}^k, \mathscr{F}) \simeq \operatorname{Ext}_U^{\bullet}(\mathscr{O}_{Z_k}, \mathscr{F})$$

involving a sign of $(-1)^{p(p+1)/2}$ in degree p. This is the unique such isomorphism which lifts the natural identification in degree 0 (for variable quasi-coherent $\mathscr{F}$ on U).

Using multiplication by $t_{i_1} \cdots t_{i_p}$ on the basis vector $e_{i_1}^{\vee} \wedge \cdots \wedge e_{i_p}^{\vee}$ of $K^p(\mathbf{t}^k, \mathscr{F})$, there is a natural map of complexes $K^{\bullet}(\mathbf{t}^k, \mathscr{F}) \to K^{\bullet}(\mathbf{t}^{k+1}, \mathscr{F})$. It is easy to check that the induced map on cohomology makes (A.2.19) compatible with respect to change in k. The resulting direct limit of cohomology $\varinjlim \operatorname{H}^{\bullet}(\mathbf{t}^k, \mathscr{F})$ is denoted $\operatorname{H}^{\bullet}((\mathbf{t}), \mathscr{F})$. Every element in these cohomology groups is killed by some power of $(t_1 \cdots t_n)|_U$, so replacing U by a smaller open affine in X around Z has no effect. In particular, we can unambiguously define $\operatorname{H}^{\bullet}((\mathbf{t}), \mathscr{F})$ for any quasi-coherent $\mathscr{O}_X$-module $\mathscr{F}$ (i.e., there is no dependence on U). Upon passing to the limit over k in (A.2.19) and using (A.2.18), we get the unique δ-functorial isomorphism between functors of quasi-coherent $\mathscr{O}_X$-modules

$$(A.2.20) \qquad \operatorname{H}^{\bullet}((\mathbf{t}), \cdot) \simeq \operatorname{H}_Z^{\bullet}(X, \cdot)$$

which is the canonical map in degree 0. This map involves a sign of $(-1)^{p(p+1)/2}$ in degree p in the identification

$$\operatorname{H}^p((\mathbf{t}^k), \mathscr{F}) \simeq \operatorname{H}^p(\operatorname{Hom}^{\bullet}(K_{\bullet}(\mathbf{t}^k), \mathscr{F}(U)))$$

for variable quasi-coherent $\mathscr{F}$ on X. Although (A.2.20) is useful in some contexts, and the universal sign in each degree is often harmless when verifying certain kinds of commutative diagrams in a fixed degree, we prefer to avoid the introduction of this extra sign (for a reason to be given shortly). Following Berthelot [**Be**, VI, (3.1.5)] and following our convention for how to compute Ext's via projective resolutions in the first variable, we *define* (A.2.17) to be the special case $\mathscr{F} = \omega_{X/Y}$ in the composite

$$(A.2.21) \quad \Gamma(Z, i^*\mathscr{F}) = H^n(\mathrm{Hom}^\bullet(K_\bullet(\mathbf{t}), \mathscr{F})) = \mathrm{Ext}_X^n(\mathscr{O}_Z, \mathscr{F}) \to H_Z^n(X, \mathscr{F}),$$

where the first equality is defined *without* the intervention of signs. Of course, (A.2.21) depends on the *ordered* set $\{t_1, \ldots, t_n\}$.

Using (A.2.21) (= (A.2.17)), for any $s \in \Gamma(X, \mathscr{F})$ we denote (out of analogy with [**Be**, p. 532]) the image of

$$i^*s \in \Gamma(Z, i^*\mathscr{F})$$

in $H_Z^n(X, \mathscr{F})$ by $s/(t_1 \cdots t_n)$. The scheme structure on Z clearly matters here, as does the *ordering* of the t_i's, so the symbol $s/t_1 \cdots t_n$ is in general *not* invariant under permutation of the t_j's. However, this 'fraction' notation is reasonable because for any positive integers $k_1, \ldots, k_n$ and non-negative integers $r_1, \ldots, r_n$,

$$s/(t_1^{k_1} \cdots t_n^{k_n}) = (t_1^{r_1} \cdots t_n^{r_n} s)/(t_1^{k_1 + r_1} \cdots t_n^{k_n + r_n})$$

as elements in $H_Z^n(X, \mathscr{F})$ (this is an easy consequence of the definitions). Moreover, it is clear from (A.2.20) that if X is affine, then as we vary $s \in \Gamma(X, \mathscr{F})$ and the positive integers $k_1, \ldots, k_n$, the cohomology class of $s/(t_1^{k_1} \cdots t_n^{k_n})$ runs through all elements in $H_Z^n(X, \mathscr{F})$.

The general relation between residues and cohomology with supports is given by the following 'corrected' lemma:

LEMMA A.2.1. [**Be**, VII, 1.2.5] *For* $f : X \to Y$, $t_1, \ldots, t_n$, $i : Z \hookrightarrow X$, ω *as above* (*in particular, assume there is an open affine in* X *containing* Z), *with* $Y = \mathrm{Spec}(R)$ *a local Gorenstein artin scheme, we have*

$$\mathrm{Tr}_{f,Z}(\omega/(t_1 \cdots t_n)) = \mathrm{Res}_{X/Y} \begin{bmatrix} \omega \\ t_1, \ldots, t_n \end{bmatrix}$$

in R.

Before giving the proof, we make several remarks. First of all, Berthelot uses the (incorrect) definition of the residue symbol as given in [**RD**], which is ours without the extra sign of $(-1)^{n(n-1)/2}$. Fortunately, this will be cancelled out by our decision (following Berthelot) to define (A.2.17) *without* the interference of signs, bypassing (A.2.20). The point is that the definition of the fundamental local isomorphism (2.5.1) involves a sign of $(-1)^{n(n+1)/2}$ (which was needed to make Theorem 2.5.1 hold) and this fundamental local isomorphism is lurking in the definition of (A.2.16) (it is used in (A.2.14), the definition of which uses Theorem 3.4.1, whose lengthy proof in [**RD**] we have not discussed). Thus, avoiding the sign of $(-1)^{n(n+1)/2}$ in the definition of (A.2.17) will turn out to contribute a factor of $(-1)^{n(n+1)/2}$. Meanwhile, the compatibility of (2.5.3) with respect to the m-fold translation functor $[m]$ *does* involve an intervention of the sign $(-1)^{nm}$. The relevant case for us will be $m = n$. Putting this all together, our modification of Berthelot's calculations will have an extra sign of $(-1)^{n(n-1)/2}(-1)^{n(n+1)/2}(-1)^{n^2} = 1$. Strictly speaking, Berthelot considers the case where the Gorenstein artin base ring is an artinian quotient of a discrete valuation ring, but his arguments carry over (with the above sign modifications) to the situation we are considering. Thus, the proof we will give is essentially the same as the proof of [**Be**, VII, 1.2.5].

We note here that the absence of a sign in Lemma A.2.1 is irrelevant in what follows; what matters is that Lemma A.2.1 is true up to a universal sign that depends only on n (when verifying the commutativity of certain kinds of diagrams, such a sign is often harmless). In particular, if the reader prefers to define (A.2.17) by using (A.2.20) and the canonical isomorphism

$$\Gamma(Z, i^* \mathscr{F}) \simeq \mathrm{H}^n(\mathbf{t}, \mathscr{F})$$

(defined without the intervention of signs), then Lemma A.2.1 should have a sign of $(-1)^{n(n+1)/2}$; this change wouldn't affect anything that follows. Although we have not formulated Lemma A.2.1 in a manner for which it makes sense to consider compatibility with respect to composites in f, the fact that both residue symbols and residual complex trace maps are compatible with composites (by (R4) and Theorem 3.4.1(2) respectively) suggests that it is probably better to define (A.2.17) to get no sign in Lemma A.2.1 rather than to get a sign of $(-1)^{n(n+1)/2}$.

PROOF. (of Lemma A.2.1)

Recall that the geometric setup we are trying to study is

(A.2.22)

$$\begin{array}{ccc} Z & \xrightarrow{\ i\ } & X \\ & {\scriptstyle g}\searrow & \downarrow{\scriptstyle f} \\ & & Y \end{array}$$

We first want to reformulate the map (A.2.21) in terms of maps of sheaves. For any $\mathscr{O}_X$-module $\mathscr{F}$, the diagram of δ-functors

$$\begin{array}{ccc} \mathrm{Ext}_X^\bullet(\mathscr{O}_Z, \mathscr{F}) & \longrightarrow & \mathrm{H}^0(X, \mathscr{E}xt_X^\bullet(\mathscr{O}_Z, \mathscr{F})) \\ \downarrow & & \downarrow \\ \mathrm{H}_Z^\bullet(X, \mathscr{F}) & \longrightarrow & \mathrm{H}^0(X, \underline{\mathrm{H}}_Z^\bullet(\mathscr{F})) \end{array}$$

is commutative, where $\underline{\mathrm{H}}_Z^\bullet(\cdot)$ denotes the derived functor of

$$\mathscr{F} \rightsquigarrow \ker(\mathscr{F} \to j_* j^* \mathscr{F}),$$

with $j : X - Z \to X$ the canonical open immersion. Note that the horizontal maps are *isomorphisms* when $\mathscr{F}$ is quasi-coherent. The Koszul construction implicit in the equalities in (A.2.21) sheafifies to define a natural map

$$i_* i^* \mathscr{F} \to \mathscr{E}xt_X^n(\mathscr{O}_Z, \mathscr{F})$$

for quasi-coherent $\mathscr{F}$ on X (which can equivalently be viewed as a natural map

$$(A.2.23) \qquad i^*\mathscr{F} \to \mathscr{E}xt_X^n(\mathscr{O}_Z, \mathscr{F})$$

if we view the $\mathscr{O}_X$-modules $\mathscr{E}xt_X^\bullet(\mathscr{O}_Z, \cdot)$ as $\mathscr{O}_Z$-modules), so we can identify (A.2.21) with $\mathrm{H}^0(X, \cdot)$ applied to the composite

$$(A.2.24) \qquad i_*i^*\mathscr{F} \to \mathscr{E}xt_X^n(\mathscr{O}_Z, \mathscr{F}) \to \underline{\mathrm{H}}_Z^n(\mathscr{F}) = \bigoplus_{z \in Z} \underline{\mathrm{H}}_{\{z\}}^n(\mathscr{F}).$$

When $\mathscr{F} = \omega_{X/Y}$, the composite of (A.2.24) with the natural map

$$(A.2.25) \qquad \underline{\mathrm{H}}_Z^n(\omega_{X/Y}) \to \bigoplus_{x \in X^0} \underline{\mathrm{H}}_{\{x\}}^n(\omega_{X/Y}) = (f^\Delta \mathscr{O}_Y)^0$$

to the degree 0 term of the residual complex $f^\Delta \mathscr{O}_Y$ certainly factors through the subsheaf killed by the ideal sheaf of Z.

Now recall from (3.3.17) (also see (3.4.5)) that for the residual complex $f^\Delta \mathscr{O}_Y$ we have the equality of complexes of $\mathscr{O}_X$-modules

$$\mathrm{Hom}_X(\mathscr{O}_Z, f^\Delta \mathscr{O}_Y) = i_*i^\Delta f^\Delta \mathscr{O}_Y.$$

Finally, by Theorem 3.4.1(3), we have a commutative diagram of complexes

$$(A.2.26)$$

$$
\begin{array}{ccc}
\mathscr{H}om_X(\mathscr{O}_Z, f^\Delta \mathscr{O}_Y) & \longrightarrow & f^\Delta \mathscr{O}_Y \\
\| & & \| \\
i_*i^\Delta f^\Delta \mathscr{O}_Y & \xrightarrow{\mathrm{Tr}_{i,f^\Delta \mathscr{O}_Y}} & f^\Delta \mathscr{O}_Y
\end{array}
$$

where the top horizontal map is "evaluate at 1" (keep in mind that $f^\Delta \mathscr{O}_Y$ is a complex of injectives, so the upper left entry coincides with $i_*i^\flat(f^\Delta \mathscr{O}_Y)$). Putting together (A.2.24), (A.2.25), (A.2.26), we arrive at a map of complexes

$$(A.2.27) \qquad i_*i^*\omega_{X/Y} \longrightarrow \mathscr{E}xt_X^n(\mathscr{O}_Z, \omega_{X/Y}) \longrightarrow \mathscr{H}om_X(\mathscr{O}_Z, f^\Delta \mathscr{O}_Y)$$

$$f^\Delta \mathscr{O}_Y \xleftarrow{\mathrm{Tr}_{i,f^\Delta \mathscr{O}_Y}} i_*i^\Delta f^\Delta \mathscr{O}_Y$$

in which the top row is the concatenation of (A.2.24) and (A.2.25). Applying f_* to (A.2.27) and composing with the residual complex trace map $\mathrm{Tr}_{f,\mathscr{O}_Y}$ in (A.2.14) recovers Berthelot's trace map (A.2.16), up to the identification of a $\mathscr{O}_Y$-modules with R-modules. Applying Theorem 3.4.1(2) to identify $\mathrm{Tr}_{f,\mathscr{O}_Y} \circ f_*(\mathrm{Tr}_{i,f^\Delta \mathscr{O}_Y})$ with the composite

$$f_*i_*i^\Delta f^\Delta \mathscr{O}_Y \xup~{c_{i,f}} g_*g^\Delta \mathscr{O}_Y \xrightarrow{\mathrm{Tr}_{g,\mathscr{O}_Y}} \mathscr{O}_Y$$

we conclude that for $\omega \in \Gamma(X, \omega_{X/Y})$, the global section $i^*\omega$ of $g_*i^*\omega_{X/Y} = f_*i_*i^*\omega_{X/Y}$ is taken to

$$\mathrm{Tr}_{f,Z}(\omega/(t_1 \cdots t_n)) \in \mathrm{H}^0(Y, \mathscr{O}_Y) = R$$

under the degree 0 part of the composite map of complexes

$$(A.2.28) \quad f_* i_* i^* \omega_{X/Y} \longrightarrow f_* \mathscr{E}xt^n_X(\mathcal{O}_Z, \omega_{X/Y}) \longrightarrow f_* \mathscr{H}om_X(\mathcal{O}_Z, f^{\triangle}\mathcal{O}_Y)$$

$$\mathcal{O}_Y \xleftarrow[\mathrm{Tr}_{g,\mathcal{O}_Y}]{} g_* g^{\triangle}\mathcal{O}_Y \xrightarrow[c_{i,f}]{\simeq} g_* i^{\triangle} f^{\triangle}\mathcal{O}_Y$$

We remind the reader that the top row is defined as the concatenation of (A.2.24) and (A.2.25).

Since the composite sheaf map in degree 0 in (A.2.28) is recovered by applying H^0 throughout, there is no harm in viewing the diagram (A.2.28) in the derived category $\mathbf{D}(Y)$. Since i_* and g_* are exact functors on $\mathcal{O}_Z$-modules (note that g is a map between finite discrete spaces), we may identify $g_* = \mathbf{R}g_*$, $i_* = \mathbf{R}i_*$ on $\mathbf{D}(Z)$. The advantage of working in the derived category is that we can then use Theorem 3.2.2(1) and Theorem 3.4.1(3) to obtain the commutativity in $\mathbf{D}(Y)$ of the diagram

$$
\begin{array}{ccccc}
g_* i^{\triangle} f^{\triangle}\mathcal{O}_Y & \xleftarrow[\simeq]{c_{i,f}} & g_* g^{\triangle}\mathcal{O}_Y & \xrightarrow{\mathrm{Tr}_{g,\mathcal{O}_Y}} & \mathcal{O}_Y \\
\simeq \downarrow & & \simeq \downarrow & & \parallel \\
g_* i^b f^\sharp \mathcal{O}_Y & \xleftarrow[\psi_{i,f}]{\simeq} & g_* g^b \mathcal{O}_Y & \xrightarrow{\mathrm{Trf}_g} & \mathcal{O}_Y
\end{array}
$$

where the map Trf_g is identified with the "evaluate at 1" map

$$\mathscr{H}om_Y(g_*\mathcal{O}_Z, \mathcal{O}_Y) \to \mathcal{O}_Y$$

since g is finite flat. But this high-brow viewpoint on the "evaluate at 1" map also implies, by the very *definition* of the residue symbol (see (A.1.3) and (A.1.4)), that

$$\mathrm{Res}_{X/Y}\begin{bmatrix} \omega \\ t_1, \ldots, t_n \end{bmatrix}$$

is $(-1)^{n(n-1)/2}$ times the image of

$$t^{\vee} \otimes i^*\omega \in \Gamma(Z, H^0(g_*(\omega_{Z/X}[-n] \otimes i^*\omega_{X/Y}[n])))$$

(with $t^{\vee} \overset{\mathrm{def}}{=} t_1^{\vee} \wedge \cdots \wedge t_n^{\vee}$) under the 0th cohomology map of the derived category map

$$g_*(\omega_{Z/X}[-n] \otimes i^*\omega_{X/Y}[n]) \xleftarrow[\simeq]{g_*(\eta_i)} g_* i^b f^\sharp \mathcal{O}_Y \xrightarrow[\simeq]{\psi_{i,f}} g_* g^b \mathcal{O}_Y \xrightarrow{\mathrm{Trf}_g} \mathcal{O}_Y$$

Thus, we are led to consider the outer part of the diagram in $\mathbf{D}(Z)$

$$
(A.2.29) \quad
\begin{array}{ccccc}
i^*\omega_{X/Y} & \xrightarrow[\text{no signs!}]{t^{\vee} \otimes (\cdot)} & \omega_{Z/X}[-n] \otimes i^*\omega_{X/Y}[n] & \xleftarrow{\eta_i} & i^b f^\sharp \mathcal{O}_Y \\
(A.2.23) \downarrow & & & & \simeq \uparrow \\
\mathscr{E}xt^n_X(\mathcal{O}_Z, \omega_{X/Y}) & \longrightarrow & \mathscr{H}om_X(\mathcal{O}_Z, f^{\triangle}\mathcal{O}_Y) & =\!=\!= & i^{\triangle} f^{\triangle}\mathcal{O}_Y
\end{array}
$$

where $\mathscr{E}xt_X^n(\mathcal{O}_Z, \omega_{X/Y})$ and $\mathscr{H}om_X(\mathcal{O}_Z, f^\Delta\mathcal{O}_Y)$ are viewed as complexes of $\mathcal{O}_Z$-modules rather than as complexes of $\mathcal{O}_X$-modules. Before defining the long diagonal map, we remind the reader again that the left map in the bottom row is defined in terms of (A.2.24) and (A.2.25). The above analysis shows that if the outside edge of (A.2.29) commutes up to a universal sign ϵ_n, then applying $g_* = \mathbf{R}g_*$ throughout and passing to H^0's yields the assertion of Lemma A.2.1 with a sign of $(-1)^{n(n-1)/2}\epsilon_n$ (so we are hoping to get $\epsilon_n = (-1)^{n(n-1)/2}$). Once we define the long diagonal map in (A.2.29), we will analyze the two 'halves' of (A.2.29) separately. The part below the diagonal will be seen to commute for quite conceptual reasons, while the commutativity of the part above the diagonal up to some universal sign will be seen to be sensitive to our choice of how we use Koszul complexes in the definition of (A.2.21); note also the presence of the fundamental local isomorphism η_i in this top part (whose definition involves Koszul complexes and signs issues).

Now we define the long diagonal map in (A.2.29). If $\mathscr{I}^\bullet$ is an injective resolution of $\omega_{X/Y}$ used to compute $\mathscr{E}xt_X^\bullet(\mathcal{O}_Z, \omega_{X/Y})$, then take $\mathscr{I}^\bullet[n]$ as an injective resolution of $\omega_{X/Y}[n] = f^\sharp\mathcal{O}_Y$. By using the fundamental local isomorphism η_i and the X-flatness of $\omega_{X/Y}$, $\mathscr{E}xt_X^{j+n}(\mathcal{O}_Z, \omega_{X/Y}) = 0$ for $j \neq 0$. Thus, there is a canonical *isomorphism* in $\mathbf{D}(Z)$

$$
\begin{aligned}
\mathscr{E}xt_X^n(\mathcal{O}_Z, \omega_{X/Y})[0] &= \mathrm{H}^n(\mathscr{H}om_X(\mathcal{O}_Z, \mathscr{I}^\bullet))[0] \\
&= \mathrm{H}^0(\mathscr{H}om_X(\mathcal{O}_Z, \mathscr{I}^\bullet[n]))[0] \\
&= \mathrm{H}^0(i^\flat f^\sharp\mathcal{O}_Y)[0] \\
&\simeq i^\flat f^\sharp\mathcal{O}_Y.
\end{aligned}
$$

This defines the long diagonal map in (A.2.29). Of course, the commutativity of the lower part will rely on the conventions in this definition.

To be more explicit, note that we can choose $\mathscr{I}^\bullet = E(\omega_{X/Y})$ as an injective resolution of $\omega_{X/Y}$ (concentrated in degrees from 0 to n), and then

$$
\mathscr{I}^\bullet[n] = E(\omega_{X/Y})[n] = f^\Delta\mathcal{O}_Y
$$

is taken as the injective resolution of $\omega_{X/Y}[n]$ for the definition of the long diagonal map in (A.2.29). This converts the right column in (A.2.29) into the identity map. In order to exploit this special choice of injective resolution, consider the canonical augmentation map

(A.2.30) $\mathscr{H}om_X(\mathcal{O}_Z, f^\Delta\mathcal{O}_Y) \longrightarrow \mathrm{H}^0(\mathbf{R}\mathscr{H}om_X(\mathcal{O}_Z, \omega_{X/Y}[n]))$

$$
\|
$$

$$
\mathrm{H}^n(\mathbf{R}\mathscr{H}om_X(\mathcal{O}_Z, \omega_{X/Y}))
$$

$$
\|
$$

$$
\mathscr{E}xt_X^n(\mathcal{O}_Z, \omega_{X/Y})
$$

where the top map is a surjection, the final $\mathscr{E}xt$ term is computed with the injective resolution $(f^\Delta\mathcal{O}_Y)[-n] = E(\omega_{X/Y})$, and the equalities do not involve any intervention of signs (exactly in accordance with our conventions for translation-compatibility of total derived functors). This is even a quasi-isomorphism, since $\mathbf{R}\mathscr{H}om_X(\mathscr{O}_Z, \omega_{X/Y}[n]) \simeq i^\Delta f^\Delta \mathscr{O}_Y \simeq g^\Delta \mathscr{O}_Y$ has vanishing cohomology outside of degree 0 and

$$\mathscr{H}om_X(\mathscr{O}_Z, (f^\Delta \mathscr{O}_Y)^j) \simeq \mathscr{H}om_X(\mathscr{O}_Z, E(\omega_{X/Y})^{n+j}) = 0$$

for $j < 0$ (as $E(\omega_{X/Y})$ is a Cousin complex with respect to the codimension filtration and Z is supported in codimension n).

By the definition we see that (A.2.30) is the inverse of the long diagonal map in (A.2.29). Thus, the commutativity of the lower part of (A.2.29) will follow if we can show that the two derived category maps

(A.2.31) $$\mathscr{E}xt_X^n(\mathscr{O}_Z, \omega_{X/Y}) \to \mathscr{H}om_X(\mathscr{O}_Z, f^\Delta \mathscr{O}_Y),$$

$$\mathscr{H}om_X(\mathscr{O}_Z, f^\Delta \mathscr{O}_Y) \to \mathrm{H}^n(\mathbf{R}\mathscr{H}om_X(\mathscr{O}_Z, \omega_{X/Y}))$$

are *inverses* to each other (the first of these maps being the left map in the bottom row in (A.2.29)). Since the second map in (A.2.31) is already known to be an isomorphism, it suffices to check that the composite map of complexes

(A.2.32)
$$\mathscr{H}om_X(\mathscr{O}_Z, f^\Delta \mathscr{O}_Y) \twoheadrightarrow \mathscr{E}xt_X^n(\mathscr{O}_Z, \omega_{X/Y}) \to \mathscr{H}om_X(\mathscr{O}_Z, f^\Delta \mathscr{O}_Y) \hookrightarrow f^\Delta \mathscr{O}_Y$$

is the canonical inclusion (i.e., the "evaluate at 1" map).

By the very *definition* of the middle map in (A.2.32), the composite of the last two maps in degree 0 is the composite of the canonical map

(A.2.33) $$\mathscr{E}xt_X^n(\mathscr{O}_Z, \omega_{X/Y}) \to \underline{\mathrm{H}}_Z^n(\omega_{X/Y})$$

and the inclusion map

$$\underline{\mathrm{H}}_Z^n(\omega_{X/Y}) \hookrightarrow \bigoplus_{x \in X^0} \underline{\mathrm{H}}_{\{x\}}^n(\omega_{X/Y})$$

to the degree 0 term of $f^\Delta \mathscr{O}_Y$. Since (A.2.33) is a special case of a general δ-functorial construction, it coincides with the H^0 map of

$$\mathbf{R}\mathscr{H}om_X(\mathscr{O}_Z, \omega_{X/Y}[n]) \to \mathbf{R}\underline{\Gamma}_Z(\omega_{X/Y}[n])$$

(there is no issue of signs here, since all linear maps commute with -1). Thus, (A.2.32) is exactly the composite of

(A.2.34) $$\mathscr{H}om_X(\mathscr{O}_Z, f^\Delta \mathscr{O}_Y) \longrightarrow \mathrm{H}^0(\mathbf{R}\mathscr{H}om_X(\mathscr{O}_Z, \omega_{X/Y}[n]))$$

$$\underline{\mathrm{H}}_Z^0(\omega_{X/Y}[n]) =\!=\!=\!=\!=\!= \mathrm{H}^0(\mathbf{R}\underline{\Gamma}_Z(\omega_{X/Y}[n]))$$

with the inclusion of $\underline{\mathrm{H}}_Z^0(\omega_{X/Y}[n])$ into the degree 0 term

$$\bigoplus_{x \in X^0} \underline{\mathrm{H}}_{\{x\}}^0(\omega_{X/Y}[n])$$

of $f^\Delta \mathscr{O}_Y = E(\omega_{X/Y}[n])$. This final formulation has the advantage that it only depends upon $\omega_{X/Y}[n]$ as an object in $\mathbf{D}(X)$! In particular, by *functoriality* in $\mathbf{D}(X)$ we can calculate all of the maps involved by replacing $\omega_{X/Y}[n]$ with its injective resolution $f^\Delta \mathscr{O}_Y$ (which is what we did in order to *define* the first map in (A.2.34)).

Keeping in mind that $f^\Delta \mathscr{O}_Y$ is supported in degrees ≤ 0 (as was required for the definition of the first map in (A.2.34)) and has degree 0 term

$$(f^\Delta \mathscr{O}_Y)^0 = \bigoplus_{x \in X^0} \underline{H}^0_{\{x\}}(\omega_{X/Y}[n]),$$

the problem becomes verifying the following two points:

- the complex $\underline{\Gamma}_Z(f^\Delta \mathscr{O}_Y)$ vanishes in degree < 0 and has degree 0 term inside of

$$\bigoplus_{z \in Z} \underline{H}^0_{\{z\}}(\omega_{X/Y}[n]) \subseteq (f^\Delta \mathscr{O}_Y)^0,$$

- for $z \in Z$, the map

$$\begin{aligned} H^0(\underline{\Gamma}_{\{z\}}(f^\Delta \mathscr{O}_Y)) &= \underline{H}^0_{\{z\}}(f^\Delta \mathscr{O}_Y) \\ &= \underline{H}^0_{\{z\}}(\omega_{X/Y}[n]) \\ &\hookrightarrow \bigoplus_{x \in X^0} \underline{H}^0_{\{x\}}(\omega_{X/Y}[n]) \\ &= (f^\Delta \mathscr{O}_Y)^0 \end{aligned}$$

is the canonical map (which makes sense, by the first point).

The first of these points is an immediate consequence of the definition of the Cousin complex $f^\Delta \mathscr{O}_Y = E(\omega_{X/Y}[n])$ with respect to the $[n]$-shift of the codimension filtration (see (3.1.6) and the surrounding discussion there), since a constant sheaf with support on an irreducible closed subscheme W of X cannot have a non-zero section supported inside of Z unless $W \subseteq Z$. The second point is then trivial. This completes the proof that the part of (A.2.29) below the diagonal is commutative in $\mathbf{D}(Z)$.

For the part of (A.2.29) above the diagonal, we must carefully look at the fundamental local isomorphism η_i in that part of the diagram. In accordance with the definition of the long diagonal map, we need to compute η_i with $f^\sharp \mathscr{O}_Y$ represented by $\mathscr{I}^\bullet[n]$, where $\mathscr{I}^\bullet$ is a choice of injective resolution for the computation of $\mathscr{E}xt^n_X(\mathscr{O}_Z, \omega_{X/Y})$. Observe that since $\mathscr{I}^\bullet[n]$ is *not* the same as $\mathscr{I}^{\bullet+n}$, we *cannot* use the explication of η_i contained in (2.7.3). We will have to appeal to the translation-compatibility of η_i in order to see what is happening. Recall that the formulation of the general fundamental local isomorphism as a map

$$(A.2.35) \qquad \eta_i : i^b(\mathscr{F}^\bullet) \simeq \omega_{Z/X}[-n] \overset{\mathbf{L}}{\otimes} \mathbf{L}i^*(\mathscr{F}^\bullet)$$

for $\mathscr{F}^\bullet \in \mathbf{D}(X)$ is the form in which we have compatibility with respect to translations, where the compatibility of the right side of (A.2.35) with respect to $\mathscr{F}^\bullet \rightsquigarrow \mathscr{F}^\bullet[m]$ involves a sign of $(-1)^{nm}$, as in (1.3.6).

We apply this in the case $\mathscr{F}^\bullet = \mathscr{I}^\bullet[n]$ for an injective resolution

$$\omega \to \mathscr{I}^\bullet$$

used to calculate $\mathscr{E}xt^\bullet_X(\cdot, \omega_{X/Y})$'s. Using a shift of $[-n]$ allows us to conclude with the help of Corollary 2.1.2 (which ensures that (2.5.3) recovers (2.5.1) when $\mathscr{F}^\bullet = \mathscr{F}[0]$) that applying H^0 to the derived category map

$$\mathscr{E}xt^n_X(\mathcal{O}_Z, \omega_{X/Y}) \xlongequal{\quad} i^b f^\sharp \mathcal{O}_Y \xrightarrow{\eta_i} \omega_{Z/X}[-n] \otimes i^* \omega_{X/Y}[n]$$

recovers $(-1)^{n^2} = (-1)^n$ times the map (2.5.1) (for the closed immersion $i : Z \hookrightarrow X$ and the sheaf $\mathscr{F} = \omega_{X/Y}$, though of course the scheme denoted Y in (2.5.1) is the scheme Z in the present setting). But in terms of the notation used near (1.3.28), the definition of (2.5.1) uses the isomorphism $\mathrm{Ext}^n(A/J, M) \simeq M/JM$ from (1.3.28), which arises from applying H^n to the Koszul calculation

$$\mathrm{Ext}^\bullet_A(A/J, M) \simeq \mathrm{H}^\bullet(\mathrm{Hom}^\bullet_A(K_\bullet(\mathbf{f}), M)) \simeq \mathrm{H}^\bullet(\mathbf{f}, M)$$

that involves a sign of $(-1)^{p(p+1)/2}$ in degree p, whereas the definition of the map (A.2.23) that appears in (A.2.29) uses the Koszul calculation

$$\mathrm{Ext}^n_A(A/J, M) \simeq \mathrm{H}^n(\mathrm{Hom}^\bullet_A(K_\bullet(\mathbf{f}), M)) = M/JM$$

that does *not* involve this extra sign. It therefore follows from the *definition* of (2.5.1) that going around the top part of (A.2.29) via the long route from $i^* \omega_{X/Y}$ to $\omega_{Z/X}[-n] \otimes i^* \omega_{X/Y}[n]$ is off by a sign of

$$\epsilon_n = (-1)^{n(n+1)/2}(-1)^n = (-1)^{n(n-1)/2}$$

from the short route map.

∎

The properties (R1) and (R10) will be easy to prove by using Lemma A.2.1, but the proof of (R9) will require an additional lemma motivated by Berthelot's proof of [Be, VII, 1.2.6].

PROOF. (of (R1)) Let $j : Z' \hookrightarrow X$ be cut out by the s_i's, so $Z \subseteq Z'$ as closed subschemes of X. There is a unique δ-functorial map

(A.2.36) $$\mathrm{H}^\bullet_Z(X, \cdot) \to \mathrm{H}^\bullet_{Z'}(X, \cdot)$$

lifting the obvious map in degree 0. In terms of the isomorphisms

$$\mathrm{H}^\bullet_Z(X, \cdot) \simeq \bigoplus_{z \in Z} \mathrm{H}^\bullet_{\{z\}}(X, \cdot), \quad \mathrm{H}^\bullet_{Z'}(X, \cdot) \simeq \bigoplus_{z' \in Z'} \mathrm{H}^\bullet_{\{z'\}}(X, \cdot),$$

the δ-functorial map (A.2.36) is the obvious map. Using the definition of (A.2.17) in terms of Koszul complexes, it is easy to check the commutativity of the diagram

(A.2.37)
$$
\begin{array}{ccccc}
\Gamma(X, \omega_{X/Y}) & \longrightarrow & \Gamma(Z, i^* \omega_{X/Y}) & \longrightarrow & \mathrm{H}^n_Z(X, \omega_{X/Y}) \\
{\scriptstyle \det(c_{ij})} \downarrow & & & & \downarrow \\
\Gamma(X, \omega_{X/Y}) & \longrightarrow & \Gamma(Z', j^* \omega_{X/Y}) & \longrightarrow & \mathrm{H}^n_{Z'}(X, \omega_{X/Y})
\end{array}
$$

so the right column of (A.2.37) takes $\omega/(t_1 \cdots t_n)$ to $(\det(c_{ij})\omega)/(s_1 \cdots s_n)$.

The definition of (A.2.16) in terms of (A.2.15) makes it obvious that the diagram

(A.2.38)

$$\begin{array}{ccc} H^n_Z(X, \omega_{X/Y}) & \xrightarrow{\mathrm{Tr}_{f,z}} & R \\ \downarrow & \nearrow{\mathrm{Tr}_{f,z'}} & \\ H^n_{Z'}(X, \omega_{X/Y}) & & \end{array}$$

commutes. Thus, by Lemma A.2.1, we deduce (R1). ∎

PROOF. (of (R10))

By Theorem 3.4.1(2), the diagram

(A.2.39)

$$\begin{array}{ccc} (fg)_*(fg)^\Delta \mathcal{O}_Y & \xrightarrow{\mathrm{Tr}_{fg,\mathcal{O}_Y}} & \mathcal{O}_Y \\ \simeq \downarrow & & \uparrow{\mathrm{Tr}_{f,\mathcal{O}_Y}} \\ f_*g_*g^\Delta f^\Delta \mathcal{O}_Y & \xrightarrow[\mathrm{Tr}_{g,f^\Delta \mathcal{O}_Y}]{} & f_*f^\Delta \mathcal{O}_Y \end{array}$$

commutes. Without loss of generality, X is affine. Since $g_*\mathcal{O}_{X'}$ is locally free of finite rank, (2.7.38) yields an isomorphism

(A.2.40)

$$\begin{array}{ccc} H^n_{Z'}(X', \omega_{X'/Y}) & \xrightarrow{\ \simeq\ } & H^n_Z(X, g_*\omega_{X'/Y}) \\ & & \downarrow{\simeq} \\ & & H^n_Z(X, \mathcal{H}om_X(g_*\mathcal{O}_{X'}, \omega_{X/Y})) \\ & & \downarrow{\simeq} \\ & & \mathrm{Hom}_X(g_*\mathcal{O}_{X'}, \underline{H}^n_Z(\omega_{X/Y})), \end{array}$$

where $\underline{H}^\bullet_Z$ is the derived functor of the functor on abelian sheaves on X given by

$$\underline{\Gamma}_Z : \mathcal{F} \rightsquigarrow \ker(\mathcal{F} \to \iota_*(\mathcal{F}|_{X-Z}))$$

(with $\iota : X - Z \hookrightarrow X$ the canonical open immersion).

Combining (A.2.39) and (A.2.40), we deduce the commutativity of the diagram of global sections

(A.2.41)

$$\begin{array}{ccc} H^n_{Z'}(X', \omega_{X'/Y}) & \xrightarrow{\mathrm{Tr}_{fg,z'}} & R \\ \simeq \downarrow & & \uparrow{\mathrm{Tr}_{f,z}} \\ H^n_Z(X, g_*\omega_{X'/Y}) & \xrightarrow[H^n_Z(\mathrm{Tr}_g)]{} & H^n_Z(X, \omega_{X/Y}) \end{array}$$

where the bottom row uses the trace map on differentials (2.7.36) (and its definition via (2.7.38)). The composite map along the left column and bottom row of (A.2.41) takes $\omega'/(t_1' \cdots t_n')$ to $\mathrm{Tr}_g(\omega')/(t_1 \cdots t_n)$. By Lemma A.2.1, this implies (R10).   ∎

Before we can prove (R9), we need a lemma inspired by [**Be**, VII, 1.2.6] (which considers a closely related, but different, situation). We warn the reader that the proof of [**Be**, VII, 1.2.6] *uses* (R9), so we must give a direct proof of what we need.

LEMMA A.2.2. *Let* $Y = \mathrm{Spec}(R)$ *be a local Gorenstein artin ring,* X *a smooth* Y-*scheme with pure dimension* n, *and* Z *a closed subscheme of* X *which is finite over* Y *and cut out by* $t_1, \ldots, t_n \in \Gamma(X, \mathcal{O}_X)$. *Assume that* Z *lies inside of an open affine* V *in* X. *Let* $\mathrm{d}_{X/Y}^{n-1} : \Omega_{X/Y}^{n-1} \to \Omega_{X/Y}^n$ *denote the usual differentiation map. The composite*

$$(A.2.42) \qquad \mathrm{H}_Z^n(X, \Omega_{X/Y}^{n-1}) \xrightarrow{\mathrm{d}_{X/Y}^{n-1}} \mathrm{H}_Z^n(X, \Omega_{X/Y}^n) \xrightarrow{\mathrm{Tr}_{I,Z}} R$$

vanishes if and only if (R9) *holds for* V *in place of* X, *all* $\eta \in \Gamma(U, \Omega_{X/Y}^{n-1})$, *and all positive integers* $k_1, \ldots, k_n$.

PROOF. Without loss of generality, $n > 0$ and $V = X$. Since X is affine, as we vary η and $k_1, \ldots, k_n$, $\eta/(t_1^{k_1} \cdots t_n^{k_n})$ runs through all elements in $\mathrm{H}_Z^n(X, \Omega_{X/Y}^{n-1})$. Thus, it suffices to prove that for all η and $k_1, \ldots, k_n$,

$$(A.2.43) \quad \mathrm{H}_Z^n(\mathrm{d}_{X/Y}^{n-1})(\eta/(t_1^{k_1} \cdots t_n^{k_n})) = \frac{\mathrm{d}\eta}{t_1^{k_1} \cdots t_n^{k_n}} - \sum_{i=1}^n k_i \frac{\mathrm{d}t_i \wedge \eta}{t_1^{k_1} \cdots t_i^{k_i+1} \cdots t_n^{k_n}}.$$

Since

$$\eta/(t_1^{k_1} \cdots t_n^{k_n}) = (t_1^{k-k_1} \cdots t_n^{k-k_n}\eta)/(t_1^k \cdots t_n^k)$$

for any $k \geq \max k_i$, we may easily reduce (A.2.43) to the case in which all k_i's are equal to a common $k \geq 1$.

The right side of (A.2.43) is equal to

$$(t_1 \cdots t_n \, \mathrm{d}\eta - \sum k_i t_1 \cdots \hat{t}_i \cdots t_n \, \mathrm{d}t_i \wedge \eta)/(t_1^{k_1+1} \cdots t_n^{k_n+1}),$$

so we are motivated to define a map of 'dual' Koszul complexes

$$\tilde{\mathrm{d}}_k^\bullet : K^\bullet(\mathbf{t}^k, \Omega_{X/Y}^{n-1}) \to K^\bullet(\mathbf{t}^{k+1}, \Omega_{X/Y}^n)$$

by

$$\tilde{\mathrm{d}}_k^p(\eta e_I^\vee) = (t_I \, \mathrm{d}\eta - k \sum_{i \in I} t_{I-\{i\}} \, \mathrm{d}t_i \wedge \eta)e_I^\vee,$$

where $e_I^\vee = e_{i_1}^\vee \wedge \cdots \wedge e_{i_p}^\vee$, $t_I = t_{i_1} \cdots t_{i_p}$ for $I = \{i_1, \ldots, i_p\}$ with

$$1 \leq i_1 < \cdots < i_p \leq n.$$

In particular, $\tilde{\mathrm{d}}_k^0 : \Omega_{X/Y}^{n-1} \to \Omega_{X/Y}^n$ is the natural differentiation map for all k. It is easy to check that $\tilde{\mathrm{d}}_k^\bullet$ is a map of complexes and is compatible with change in k, so we get a map on the direct limit of cohomology

$$(A.2.44) \qquad \tilde{h}^\bullet = \varinjlim \mathrm{H}^\bullet(\tilde{\mathrm{d}}_k^\bullet) : \mathrm{H}^\bullet((\mathbf{t}), \Omega_{X/Y}^{n-1}) \to \mathrm{H}^\bullet((\mathbf{t}), \Omega_{X/Y}^n).$$

It suffices to show that under the δ-functorial identification (A.2.20), $\widetilde{h}^n$ is identified with the map $H_Z^n(d_{X/Y}^{n-1})$.

Let $U = X - Z$, so for all p there is a commutative diagram

(A.2.45)
$$
\begin{array}{ccc}
H^p(U, \Omega_{X/Y}^{n-1}) & \xrightarrow{\;\delta\;} & H_Z^{p+1}(X, \Omega_{X/Y}^{n-1}) \\
{\scriptstyle H^p(U, d_{X/Y}^{n-1})} \downarrow & & \downarrow {\scriptstyle H_Z^{p+1}(d_{X/Y}^{n-1})} \\
H^p(U, \Omega_{X/Y}^{n}) & \xrightarrow[\;\delta\;]{} & H_Z^{p+1}(X, \Omega_{X/Y}^{n})
\end{array}
$$

in which the rows are *surjective* (and even isomorphisms for $p \geq 1$). By using (A.2.20), it suffices to check that replacing $H_Z^{p+1}(d_{X/Y}^{n-1})$ in (A.2.45) with $\widetilde{h}^{p+1}$ still gives a commutative diagram.

By the Čech theory for *abelian* sheaves and [**EGA**, III$_1$, 1.4.1], the ordered open affine covering
$$
\mathfrak{U} = \{X_{t_1}, \ldots, X_{t_n}\}
$$
of U gives rise to a commutative diagram

(A.2.46)
$$
\begin{array}{ccc}
H^p(\mathfrak{U}, \Omega_{X/Y}^{n-1}) & \xrightarrow{\;\simeq\;} & H^p(U, \Omega_{X/Y}^{n-1}) \\
{\scriptstyle d} \downarrow & & \downarrow {\scriptstyle d} \\
H^p(\mathfrak{U}, \Omega_{X/Y}^{n}) & \xrightarrow[\;\simeq\;]{} & H^p(U, \Omega_{X/Y}^{n})
\end{array}
$$

for all p, with isomorphisms in the horizontal direction. An easy calculation and [**EGA**, III$_1$, (1.2.2.3)] yield a commutative diagram with surjective maps (even isomorphisms for $p \geq 1$) in the horizontal direction

(A.2.47)
$$
\begin{array}{ccc}
H^p(\mathfrak{U}, \Omega_{X/Y}^{n-1}) & \longrightarrow & H^{p+1}((\mathbf{t}), \Omega_{X/Y}^{n-1}) \\
{\scriptstyle d} \downarrow & & \downarrow {\scriptstyle \widetilde{h}^{p+1}} \\
H^p(\mathfrak{U}, \Omega_{X/Y}^{n}) & \longrightarrow & H^{p+1}((\mathbf{t}), \Omega_{X/Y}^{n})
\end{array}
$$

Finally, a little thought enables one to generalize [**SGA2**, II, Prop 5] to get a diagram of functors on quasi-coherent sheaves

(A.2.48)
$$
\begin{array}{ccc}
H^p(\mathfrak{U}, \cdot) & \longrightarrow & H^{p+1}((\mathbf{t}), \cdot) \\
{\scriptstyle \simeq} \downarrow & & \downarrow {\scriptstyle \simeq} \\
H^p(U, \cdot) & \longrightarrow & H_Z^{p+1}(X, \cdot)
\end{array}
$$

with surjective horizontal maps and commutativity up to a universal sign ϵ_p that depends only on p. Using (A.2.48) for $\Omega_{X/Y}^{n-1}$, $\Omega_{X/Y}^{n}$ and the diagrams (A.2.45), (A.2.46), (A.2.47), we get (for a fixed choice of p) a cube diagram with (A.2.48) along the top and bottom faces. These two faces commute up to the *same* universal sign ϵ_p, and all other faces commute except for possibly the face

(A.2.49)
$$H^{p+1}((t), \Omega_{X/Y}^{n-1}) \xrightarrow{\simeq} H_Z^{p+1}(X, \Omega_{X/Y}^{n-1})$$

$$\bar{h}^{p+1} \downarrow \qquad\qquad\qquad \downarrow d$$

$$H^{p+1}((t), \Omega_{X/Y}^{n}) \xrightarrow{\simeq} H_Z^{p+1}(X, \Omega_{X/Y}^{n})$$

The surjectivity of all horizontal maps in (A.2.45), (A.2.46), (A.2.47), (A.2.48) forces the commutativity of (A.2.49). Taking $p = n - 1 \geq 0$ completes the proof of (A.2.43), and hence of the lemma.

∎

With the proof of Lemma A.2.2 complete, we can give the proof of (R9). The basic idea is that if X is a smooth scheme with pure relative dimension n over a local artin scheme Y and $x \in X$ is a closed point with residue field equal to that of Y (e.g., if the residue field of Y is algebraically closed), then the functor $H_{\{x\}}^n$ based at $x \in X$ should be identified with functor $H_{\{0\}}^n$ based at the origin on affine space. Thus, Lemma A.2.2 should enable us to reduce the proof of (R9) to the special case when $X = \mathbf{A}_Y^n$ and the t_j's are the coordinate functions. This special case can be attacked by a direct calculation. We will need to be careful, because the differential $\Omega_{X/Y}^{n-1} \to \Omega_{X/Y}^n$ is not $\mathcal{O}_X$-linear.

PROOF. (of (R9)) First consider the special case when $X = \mathbf{A}_Y^n$ and the t_j's are the coordinate functions. In this case, it is enough to consider $\omega = f \, dT_1 \wedge \cdots \wedge \widehat{dT_i} \wedge \ldots dT_n$ with $f = T_1^{r_1} \cdots T_n^{r_n}$ a monomial. It suffices to prove that if $r_i \geq 1$, then

$$\operatorname{Res}_{\mathbf{A}_Y^n/Y} \left[\begin{matrix} r_i T_1^{r_1} \cdots T_i^{r_i-1} \cdots T_n^{r_n} \, dT_1 \wedge \cdots \wedge dT_n \\ T_1^{k_1}, \ldots, T_n^{k_n} \end{matrix} \right]$$

is equal to

$$k_i \operatorname{Res}_{\mathbf{A}_Y^n/Y} \left[\begin{matrix} T_1^{r_1} \cdots T_n^{r_n} \, dT_1 \wedge \cdots \wedge dT_n \\ T_1^{k_1}, \ldots, T_i^{k_i+1}, \ldots T_n^{k_n} \end{matrix} \right].$$

By (R4), we are reduced to showing

(A.2.50)
$$\operatorname{Res}_{\mathbf{A}_Y^1/Y} \left[\begin{matrix} r T^{r-1} \, dT \\ T^k \end{matrix} \right] = k \operatorname{Res}_{\mathbf{A}_Y^1/Y} \left[\begin{matrix} T^r \, dT \\ T^{k+1} \end{matrix} \right]$$

for all positive integers $k \geq 1$ and $r \geq 1$. By (R8) (resp. (R1)), both sides of (A.2.50) vanish for $r > k$ (resp. $r < k$). For $r = k$, we want to prove

$$\operatorname{Res}_{\mathbf{A}_Y^1/Y} \left[\begin{matrix} T^k \, dT \\ T^{k+1} \end{matrix} \right] = 1$$

for all $k \geq 1$. This follows from (R1) and (R6).

We now reduce the general case to the special case just treated. Without loss of generality, $Y = \operatorname{Spec}(R)$ for a local Gorenstein artin ring R with an *algebraically closed* residue field, X is smooth over Y with pure dimension n, and $Z = \{z_0\}$ is a single closed point on X. By Lemma A.2.2, it suffices to prove that the composite

$$(A.2.51) \qquad H^n_{\{z_0\}}(X, \Omega^{n-1}_{X/Y}) \xrightarrow{d} H^n_{\{z_0\}}(X, \Omega^n_{X/Y}) \xrightarrow{\mathrm{Tr}_{z_0}} R$$

is zero. In particular, the scheme structure on the underlying space $\{z_0\}$ does not matter. By the special case just treated, we know that the composite

$$(A.2.52) \qquad H^n_{\{0\}}(\mathbf{A}^n_Y, \Omega^{n-1}_{\mathbf{A}^n_Y/Y}) \xrightarrow{d} H^n_{\{0\}}(\mathbf{A}^n_Y, \Omega^n_{\mathbf{A}^n_Y/Y}) \xrightarrow{\mathrm{Tr}_0} R$$

vanishes.

The unique point of Y has an algebraically closed residue field, so we may choose a section in $X(Y)$ based at the point z_0. Shrinking X around this section, we may assume that $Z = Y$ and that there is an étale map $\pi : X \to \mathbf{A}^n_Y$ taking the section $Z = Y$ to the origin. By shrinking X some more, we can assume that $\pi^{-1}(0) = Z$ as schemes (since the residue field of Y is separably closed and π is étale). It suffices to consider the following more general situation. Let $\pi : X' \to X$ be an étale map between smooth schemes with pure relative dimension $n \geq 1$ over a local Gorenstein artin scheme $Y = \mathrm{Spec}(R)$ (such as $X \to \mathbf{A}^n_Y$ above). Let $x \in X$ be a closed point on X and assume that $\pi^{-1}(x) = \{x'\}$, with the residue fields at x and x' equal to that of Y. In order to deduce the vanishing of (A.2.51) from the vanishing of (A.2.52), and thereby complete the proof of (R9), it suffices to prove the following three general facts:

- the canonical pullback maps

$$(A.2.53) \qquad H^n_{\{x\}}(X, \Omega^j_{X/Y}) \to H^n_{\{x'\}}(X', \Omega^j_{X'/Y})$$

  are isomorphisms for all $j \geq 0$,
- the diagram

$$(A.2.54) \qquad \begin{array}{ccc} H^n_{\{x\}}(X, \Omega^{j-1}_{X/Y}) & \xrightarrow{\ d\ } & H^n_{\{x\}}(X, \Omega^j_{X/Y}) \\ \downarrow & & \downarrow \\ H^n_{\{x'\}}(X', \Omega^{j-1}_{X'/Y}) & \xrightarrow{\ d\ } & H^n_{\{x'\}}(X', \Omega^j_{X'/Y}) \end{array}$$

  commutes for all $j \geq 1$,
- the diagram

$$(A.2.55) \qquad \begin{array}{c} H^n_{\{x\}}(X, \Omega^n_{X/Y}) \xrightarrow{\mathrm{Tr}_{f,\{x\}}} R \\ \downarrow \qquad \nearrow_{\mathrm{Tr}_{f \circ \pi, \{x'\}}} \\ H^n_{\{x'\}}(X', \Omega^n_{X'/Y}) \end{array}$$

  commutes, where $f : X \to Y$ is the structure map.

Since Y is a local artin scheme, X is smooth over Y, and $k(x) = k(Y)$, we can choose a section $s : Y \hookrightarrow X$ based at the point x. Shrinking X around x, we may assume that X, X' are separated and s is cut out by n global functions $t_1, \ldots, t_n$ on X. By our hypotheses, the base change subscheme $s' : Y' = Y \times_X X' \hookrightarrow X'$

cut out by the functions $t'_j = \pi^* t_j$ projects isomorphically to Y, so s' can be viewed as a section to $X' \to Y$, based at the point x'. Let Y_k denote the infinitesimal neighborhood of Y in X defined by $t_1^k, \ldots, t_n^k$, and likewise define Y'_k in X'. The scheme diagram of interest to us is

(A.2.56)

$$
\begin{array}{ccc}
Y'_k & \xrightarrow{s'_k} & X' \\
\| & & \downarrow{\scriptstyle \pi} \\
Y_k & \xrightarrow{s_k} & X \\
& \searrow{\scriptstyle h_k} & \downarrow{\scriptstyle f} \\
& & Y
\end{array}
$$

where the top square is cartesian and $h_k = f \circ s_k$ is finite flat.

Since π is étale, the natural map $\pi^* \Omega^1_{X/Y} \to \Omega^1_{X'/Y}$ is an isomorphism, so (A.2.53) is a special case of the natural pullback map

(A.2.57) $$\mathrm{H}^n_{\{x\}}(X, \mathscr{F}) \to \mathrm{H}^n_{\{x'\}}(X', \pi^* \mathscr{F})$$

for an arbitrary quasi-coherent $\mathscr{O}_X$-module $\mathscr{F}$. We will now prove that (A.2.57) is an isomorphism. Replacing X by an open affine around x and X' by an open affine around x', we may assume that X and X' are affine. Using (A.2.20), we get a diagram of functors (on quasi-coherent $\mathscr{O}_X$-modules)

$$
\begin{array}{ccc}
\mathrm{H}^\bullet((\mathbf{t}), \cdot) & \xrightarrow{\ \simeq\ } & \mathrm{H}^\bullet_{\{x\}}(X, \cdot) \\
\downarrow & & \downarrow \\
\mathrm{H}^\bullet((\mathbf{t}'), \pi^*(\cdot)) & \xrightarrow{\ \simeq\ } & \mathrm{H}^\bullet_{\{x'\}}(X', \pi^*(\cdot))
\end{array}
$$

whose commutativity is easy to prove (interpret in terms of Ext's). Since the natural map $Y'_k \to Y_k$ is an isomorphism for all k, it is clear that

$$\mathrm{H}^n((\mathbf{t}), \mathscr{F}) \to \mathrm{H}^n((\mathbf{t}'), \pi^* \mathscr{F})$$

is an isomorphism. Similarly, we can reduce the commutativity of (A.2.54) to an easy commutativity claim involving Koszul complexes and a variant of the map $\tilde{h}^\bullet$ in (A.2.44) (defined for $\Omega^{j-1}_{X/Y}$, $\Omega^j_{X/Y}$ instead of $\Omega^{n-1}_{X/Y}$, $\Omega^n_{X/Y}$).

Finally, we need to prove the commutativity of (A.2.55). Our first step is to remove any reference to abstract trace maps. For all $k \geq 1$, we have a composite injection of $\mathscr{O}_Y$-modules

$$
\begin{aligned}
\mathscr{H}om_Y(\mathscr{O}_{Y_k}, \mathscr{O}_Y) &= \mathrm{H}^0(h_k^\flat(\mathscr{O}_Y)) \\
&\simeq \mathrm{H}^0(s_k^\flat f^\Delta \mathscr{O}_Y) \\
&= \mathscr{H}om_X(\mathscr{O}_{Y_k}, \underline{\mathrm{H}}^n_{\{x\}}(\omega_{X/Y})) \\
&\hookrightarrow \underline{\mathrm{H}}^n_{\{x\}}(\omega_{X/Y})
\end{aligned}
$$

which induces an injection on global sections

(A.2.58) $$\mathrm{Hom}_Y(\mathscr{O}_{Y_k}, \mathscr{O}_Y) \hookrightarrow \mathrm{H}^n_{\{x\}}(X, \omega_{X/Y}).$$

This map is compatible with change in k and passing to the direct limit yields an isomorphism. Also, by the definition of (A.2.16), the composite of (A.2.58) and $\mathrm{Tr}_{f,\{x\}}$ is just the canonical 'evaluate at 1' map

$$\mathrm{Tr}_{h_k,\{x\}} : \mathrm{Hom}_Y(\mathscr{O}_{Y_k}, \mathscr{O}_Y) \to R.$$

Thus, instead of studying (A.2.55), it suffices to show the commutativity of the diagram

$$(\mathrm{A.2.59}) \qquad \begin{array}{ccc} \mathrm{Hom}_Y(\mathscr{O}_{Y_k}, \mathscr{O}_Y) & \longrightarrow & \mathrm{H}^n_{\{x\}}(X, \omega_{X/Y}) \\ {\scriptstyle \simeq}\downarrow & & \downarrow \\ \mathrm{Hom}_Y(\mathscr{O}_{Y'_k}, \mathscr{O}_Y) & \longrightarrow & \mathrm{H}^n_{\{x'\}}(X', \omega_{X'/Y}) \end{array}$$

for all $k \geq 1$. Note that this diagram does not involve any abstract trace maps.

By definition, (A.2.58) factorizes as a composite

$$\begin{aligned} \mathrm{Hom}_Y(\mathscr{O}_{Y_k}, \mathscr{O}_Y) &\simeq \mathrm{Ext}^n_Y(\mathscr{O}_{Y_k}, \omega_{X/Y}) \\ &\simeq \mathrm{Hom}_X(\mathscr{O}_{Y_k}, \underline{\mathrm{H}}^n_{\{x\}}(\omega_{X/Y})) \\ &\hookrightarrow \mathrm{H}^n_{\{x\}}(X, \omega_{X/Y}) \end{aligned}$$

Since the pullback $\mathrm{H}^\bullet_{\{x\}}(X, \cdot) \to \mathrm{H}^\bullet_{\{x'\}}(X', \pi^*(\cdot))$ can be computed in terms of Koszul complexes and the pullback $\mathrm{Ext}^\bullet_X(\mathscr{O}_{Y_k}, \cdot) \to \mathrm{Ext}^\bullet_X(\mathscr{O}_{Y'_k}, \pi^*(\cdot))$ can be computed in terms of Koszul resolutions of $\mathscr{O}_{Y_k}$ and $\mathscr{O}_{Y'_k}$, it is easy to check that the diagram

$$(\mathrm{A.2.60}) \qquad \begin{array}{ccc} \mathrm{Ext}^n_X(\mathscr{O}_{Y_k}, \omega_{X/Y}) & \longrightarrow & \mathrm{H}^n_{\{x\}}(X, \omega_{X/Y}) \\ {\scriptstyle \simeq}\downarrow & & \downarrow{\scriptstyle \simeq} \\ \mathrm{Ext}^n_{X'}(\mathscr{O}_{Y'_k}, \omega_{X'/Y}) & \longrightarrow & \mathrm{H}^n_{\{x'\}}(X', \omega_{X'/Y}) \end{array}$$

commutes. Meanwhile, the isomorphism $\mathrm{Hom}_Y(\mathscr{O}_{Y_k}, \mathscr{O}_Y) \simeq \mathrm{Ext}^n_X(\mathscr{O}_{Y_k}, \omega_{X/Y})$ is H^0 of the isomorphism $\psi_{s_k,f} : h^\flat_k \simeq s^\flat_k f^\sharp$ applied to $\mathscr{O}_Y$. Thus, the commutativity of the diagram

$$\begin{array}{ccc} h^\flat_k & \xrightarrow{\ \psi_{s_k,f}\ } & s^\flat_k f^\sharp \\ {\scriptstyle \psi_{s'_k,f\pi}}\downarrow & & \diagdown \\ s'^{\,\flat}_k(f\pi)^\sharp & =\!=\!= & s'^{\,\flat}_k \pi^\sharp f^\sharp \xleftarrow[\ \simeq\ ]{} s'^{\,\flat}_k \pi^\flat f^\sharp \end{array}$$

(see Theorem 2.8.1) yields a commutative diagram of $\mathscr{O}_Y$-modules

$$(\mathrm{A.2.61}) \qquad \begin{array}{ccc} \mathscr{H}om_Y(\mathscr{O}_{Y_k}, \mathscr{O}_Y) & \xrightarrow{\ \simeq\ } & \mathscr{E}xt^n_X(\mathscr{O}_{Y_k}, \omega_{X/Y}) \\ \| & & \downarrow{\scriptstyle \simeq} \\ \mathscr{H}om_Y(\mathscr{O}_{Y'_k}, \mathscr{O}_Y) & \xrightarrow[\ \simeq\ ]{} & \mathscr{E}xt^n_{X'}(\mathscr{O}_{Y'_k}, \omega_{X'/Y}) \end{array}$$

Passing to global sections and 'gluing' (A.2.60) and (A.2.61) will yield the commutativity of (A.2.59), provided we verify that the right column in (A.2.61) is the canonical base change map for $\mathscr{E}xt$. Since the henselizations of $\mathcal{O}_{X,x}$ and $\mathcal{O}_{X',x'}$ are the same, we can find a pointed étale neighborhood (U, u) of (X, x) with $k(u) = k(x)$ such that $\pi \times_X U$ is an *isomorphism* near $u' = (x', u) \in U' = X' \times_X U$. Thus, the right column of (A.2.61) is easily seen to be the expected map. ∎

Trace Map on Smooth Curves

B.1. Motivation

The most classical instance of Grothendieck-Serre duality is the following. Let $f : X \to \mathrm{Spec}(k)$ be a proper, smooth, connected curve over an algebraically closed field k. Then $\mathrm{H}^1(X, \Omega^1_{X/k})$ is a 1-dimensional k-vector space, by the last part of Corollary 3.6.6, and for any coherent sheaf $\mathscr{F}$ on X, the pairing of finite-dimensional k-vector spaces

$$\mathrm{Hom}_X(\mathscr{F}, \Omega^1_{X/k}) \times \mathrm{H}^1(X, \mathscr{F}) \to \mathrm{H}^1(X, \Omega^1_{X/k})$$

is a perfect pairing (by Corollary 5.1.3). In order to make this into a genuine duality pairing, one wants a *canonical* isomorphism $\mathrm{H}^1(X, \Omega^1_{X/k}) \simeq k$. In Grothendieck's theory this is given by the trace map γ_f, while in the classical theory it is given by the theory of residues. More precisely, if we let $K = k(X)$ be the function field of X, then there is a quasi-coherent injective resolution

$$0 \to \Omega^1_{X/k} \to \underline{\Omega}^1_{K/k} \to \bigoplus_{x \in X^0} i_{x*}(\Omega^1_{K/k} / \Omega^1_{\mathscr{O}_{X,x}/k}) \to 0,$$

where X^0 denotes the set of closed points of X, $\underline{\Omega}^1_{K/k}$ is the (constant) quasi-coherent sheaf attached to $\Omega^1_{K/k}$, and

$$i_x : \mathrm{Spec}(\mathscr{O}_{X,x}) \to X$$

is the canonical map of schemes (note that for $x \in X^0$, the $\mathscr{O}_{X,x}$-module $\Omega^1_{K/k} / \Omega^1_{\mathscr{O}_{X,x}/k}$ consists entirely of $\mathfrak{m}_x$-power torsion, so it is supported at $\{x\}$). We then get a long exact cohomology sequence

(B.1.1) $$\Omega^1_{K/k} \to \bigoplus_{x \in X^0} \Omega^1_{K/k} / \Omega^1_{\mathscr{O}_{X,x}/k} \to \mathrm{H}^1(X, \Omega^1_{X/k}) \to 0.$$

The classical theory of residues provides k-linear maps

$$\mathrm{res}_x : \Omega^1_{K/k} \to k$$

for all $x \in X^0$, uniquely determined by the condition that

$$\mathrm{res}_x(\eta) = a_{-1}$$

where, for a choice of uniformizer $t_x \in \mathscr{O}_{X,x}$, the image of η under

$$\Omega^1_{K/k} \hookrightarrow \Omega^1_{K/k} \otimes_{\mathscr{O}_{X,x}} \widehat{\mathscr{O}}_{X,x} = k((t_x)) \, dt_x$$

is $\sum a_n t_x^n dt_x$. Of course, it is by no means obvious that such *well-defined* maps res_x exist (i.e., that the choice of the t_x's doesn't matter). One also has the *residue theorem*: the composite map

$$\Omega^1_{K/k} \longrightarrow \bigoplus_{x \in X^0} \Omega^1_{K/k}/\Omega^1_{\mathscr{O}_{X,x},k} \xrightarrow{\sum \mathrm{res}_x} k$$

is zero. Granting these facts, the exact sequence (B.1.1) then provides us with a (visibly non-zero) k-linear map

(B.1.2) $\mathrm{res}_X : \mathrm{H}^1(X, \Omega^1_{X/k}) \to k$

which must then be an isomorphism, thereby giving us the desired 'duality theorem'.

In the foundations of Grothendieck's general theory, it is very important that for a proper map $f : X \to Y$ of noetherian schemes with finite Krull dimension and a residual complex $\mathscr{K}^\bullet$ on Y, the map of graded sheaves $\mathrm{Tr}_{f,\mathscr{K}^\bullet}$ is a map of complexes when f is *proper* (see Theorem 3.4.1). This is proven in [**RD**, VII, 2.1], where it is called the 'residue theorem'. The reason for this name is that the method of proof is to use the general theory and *construction* of the 'residual complex trace map' to reduce to the case of finite schemes over $\mathbf{P}^1_A$ with A an artin local ring having algebraically closed residue field, and to then ultimately reduce to the case of $\mathbf{P}^1_A \to \mathrm{Spec}(A)$ for such A. In general, for any proper, smooth, connected curve $f : X \to \mathrm{Spec}(A)$ over such A, the trace maps $\mathrm{Tr}_{f,\mathscr{K}^\bullet}$ can be expressed in terms of 'residues' [**RD**, VII, 1.3] (in particular, one *derives* from Grothendieck's theory the existence and well-definedness of the notion of residue of a meromorphic differential at a point on such a curve). We will explain this below. Once this is done for the curve $f : X \to \mathrm{Spec}(A)$, the assertion for such f that $\mathrm{Tr}_{f,\mathscr{K}^\bullet}$ is a map of complexes is *exactly* the classical statement that the sum of the residues of a meromorphic differential on X is always 0. In the case $X = \mathbf{P}^1_A$, this is proven by a direct calculation [**RD**, VII, 1.5]. Then, after one uses reduction to the special case of the projective line to complete the proof of the *general* result that $\mathrm{Tr}_{f,\mathscr{K}^\bullet}$ is a map of complexes for proper f [**RD**, VII, 2.1], one can go back to conclude the classical residue theorem on any proper, smooth, connected curve over an algebraically closed field (or more generally, over an artin local ring with algebraically closed residue field).

However, there is a subtle point here: for a proper, smooth, connected curve $f : X \to A$ as above, the residual complex trace map $\mathrm{Tr}_{f,\mathscr{K}^\bullet}$ turns out to be essentially the *negative* of the map one defines using residues. The reason for this sign problem is hidden in the proof of the key lemma [**RD**, VII, 1.2], but the proof of this lemma is omitted in [**RD**], left to the reader as a "good exercise in definition-chasing." We will give the proof of [**RD**, VII, 1.2] in order to eliminate any possibility of sign ambiguity. The purpose of this appendix is two-fold:

- state and prove [**RD**, VII, 1.2], which ensures the existence of the theory of residues on smooth curves,

- use [**RD**, VII, 1.2] to prove that Grothendieck's trace map (3.4.11) on proper smooth curves is the *negative* of the map one naturally defines using residues (thereby settling the smooth case of Theorem 5.2.3).

We also include a discussion of the relationship between Grothendieck duality on a curve and duality on its Jacobian, as this is rather important in applications but the justification is not immediately obvious.

B.2. Preparations

Before we can formulate the theory of residues [**RD**, VII, 1.2], some preparations are necessary. Throughout our discussion, we often identify an object in an abelian category with a complex concentrated in degree 0 (or the associated object in the derived category). Let A be an artin local ring with algebraically closed residue field and let $f : X \to Y = \mathrm{Spec}(A)$ be a smooth, connected curve over A. Let ξ, X^0, $K = \mathcal{O}_{X,\xi}$ be as in Lemma 3.1.1. Fix an injective hull I for k over A, so the complex $\tilde{I} = \tilde{I}[0]$ on Y is a residual complex with associated filtration equal to the codimension filtration on the one point space Y. Choose a closed point $x_0 \in X^0$ and a closed subscheme $i : Z = \mathrm{Spec}(B) \hookrightarrow X$ supported at x_0. Thus, B is a local artin ring and the natural composite map $g = f \circ i : Z \to Y$ makes B a finite A-algebra. We want to unwind the so-called 'residue isomorphism'

$$(B.2.1) \qquad g^y(\tilde{I}) \simeq i^y f^z(\tilde{I})$$

from (3.2.5).

By definition, we have that on Z

$$g^y(\tilde{I}) = E(g^\flat(\tilde{I})) = E(\mathrm{Hom}_A(B, I)^\sim) = \mathrm{Hom}_A(B, I)^\sim,$$

where the final equality is an obvious consequence of the rather simple nature of the codimension filtration on the 0-dimensional Z. We also have by Corollary 3.1.2 with $M = I$ and $\mathscr{L} = \Omega^1_{X/A}$ that

$$i^y f^z(\tilde{I}) \simeq E(i^\flat(E(f^\sharp(\tilde{I})))) = E(i^\flat(\mathscr{K}^\bullet)),$$

where the *residual complex* $\mathscr{K}^\bullet = f^z(\tilde{I}) = E(f^\sharp(\tilde{I}))$ is the two-term complex in degrees -1, 0 given by

$$(B.2.2) \qquad i_{\xi*}(I \otimes_A \Omega^1_{X/A,\xi}) \to \bigoplus_{x \in X^0} i_{x*}(I \otimes_A \Omega^1_{X/A,\xi}/\Omega^1_{X/A,x}),$$

with localization at any $x \in X^0$ yielding a map

$$(B.2.3) \qquad I \otimes_A \Omega^1_{X/A,\xi} \to I \otimes_A \Omega^1_{X/A,\xi}/\Omega^1_{X/A,x}$$

which is the canonical projection. Recall, as was noted after Corollary 3.1.2, that the sign in (1.3.28) plays a role in the proof that (B.2.3) is the canonical projection rather than its negative.

Since the residual complex $\mathscr{K}^\bullet$ is a complex of injectives on X, we may compute $i^\flat(\mathscr{K}^\bullet)$ by simply applying the functor $\mathscr{H}om_X(\mathcal{O}_Z, \cdot)$, which takes injectives on X to injectives on Z. Using (B.2.2), we see that $\mathscr{H}om_X(\mathcal{O}_Z, \mathscr{K}^{-1}) = 0$, since $\Omega^1_{X/A,\xi}$ is a K-module and K is the total ring of fractions of $\mathcal{O}_{X,x_0}$

(whereas B is an artinian quotient of $\mathcal{O}_{X,x_0}$). It is likewise clear that applying $\mathscr{H}om_X(\mathcal{O}_Z, \cdot)$ to $\mathscr{K}^0$ kills off the terms in the part of the direct sum in (B.2.2) indexed by $x \in X^0$ with $x \notin \operatorname{Supp}(Z) = \{x_0\}$. Thus, $i^{\flat}(\mathscr{K}^\bullet)$ is the complex in degree 0 on Z given by the quasi-coherent sheaf

$$\mathscr{H}om_X(\mathcal{O}_Z, i_{x_0*}(I \otimes_A \Omega^1_{X/A,\xi}/\Omega^1_{X/A,x_0})),$$

or equivalently, in terms of B-modules,

$$\operatorname{Hom}_{\mathcal{O}_{x_0}}(B, I \otimes_A \omega_{x_0} \otimes_{\mathcal{O}_{x_0}} K/\mathcal{O}_{x_0}),$$

where $\mathcal{O}_{x_0} = \mathcal{O}_{X,x_0}$, $\omega_{x_0} = \Omega^1_{X/A,x_0}$. Applying E has no effect on this (since Z is a 1-point space), so we arrive at

$$i^y f^z(\tilde{I}) \simeq \operatorname{Hom}_{\mathcal{O}_{x_0}}(B, I \otimes_A \omega_{x_0} \otimes_{\mathcal{O}_{x_0}} K/\mathcal{O}_{x_0})^{\sim}$$

on $Z = \operatorname{Spec}(B)$.

The residue isomorphism (B.2.1) on Z therefore amounts to a B-linear isomorphism

(B.2.4) $$\operatorname{Hom}_A(B, I) \simeq \operatorname{Hom}_{\mathcal{O}_{x_0}}(B, I \otimes_A \omega_{x_0} \otimes_{\mathcal{O}_{x_0}} K/\mathcal{O}_{x_0}).$$

Choose any $t \in \mathcal{O}_{x_0}$ which cuts out a section in $X(A)$ supported at x_0, so $\widehat{\mathcal{O}}_{x_0} \simeq A[\![t]\!]$ and ω_{x_0} is free over $\mathcal{O}_{x_0}$ on the basis dt. We call any such t a *uniformizer* at x_0, out of analogy with the case $A = k$. Clearly $\omega_{x_0} \otimes_{\mathcal{O}_{x_0}} K/\mathcal{O}_{x_0}$ is free over $\mathcal{O}_{x_0}$ on the basis $\{t^{-1}dt, t^{-2}dt, t^{-3}dt, \dots\}$. Let b_0 be the (nilpotent) image of t under the canonical surjection $\mathcal{O}_{x_0} \twoheadrightarrow B$ corresponding to the closed immersion $i : Z \hookrightarrow X$. The explicit description of (B.2.4), which also explains the reason it is called the residue isomorphism, is:

THEOREM B.2.1. [**RD**, VII, 1.2] *If $\varphi \mapsto \psi$ under (B.2.4), then*

$$\psi(b) = \sum_{r \geq 0} -\varphi(bb_0^r)t^{-r-1}dt,$$

so $\varphi(b) \in I \subseteq I \otimes_A \mathcal{O}_{x_0}$ is the negative of the coefficient of $t^{-1}dt$ in $\psi(b)$.

The proof of this theorem will be given in §B.3. From the proof, it will be clear that the sign in Theorem B.2.1 is unaffected by omitting the sign in (1.3.28) (which is used in the definition of (2.5.3) and plays a role in the proof of Corollary 3.1.2). In [**RD**, VII, 1.2], the sign in Theorem B.2.1 is missing. Right now we record some consequences. Using $f^\Delta(\tilde{I}) = f^z(\tilde{I})$ and the description of $\mathscr{K}^\bullet = f^z(\tilde{I})$ via (B.2.2), the proof of [**RD**, VII, 1.3] explains how to use Theorem B.2.1 to show that when f is quasi-compact (so f_* commutes with direct sums) the map of graded sheaves

$$\operatorname{Tr}_{f,\tilde{I}} : f_* f^\Delta(\tilde{I}) \to \tilde{I} = \tilde{I}[0],$$

which certainly kills the degree -1 term on the left, is given in degree 0 by the B-linear map

(B.2.5) $$\bigoplus_{x \in X^0} -\operatorname{res}_x : \bigoplus_{x \in X^0} (I \otimes_A \omega_x \otimes_{\mathcal{O}_x} K/\mathcal{O}_x) \to I$$

in which $-\mathrm{res}_x$ (by definition) sends

$$u = \sum_{r<0} a_r t_x^r dt_x \in I \otimes_A (\omega_x \otimes_{\mathcal{O}_x} K/\mathcal{O}_x)$$

to $-a_{-1}$ (where t_x is a uniformizer at x, all $a_r \in I$, and $a_r = 0$ for sufficiently negative r). In particular, this justifies the fact that taking 'residues at $x \in X^0$' (with values in I) is independent of the choice of uniformizer t_x at $x \in X^0$. Moreover, the assertion that $\mathrm{Tr}_{f,\overline{f}}$ is a map of *complexes* is obviously equivalent to a 'residue theorem' for elements in $I \otimes_A \Omega^1_{X/A,\xi}$ (specializing to the classical residue theorem when $A = k$, $I = k$). We refer to [**RD**, p.364ff] for a development of the theory of residues on smooth curves on smooth curves over artin rings and its application to the proof of the final assertion in Theorem 3.4.1 that 'residual complex trace maps' attached to *proper* morphisms are always maps of complexes, conditional on Theorem B.2.1.

Granting these results, when the smooth connected curve $f : X \to \mathrm{Spec}(A)$ is *proper* we want to give an explicit description of Grothendieck's trace map $\gamma_f : \mathrm{H}^1(X, \Omega^1_{X/A}) \to A$. We have already observed that the canonical sequence (without signs)

$$0 \to I \otimes_A \Omega^1_{X/A} \to i_{\xi_*}(I \otimes_A \Omega^1_{X/A,\xi}) \to \bigoplus_{x \in X^0} i_{x_*}(I \otimes_A \Omega^1_{X/A,\xi}/\Omega^1_{X/A,x}) \to 0$$

is a flasque (even injective) resolution of $I \otimes_A \Omega^1_{X/A}$, so we get an exact sequence of A-modules

$$I \otimes_A \Omega^1_{X/A,\xi} \to \bigoplus_{x \in X^0} I \otimes_A \Omega^1_{X/A,\xi}/\Omega^1_{X/A,x} \to \mathrm{H}^1(X, \Omega^1_{X/A}) \to 0.$$

By the 'residue theorem', the sum of the maps res_x for $x \in X^0$ kills $I \otimes_A \Omega^1_{X/A,\xi}$, so there is an induced map

(B.2.6) $$\bigoplus_{x \in X^0} \mathrm{res}_x : \mathrm{H}^1(X, I \otimes_A \Omega^1_{X/A}) \to I.$$

Suppose A is Gorenstein, so we can take $I = A$ and therefore get a 'residue map'

(B.2.7) $$\mathrm{res}_{X/A} : \mathrm{H}^1(X, \Omega^1_{X/A}) \to A.$$

A special case of the relation between residues and Grothendieck's theory is:

THEOREM B.2.2. *For a local Gorenstein artin ring A with residue field which is algebraically closed, we have* $\mathrm{res}_{X/A} = -\gamma_f$.

After the proof, we will see by a *base change* argument that the same assertion is intrinsically meaningful and true without the Gorenstein hypothesis. We emphasize that the sign in Theorem B.2.2 rests not only on Theorem B.2.1, but also on the definition of (3.4.13) (which is used in the definition of γ_f). If (3.4.13) were changed by a sign of $(-1)^1 = -1$, then there would be no sign in Theorem B.2.2 but there would be a sign in Theorem B.4.1.

PROOF. Let $Z^\bullet$ denote the codimension filtration on X. Since $\mathscr{O}_Y$ is a residual complex on $Y = \mathrm{Spec}(A)$ with associated filtration equal to the codimension filtration on Y, the description (B.2.5) of Grothendieck's trace map $\mathrm{Tr}_{f,\tilde{I}} = \mathrm{Tr}_{f,\mathscr{O}_Y} = \gamma_f$ via the residual complex $\mathscr{K}^\bullet = E_{Z^\bullet[1]}(f^\sharp(\mathscr{O}_Y))$ from (B.2.2) is given by almost exactly the same recipe as the one used to define the residue map $\mathrm{res}_{X/A}$. Grothendieck's construction uses the flasque resolution of $\Omega^1_{X/A}[1]$ given by the two-term complex in degrees $-1, 0$

$$(B.2.8) \qquad i_{\xi*}(\Omega^1_{X/A,\xi}) \to \bigoplus_{x \in X^0} i_{x*}(\Omega^1_{X/A,\xi}/\Omega^1_{X/A,x})$$

which localizes at $x \in X^0$ to the canonical projection

$$\Omega^1_{X/A,\xi} \to \Omega^1_{X/A,\xi}/\Omega^1_{X/A,x},$$

whereas the residue map uses the *same* complex, viewed in degrees 0, 1 (without the intervention of signs). Also, *both* constructions use the same canonical augmentation map $\Omega^1_{X/A} \to i_{\xi*}(\Omega^1_{X/A,\xi})$.

Since the isomorphism $\mathrm{H}^0(\mathbf{R}f_*(\Omega^1_{X/A}[1])) \simeq \mathrm{H}^1(X, \Omega^1_{X/A})$ is defined in accordance with (3.4.13) and (3.6.13) by shifting the injective resolution (B.2.8) *without* changing signs in the differentials, Grothendieck's definition of γ_f computes $\mathrm{H}^1(X, \Omega^1_{X/A})$ by using the flasque resolution (B.2.8), as does the definition of $\mathrm{res}_{X/A}$. Thus, due the sign in (B.2.5) which is not in (B.2.6), the resulting maps $\mathrm{H}^1(X, \Omega^1_{X/A}) \to A$ are negatives of each other.

∎

When $A = k$, Theorem B.2.2 implies the 'smooth connected' case of Theorem 5.2.3. Let us now explain why Theorem B.2.2 is true without a Gorenstein assumption. We need to use the existence of a general theory of residues with values in an A-module, including the residue theorem (the details of this are given in [**RD**, p.364ff]). Since flasque resolutions compute cohomology, the construction of a map

$$\mathrm{res}_{X/A} : \mathrm{H}^1(X, \Omega^1_{X/A}) \to A$$

can then be carried out for a general local artin A as above, with I replaced by A. It then makes sense to ask if this is the negative of Grothendieck's trace map. Since *both* constructions are compatible with base change to a quotient of A and A is a quotient of a Gorenstein artin ring (with the same residue field k), by using Grothendieck's fundamental theorem that proper smooth curves always lift through a nilpotent thickening of the affine base [**SGA1**, III, Thm 6.3] we immediately reduce to the case where A is Gorenstein. But this is the content of Theorem B.2.2. It is now natural to ask if the abstract map (B.2.6) is just $I \otimes_A (\cdot)$ applied to $-\gamma_f$ using the natural isomorphism

$$I \otimes_A \mathrm{H}^1(X, \Omega^1_{X/A}) \simeq \mathrm{H}^1(X, I \otimes_A \Omega^1_{X/A}).$$

This follows by a functoriality argument.

B.3. The Proof

Before we begin the proof of Theorem B.2.1, note that if we can show $-\varphi(b)$ is the coefficient of $t^{-1}dt$ in $\psi(b)$, then we deduce the entire theorem as follows. The $\mathscr{O}_{x_0}$-linearity of (B.2.4) implies that $\psi(bb_0^r) = t^r\psi(b)$ for all $r \geq 0$, so $-\varphi(bb_0^r)$ is the coefficient of $t^{-1}dt$ in $t^r\psi(b)$. In other words, $-\varphi(bb_0^r)$ is the coefficient of $t^{-r-1}dt$ in $\psi(b)$, as desired. Thus, we may now focus our attention on computing the coefficient of $t^{-1}dt$ in $\psi(b)$, where $\varphi \mapsto \psi$ under (B.2.4).

The essential problem is to unwind the abstract isomorphisms that are used to construct (B.2.4). The relevant scheme diagram is

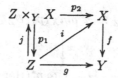

By definition, (B.2.4) is the result of applying H^0 to the composite of isomorphisms

$$\text{(B.3.1)} \qquad g^b(\widetilde{I}) \xrightarrow[\simeq]{\psi_{j,p_1}} j^b p_1^\sharp g^b(\widetilde{I}) \xrightarrow{\simeq} j^b p_2^b f^\sharp(\widetilde{I}) \xleftarrow{\simeq} i^b f^\sharp(\widetilde{I})$$

where we use $f^\sharp(\widetilde{I}) = \Omega^1_{X/A}[1] \simeq f^z\widetilde{I}$ via (B.2.8). Let's first look at the derived category isomorphism $p_1^\sharp g^b(\widetilde{I}) \simeq p_2^b f^\sharp(\widetilde{I})$ in the middle of (B.3.1). Since residual complexes (such as $f^z(\widetilde{I})$) consist of injective terms, this *derived category* isomorphism is represented by the map of (vertical) complexes in degrees $-1, 0$

$$\begin{array}{ccc}
\omega_{p_1} \otimes p_1^*(\mathrm{Hom}_A(B,I)^\sim) & \longrightarrow & \omega_{p_1} \otimes \mathscr{H}om_X(p_{2*}\mathscr{O}, i_{\xi_*}(I \otimes_A K)) \\
\downarrow & & \downarrow \\
0 & \longrightarrow & \bigoplus_{x \in X^0} \omega_{p_1} \otimes \mathscr{H}om_X(p_{2*}\mathscr{O}, i_{x*}\mathrm{H}^1_x(I \otimes_A \mathscr{O}_{X,x}))
\end{array}$$

in which $\omega_{p_1} = \omega_{Z \times_Y X/Z}$, $\mathscr{O} = \mathscr{O}_{Z \times_Y X}$, and all maps are the canonical ones (with the right column arising from the $E_{Z^\bullet[1]}$ spectral sequence construction on X, with $Z^\bullet$ the codimension filtration). This is a quasi-isomorphism of complexes since it is an isomorphism in the derived category, so the right column is an *injective* resolution of the left side.

In fact, this is just a special case of Corollary 3.1.2 (with $M = I$, $\mathscr{L} = \mathscr{O}_X$), so it can be reformulated as

$$\text{(B.3.2)}$$
$$\begin{array}{ccc}
\omega_{p_1} \otimes p_1^*(\mathrm{Hom}_A(B,I)^\sim) & \longrightarrow & \omega_{p_1} \otimes \mathscr{H}om_X(p_{2*}\mathscr{O}, i_{\xi_*}(I \otimes_A K)) \\
\downarrow & & \downarrow \\
0 & \longrightarrow & \bigoplus_{x \in X^0} \omega_{p_1} \otimes \mathscr{H}om_X(p_{2*}\mathscr{O}, i_{x*}(I \otimes_A K/\mathscr{O}_{X,x}))
\end{array}$$

with the differential on the right corresponding to the canonical projection maps. Since the right column is an injective resolution of the left column, applying $j^\flat$ amounts to just applying $\mathcal{H}om_{Z\times_Y X}(\mathcal{O}_Z, \cdot)$ to the right column. This kills all terms in the direct sum indexed by $x \neq x_0$ and thereby gives rise to a canonical coboundary map isomorphism of $\mathcal{O}_Z$-modules

$$\mathcal{H}om_{Z\times_Y X}(\mathcal{O}_Z, \omega_{Z\times_Y X/Z} \otimes \mathcal{H}om_X(p_{2*}\mathcal{O}_{Z\times_Y X}, i_{x_0*}(I \otimes_A K/\mathcal{O}_{x_0})))$$

$$\simeq \Big\downarrow \delta$$

$$\mathcal{E}xt^1_{Z\times_Y X}(\mathcal{O}_Z, \omega_{Z\times_Y X} \otimes p_1^*(\mathrm{Hom}_A(B, I)^\sim))$$

Meanwhile, the isomorphism $j^\flat p_1^\sharp \simeq 1$ at the beginning of (B.3.1) is the composite of the derived category isomorphisms

$$j^\flat(\omega_{Z\times_Y X/Z}[1] \otimes p_1^*(\cdot)) \xrightarrow{\ \eta_j\ } \omega_{Z/Z\times_Y X}[-1] \overset{\mathbf{L}}{\otimes} \mathbf{L}j^*(\omega_{Z\times_Y X/Z}[1] \otimes p_1^*(\cdot))$$

$$\Big\|$$

$$(\omega_{Z/Z\times_Y X}[-1] \otimes j^*\omega_{Z\times_Y X/Z}[1]) \otimes (\cdot)$$

$$\Big\uparrow \zeta'_{j,p_1}$$

$$(\cdot)$$

where the map η_j can be computed without using flat resolutions to compute $\mathbf{L}j^*$ because $p_1^*(\cdot)$ is always j^*-acyclic (as p_1 and $p_1 \circ j = 1$ are flat); see (2.7.3) for an explication of this *without* signs when $(\cdot) = \mathcal{G}[m]$ (we only need the case $m = 0$).

Putting this all together, the *inverse* of (B.2.4) is the composite $\mathcal{O}_Z$-linear isomorphism

(B.3.3)
$$\mathrm{Hom}_{\mathcal{O}_{x_0}}(B, I \otimes_A \omega_{x_0} \otimes_{\mathcal{O}_{x_0}} K/\mathcal{O}_{x_0})^{\sim}$$

$$\parallel$$

$$\mathcal{H}om_{Z\times_Y X}(\mathcal{O}_Z, \mathcal{H}om_X(\mathcal{O}_{Z\times_Y X}, i_{x_0*}(I \otimes_A \omega_{x_0} \otimes_{\mathcal{O}_{x_0}} K/\mathcal{O}_{x_0})))$$

$$\parallel$$

$$\mathcal{H}om_{Z\times_Y X}(\mathcal{O}_Z, \omega_{Z\times_Y X/Z} \otimes \mathcal{H}om_X(\mathcal{O}_{Z\times_Y X}, i_{x_0*}(I \otimes_A K/\mathcal{O}_{x_0})))$$

$$\simeq \Big\downarrow \delta$$

$$\mathcal{E}xt^1_{Z\times_Y X}(\mathcal{O}_Z, \omega_{Z\times_Y X} \otimes p_1^*(\mathrm{Hom}_A(B,I)^{\sim}))$$

$$\simeq \Big\downarrow \eta_j$$

$$\omega_{Z/Z\times_Y X} \otimes j^*\omega_{Z\times_Y X/Z} \otimes j^*p_1^*(\mathrm{Hom}_A(B,I)^{\sim})$$

$$\Big\Vert \varsigma'_{j,p_1}$$

$$\mathrm{Hom}_A(B,I)^{\sim}$$

with η_j computed as in (2.5.1), using Koszul complexes. The reason that the map δ (rather than $-\delta$) is the correct one to use in (B.3.3) is based on two facts. First, quite generally in an abelian category $\mathscr{A}$ with enough injectives and an injective resolution $M \to I^\bullet$, the short exact sequence

$$0 \to M \to I^0 \xrightarrow{\ d^0\ } \ker(d^1) \to 0$$

gives rise to a coboundary map $\delta : \mathrm{Hom}(N, \ker d^1) \to \mathrm{Ext}^1(N, M)$ which is exactly the canonical map $\mathrm{Hom}(N, \ker d^1) \to \mathrm{H}^1(\mathrm{Hom}(N, I^\bullet))$ when we compute $\mathrm{Ext}^1(N, M)$ using the injective resolution $I^\bullet$ of M. Second, according to the conventions on injective resolutions in the explication (2.7.3) of ψ_{j,p_1}, we are required to calculate the above $\mathcal{E}xt^1$ term using exactly the injective resolution arising from (B.3.2), without the introduction of signs.

The sign (of $(-1)^{1(1+1)/2} = -1$) implicit in the definition of the map η_j is the same as the sign used in the proof of Corollary 3.1.2 to obtain the canonical projection in the right column of (B.3.2). Changing this sign would change both η_j and δ in (B.3.3) by the same sign, so the composite map in (B.3.3) is 'independent' of the sign in (2.5.1). This is the 'cancellation' referred to after Corollary 3.1.2.

We now summarize the above analysis, and in particular the diagram (B.3.3) of quasi-coherent $\mathcal{O}_Z$-modules, in terms of B-modules. To start off, there is the injective resolution of $B \otimes_A \mathcal{O}_{x_0}$-modules

$$0 \longrightarrow \operatorname{Hom}_A(B, I) \otimes_A \omega_{x_0} \longrightarrow \operatorname{Hom}_{\mathcal{O}_{x_0}}(B \otimes_A \mathcal{O}_{x_0}, I \otimes_A \omega_{x_0} \otimes_{\mathcal{O}_{x_0}} K)$$

$$0 \longleftarrow \operatorname{Hom}_{\mathcal{O}_{x_0}}(B \otimes_A \mathcal{O}_{x_0}, I \otimes_A \omega_{x_0} \otimes_{\mathcal{O}_{x_0}} K/\mathcal{O}_{x_0})$$

which we can equivalently write as

$$(B.3.4) \quad 0 \longrightarrow \operatorname{Hom}_A(B, I) \otimes_A \omega_{x_0} \longrightarrow \operatorname{Hom}_A(B, I \otimes_A \omega_{x_0} \otimes_{\mathcal{O}_{x_0}} K)$$

$$0 \longleftarrow \operatorname{Hom}_A(B, I \otimes_A \omega_{x_0} \otimes_{\mathcal{O}_{x_0}} K/\mathcal{O}_{x_0})$$

and this resolution uses the canonical projection $K \to K/\mathcal{O}_{x_0}$. Also, the surjection $\mathcal{O}_{x_0} \twoheadrightarrow B$ over A corresponding to $Z \hookrightarrow X$ takes t onto b_0 and induces a surjection $j^* : B \otimes_A \mathcal{O}_{x_0} \twoheadrightarrow B$ with kernel generated by $\tau \overset{\text{def}}{=} 1 \otimes t - b_0 \otimes 1$ (since $\mathcal{O}_{x_0}/t \simeq A$ over A and b_0 is nilpotent). Combining all this, we find that the inverse (B.3.3) of (B.2.4) is equal to the B-linear composite

$$(B.3.5) \qquad \operatorname{Hom}_{\mathcal{O}_{x_0}}(B, I \otimes_A \omega_{x_0} \otimes_{\mathcal{O}_{x_0}} K/\mathcal{O}_{x_0})$$

$$\Big\uparrow {\scriptstyle\simeq}$$

$$\operatorname{Hom}_{B \otimes_A \mathcal{O}_{x_0}}(B, \operatorname{Hom}_{\mathcal{O}_{x_0}}(B \otimes_A \mathcal{O}_{x_0}, I \otimes_A \omega_{x_0} \otimes_{\mathcal{O}_{x_0}} K/\mathcal{O}_{x_0}))$$

$$\Big\|$$

$$\operatorname{Hom}_{B \otimes_A \mathcal{O}_{x_0}}(B, \operatorname{Hom}_A(B, I \otimes_A \omega_{x_0} \otimes_{\mathcal{O}_{x_0}} K/\mathcal{O}_{x_0}))$$

$$\Big\downarrow {\scriptstyle\delta}$$

$$\operatorname{Ext}^1_{B \otimes_A \mathcal{O}_{x_0}}(B, \operatorname{Hom}_B(A, I) \otimes_A \omega_{x_0})$$

$$\simeq \Big\downarrow {\scriptstyle\eta_j}$$

$$(\operatorname{Hom}_A(B, I) \otimes_A \omega_{x_0})/\tau$$

$$\simeq \Big\downarrow {\scriptstyle\zeta}$$

$$\operatorname{Hom}_A(B, I)$$

where δ is the coboundary map arising from (B.3.4), the final isomorphism ζ uses $\omega_{x_0} \simeq \mathcal{O}_{x_0}$ via the basis dt in order to define

$$
\begin{aligned}
(\operatorname{Hom}_A(B, I) \otimes_A \omega_{x_0})/\tau &\simeq \operatorname{Hom}_A(B, I) \otimes_B (B \otimes_A \omega_{x_0})/\tau \\
&\simeq \operatorname{Hom}_A(B, I) \otimes_B (B \otimes_A \mathcal{O}_{x_0})/\tau \\
&\simeq \operatorname{Hom}_A(B, I),
\end{aligned}
$$

and the map η_j is the isomorphism given by the first map in (2.5.2) with the ring $B \otimes_A \mathcal{O}_{x_0}$, the module $\text{Hom}_A(B, I) \otimes_A \mathcal{O}_{x_0}$, and the regular sequence $\{\tau\}$. The link between t and τ (as opposed to $-\tau$, for example) underlying the correctness of the description of ζ and η_j in (B.3.5) is the equality $d\tau = 1 \otimes dt$ with respect to the isomorphism

$$\Omega^1_{B \otimes_A \mathcal{O}_{x_0}/B} \simeq B \otimes_A \Omega^1_{\mathcal{O}_{x_0}/A}.$$

How can we compute (B.3.5)? To start off, let

$$0 \to \text{Hom}_A(B, I) \otimes_A \omega_{x_0} \to I^0 \xrightarrow{d} I^1 \to 0$$

denote the $B \otimes_A \mathcal{O}_{x_0}$-module injective resolution (B.3.4) and let

$$0 \to P^{-1} \xrightarrow{\tau} P^0 \to B \to 0$$

denote the $B \otimes_A \mathcal{O}_{x_0}$-module Koszul (projective) resolution $K_\bullet(\tau, B)$ with $P^0 = P^{-1} = B \otimes_A \mathcal{O}_{x_0}$ as used in the definition of η_j. From the first two steps in (B.3.5), we extract (using $\omega_{x_0} = \mathcal{O}_{x_0} dt \simeq \mathcal{O}_{x_0}$) the natural 'evaluate at 1' isomorphism

(B.3.6)

$$\text{Hom}_{B \otimes_A \mathcal{O}_{x_0}}(B, I^1) = \text{Hom}_{B \otimes_A \mathcal{O}_{x_0}}(B, \text{Hom}_{\mathcal{O}_{x_0}}(B \otimes_A \mathcal{O}_{x_0}, I \otimes_A K/\mathcal{O}_{x_0}))$$

$$\downarrow \simeq$$

$$\text{Hom}_{\mathcal{O}_{x_0}}(B, I \otimes_A K/\mathcal{O}_{x_0})$$

More generally, one can check directly that in our situation

$$\text{Hom}_{B \otimes_A \mathcal{O}_{x_0}}(B, \text{Hom}_A(B, N)) \to \text{Hom}_{\mathcal{O}_{x_0}}(B, N)$$

is an isomorphism for any $\mathcal{O}_{x_0}$-module N. Using the usual 'Hom double complex' whose total complex $\text{Hom}^\bullet_{B \otimes_A \mathcal{O}_{x_0}}(P^\bullet, I^\bullet)$ computes Ext (according to our definition of Ext in §1.3), we want to 'snake' our way through the diagram with exact rows and columns (and anti-commutative right square)

(B.3.7)

$$\begin{array}{ccccccc}
\text{Hom}_A(B,I) \otimes_A \omega_{x_0} & \longrightarrow & \text{Hom}_{B \otimes_A \mathcal{O}_{x_0}}(P^{-1}, I^0) & \xrightarrow{d} & \text{Hom}_{B \otimes_A \mathcal{O}_{x_0}}(P^{-1}, I^1) & \longrightarrow & 0 \\
{\scriptstyle -\tau}\uparrow & & {\scriptstyle -\tau}\uparrow & & {\scriptstyle \tau}\uparrow & & \\
\text{Hom}_A(B,I) \otimes_A \omega_{x_0} & \longrightarrow & \text{Hom}_{B \otimes_A \mathcal{O}_{x_0}}(P^0, I^0) & \xrightarrow{d} & \text{Hom}_{B \otimes_A \mathcal{O}_{x_0}}(P^0, I^1) & \longrightarrow & 0 \\
& & & & {\scriptstyle}\uparrow & & \\
& & & & \text{Hom}_{B \otimes_A \mathcal{O}_{x_0}}(B, I^1) & &
\end{array}$$

going from $\text{Hom}_{B \otimes_A \mathcal{O}_{x_0}}(B, I^1) \simeq \text{Hom}_{\mathcal{O}_{x_0}}(B, I \otimes_A \omega_{x_0} \otimes_{\mathcal{O}_{x_0}} K/\mathcal{O}_{x_0})$ over into the cokernel of the left column.

To make this diagram clearer, let M denote the B-module $\text{Hom}_A(B, I)$. Since

$$K/\mathcal{O}_{x_0} = \varinjlim \mathcal{O}_{x_0}/t^n$$

is A-flat (or even better, A-free), the natural map of $B \otimes_A \mathscr{O}_{x_0}$-modules

(B.3.8) $M \otimes_A K/\mathscr{O}_{x_0} \simeq \mathrm{Hom}_A(B, I \otimes_A K/\mathscr{O}_{x_0})$

is an isomorphism. Thus, we have a natural isomorphism of $B \otimes_A \mathscr{O}_{x_0}$-modules

(B.3.9) $\mathrm{Hom}_{B \otimes_A \mathscr{O}_{x_0}}(B, M \otimes_A K/\mathscr{O}_{x_0}) \simeq \mathrm{Hom}_{\mathscr{O}_{x_0}}(B, I \otimes_A K/\mathscr{O}_{x_0})$

induced by the A-linear 'evaluation at 1' map

$$M = \mathrm{Hom}_A(B, I) \to I.$$

This is a restatement of (B.3.6). Combining this with the isomorphism $\omega_{x_0} \simeq \mathscr{O}_{x_0}$ arising from the basis dt, we may identify (B.3.7) with the diagram (with anti-commutative right square)

(B.3.10)

$$
\begin{array}{ccccccc}
M \otimes_A \mathscr{O}_{x_0} & \longrightarrow & M \otimes_A K & \longrightarrow & M \otimes_A K/\mathscr{O}_{x_0} & \longrightarrow & 0 \\
\uparrow{\scriptstyle -\tau} & & \uparrow{\scriptstyle -\tau} & & \uparrow{\scriptstyle \tau} & & \\
M \otimes_A \mathscr{O}_{x_0} & \longrightarrow & M \otimes_A K & \longrightarrow & M \otimes_A K/\mathscr{O}_{x_0} & \longrightarrow & 0 \\
& & & & \uparrow & & \\
& & & & \mathrm{Hom}_{B \otimes_A \mathscr{O}_{x_0}}(B, M \otimes_A K/\mathscr{O}_{x_0}) & &
\end{array}
$$

where the augmentation at the bottom is $\psi \mapsto \psi(1)$ and the map $K \to K/\mathscr{O}_{x_0}$ used in both rows is the canonical projection.

We want to compute how the 'snake lemma' method for computing the cohomology of a total complex sends an element ψ in the bottom object

(B.3.11) $\mathrm{Hom}_{\mathscr{O}_{x_0}}(B, I \otimes_A K/\mathscr{O}_{x_0}) \simeq \mathrm{Hom}_{B \otimes_A \mathscr{O}_{x_0}}(B, \mathrm{Hom}_A(B, I \otimes_A K/\mathscr{O}_{x_0}))$

(see (B.3.8), (B.3.9)) in the right column of (B.3.10) over to an element φ in the cokernel

$$(M \otimes_A \mathscr{O}_{x_0})/\tau = M \otimes_B (B \otimes_A \mathscr{O}_{x_0})/\tau \simeq M = \mathrm{Hom}_A(B, I)$$

of the left column. Due to the sign of -1 in the definition of η_j, our goal is to show that $\varphi \in \mathrm{Hom}_A(B, I)$ is the A-linear map which sends $b \in B$ to the coefficient (in I) of t^{-1} in $\psi(b) \in I \otimes_A K/\mathscr{O}_{x_0}$, relative to the A-basis $\{t^{-1}, t^{-2}, \dots\}$ of $K/\mathscr{O}_{x_0}$. It is easy to check that the isomorphism (B.3.11) sends ψ to the map

$$b \mapsto (b' \mapsto \psi(bb')).$$

Thus, the augmentation map

$$\mathrm{Hom}_{\mathscr{O}_{x_0}}(B, I \otimes_A K/\mathscr{O}_{x_0}) \hookrightarrow M \otimes_A K/\mathscr{O}_{x_0} \simeq \bigoplus_{r \geq 1} M \otimes t^{-r}$$

in the lower right of (B.3.10) sends ψ to $\sum \psi_r \otimes t^{-r}$, where $\psi_r \in M = \mathrm{Hom}_A(B, I)$ gives the coefficient of t^{-r} in ψ via $I \otimes_A K/\mathscr{O}_{x_0} \simeq \bigoplus(I \otimes t^{-r})$. The fact that ψ is $\mathscr{O}_{x_0}$-linear (not just A-linear) says that $t \cdot \psi(\cdot) = \psi(b_0(\cdot))$. This is nothing other than the τ-torsion condition on ψ, viewed as an element of $M \otimes_A K/\mathscr{O}_{x_0}$. In terms of the ψ_r's, this amounts to the relation $\psi_r(b_0(\cdot)) = \psi_{r+1}$ for $r \geq 1$ (and this makes explicit that $\psi_r = 0$ for large r, since b_0 is nilpotent).

Now we 'snake' through (B.3.10). Since the rows use the canonical projection $K \to K/\mathscr{O}_{x_0}$, we can lift the element $\sum \psi_r \otimes t^{-r} \in M \otimes_A K/\mathscr{O}_{x_0}$ to the element $\sum \psi_r \otimes t^{-r} \in M \otimes_A K$. Multiplying by $-\tau$ brings us to the element

$$-\psi_1 \otimes 1 - \sum_{r \geq 1} \psi_{r+1} \otimes t^{-r} + \sum_{r \geq 1} \psi_r(b_0(\cdot)) \otimes t^{-r} = -\psi_1 \otimes 1 \in M \otimes_A \mathscr{O}_{x_0} \subseteq M \otimes_A K.$$

Thus, in the total complex of the anti-commutative square in (B.3.10), the ordered pairs

$$(\psi_1 \otimes 1, 0), \quad (0, \sum \psi_r \otimes t^r)$$

define the same cohomology class (in degree 1). We conclude that the sought-after element

$$\varphi \in \mathrm{Hom}_A(B, I) = M \simeq (M \otimes_A \mathscr{O}_{x_0})/\tau(M \otimes_A \mathscr{O}_{x_0}),$$

using the left column of (B.3.10), is exactly ψ_1. But $\psi_1(b)$ is the coefficient of t^{-1} in $\psi(b)$, so we're done!

B.4. Duality on Jacobians

We conclude our discussion of Grothendieck duality on proper smooth curves by relating it to duality on Jacobians. In the classical situation of a proper, smooth, connected curve X over an algebraically closed field k, with Jacobian J, there are canonical isomorphisms $\mathrm{H}^0(X, \Omega^1_{X/k}) \simeq T_0(J)^\vee$ and $\mathrm{H}^1(X, \mathscr{O}_X) \simeq T_0(J)$ with the cotangent and tangent spaces at the origin of J (to be reviewed below). How does the canonical pairing between $T_0(J)$ and $T_0(J)^\vee$ relate to the Grothendieck duality pairing (or its negative, the residue pairing) between $\mathrm{H}^0(X, \Omega^1_{X/k})$ and $\mathrm{H}^1(X, \mathscr{O}_X)$? Since the pairing on J is very "local" and the pairing on X is very "global", it is not immediately obvious how to relate these two pairings.

For conceptual clarity, we work in a relative situation. Let $f : X \to S$ be a proper smooth map with geometrically connected fibers of pure dimension 1. The genus of the fibers is a locally constant function on the base, so there is no serious loss of generality in assuming (as we now do) that the fibers have constant genus g. Let $J = \mathrm{Pic}^0_{X/S}$, an abelian scheme of relative dimension g. This abelian scheme is a retrocompact open in the smooth, separated algebraic space $\mathrm{Pic}_{X/S}$ and coincides with the connected component of the identity section on fibers. When $g \geq 2$, the dualizing sheaf $\omega_f = \Omega^1_{X/S}$ is relatively ample (by the Riemann-Roch theorem on fibers and [EGA, IV$_3$, 9.6.5]), so $\mathrm{Pic}_{X/S}$ exists as a scheme, via Grothendieck's projective methods. When $X(S) \neq \emptyset$, there is (for any g) a universal line rigidified line bundle on $X \times_S J$, so in such cases J can be viewed as representing isomorphism classes of certain line bundles. When $X(S) = \emptyset$, there is no universal line bundle on $X \times_S J$. See [BLR, §8.4, Ch 9] for an elegant exposition of the basic theory of Jacobians of relative curves.

Let $\mathscr{I}$ denote the ideal sheaf of the zero section $e : S \to J$, so $e^*(\mathscr{I}/\mathscr{I}^2)$ is locally free of rank g on S. There is a canonical exact sequence of abelian sheaves on $X_{\text{ét}}$

$$0 \to \mathscr{O}_X \to \mathscr{O}^\times_{X[\epsilon]} \to \mathscr{O}^\times_X \to 1$$

where $X[\epsilon] = X \times_S S[\epsilon]$, with $\epsilon^2 = 0$. The long exact sequence of étale higher direct image sheaves gives an isomorphism of abelian sheaves (for both the Zariski and étale topologies)

$$R^1 f_*(\mathscr{O}_X) \simeq \ker(R^1 f_*(\mathscr{O}^\times_{X[\epsilon]}) \to R^1 f_*(\mathscr{O}^\times_X)).$$

Since $X \to S$ *does* admit sections locally for the étale topology on S, we can identify the right side with the relative tangent space $e^*(\mathscr{I}/\mathscr{I}^2)^\vee$ at the origin of $J = \mathrm{Pic}^0_{X/S}$ (or better, of $\mathrm{Pic}_{X/S}$) once we make a universal functorial *choice* of isomorphism

(B.4.1) $\mathrm{Pic}_{\text{ét}}(T) = \mathrm{Pic}(T) \simeq \check{\mathrm{H}}^1(T, \mathscr{O}^\times_T) = \mathrm{H}^1(T, \mathscr{O}^\times_T)$

for all schemes T.

The point is that the isomorphism $\mathrm{Pic} \simeq \check{\mathrm{H}}^1(\mathbf{G}_m)$ involves a choice of sign. More precisely, if $\mathscr{L}$ is an invertible sheaf on T and $\mathfrak{U} = \{U_i\}$ is an ordered open covering such that $\mathscr{L}|_{U_i} \simeq \mathscr{O}_{U_i}$ for all i, then there are two canonical ways to associate 'transition data' $c_{ij} \in \Gamma(U_i \cap U_j, \mathscr{O}^\times_T)$ for $i < j$, depending on whether one considers the isomorphism

$$\varphi_{ij} : \mathscr{O}_{U_i \cap U_j} \simeq (\mathscr{L}|_{U_i})|_{U_i \cap U_j} = (\mathscr{L}|_{U_j})|_{U_i \cap U_j} \simeq \mathscr{O}_{U_i \cap U_j}$$

or its inverse. The isomorphisms φ_{ij} appear to be simpler to use in the setting of vector bundles of higher rank (corresponding to non-abelian H^1), since the identity

$$\varphi_{ik} = \varphi_{ij} \circ \varphi_{jk}$$

holds in $\mathrm{GL}_n(\mathscr{O}_{U_i \cap U_j \cap U_k})$ when $\mathscr{L}$ is replaced by an arbitrary vector bundle of rank n. Thus, we use these isomorphisms φ_{ij} (rather than their inverses) to define (B.4.1). Now we have unambiguously defined an isomorphism of abelian sheaves

(B.4.2) $e^*(\mathscr{I}/\mathscr{I}^2)^\vee \simeq R^1 f_*(\mathscr{O}_X),$

and from the construction one checks this is $\mathscr{O}_S$-linear and is compatible with base change on S. When $S = \mathrm{Spec}(k)$ for a field k, this is the classical isomorphism $T_0(J) \simeq \mathrm{H}^1(X, \mathscr{O}_X)$, conditional on our choice of sign in the definition of (B.4.1).

One can likewise relate the relative cotangent space $e^*(\mathscr{I}/\mathscr{I}^2)$ of J with cohomology of the curve, as follows. Let $F : J \to S$ be the structure map. Since $\mathscr{O}_S \simeq F_*(\mathscr{O}_J)$, we conclude from [**BLR**, 4.3/2] that there is a canonical isomorphism, compatible with base change on S:

$$e^*(\mathscr{I}/\mathscr{I}^2) \simeq F_*(\Omega^1_{J/S})$$

(i.e., any global 1-form on the group scheme J is uniquely determined by its value in the cotangent space at the origin). Thus, $F_*\Omega^1_{J/S}$ is locally free of rank g and commutes with base change, so $\underline{\mathrm{Aut}}_S(F_*\Omega^1_{J/S})$ is an *affine* S-group scheme of finite type. It follows that the 'translation' map $J \to \underline{\mathrm{Aut}}_S(F_*\Omega^1_{J/S})$ is trivial on geometric fibers and therefore is trivial by [**GIT**, Cor 6.2]. In other words, all global relative 1-forms on J are translation-invariant. When $X(S) \neq \emptyset$, any $P \in X(S)$ defines a map

(B.4.3) $$j_P : X \to J$$

via "$x \mapsto \mathcal{O}(x) \otimes \mathcal{O}(P)^{-1}$", and when $g > 0$ this is a proper monomorphism. This says that if $x_1, x_2 \in X(S)$ and there is an isomorphism $\alpha : \mathcal{O}(x_1) \simeq \mathcal{O}(x_2)$ of $\mathcal{O}_S$-modules, then $x_1 = x_2$. To prove this, we just need the subsheaves $\mathcal{O}(x_1), \mathcal{O}(x_2) \subseteq \mathcal{O}(x_1 + x_2)$ to coincide. Without loss of generality S is locally noetherian, so since $g > 0$ the theory of cohomology and base change ensures that

$$f_* \mathcal{H}om_X(\mathcal{O}(x_j), \mathcal{O}(x_1 + x_2)) = f_* \mathcal{O}(x_{3-j})$$

is an invertible sheaf of formation compatible with base change. The map

$$\mathcal{O}_S \to f_* \mathcal{H}om_X(\mathcal{O}(x_j), \mathcal{O}(x_1 + x_2))$$

corresponding to the canonical inclusion $\mathcal{O}(x_j) \hookrightarrow \mathcal{O}(x_1 + x_2)$ is non-zero on fibers and therefore is an isomorphism of $\mathcal{O}_S$-modules. This forces the diagram

(B.4.4)
$$
\begin{array}{ccc}
\mathcal{O}(x_1) & \longrightarrow & \mathcal{O}(x_1 + x_2) \\
{\scriptstyle \simeq} \downarrow {\scriptstyle \alpha} & \nearrow & \\
\mathcal{O}(x_2) & &
\end{array}
$$

to commute up to an element of $\mathrm{H}^0(S, \mathcal{O}_S^\times)$, so the $\mathcal{O}(x_j)$'s coincide as subsheaves of $\mathcal{O}(x_1 + x_2)$, as desired.

Since proper monomorphisms are closed immersions [**EGA**, IV$_4$, 18.12.6], we conclude that j_P is a closed immersion when $g > 0$. Using pullback, we get an induced map

(B.4.5) $$F_* \Omega^1_{J/S} \to f_* \Omega^1_{X/S}$$

which is *independent* of P (since all global relative 1-forms on J are translation-invariant) and compatible with base change on S. Since the fppf S-scheme X acquires sections fppf-locally over S, by descent we can therefore define (B.4.5) even when $X(S) = \emptyset$. To see that the map (B.4.5) is an isomorphism, first note that both sides are locally free of rank g and commute with base change; for the right side, this follows from the local (even global) freeness of $\mathrm{R}^1 f_* \Omega^1_{X/S}$ (see Corollary 3.6.6) and Grothendieck's theory of cohomology and base change. Thus, we may base change to geometric fibers, where it suffices (for dimension reasons) to check injectivity. This injectivity follows from the classical description of J in terms of the g-fold symmetric product of X [**BLR**, 9.3/6]. Putting this all together, we get a map of $\mathcal{O}_S$-modules

(B.4.6) $$e^*(\mathscr{I}/\mathscr{I}^2) \to f_*(\Omega^1_{X/S})$$

of formation compatible with base change on S, and this map is an isomorphism. When $S = \mathrm{Spec}(k)$ for a field k, this is the classical isomorphism $T_0(J)^\vee = \mathrm{H}^0(J, \Omega^1_{J/k}) \simeq \mathrm{H}^0(X, \Omega^1_{X/k})$. Note that even if we didn't know (B.4.5) were an isomorphism, (B.4.6) would still exist as a natural map.

Looking at (B.4.2) and (B.4.6), a subtle point of compatibility arises, which we discussed above when the base is an algebraically closed field. There is an obvious duality between $e^*(\mathscr{I}/\mathscr{I}^2)$ and $e^*(\mathscr{I}/\mathscr{I}^2)^\vee$. Meanwhile, Grothendieck duality gives a perfect pairing between $f_*\Omega^1_{X/S}$ and $\mathrm{R}^1 f_*(\mathcal{O}_X)$. Are these pairings compatible? Of course, since we had to make a choice of sign in the definition of (B.4.1), one really has to be careful here. Moreover, in view of Theorem 5.2.3, the choice of sign in (B.4.1) that is compatible with Grothendieck duality is the negative of the one that is compatible with the residue pairing on geometric fibers. In any case, it is rather important that the isomorphisms in (B.4.2) and (B.4.6) are compatible with the canonical pairings between the left sides and between the right sides, at least up to a universal sign. This compatibility is essentially the content of [**Maz**, Lemma 2.1], but we want to remove the sign ambiguity in [**Maz**] and fill in the details omitted there. Since Grothendieck duality makes sense over an arbitrary base, we state the compatibility in terms of Grothendieck duality (though the first step in the proof is to reduce to a statement about the more concrete residue pairing over algebraically closed fields).

THEOREM B.4.1. *Fix* $f : X \to S$ *as above. The maps* $e^*(\mathscr{I}/\mathscr{I}^2) \to f_*(\Omega^1_{X/S})$ *and* $e^*(\mathscr{I}/\mathscr{I}^2)^\vee \simeq \mathrm{R}^1 f_*(\mathcal{O}_X)$ *are compatible with the canonical pairing between the left sides and the Grothendieck duality pairing between the right sides.*

As we have mentioned already, it is not obvious how to begin the proof because the tangent space and cotangent space on the Jacobian are very "local", whereas the cohomology groups on the curve are very "global". In [**Mi**, Prop 2.2], Theorem B.4.1 (with S the spectrum of a field) is used to prove that (B.4.5) is an isomorphism, but unfortunately the proof of Theorem B.4.1 is omitted there as a "rather complicated" exercise. We mention this point because the proof we give for Theorem B.4.1 rests on knowing *a priori* that (B.4.5) is an isomorphism; we do not know of a proof that avoids this, even if the base is an algebraically closed field (which, as we shall see, is the essential case).

PROOF. Since everything is compatible with base change, we may use direct limit arguments to reduce to the case in which S is local noetherian, and then even local artinian with algebraically closed residue field. In particular, $X(S) \neq \emptyset$, so X admits an ample line bundle. Any local artin ring is the quotient of a complete regular local ring and a proper smooth curve over an affine base can always be lifted across nilpotent thickenings of the affine base [**SGA1**, III, Thm 6.3]. Thus, by Grothendieck's algebraization theorem for proper formal schemes [**EGA**, III$_1$, 5.4.5], we can lift our X to a proper scheme X' over a complete regular local ring R such that all of the infinitesimal fibers over the closed point are smooth (and therefore flat). By the local flatness criterion [**Mat**, 22.3], the R-flat locus of X' contains the closed fiber. But X' is proper over R and the R-flat locus in open in X', so X' is R-flat. Since the closed fiber is geometrically connected and smooth of dimension 1, the same holds for all fibers by [**EGA**, IV$_3$, 12.2.1, 12.2.4]. Thus, we have reduced to the case $S = \mathrm{Spec}(R)$ for a regular local ring R. We may then base change to an algebraic closure of the fraction field of R, so we are finally in the case of an algebraically closed base field k. In this situation, Theorem 5.2.3 translates the present theorem into the assertion that the isomorphisms of k-vector spaces

$$(B.4.7) \qquad T_0(J) \simeq \mathrm{H}^1(X, \mathscr{O}_X), \ T_0(J)^\vee \simeq \mathrm{H}^0(X, \Omega^1_{X/k})$$

are compatible with the duality between $T_0(J), T_0(J)^\vee$ and the negative of the residue pairing between $\mathrm{H}^1(X, \mathscr{O}_X), \mathrm{H}^0(X, \Omega^1_{X/k})$. Also, the case $g = 0$ is trivial, so we may (and do) assume $g > 0$.

I am grateful to Gabber for pointing out to me how to compare these two pairings: since we are trying to relate two bilinear pairings, it suffices to study a set of spanning vectors in each space! For $x_0 \in X(k)$, the map $j_{x_0} : X \hookrightarrow J$ defined by (B.4.3) is a closed immersion (since $g > 0$). The tangent space map $dj_{x_0} : T_{x_0}(X) \to T_0(J)$ is therefore an injection, so its image is a line in $T_0(J)$. The key point is that as we vary x_0, the resulting lines in $T_0(J)$ span the entire space. Indeed, consider an $\omega \in \mathrm{H}^0(J, \Omega^1_{J/k}) \simeq T_0(J)^\vee$ which pairs to 0 with all $dj_{x_0}(v)$ for all $v \in T_{x_0}(X), x_0 \in X(k)$. We want to prove $\omega = 0$. The hypothesis on ω forces $j^*_{x_0}(\omega)|_{x_0} \in T_{x_0}(X)^\vee$ to vanish for all $x_0 \in X(k)$. But $j^*_{x_0}(\omega)$ is exactly the image of ω under the canonical pullback *isomorphism*

$$\mathrm{H}^0(J, \Omega^1_{J/k}) \simeq \mathrm{H}^0(X, \Omega^1_{X/k})$$

from (B.4.5), which is *independent* of x_0. Thus, the global 1-form on X corresponding to ω vanishes in the cotangent space at all points in $X(k)$ and so is zero. This forces $\omega = 0$, as desired.

We are now reduced to checking that for $x \in X(k)$, $v = \partial_{t_x}|_x = (dt_x|_x)^\vee \in T_x(X)$ —with $t_x \in \mathfrak{m}_x$ any desired fixed choice of uniformizer at x— and $\omega \in \mathrm{H}^0(J, \Omega^1_{J/k})$, there is an equality

$$\langle dj_x(v), \omega|_0 \rangle = -\mathrm{res}_{X/k}(\tilde{v} \cup j^*_x(\omega)),$$

where $\tilde{v} \in \mathrm{H}^1(X, \mathscr{O}_X)$ corresponds to $dj_x(v) \in T_0(J)$ under (B.4.7), the map $\mathrm{res}_{X/k} : \mathrm{H}^1(X, \Omega^1_{X/k}) \simeq k$ is the residue map, and $\langle , \rangle$ denotes the pairing between $T_0(J), T_0(J)^\vee$. Recall that the definition of $\tilde{v}$ uses the choice of sign in (B.4.1). Since $\langle dj_x(v), \omega|_0 \rangle = \langle v, j^*_x(\omega)|_x \rangle$, with the latter pairing between $T_x(X)$ and $T_x(X)^\vee$, this entire situation can be phrased entirely in terms of X without mentioning J at all, as we now explain. Choose $x \in X(k)$ and a uniformizer t_x at x such that the zero locus of the rational function t_x on X consists of exactly x and one other point y on X. The role of y will be to simplify a certain Čech calculation later on. Let $v = \partial_{t_x}|_x \in T_x(X) \subseteq X(k[\epsilon]) = X[\epsilon](k[\epsilon])$, so the line bundle $j_x(v) = \mathscr{O}(v) \otimes \mathscr{O}(x)^{-1}$ on $X[\epsilon]$ gives rise to an element

$$\tilde{v} \in \ker(\mathrm{Pic}(X[\epsilon]) \to \mathrm{Pic}(X)) \simeq \mathrm{H}^1(X, \mathscr{O}_X)$$

using (B.4.1). We must show that for any $\omega \in \mathrm{H}^0(X, \Omega^1_{X/k})$, the elements $\mathrm{res}_{X/k}(\tilde{v} \cup \omega) \in k$ and $-\langle v, \omega|_x \rangle \in k$ are equal.

Explicitly, if we write $\omega = f dt_x$ for $f \in \mathscr{O}_{X,x}$, we want $\mathrm{res}_{X/k}(\widetilde{v} \cup \omega) = -f(0)$. Since $f(0) = \mathrm{res}_x(\omega/t_x)$, our assertion is

$$-\mathrm{res}_x\left(\frac{\omega}{t_x}\right) \stackrel{?}{=} \mathrm{res}_{X/k}(\widetilde{v} \cup \omega).$$

The first step is to compute

$$\widetilde{v} \in \mathrm{H}^1(X, \mathscr{O}_X) \simeq \check{\mathrm{H}}^1(X, \mathscr{O}_X) = \ker(\check{\mathrm{H}}^1(X, \mathscr{O}^{\times}_{X[\epsilon]}) \to \check{\mathrm{H}}^1(X, \mathscr{O}^{\times}_X)).$$

Let $\mathfrak{U} = \{U_0, U_1\}$ be an ordered open affine covering of X with $U_0 = X - \{x\}$, $U_1 = X - \Sigma$, with Σ a non-empty finite subset of $X(k) - \{x\}$ containing all zeros and poles of t_x aside from x. Viewing $x, v \in X(k[\epsilon]) = (X[\epsilon])(k[\epsilon])$, we want to compute Čech 1-cocycles in $\Gamma(U_0 \cap U_1, \mathscr{O}^{\times}_{X[\epsilon]})$ which represent the inverse ideal sheaves $\mathscr{O}(x) = \mathscr{I}^{-1}_{\{x\}}$, $\mathscr{O}(v) = \mathscr{I}^{-1}_{\{v\}}$ in $\mathrm{Pic}(X[\epsilon]) \simeq \check{\mathrm{H}}^1(X, \mathscr{O}^{\times}_{X[\epsilon]})$. Since $\mathscr{O}(x)|_{U_0}$ has basis t_x^{-1} and $\mathscr{O}(x)|_{U_1}$ has basis 1, a representative of $\mathscr{O}(x)$ is $1/(t_x)^{-1} = t_x \in \Gamma(U_0 \cap U_1, \mathscr{O}^{\times}_{X[\epsilon]})$. Here we have used our choice in the definition of (B.4.1). The analogous calculation for v requires a little more care. The map $v = \partial_{t_x}|_x : \mathscr{O}_x[\epsilon] \twoheadrightarrow k[\epsilon]$ is defined by

$$a + b\epsilon \mapsto a(x) + \left(b(x) + \frac{\partial a}{\partial t_x}(x)\right)\epsilon,$$

which has kernel $(t_x - \epsilon, t_x^2, t_x \epsilon)$. Since $(t_x - \epsilon)\epsilon = t_x \epsilon$ and $(t_x - \epsilon)(t_x + \epsilon) = t_x^2$, the kernel of v is generated by $t_x - \epsilon$. Thus, $\mathscr{O}(v)|_{U_0}$ has basis $(t_x - \epsilon)^{-1}$ and $\mathscr{O}(v)|_{U_1}$ has basis 1, so $\mathscr{O}(v)$ has representative Čech cocycle $t_x - \epsilon \in \Gamma(U_0 \cap U_1, \mathscr{O}^{\times}_{X[\epsilon]})$. We conclude that $\mathscr{O}(v) \otimes \mathscr{O}(x)^{-1} \in \mathrm{Pic}(X[\epsilon]) \simeq \check{\mathrm{H}}^1(X, \mathscr{O}^{\times}_{X[\epsilon]})$ is represented by the Čech 1-cocycle

$$(t_x - \epsilon)t_x^{-1} = 1 - \frac{\epsilon}{t_x} \in \Gamma(U_0 \cap U_1, \mathscr{O}^{\times}_{X[\epsilon]}).$$

Thus, when viewed as an element $\widetilde{v} \in \mathrm{H}^1(X, \mathscr{O}_X) \subseteq \mathrm{H}^1(X, \mathscr{O}^{\times}_{X[\epsilon]})$, $\mathscr{O}(v) \otimes \mathscr{O}(x)^{-1}$ is the image of

$$-\frac{1}{t_x} \in \Gamma(U_0 \cap U_1, \mathscr{O}_X) \twoheadrightarrow \check{\mathrm{H}}^1(\mathfrak{U}, \mathscr{O}_X) \simeq \check{\mathrm{H}}^1(X, \mathscr{O}_X) \simeq \mathrm{H}^1(X, \mathscr{O}_X).$$

This computes the desired representative for $\widetilde{v}$.

It is now clear that the cup product $\widetilde{v} \cup \omega \in \mathrm{H}^1(X, \Omega^1_{X/k})$ is represented by

$$-\frac{\omega}{t_x} \in \Gamma(U_0 \cap U_1, \Omega^1_{X/k}) \twoheadrightarrow \check{\mathrm{H}}^1(\mathfrak{U}, \Omega^1_{X/k}) \simeq \check{\mathrm{H}}^1(X, \Omega^1_{X/k}) \simeq \mathrm{H}^1(X, \Omega^1_{X/k}).$$

We want to show that the composite

$$\Gamma(U_0 \cap U_1, \Omega^1_{X/k}) \twoheadrightarrow \check{\mathrm{H}}^1(\mathfrak{U}, \Omega^1_{X/k}) \simeq \mathrm{H}^1(X, \Omega^1_{X/k}) \xrightarrow{\mathrm{res}_{X/k}} k$$

sends $-\omega/t_x$ to $-\mathrm{res}_x(\omega/t_x)$, or equivalently sends ω/t_x to $\mathrm{res}_x(\omega/t_x)$. Now recall that we chose t_x to have zero locus $\{x, y\}$ on $X(k)$. Define $U_0' = X - \{y\} \supseteq U_0$ and $\mathfrak{U}' = \{U_0', U_1\}$, so $\mathfrak{U}'$ is an ordered open covering of X which is refined by $\mathfrak{U}$. The natural refinement map $\check{\mathrm{H}}^1(\mathfrak{U}', \Omega^1_{X/k}) \to \check{\mathrm{H}}^1(\mathfrak{U}, \Omega^1_{X/k})$ takes the class of $\omega/t_x \in \Gamma(U_0' \cap U_1, \Omega^1_{X/k})$ to the class of $\omega/t_x \in \Gamma(U_0 \cap U_1, \Omega^1_{X/k})$.

We are now reduced to showing that if $\mathfrak{V} = \{V_0, V_1\}$ is an ordered open covering of X with $V_i = X - \{x_i\}$ for distinct points $x_0, x_1 \in X(k)$ and $\eta \in \Gamma(V_0 \cap V_1, \Omega^1_{X/k})$ is a meromorphic differential on X with pole set inside of $\{x_0, x_1\}$, then the composite map

$$(B.4.8) \qquad \check{H}^1(\mathfrak{V}, \Omega^1_{X/k}) \simeq H^1(X, \Omega^1_{X/k}) \overset{\mathrm{res}_{X/k}}{\longrightarrow} k$$

takes the class of η to $\mathrm{res}_{x_1}(\eta)$. Note that since $\mathfrak{V}$ is ordered, there is an a priori distinction between x_0 and x_1. The first map in (B.4.8) is the edge map in the Čech to derived functor cohomology spectral sequence.

In our situation, $\Omega^1_{X/k}$ has a rather concrete injective resolution given by

$$(B.4.9) \qquad \Omega^1_{X/k} \to \underline{\Omega}^1_{K/k} \to \bigoplus_{x \in X^0} i_{x*}(\Omega^1_{K/k}/\Omega^1_{\mathcal{O}_{X,x}/k}) \to 0,$$

defined without an intervention of signs. If $j_i : V_i \hookrightarrow X$, $j : V_0 \cap V_1 \hookrightarrow X$ are the canonical open immersions, then the Čech resolution $\mathscr{C}^\bullet(\mathfrak{V}, \Omega^1_{X/k})$ relative to $\mathfrak{V}$ is

$$(B.4.10) \qquad \Omega^1_{X/k} \to j_{0*}\Omega^1_{V_0/k} \oplus j_{1*}\Omega^1_{V_1/k} \to j_*\Omega^1_{V_0 \cap V_1/k} \to 0,$$

with the second map given by "$(\eta_0, \eta_1) \mapsto \eta_1 - \eta_0$". We can define a map of complexes

$$(B.4.11) \quad \begin{array}{ccccc} \Omega^1_{X/k} & \longrightarrow & j_{0*}\Omega^1_{V_0/k} \oplus j_{1*}\Omega^1_{V_1/k} & \longrightarrow & j_*\Omega^1_{V_0 \cap V_1/k} \\ \| & & \downarrow & & \downarrow \\ \Omega^1_{X/k} & \longrightarrow & \underline{\Omega}^1_{K/k} & \longrightarrow & \bigoplus_{x \in X^0} i_{x*}(\Omega^1_{K/k}/\Omega^1_{\mathcal{O}_{X,x}/k}) \end{array}$$

extending the identity on $\Omega^1_{X/k}$, where the map in the middle column sends (η_0, η_1) to η_1 and the map in the right column sends η to the element whose x_1th coordinate is represented by η and whose xth coordinate is 0 for all $x \neq x_1$. The diagram (B.4.11) computes the edge map in (B.4.8), thanks to the explicit description of (1.3.23) via (1.3.24). By the *definition* of $\mathrm{res}_{X/k}$ in terms of (B.4.9), the map (B.4.8) must therefore send the class of η to $\mathrm{res}_{x_1}(\eta) \in k$, as desired. ∎

Bibliography

[LLT] L. Alonso, A. Jeremías, J. Lipman, *Studies in duality on noetherian formal schemes and non-noetherian ordinary schemes*, Contemporary Math. **244**, AMS, 1999.

[AK1] A. Altman, S. Kleiman, *Introduction to Grothendieck duality theory*, Lecture Notes in Mathematics **146**, Springer-Verlag, New York, 1970.

[AK2] A. Altman, S. Kleiman, *Compactification of the Picard scheme*, Advances in Mathematics, **35** (1980), pp. 53–112.

[AM] M. Atiyah, I. MacDonald, *Introduction to commutative algebra*, Addison-Wesley, Reading, 1969.

[AFH] L. Avramov, H. Foxby, B. Herzog, *Structure of local homomorphisms*, Journal of Algebra, **164** (1994), pp. 124-145.

[Be] P. Berthelot, *Cohomologie cristalline des schémas de caractéristique p > 0*, Lecture Notes in Mathematics **407**, Springer-Verlag, New York, 1974.

[BBM] P. Berthelot, L. Breen, W. Messing, *Théorie de Dieudonné cristalline II*, Lecture Notes in Mathematics **930**, Springer-Verlag, New York, 1982.

[BLR] S. Bosch, W. Lütkebohmert, M. Raynaud, *Néron models*, Springer-Verlag, New York, 1980.

[CE] H. Cartan, S. Eilenberg, *Homological algebra*, Princeton Univ. Press, Princeton, 1956.

[D] P. Deligne, *Intégration sur un cycle évanescent*, Inv. Math., **76** (1983), pp. 129–143.

[DM] P. Deligne, D. Mumford, *The irreducibility of the space of curves of a given genus*, Publ. Math. IHES, **36**, 1969, pp. 75-110.

[DR] P. Deligne, M. Rapoport, *Les schémas de modules de courbes elliptiques*, Lecture Notes in Mathematics **349**, Springer-Verlag, New York, 1973.

[EGA] J. Dieudonné, A. Grothendieck, *Éléments de géométrie algébrique*, Publ. Math. IHES, **4, 8, 11, 17, 20, 24, 28, 32**, 1960-7.

[FK] E. Freitag, R. Kiehl, *Étale cohomology and the Weil conjecture*, Springer-Verlag, New York, 1988.

[Tohoku] A. Grothendieck, *Sur quelques points d'algebre homologique*, Tôhoku Math Journal, **9** (1957), pp. 119–221.

[SGA1] A. Grothendieck, *Revêtments étales et groupe fondamental*, Lecture Notes in Mathematics **224**, Springer-Verlag, 1971.

[SGA2] A. Grothendieck, *Cohomologie locale de faisceaux coherents et theoremes de Lefschetz locaux et globaux*, North-Holland Publ. Co., Amsterdam, 1962.

[SGA4] A. Grothendieck, *Théorie des topos et cohomologie étale des schémas* Tome 3, Lecture Notes in Mathematics **305**, Springer-Verlag, New York, 1973.

[SGA6] A. Grothendieck, *Théorie des Intersections et Théroème de Riemann-Roch*, Lecture Notes in Mathematics **225**, Sprinfer-Verlag, 1971.

[RD] R. Hartshorne, *Residues and duality*, Lecture Notes in Mathematics **20**, Springer-Verlag, New York, 1966.

[H] R. Hartshorne, *Algebraic geometry*, Springer-Verlag, New York, 1977.

[K] S. Kleiman, *Relative duality for quasi-coherent sheaves*, Compositio Math., **41** (1980), pp. 39-60.

[L] J. Lipman, *Residues and traces of differential forms via Hochshild homology*, Contemporary Math. **61**, AMS, 1987.

[Mac] S. MacLane, *Categories for the working mathematician*, GTM **5**, Springer-Verlag, New York, 1971.

[Mat] H. Matsumura, *Commutative ring theory*, Cambridge Univ. Press, Cambridge, 1986.

[M] B. Mazur, *Modular curves and the Eisenstein ideal*, Publ. Math. IHES, **47** (1977), pp. 33–186.

[Maz] B. Mazur, *Rational isogenies of prime degree*, Inv. Math. **44** (1978), pp. 179–162.

[MR] B. Mazur, L. Roberts, *Local Euler characteristics*, Inv. Math. **9** (1970), pp. 201–234.

[Mi] J. Milne, "Jacobian Varieties" in *Arithmetic geometry* (ed. Cornell, Silverman), Springer-Verlag, New York, 1986.

[GIT] D. Mumford, J. Fogarty, F. Kirwan, *Geometric invariant theory* 3rd. ed., Springer-Verlag, New York, 1994.

[R] M. Raynaud, *Revêtements de la droite affine en caractérisque p > 0 et conjecture d'Abhyanakar*, Inv. Math. **116** (1994), pp. 425–462.

[Sp] N. Spaltenstein, *Resolution of unbounded complexes*, Compositio Math. **65** (1988), pp. 121–154.

[Verd] J.-L. Verdier, *Base change for twisted inverse image of coherent sheaves*, International Colloquium on Algebraic Geometry, Bombay, 1968, pp. 393-408.

[W] C. Weibel, *An introduction to homological algebra*, Cambridge Univ. Press, Cambridge, 1994.

Index

Vol. 1654: R. W. Ghrist, P. J. Holmes, M. C. Sullivan, Knots and Links in Three-Dimensional Flows. X, 208 pages. 1997.

Vol. 1655: J. Azéma, M. Emery, M. Yor (Eds.), Séminaire de Probabilités XXXI. VIII, 329 pages. 1997.

Vol. 1656: B. Biais, T. Björk, J. Cvitanic, N. El Karoui, E. Jouini, J. C. Rochet, Financial Mathematics. Bressanone, 1996. Editor: W. J. Runggaldier. VII, 316 pages. 1997.

Vol. 1657: H. Reimann, The semi-simple zeta function of quaternionic Shimura varieties. IX, 143 pages. 1997.

Vol. 1658: A. Pumarino, J. A. Rodrıguez, Coexistence and Persistence of Strange Attractors. VIII, 195 pages. 1997.

Vol. 1659: V, Kozlov, V. Maz'ya, Theory of a Higher-Order Sturm-Liouville Equation. XI, 140 pages. 1997.

Vol. 1660: M. Bardi, M. G. Crandall, L. C. Evans, H. M. Soner, P. E. Souganidis, Viscosity Solutions and Applications. Montecatini Terme, 1995. Editors: I. Capuzzo Dolcetta, P. L. Lions. IX, 259 pages. 1997.

Vol. 1661: A. Tralle, J. Oprea, Symplectic Manifolds with no Kähler Structure. VIII, 207 pages. 1997.

Vol. 1662: J. W. Rutter, Spaces of Homotopy Self-Equivalences – A Survey. IX, 170 pages. 1997.

Vol. 1663: Y. E. Karpeshina; Perturbation Theory for the Schrödinger Operator with a Periodic Potential. VII, 352 pages. 1997.

Vol. 1664: M. Väth, Ideal Spaces. V, 146 pages. 1997.

Vol. 1665: E. Giné, G. R. Grimmett, L. Saloff-Coste, Lectures on Probability Theory and Statistics 1996. Editor: P. Bernard. X, 424 pages, 1997.

Vol. 1666: M. van der Put, M. F. Singer, Galois Theory of Difference Equations. VII, 179 pages. 1997.

Vol. 1667: J. M. F. Castillo, M. González, Three-space problems in Banach Space Theory. XII, 267 pages. 1997.

Vol. 1668: D. B. Dix, Large-Time Behavior of Solutions of Linear Dispersive Equations. XIV, 203 pages. 1997.

Vol. 1669: U. Kaiser, Link Theory in Manifolds. XIV, 167 pages. 1997.

Vol. 1670: J. W. Neuberger, Sobolev Gradients and Differential Equations. VIII, 150 pages. 1997.

Vol. 1671: S. Bouc, Green Functors and G-sets. VII, 342 pages. 1997.

Vol. 1672: S. Mandal, Projective Modules and Complete Intersections. VIII, 114 pages. 1997.

Vol. 1673: F. D. Grosshans, Algebraic Homogeneous Spaces and Invariant Theory. VI, 148 pages. 1997.

Vol. 1674: G. Klaas, C. R. Leedham-Green, W. Plesken, Linear Pro-p-Groups of Finite Width. VIII, 115 pages. 1997.

Vol. 1675: J. E. Yukich, Probability Theory of Classical Euclidean Optimization Problems. X, 152 pages. 1998.

Vol. 1676: P. Cembranos, J. Mendoza, Banach Spaces of Vector-Valued Functions. VIII, 118 pages. 1997.

Vol. 1677: N. Proskurin, Cubic Metaplectic Forms and Theta Functions. VIII, 196 pages. 1998.

Vol. 1678: O. Krupková, The Geometry of Ordinary Variational Equations. X, 251 pages. 1997.

Vol. 1679: K.-G. Grosse-Erdmann, The Blocking Technique. Weighted Mean Operators and Hardy's Inequality. IX, 114 pages. 1998.

Vol. 1680: K.-Z. Li, F. Oort, Moduli of Supersingular Abelian Varieties. V, 116 pages. 1998.

Vol. 1681: G. J. Wirsching, The Dynamical System Generated by the 3n+1 Function. VII, 158 pages. 1998.

Vol. 1682: H.-D. Alber, Materials with Memory. X, 166 pages. 1998.

Vol. 1683: A. Pomp, The Boundary-Domain Integral Method for Elliptic Systems. XVI, 163 pages. 1998.

Vol. 1684: C. A. Berenstein, P. F. Ebenfelt, S. G. Gindikin, S. Helgason, A. E. Tumanov, Integral Geometry, Radon Transforms and Complex Analysis. Firenze, 1996. Editors: E. Casadio Tarabusi, M. A. Picardello, G. Zampieri. VII, 160 pages. 1998

Vol. 1685: S. König, A. Zimmermann, Derived Equivalences for Group Rings. X, 146 pages. 1998.

Vol. 1686: J. Azéma, M. Émery, M. Ledoux, M. Yor (Eds.), Séminaire de Probabilités XXXII. VI, 440 pages. 1998.

Vol. 1687: F. Bornemann, Homogenization in Time of Singularly Perturbed Mechanical Systems. XII, 156 pages. 1998.

Vol. 1688: S. Assing, W. Schmidt, Continuous Strong Markov Processes in Dimension One. XII. 137 page. 1998.

Vol. 1689: W. Fulton, P. Pragacz, Schubert Varieties and Degeneracy Loci. XI, 148 pages. 1998.

Vol. 1690: M. T. Barlow, D. Nualart, Lectures on Probability Theory and Statistics. Editor: P. Bernard. VIII, 237 pages. 1998.

Vol. 1691: R. Bezrukavnikov, M. Finkelberg, V. Schechtman, Factorizable Sheaves and Quantum Groups. X, 282 pages. 1998.

Vol. 1692: T. M. W. Eyre, Quantum Stochastic Calculus and Representations of Lie Superalgebras. IX. 138 pages. 1998.

Vol. 1694: A. Braides, Approximation of Free-Discontinuity Problems. XI, 149 pages. 1998.

Vol. 1695: D. J. Hartfiel, Markov Set-Chains. VIII. 131 pages. 1998.

Vol. 1696: E. Bouscaren (Ed.): Model Theory and Algebraic Geometry. XV, 211 pages. 1998.

Vol. 1697: B. Cockburn, C. Johnson, C.-W. Shu, E. Tadmor, Advanced Numerical Approximation of Nonlinear Hyperbolic Equations. Cetraro, Italy, 1997. Editor: A. Quarteroni. VII, 390 pages. 1998.

Vol. 1698: M. Bhattacharjee, D. Macpherson, R. G. Möller, P. Neumann, Notes on Infinite Permutation Groups. XI, 202 pages. 1998.

Vol. 1699: A. Inoue,Tomita-Takesaki Theory in Algebras of Unbounded Operators. VIII, 241 pages. 1998.

Vol. 1700: W. A. Woyczyński, Burgers-KPZ Turbulence,XI, 318 pages. 1998.

Vol. 1701: Ti-Jun Xiao, J. Liang, The Cauchy Problem of Higher Order Abstract Differential Equations. XII, 302 pages. 1998.

Vol. 1702: J. Ma, J. Yong, Forward-Backward Stochastic Differential Equations and Their Applications. XIII, 270 pages. 1999.

Vol. 1703: R. M. Dudley, R. Norvaiša, Differentiability of Six Operators on Nonsmooth Functions and p-Variation. VIII, 272 pages. 1999.

Vol. 1704: H. Tamanoi, Elliptic Genera and Vertex Operator Super-Algebras. VII, 390 pages. 1999.

Vol. 1705: I. Nikolaev, E. Zhuzhoma, Flows in 2-dimensional Manifolds. XIX, 294 pages. 1999.

4. Lecture Notes are printed by photo-offset from the master-copy delivered in camera-ready form by the authors. Springer-Verlag provides technical instructions for the preparation of manuscripts. Macro packages in T_EX, L^AT_EX2e, $L^AT_EX2.09$ are available from Springer's web-pages at

http://www.springer.de/math/authors/b-tex.html.

Careful preparation of the manuscripts will help keep production time short and ensure satisfactory appearance of the finished book.

The actual production of a Lecture Notes volume takes approximately 12 weeks.

5. Authors receive a total of 50 free copies of their volume, but no royalties. They are entitled to a discount of 33.3 % on the price of Springer books purchase for their personal use, if ordering directly from Springer-Verlag.

Commitment to publish is made by letter of intent rather than by signing a formal contract. Springer-Verlag secures the copyright for each volume. Authors are free to reuse material contained in their LNM volumes in later publications: A brief written (or e-mail) request for formal permission is sufficient.

Addresses:

Professor Jean-Michel Morel
CMLA, École Normale Supérieure de Cachan
61 Avenue du Président Wilson
94235 Cachan Cedex France
E-mail: Jean-Michel.Morel@cmla.ens-cachan.fr

Professor B. Teissier, DMI, École Normale Supérieure
45, rue d'Ulm
F-7500 Paris, France
E-mail: Teissier@ens.fr

Professor F. Takens, Mathematisch Instituut
Rijksuniversiteit Groningen, Postbus 800
9700 AV Groningen, The Netherlands
E-mail: F.Takens@math.rug.nl

Springer-Verlag, Mathematics Editorial, Tiergartenstr. 17
D-69121 Heidelberg, Germany
Tel.: *49 (6221) 487-701
Fax: *49 (6221) 487-355
E-mail: lnm@Springer.de